METHODS OF BIOCHEMICAL ANALYSIS

Volume 24

Advisory Board

METHODS OF BIOCHEMICAL ANALYSIS

Edited by DAVID GLICK
Life Sciences Division
Stanford Research Institute
Menlo Park, California

VOLUME 24

An Interscience® Publication
JOHN WILEY & SONS, New York · London · Sydney · Toronto

An Interscience® Publication

Library of Congress Catalogue Card Number: 54-7232

ISBN 0-471-02764-2

Printed in the United States of America.

10 9 8 7 6 5 4 3 2 1

PREFACE

Annual review volumes dealing with many different fields of science have proved their value repeatedly and are now widely used and well established. These reviews have been concerned primarily with the results of the developing fields, rather than with the techniques and methods employed, and they have served to keep the ever-expanding scene within the view of the investigator, the applier, the teacher, and the student.

It is particularly important that review services of this nature should now be extended to cover methods and techniques, because it is becoming increasingly difficult to keep abreast of the manifold experimental innovations and improvements which constitute the limiting factor in many cases for the growth of the experimental sciences. Concepts and vision of creative scientists far outrun that which can actually be attained in present practice. Therefore an emphasis on methodology and instrumentation is a fundamental need in order for material achievement to keep in sight of the advance of useful ideas.

The volumes in this series are designed to try to meet the need in the field of biochemical analysis. The topics to be included are chemical, physical, microbiological, and if necessary, animal assays, as well as basic techniques and instrumentation for the determination of enzymes, vitamins, hormones, lipids, carbohydrates, proteins and their products, minerals, antimetabolites, etc.

Certain chapters will deal with well-established methods or techniques which have undergone sufficient improvement to merit recapitulation, reappraisal, and new recommendations. Other chapters will be concerned with essentially new approaches which bear promise of great usefulness. Relatively few subjects can be included in any single volume, but as they accumulate these volumes should comprise a self-modernizing encyclopedia of methods of biochemical analysis. By judicious selection of topics it is planned that most subjects of current importance will receive treatment in these volumes.

The general plan followed in the organization of the individual chapters is a discussion of the background and previous work, a critical evaluation of the various approaches, and a presentation of the procedural details of the method or methods recommended by the author.

The presentation of the experimental details is to be given in a manner that will furnish the laboratory worker with the complete information required to carry out the anaylsis.

Within this comprehensive scheme the reader may note that the treatments vary widely with respect to taste, style, and point of view. It is the Editor's policy to encourage individual expression in these presentations because it is stifling to originality and justifiably annoying to many authors to submerge themselves in a standard mold. Scientific writing need not be as dull and uniform as it too often is. In certain technical details, a consistent pattern is followed for the sake of convenience, as in the form used for reference citations and indexing.

The success of the treatment of any topic will depend primarily on the experience, critical ability, and capacity to communicate of the author. Those invited to prepare the respective chapters are scientists who either have originated the methods they discuss or have had intimate personal experience with them.

It is the wish of the Advisory Board and the Editor to make this series of volumes as useful as possible and to this end suggestions will be always welcome.

DAVID GLICK

CONTENTS

METHODS OF BIOCHEMICAL ANALYSIS

Volume 24

Analysis of Morphine and Related Analgesics by Gas Phase Methods

JEAN-PAUL G. THÉNOT AND KLAUS D. HAEGELE,* *Institute for Lipid Research, Baylor College of Medicine, Houston, Texas*

*Present address: Department of Pharmacology, University of Texas, Health Science Center at San Antonio.

I. INTRODUCTION

The analgesic virtue of opium was known to early civilizations. Roman literature contains accurate descriptions of its pharmacological activity. In 1804 Sertürner isolated morphine as the main alkaloid from *Papaver sommiferum* (1). Seventy years later, Wright (2) prepared a series of morphine derivatives including the diacetyl, but the name "heroin" for diacetylmorphine was coined by Dreser (3) of the Bayer Company in Germany. Heroin became very popular for the treatment of cough, bronchitis, and asthma, for it was thought to lack the undesirable side effects of morphine such as nausea and vomiting. Although Dreser first claimed that heroin was not habit-forming, Morel-Lavallee (4) warned against its addictive properties as early as 1902. Because of its addictive potential, heroin is no longer prescribed in the United States and its preparation is illegal. The drug is still available legally in many other countries.

In searching for new analgesics, Eislieb and Schauman prepared meperidine in 1939. This analgesic is still commonly used for the relief of labor pains. Methadone, introduced in the late 1940s, is currently used for the treatment of heroin addiction (5).

Following are the main analgesics in current use that are structurally related to morphine: codeine, dihydrocodeine, dihydromorphine, and hydroxydihydrocodeinone (oxycodone). Recently Rice and Jacobson (6) reported the preparation and analgesic activity of O^3, O^6-diacetylnormorphine and O^6-acetylnormorphine.

The literature dealing with morphine analysis is very extensive. Taylor (7) reviewed methods of chemical analysis. Techniques used for morphine analysis include thin-layer chromatography (TLC) (8), spectrofluorometry (9), colorimetry, radioimmunoassay (RIA) (10), free radical assay techniques (FRAT) (11), gas chromatography (GC), and gas chromatography–mass spectrometry (GC-MS). An evaluation of current methods for heroin screening and morphine analysis has been presented (12,13). Gas chromatography is a well-established method for screening and confirmation of drugs of abuse. When both sensitivity and specificity are required, as for example in forensic sciences, gas chromatography–mass spectrometry is the method of choice. Applications of GC-MS in clinical chemistry have been reviewed recently by Roboz (14).

This chapter describes methods of analysis for morphine and related analgesics by gas phase methods. The literature of the last decade has been reviewed.

II. EXTRACTION PROCEDURES

Many biologic compounds are present in body fluids as ionized species that are not directly amenable to determination by gas chromatography, thus must be extracted as neutral molecules prior to the analysis. The extraction step also serves the purpose of purification and concentration.

Morphine-type analgesics are nitrogen-containing compounds with a secondary or tertiary amino group. Their p*K* values (15) are little affected by structural changes. Phenolic alkaloids like morphine (p*K*a 10) present an additional problem with respect to complete extraction. The pH of the aqueous phase must be maintained slightly lower than the p*K* of the phenol group with an appropriate buffer.

Many methods have been described for the extraction of morphine, most of them derived from the historical procedure of Stas (16). More than a century ago, Stas described a general method for the extraction of alkaloids from biologic tissues. The tissue proteins were precipitated upon heating with an alcoholic solution; then the filtrate was made alkaline with sodium bicarbonate and extracted with diethyl ether. This method is still valid for efficient extraction of numerous drugs. Niyogi (17), in an extensive review, has described various modifications of this original technique.

1. Continuous Solvent Extraction

The continuous extraction of a sample (liquid or solid) with a boiling solvent is a useful method, when large solid samples are analyzed; since more time is necessary for efficient extraction, however, it is not widely used. The large volumes of solvent used in this process must be concentrated. With increasingly sensitive instrumentation, the analysis of large samples is no longer a necessity.

2. Direct Solvent Extraction

In contrast to continuous extraction, the sample can be extracted for a short period of time with a relatively small volume of solvent. At physiological pH's, the drugs are ionized, and since the free base is more soluble than the salt in organic solvents, the extraction is carried out after the pH of the solution has been adjusted to 9–10. Many buffers have been reported in the literature; the most common are bicarbonate, carbonate, or borate buffers. Too high a pH will ionize the phenolic group of morphine and decrease the partition ratio between the aqueous and the organic phases. The addition of salt is often used aid in increase the extraction.

Many solvents have been used for the extraction of morphine and related compounds. Ether, used by Stas, is no longer widely employed. Chloroform is frequently used. Addition of an alcohol (ethanol, isopropanol, butanol) increases the partition ratio and also prevents adsorption to glassware. Unfortunately the recovery is not always explicitly stated in published methods.

3. Ion-Pair Extraction

In the usual extraction, basic drugs are extracted in un-ionized form. It is, however, possible to transfer ionized compounds into an organic phase if a suitable counterion is added to the aqueous phase. Depending on the nature of the molecule, a cation or an anion is necessary to complete the transfer. It is well known that morphine recovery is improved if ammonium ions are present in the water phase. Horning et al. (18) reported quantitative extraction for a variety of basic drugs including morphine, codeine, and meperidine from blood and urine with ammonium carbonate as the buffer. An ion pair between an ionized phenolic group and the ammonium ion is soluble in organic solvents, and solubility increases if the ammonium ions are tetraalkyl substituted.

The amino group of a basic drug is ionized at physiological pH's; a negatively charged counterion may be used to promote the transfer into an organic phase. Schill and Marsh (19) in an extensive study reported the extraction of some 40 alkaloids with bromothymol blue at pH 7.5. The compounds were subsequently determined photometrically. Boon and Mace

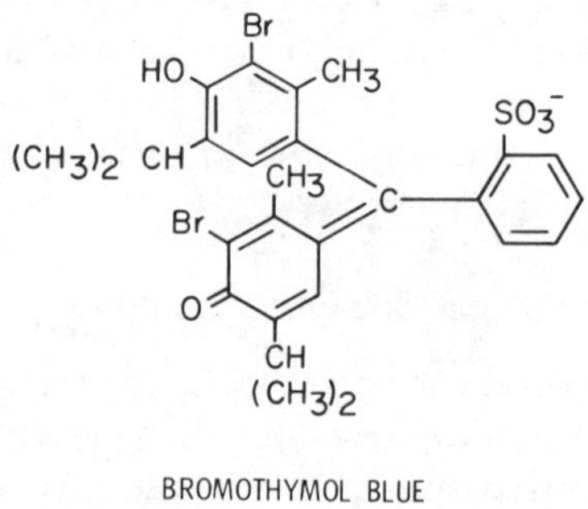

BROMOTHYMOL BLUE

(20) extracted five basic drugs including meperidine at pH 5 with chloroform after addition of a bromothymol blue solution. The ion pair was directly injected into the column and no interference was observed due to the counterion. This interesting technique could be extended to morphine.

4. Clean-up Procedures

Biologic samples such as blood and urine contain hundreds of compounds, and a simple extraction procedure is usually not sufficient to avoid interference from other endogenous compounds. It is often necessary to purify a sample before derivatization, even when high-resolution techniques such as capillary column chromatography and multiple-ion detection are used. The objective of a purification step is to remove most of the "unwanted" substances—endogenous or exogenous—but not at the expense of the recovery of the drug to be measured. This problem becomes very critical when very small amounts are monitored in biologic samples, since endogenous substances may impair the GC separation or may saturate the ion source. With instruments that can detect picogram and even femtogram samples, purification techniques are becoming the limiting factors.

A. DIFFERENTIAL EXTRACTION

Morphine and related compounds, which are bases, are easily separated from acidic and neutral substances. Several approaches are possible. In an acidic medium, neutral and acidic compounds are first removed, then the solution is adjusted to pH 9–10 and the alkaloids are extracted. Ion-pair extraction techniques can be applied for the purification step. After the removal of neutral substances, the counterion is added to the acidic solution and the ion pair is transferred into the organic phase. This method is particularly suitable for alkali-sensitive compounds. A second approach, often called back extraction, consists of extracting the basic compound in unionized form, then washing the organic phase with a solution of a mineral acid. The bases are reextracted from the acid solution after alkalinization as for a direct extraction. This method has been used extensively by many authors with different solvents and acids. The presence of ammonium ions may lead to the formation of ion pairs, which in turn result in poor yields during back extractions. Addition of hexane to the first organic extract results in quantitative transfer to the acid solution.

B. ION EXCHANGE

The basic properties of morphine and its analogs allow these substances to be separated from other endogenous substances by ion-exchange chromatography. Tompsett (21) reported the isolation of morphine alkaloids with a

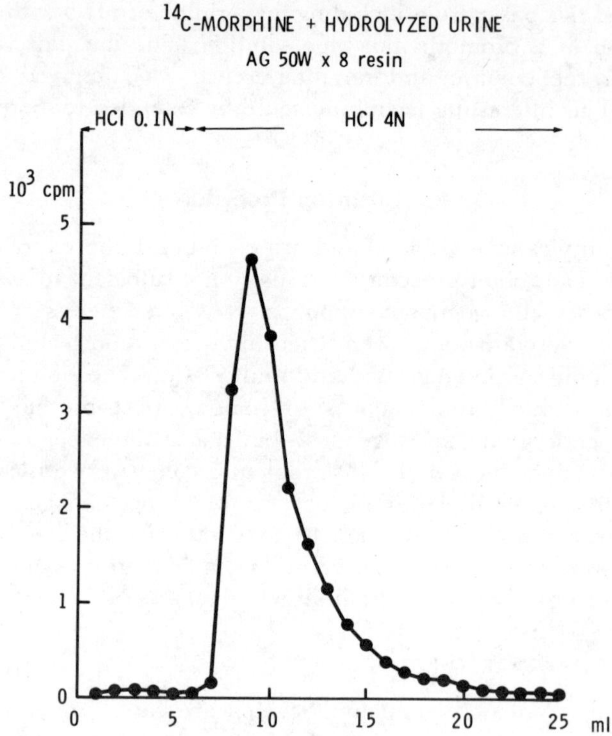

Figure 1. Elution of ^{14}C-morphine in hydrolyzed urine from a small ion-exchange column (0.3 g of AG 50W × 8) with 4*N* hydrochloric acid.

sulfonic acid resin (Dowex 50). These resins have a high capacity of binding, and a very small column is usually sufficient to retain basic compounds contained in biologic fluids. The difficulty lies in elution from the column. Efficient recovery can be obtained when column operating parameters are optimized. Figure 1 illustrates the elution of ^{14}C-morphine in hydrolyzed urine from a small ion-exchange column containing only 0.3 g of the sulfonic acid resin AG 50W × 8, 200-400 mesh (Bio Rad Laboratories, Richmond, Calif.) in a disposable Pasteur pipette. The flow rate was controlled (at the exit) at 1.5 ml/min through the use of a peristaltic pump.

The recent development of high-pressure liquid chromatography (HPLC) makes it possible to obtain good resolution and fast analyses. Opium alkaloids may be separated by HPLC, as described by several authors (22,23).

C. ADSORPTION

Charcoal has been used for many years as an adsorbent for biologic compounds. In a method for the determination of morphine by GC described by Ikekawa et al. (24), charcoal was used to retain morphine alkaloids from urine.

However synthetic adsorbents, with their more well-defined properties, are replacing natural adsorbents. Florisil, a synthetic magnesium silicate, is often used to purify biologic extracts. With this method, Stolman and Stewart (25) reported quantitative recoveries for morphine, codeine, and heroin. The mechanism of adsorption is not fully understood, and empirical conditions prevail; for example, the optimal pH is different for different specimens: for tissue extracts, between 7 and 8.5; for blood, between 8 and 9; and for urine, between 5 and 6. No rationale is apparent for these variations.

The most widely used adsorbent is currently a styrene-divinylbenzene neutral copolymer (XAD-2). The usefulness of this adsorbent lies in the capacity of the resin to adsorb organic material in aqueous solution. After washing with water, organic solvents release the adsorbed compounds. Extraction and purification are carried out in one step. Several authors (26–28) have applied this technique for the purification of drugs of abuse in human urine. Disposable prefilled columns are commercially available.

5. Conclusion

A review of the principal methods of extraction and purification of morphine alkaloids indicates that no method of extraction has pronounced advantages in all applications. The extraction procedure should be modified to suit the particular analytical problem.

Plasticizers must be kept to a minimum. It is almost impossible, for example, to obtain a blood sample free from plasticizers. Disposable plastic ware should be avoided during the sample preparation. An intense ion at m/e 149 is usually indicative of the presence of a phthalate plasticizer.

III. ANALYSIS BY GAS CHROMATOGRAPHY

The potential value of gas chromatography in the analysis of opium alkaloids was recognized in the early 1960s (29,30) immediately after Vanden Heuvel et al. (31) described the preparation of thin-film GC columns made with a thermostable liquid phase (a dimethylsiloxane polymer) and suitable for the separation of steroids. Morphine and related compounds are directly amenable to gas chromatographic analysis as free bases. Extracts have been injected, without derivatization, in the determination of opium alkaloids (32), thebaine (33) and morphine standards (34,35), legal (36,38) and illegal (39–41) preparations, and biologic samples (8,42–47).

Upon injection of a mixture of morphine and diacetylmorphine, as free bases, a transacetylation may occur in the injector zone with the formation of O^3- and O^6-acetylmorphine. This phenomenon, observed first by Brockmann-Hanssen and Swendsen (30) and Viala et al. (40), is a severe limitation on the quantitative measurement of samples containing both compounds.

In practice, it is usually desirable to prepare a suitable derivative before GC analysis or analysis by GC-MS.

1. Derivative Formation

A. ACETATE DERIVATIVES

Anders (48) described a procedure in which acetylation was performed on the column. This technique was also used by Mulé (49). If the proper acid anhydride is added, other esters may be prepared. Acetylation of morphine before injection was reported by Wallace et al. (50). Obviously, acetylation is not the method of choice if the same sample contains any combination of morphine, monoacetylmorphine, and diacetylmorphine.

It is possible to acetylate the phenolic or the allylic group of morphine in specific fashion. In 1954 Welsh (51) described the preparation of O^3-acetylmorphine. O^6-Acetylmorphine is formed by mild hydrolysis of the diacetyl derivative with hydroxylamine hydrochloride in alcohol (52).

Trifluoroacetyl derivatives, prepared by reaction with trifluoroacetic anhydride (53), have a shorter retention time than the acetyl derivatives and may be used for electron capture detection analyses.

Ebbighausen, et al. (54) recently developed an analytical procedure for the study of morphine and codeine based on the use of the trifluoracetyl derivative (morphine) and the heptafluorobutyryl derivative (codeine). Smith and Cole (55) used the O^3-trifluoracetyl derivative of O^6-acetylmorphine in a study of diacetylmorphine metabolism. Ebbighausen et al. (56) also used trifluoracetyl derivatives in a study of codeine metabolism.

B. TRIMETHYLSILYL DERIVATIVES

The most satisfactory derivative of morphine for analytical purposes is the ditrimethylsilyl ether. The phenolic group and the allylic hydroxyl group are readily converted to trimethylsilyl ethers by the usual silylating reagents. Bis-trimethylsilylacetamide, bis-trimethylsilyltrifluoracetamide, or *N*-trimethylsilylimidazole may be used; the reaction is usually catalyzed by the addition of trimethylchlorosilane. The trimethylsilyl (TMS) derivative has good gas chromatographic properties.

This derivative was employed by Wilkinson and Way (57) in an early quantitative study of morphine metabolism, and it has been used many times in later investigations. Although trimethylsilyl ethers undergo hydrolysis relatively easily, they are thermally stable and show little adsorption on GC columns. Column loss may occur if acidic conditions develop on the column packing; the best way of avoiding this circumstance is to employ an initial 1- to 2-cm zone of 10% SE-30 packing, according to the practice described by Thénot and Horning (58). The TMS derivative was employed in a recent method described by Clarke and Foltz (59). Other studies (60–66) have also been based on the use of this derivative.

Codeine forms a O^6-trimethylsilyl ether; this derivative is suitable for analytical studies. Diacetylmorphine does not require derivative formation. O^6-Acetylmorphine forms a O^3-trimethylsilyl ether. Normorphine forms a ditrimethylsilyl ether, in the same fashion as morphine. The secondary amino group also reacts with most silylating reagents, but not with *N*-trimethylsilylimidazole (67), to yield an *N*-trimethylsilyl derivative. Compounds of this type are active silylating agents, and when they are employed as derivatives it is not unusual to find both the free amine and the *N*-trimethylsilyl derivative present during the GC separation.

The phenolic group may be selectively silylated in the presence of other alcohol groups with isopropenyltrimethylsilane (67).

C. ALKYL DERIVATIVES

Peralkylation of morphine may be carried out according to a procedure first described by Corey (68,69) and later applied by Hakomori (70) for the permethylation of sugars and by Haegele et al. (71) for the peralkylation of peptides and amino acids.

Procedure. In the experimental preparation of diethylmorphine, morphine hydrochloride (10.7 mg, 0.1 mmole) was dissolved in 600 μl of dimethylsulfoxide (distilled over calcium hydride). To this solution, 150 μl of a 1*M* solution of methylsulfinylmethide carbanion was added. The reaction mixture was sonicated for 10 min to break gel particles that were formed. This was followed by the addition of 10.5 μl of ethyl iodide (eqmolar excess), and the reaction mixture was sonicated for 50 min. Ice and water were added (approximately 1 ml) and the diethylmorphine was extracted with 2 ml of chloroform. The chloroform solution was washed three times with 1-ml portions of water, and the solvent was removed with a stream of nitrogen. The reaction is conveniently carried out in a 3.5-ml screw-cap vial, which is flushed with nitrogen when reagents are added, since the carbanion solution is extremely sensitive to moisture and to oxygen. The yield was 10.9 mg (96%).

Alkyl derivatives are more stable than TMS derivatives and do not increase the molecular weight significantly. However, they have not been used for analytical purposes.

D. CARBAMATE DERIVATIVES

Chloroformates react with tertiary amines to yield carbamates.

$$R_1R_2NCH_3 + ClCOOR_3 \rightarrow R_1R_2NCOOR_3 \qquad [1]$$

The carbamates of morphine have been used in synthetic work for the preparation of the secondary amine (72–74) or for isotope labeling (75). Hartvig and Vessman (76,77) described methods for the determination of tertiary amines after demethylation to secondary amines and subsequent perfluoroacylation before electron capture gas chromatography. The reaction may be carried out in one step with the appropriate chloroformate (e.g., pentafluorobenzylchloroformate) (78).

2. Column Selection

Since morphine and morphinelike analgesics are basic compounds, it is necessary to avoid any accumulation of acidic material on the column. Repeated injections may result in erosion of the stationary phase at the head of the column; the best way of avoiding this circumstance is to employ an initial 1- to 2-cm zone of 10% SE-30 packing, according to the practice of Thénot and Horning (58).

A base-treated support may be used to prepare the packing. However in our experience tailing is not observed with a 3 to 5% coating.

A survey of the literature shows that silicone phases (OV-17, SE-30) are the most widely used of all liquid phases. Moffat (79) reported the retention index of 180 drugs on SE-30, which he called the "preferred stationary phase" for the identification of basic drugs. With a nonpolar phase like SE-30, compounds are eluted according to their size and some predictions of retention behavior can be made. Normorphine diTMS, morphine diTMS, and normorphine triTMS are eluted from a SE-30 column in that order, as expected from the size of the molecule: $H < CH_3 < Si(CH_3)_3$.

3. Detectors

A. FLAME IONIZATION DETECTOR

The flame ionization detector (FID) is the most widely used of all GC detectors because of its stability, its large dynamic response range, and its universal applicability. No special derivative is required for FID. The sensitivity is quite adequate for most applications; Ikekawa et al. (24) reported 10 ng as a detectable amount of morphine TMS.

B. ELECTRON CAPTURE DETECTOR

Several orders of magnitude in sensitivity of detection may be gained over the FID with the electron capture detector. However it is necessary to prepare special derivatives to obtain such an improvement. Perfluoroacyl derivatives are suitable for electron capture detection. With the pentafluoropropionyl derivative, Dahlström and Paalzow (80) obtained a sensitivity limit of about 5 pg for morphine: 100 pg could be determined in a 30-mg brain tissue sample.

C. NITROGEN DETECTOR

A specific response to compounds containing nitrogen (and phosphorus) may be obtained when the collector plate of a conventional flame ionization detector is replaced by an alkali-coated ceramic. The exact mechanism of the response is not fully understood.

The selective detection of nitrogen-containing compounds is an attractive technique. Previous difficulties (mainly lack of stability) have prevented the full investigation of the use of the nitrogen detector. Recent improvements in technology should result in a wider use of this detector. Applications to morphine-type compounds were reported by Riedmann (81) and Smith and Cole (82).

4. Quantification

Quantification by gas chromatography is carried out by adding a suitable internal standard. This standard should be chemically related; it should also have a similar retention behavior, and it should be added at the first steps of the sample preparation.

Tetraphenylethylene was used in the early work of Wilkinson and Way (57), but ethylmorphine (55) and nalorphine (53) are more appropriate internal standards.

IV. ANALYSIS BY GAS CHROMATOGRAPHY—MASS SPECTROMETRY—COMPUTER SYSTEMS

1. Introduction

Analytical systems based on a combination of a gas chromatograph, a mass spectrometer, and a computer, and operated as a single instrumental system, provide the most powerful and most reliable method of analysis now known for the study of complex mixtures of biologic origin. They are particularly valuable in studies of drugs and drug metabolism. The function of the gas chromatograph is to separate components of the mixtures under investigation. For example, most drugs yield multiple metabolites; some metabolites may have a physiological action related to that of the original drug, some may have toxic properties due to their specific structure, and some may be inactive. The structural differences introduced through metabolic transformations are usually such that separation of the parent drug and individual metabolites is possible with ordinary GC columns. It is usually necessary to prepare derivatives before the instrumental analysis step, since many metabolites contain polar groups that would lead to undue adsorption if derivatives were not prepared. The mass spectrometer provides an intermittent or continuous record of mass spectral data. Quantitative analyses are usually carried out by using the mass spectrometer as a specific ion detector.

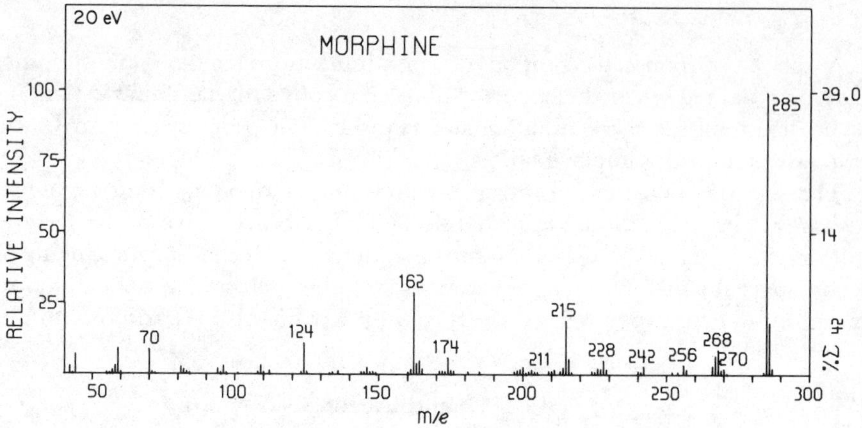

Figure 2. EI ionization (20 eV) mass spectrum of morphine.

2. Ionization Processes

A. ELECTRON IMPACT

In an electron impact (EI) source, an electron beam produced by a heated filament ionizes the sample upon collision.

$$M + e \rightarrow M.^{+} + 2e \qquad [2]$$

The molecular ion $(M.^{+})$ usually decomposes to many fragments. Within a class of compounds, similar fragmentation patterns are observed. In the morphine series, the molecular ion is always intense and is usually the base peak, as shown by the EI mass spectra of morphine (Figure 2) and diacetylmorphine (Figure 3). The fragmentation pattern of morphine alkaloids has been investigated in detail by several authors (83–87).

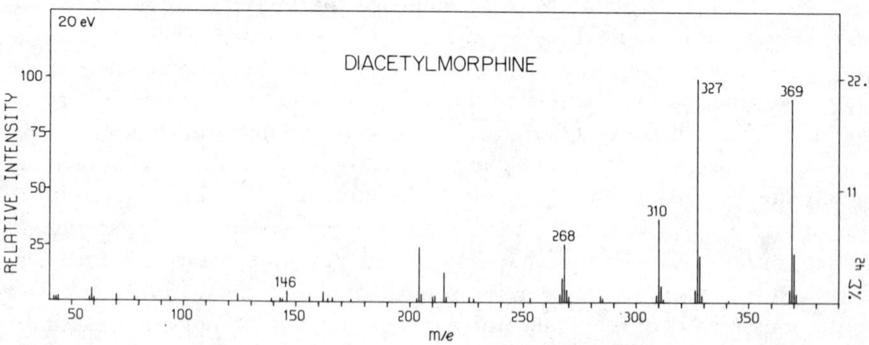

Figure 3. EI ionization (20 eV) mass spectrum of O^3,O^6-diacetylmorphine.

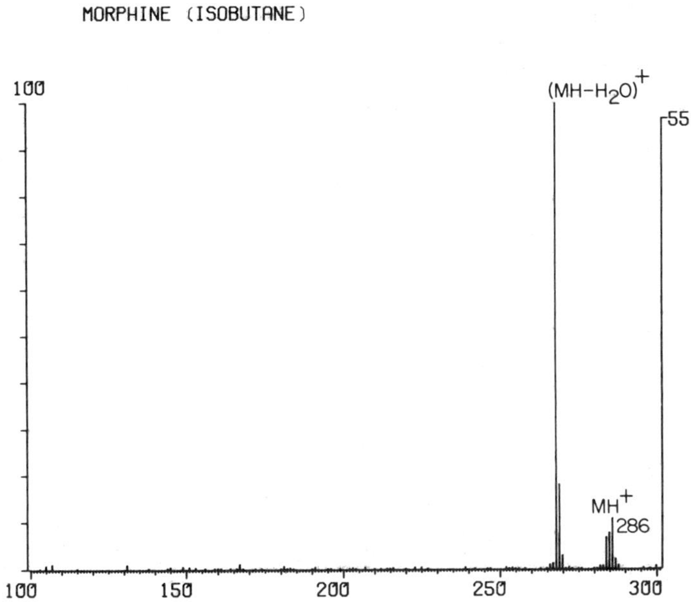

Figure 4. CI mass spectrum of morphine with isobutane as the reagent gas.

B. CHEMICAL IONIZATION

In chemical ionization (CI) mass spectrometry, the primary process is the ionization of a reagent gas by a high-energy electron beam. The sample is ionized through ion-molecule reactions with ions derived from the reagent gas.

Positive ions may be formed by proton transfer, by hydride abstraction, or by charge exchange. With methane as the reagent gas, Field and Munson (88,89) have shown that the reactive species at 0.5 to 1 torr are principally CH_5^+ and $C_2H_5^+$. Under similar conditions, isobutane produces the $C(CH_3)_3^+$ ion, which has a lower energy than CH_5^+. Chemical ionization with NH_4^+ is possible with ammonia or with 10% NH_3 in nitrogen as the reagent. Analyses of morphine and morphine-related compounds have been reported with chemical ionization using methane (59,90) or isobutane (91,92). With these reagents an MH^+ ion is formed, but the main reaction products are ions resulting from the loss of R_2OH. The mechanism is an acid-catalyzed elimination reaction occurring in the gas phase. With morphine, loss of water occurs only at C^{-6} (Figure 4); if morphine is derivatized, loss of acetic acid (Figure 5) or trimethylsilanol occurs with acetyl or TMS derivatives (Table I). The O^6-acetyl and O^6-TMS groups are eliminated.

Under chemical ionization conditions, substitution at the 6- or 3-position can be easily differentiated, as shown by the CI spectra of O^3-and O^6-acetylmorphine (Figure 6).

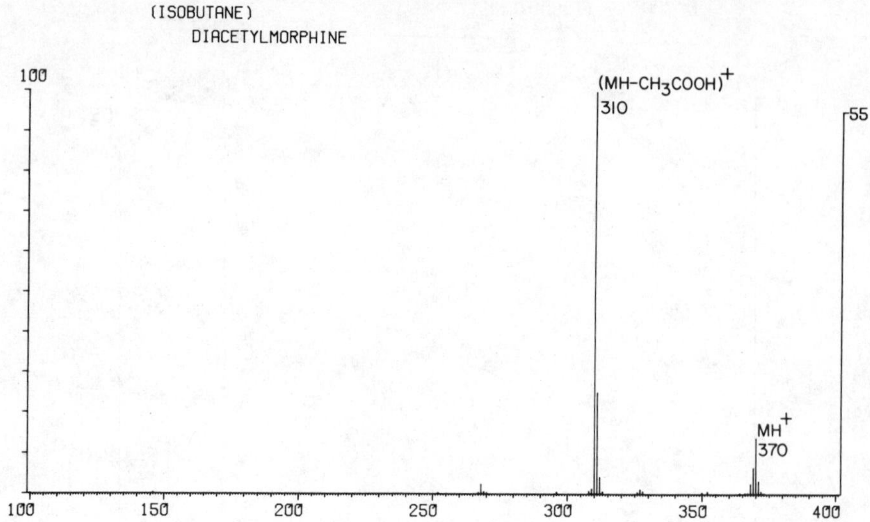

Figure 5. CI mass spectrum of O^3,O^6-diacetylmorphine with isobutane as the reagent gas.

Loss of a methyl group from TMS groups occurs when methane is the reagent gas, but not with isobutane. The ions observed under chemical ionization conditions with methane as the reagent gas, for morphine TMS, are MH^+, $(MH - CH_4)^+$, and $(MH - 90)^+$.

Ammonia, used as a 10% mixture in nitrogen, gives the same spectrum as

TABLE I

FRAGMENTATION OF MORPHINE DERIVATIVES
- CI METHANE -

R_1 (C_3)	R_2 (C_6)	MH^+	%	$(MH-R_2OH)^+$	%
CH_3	H	300	32	282	100
CH_3	TMS	372	30	282	100
CH_3	Ac	342	14	282	100
CD_3	CD_3	320	40	285	100
H	H	286	35	268	100
H	Ac	328	19	268	100
Ac	H	328	18	310	100
Ac	Ac	370	6	310	100
TMS	H	358	35	340	100
TMS	TMS	430	58	340*	100
TMS	Ac	400	25	340	100

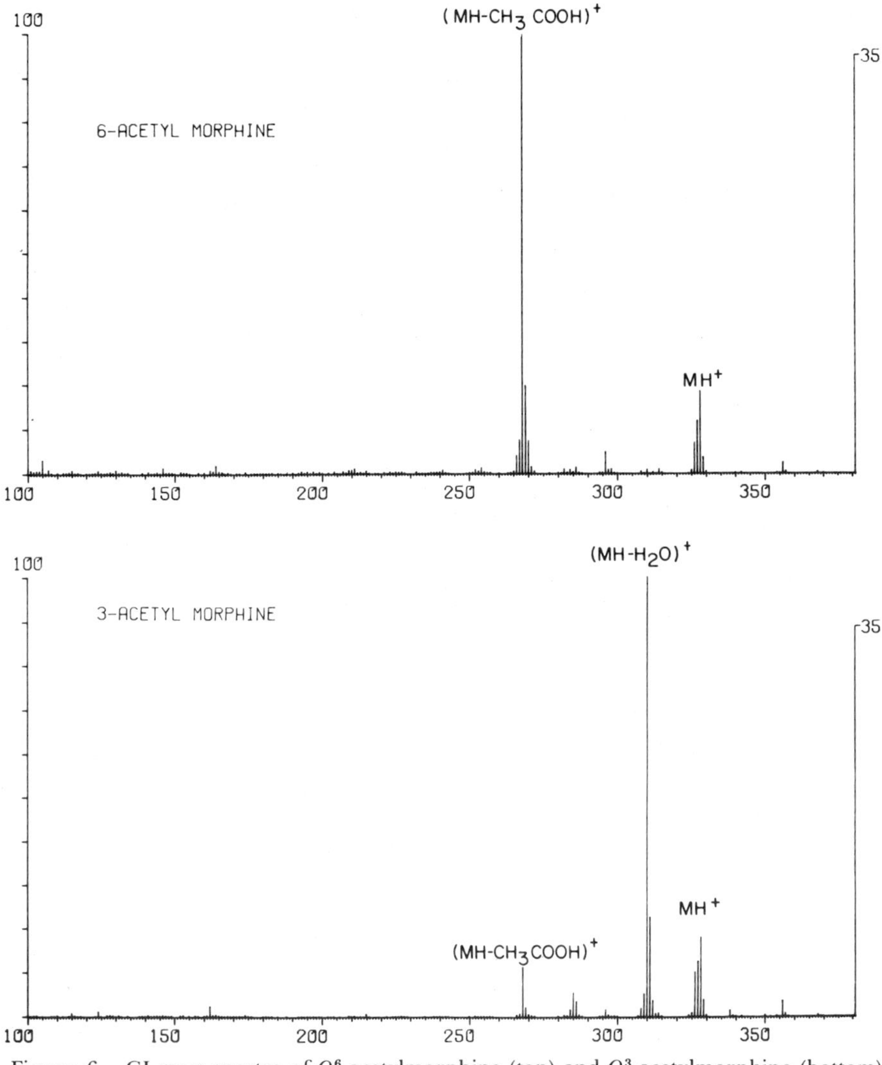

Figure 6. CI mass spectra of O^6-acetylmorphine (top) and O^3-acetylmorphine (bottom) with methane as the reagent gas. The O^3-acetylmorphine sample may have contained a small amount of O^6-acetylmorphine as an impurity, giving rise to the $(MH - CH_3COOH)^+$ ion.

found with isobutane; the ions correspond to MH^+ and $(MH - 90)^+$ for morphine TMS.

Chemical ionization conditions are usually used in quantitative work, whereas electron impact ionization usually serves for identification and structural studies.

Ionized helium (He^+) and nitrogen (N_2^+, N^+) give, through charge transfer, a spectrum similar to an EI spectrum.

Jardine and Fenselau (93,94) have shown that nitric oxide, diluted with nitrogen, gives with morphine a spectrum simpler than the one obtained under electron impact. This is the result of a relatively low-energy charge-transfer reaction. The chief ion is the molecular ion $M.^+$.

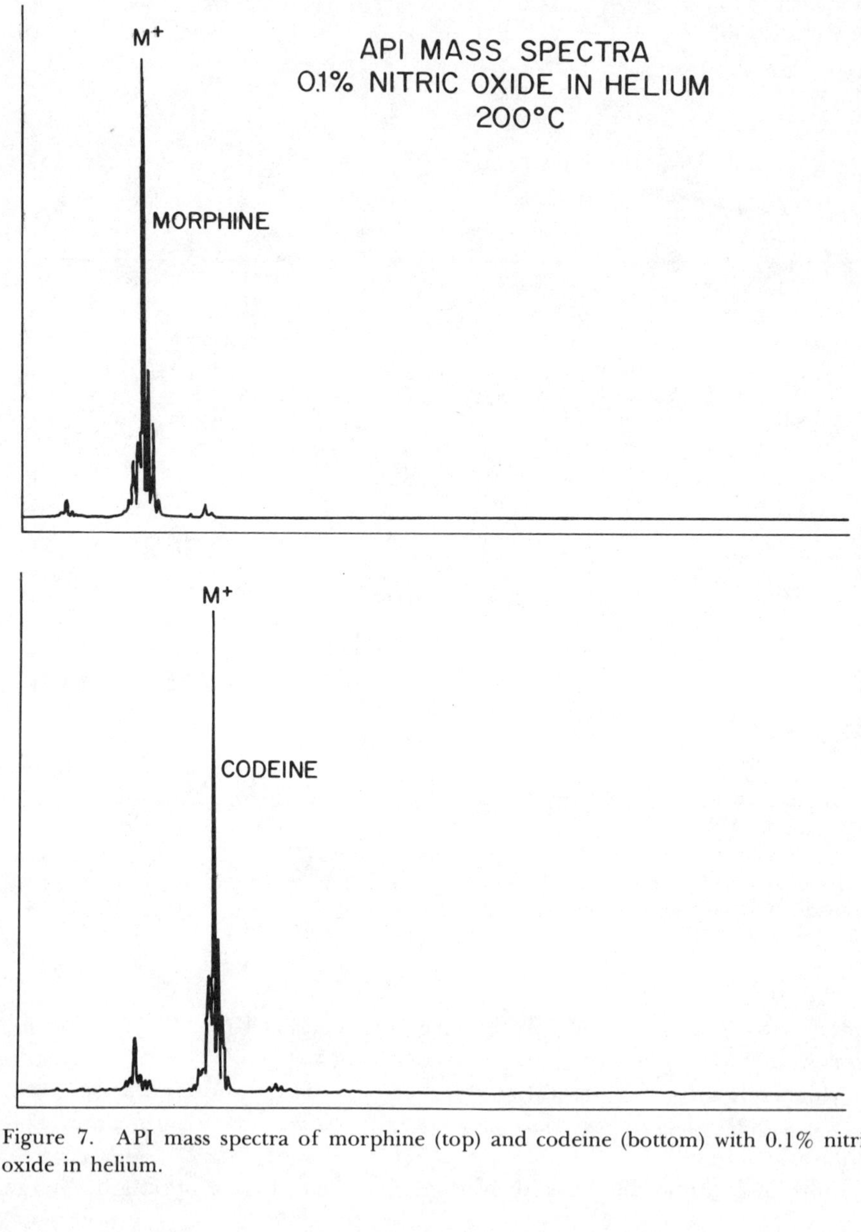

Figure 7. API mass spectra of morphine (top) and codeine (bottom) with 0.1% nitric oxide in helium.

C ATMOSPHERIC PRESSURE IONIZATION

In atmospheric pressure ionization (API) mass spectrometry (95), positive and negative ions are generated in a reaction chamber external to the low-pressure mass analyzer region of a quadrupole mass spectrometer. Through a small aperture in the wall of the reaction chamber, ions are admitted into the mass analyzer. Samples are introduced into the reaction chamber either by direct injection (95–99) or by means of a liquid chromatograph (100,101) or a gas chromatograph (102). The reaction chamber contains a ^{63}Ni source and/or a corona discharge electrode as a source of high-energy electrons, which in turn ionize a carrier gas such as nitrogen. If the nitrogen contains a low concentration of a reagent gas (e.g., 0.1% of nitric oxide), NO^+ formed by [3] is present as a reagent ion:

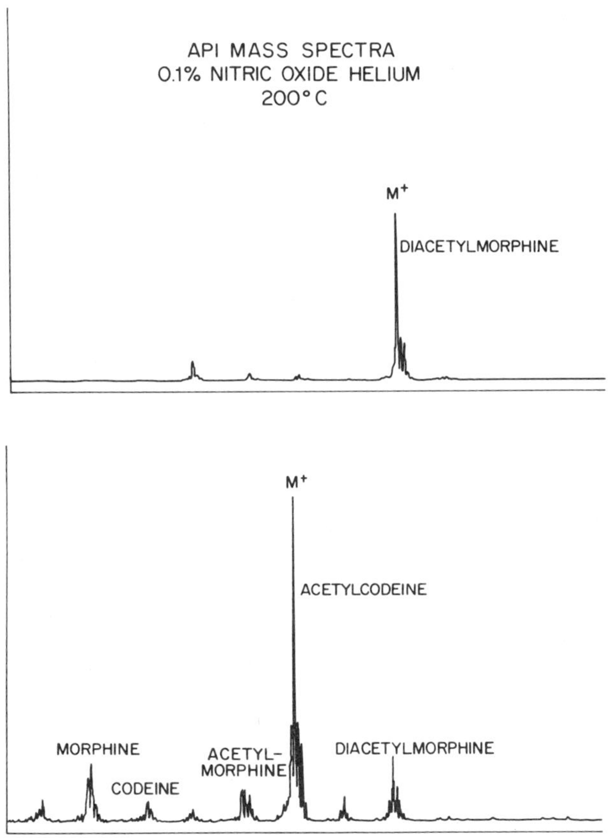

Figure 8. API mass spectra of O^3,O^6-diacetylmorphine (top) and an acetylated codeine sample (bottom) with 0.1% nitric oxide in helium. Morphine is present as an impurity in the codeine sample.

$$N_4^+ + NO \rightarrow NO^+ + 2N_2 \quad [3]$$

The NO^+ reagent ions ionize the sample by [4] and [5].

$$NO^+ + M \rightarrow (M - H)^+ + HNO \quad [4]$$

$$NO^+ + M \rightarrow M^+ + NO \quad [5]$$

An example of [5] is illustrated in Figure 7, for morphine and codeine. The advantage of using NO^+ ions for ionization and detection of these compounds by CI mass spectrometry is described in refs. 93 and 94. The advantage of using API mass spectrometry lies mainly in high sensitivity of detection in addition to versatility of the sample introduction. This technique may be used to detect impurities in morphine-related preparations. For example, a sample of acetylcodeine (O^3-methyl-O^6-acetylmorphine) was analyzed by API mass spectrometry with nitric oxide as the reagent gas. The original sample evidently contained free morphine, since the minor components were found to be diacetylmorphine, a monoacetylmorphine, and morphine, as well as codeine (Figure 8).

3. Data Acquisition and Processing

A. REPETITIVE SCANNING

If the primary purpose of the analysis is to obtain qualitative data, the system may be operated manually to record mass spectra under the GC peaks. The development of small laboratory computers made it possible to design analytical systems in which data are recorded continuously. The procedure of mass chromatography, developed by Biemann (103), involves the continuous cycling of the mass spectrometer to provide a series of mass spectra obtained throughout the separation process. Each scan may require about 2 to 6 sec, depending on the mass range selected for the scan. Some relaxation time is required between scans when the scan is obtained by magnetic field changes; if the scan involves electrical field or accelerating voltage changes, the cycling is essentially continuous. An analysis may require 5 to 10 min if only one or two compounds are under study; if multicomponent analyses are needed, the analysis time may be 30 to 60 min or more. The mass spectral data are subjected to computer-based analysis. The programs may be relatively simple, but generally a sophisticated program is required to accommodate problems of incomplete separation.

The greatest value of the repetitive scanning approach lies in its unparalleled capabilities for the detection of specific compounds, often in small amount, of interest because of their beneficial or toxic physiological action. For this reason, electron impact spectra are almost always obtained for analysis. It is also possible in some instances to employ a charge-transfer mode of operation with nitrogen as a carrier gas, but this form of operation has never been investigated in detail. The use of fragmentation spectra is advantageous because it usually allows the analyst to arrive at a unique identification when EI mass spectral data are combined with GC data. Retention behavior is a physical property that is based on the free energy of solution of the solute in the liquid phase under the conditions of the separation, whereas the fragmentation spectrum is based on the chemical structure of the compound.

Some types of isomer give virtually identical EI mass spectra, but the retention times differ. Structurally unrelated compounds may show virtually identical retention behavior with a specific column, but the EI spectra will be different. Identifications based on criteria involving both physical properties and chemical structure have high validity. The system is particularly valuable when complex biological mixtures are analyzed for drugs; the computer allows rapid identification of each drug by comparison with spectra in a library. Several collections of drug mass spectra have been published, and these data are usually furnished with commercial instruments.

B. SELECTED ION MONITORING

In the selected ion monitoring mode, the mass spectrometer is set up to monitor continuously one or several ions. This technique, originally called mass fragmentography, was developed for the detection and quantification of small amounts of sample (104). The method is widely used in drug metabolism studies, and a recent article reviews its biomedical applications (105).

4. Quantification

A. PRINCIPLE

For quantitative work, a stable isotope-labeled internal standard is added to the sample, and the analytical system is operated in the selected ion monitoring mode. Ions corresponding to the unlabeled and labeled drug are monitored.

The preferred stable isotope label is ^{13}C, since there is no discernible isotope effect in the separation or ionization processes for ^{13}C compounds. It is customary to introduce three or more ^{13}C atoms. The adsorption losses are the same for both labeled and unlabeled compounds, and the retention behavior is the same. Deuterium-labeled compounds are also used, usually satisfactorily, although there may be a recognizable difference in retention behavior (and perhaps in adsorption losses). For compounds with NCH_3 groups, the usual label is NCD_3. Homologs and analogs have also served as internal standards. They are less satisfactory than stable isotope-labeled compounds, but they have been selected for some applications. The usual practice is to monitor two or four ions (one or two each for the compound under investigation and one or two each for the internal standard), and most programs allow for the monitoring of eight ions if necessary (for multicomponent analyses).

Complicated technical problems are associated with quantitative work. From an instrumental point of view, it is necessary to employ power supplies of high stability in quadrupole instruments, and to have a means of detecting or correcting drift for both magnetic field and electrical field instruments. The peak setting is usually made to the nearest 0.1 amu, and adjustments for mass defects may be required. Derivatives should be selected to minimize adsorption losses, and the sample size should be large enough to avoid errors in ion intensity measurements.

B. PREPARATION OF INTERNAL REFERENCE COMPOUNDS LABELED WITH DEUTERIUM

Internal reference compounds labeled with stable isotopes are the most suitable standards for quantitative analysis by gas chromatography–mass spectrometer–computer techniques. Deuterium-labeled standards possess physical properties nearly identical with to those of their corresponding unlabeled analogs, but they are distinguishable by mass spectrometry. Deuterium-labeled internal standards are commonly used for the quantitative analysis of drugs and drug metabolites because they can be synthesized easily and cheaply. Since many drugs contain an *N*-methyl function, the most accessible site for the introduction of the deuterium label is by forming the *N*-demethylated compound (normorphine in this work), which in turn is then alkylated using d_3-labeled methyl iodide to form the N-d_3-labeled drug (N-d_3-morphine). This compound can also be prepared by reducing the carbamate with lithium aluminum deuteride.

a. Preparation of Normorphine: Cyanogen Bromide Method of von Braun. Diacetylmorphine acetate (215 mg, 0.5 mmole) was dissolved in 4 ml of anhydrous chloroform. A solution of 96 mg of cyanogen bromide (0.9 mmole) in 1 ml of chloroform was added to the solution, and the reaction mixture was heated under reflux for 2.5 hr. The chloroform was evaporated, and the residue was treated with 5 ml of boiling water. The solution was allowed to cool, and the precipitate was removed by filtration. After recrystallization from an ethanol-water solution, colorless needles of diacetyl-*N*-cyanonormorphine were obtained. The yield was 164 mg (86%).

Upon refluxing 121 mg (0.33 mmole) of diacetyl-*N*-cyanonormorphine for 5 min with concentrated hydrochloric acid, the two ester functions were saponified and *N*-cyanonormorphine crystallized from the cooled mixture. To complete the crystallization process, the mixture was refrigerated overnight. The product was removed by filtration. The yield was 94 mg (95%).

N-Cyanonormorphine was converted to normorphine by refluxing 94 mg (0.317 mmole) with 60 ml of 6% hydrochloric acid for 6 hr. The solvent was removed *in vacuo* and the residue was dissolved in ethanol. Normorphine hydrochloride was precipitated upon addition of *n*-pentane. Storage of the mixture (freezer) completed the precipitation process. The product was removed by filtration, washed with cold *n*-pentane, and dried. The yield was 79 mg (81%) of normorphine hydrochloride.

b. Preparation of Morphine-*N*-CD_3 (Morphine-d_3). The synthesis of morphine labeled with deuterium in the *N*-methyl group can be accomplished either through direct methylation of normorphine with d_3-methyl iodide or by the reduction of a carbamate ester of normorphine with lithium aluminum deuteride. The direct route was used by Ebbinghausen et al. (54), but as expected the yield was low and other products were obtained as well (codeine-d_6, unreacted normorphine, and norcodeine-d_3). The method of choice is therefore the utilization of a carbamate ester that does not require the preparation of normorphine as the starting compound for the introduction of the N-CD_3 group.

The most satisfactory method involves the use of ethyl chloroformate to effect *N*-demethylation of morphine, leading to formation of the corresponding normorphine carbamate as described by Elison et al. (75). The reduction of *N*-carbophenoxynormorphine, according to Abdel-Monem and Portoghese (72), and reduction of *N*-trichlorocarbethoxynormorphine, as described by Montzka et al. (73), were less satisfactory.

c. Synthesis of O^3, O^6,*N*-Tricarbethoxynormorphine. The procedure described by Elison et al. (75) was followed without major change, but the final product was identified as O^3, O^6,*N*-tricarbethoxynormorphine, and not O^3,*N*-dicarbethoxynormorphine as indicated by the authors.

Normorphine as the free base (28 mg, 0.01 mmole), 0.4 ml (4 mmole) of ethyl chloroformate, and 1 g (20m*M*) of potassium hydroxide in 6 ml of water and 10 ml of chloroform were shaken in a separatory funnel for 15 min. The chloroform layer was collected, and the aqueous phase was extracted with 2 portions of 10 ml of chloroform. The combined chloroform extracts were washed with 1*N* hydrochloric acid and with water. The chloroform solution was evaporated. The yield was 44.2 mg (88%) of a slightly yellow, resinlike material, identified by its mass spectrum as a diester carbamate.

d. Preparation of *N*-CD_3-Morphine (morphine-d_3). O^3, O^6,*N*-Tricarbethoxynormorphine (79 mg, 0.162 mmole) was dissolved in 5 ml of tetrahydrofuran, which had been freshly distilled from lithium aluminum hydride. To this solution, a suspension of 42 mg (1 mmole) of lithium aluminum deuteride in 2 ml of tetrahydrofuran was added dropwise and with stirring. After the addition was completed, the reaction mixture was heated under reflux for 2 hr. Ethyl acetate was added to destroy excess reagent. This was followed by the addition of 25 ml of 2*N* hydrochloric acid and 4 g of potassium tartrate, and the mixture was heated under reflux for 2 hr. After adjusting the pH to 8.3 with aqueous potassium hydroxide, the mixture was extracted with methylene chloride for 24 hr by using a continuous extractor. After evaporation of the solvent, 22 mg (47%) of N - CD_3-morphine (morphine-d_3) was obtained.

e. Preparation of O^6-Acetylmorphine. Hydroxylamine hydrochloride (60 mg) in water (0.3 ml) was added to 100 mg of diacetylmorphine in ethanol (3 ml). The mixture was allowed to react overnight at room temperature. O^6-Acetylmorphine crystallized upon cooling (+ 4 °C). The product was removed by filtration, washed with ice-cold water, and dried under vacuum. The sample was contaminated with less than 1% of unreacted diacetylmorphine.

The same procedure may be used for the preparation of O^6-acetylmorphine-d_3 with diacetylmorphine-N - CD_3 as the starting material.

V. APPLICATIONS

1. Analysis of Diacetylmorphine and Its Metabolites in Biological Fluids

The literature dealing with morphine and morphine-related compounds is extensive and is distributed through many disciplines. Comparatively little quantitative data relating to morphine metabolism and distribution in humans are available, however, because of the earlier lack of analytical methods with sufficient sensitivity of detection and specificity for use in human studies. The recent review of Boerner et al. (106) summarizes current knowledge of the metabolism of morphine and heroin in humans.

Diacetylmorphine (heroin) is rapidly metabolized to both O^6-acetylmorphine and O^3-acetylmorphine, but the rate of enzymic hydrolysis of the ester group at the 3-position is considerably faster than that of the corresponding group at the 6-position. As a consequence, the apparent route of metabolism is O^3,O^6-diacetylmorphine→ O^6-acetylmorphine → morphine.

Since hydrolytic enzymes are widely distributed in the body, the removal of the O^3- and O^6-acetyl groups probably begins immediately and occurs at many sites after heroin ingestion. The distribution of these three compounds, however, is likely to be somewhat different for each compound. The rate of entry of diacetylmorphine into the central nervous system is believed to be faster than that of morphine. After a very short period, however, the metabolic processes that occur are those of morphine itself. The concentration of morphine in blood falls relatively rapidly after a single dose of either heroin or morphine, but low concentrations of morphine persist for a long time.

From a mass transfer point of view, the principal metabolite of morphine is the 3-glucuronide, a urinary excretion is the principal pathway of excretion. The O^6-glucuronide is also a human (and animal) metabolite, but the rate of formation of the O^6-glucuronide is very much slower than that of the O^3-isomer.

The metabolic problems of interest, particularly in terms of physiologically active compounds, involve reactions other than conjugation. Two routes of considerable interest are *N*-demethylation to form normorphine and *O*-methylation to form codeine.

Recognized major and minor pathways, accounting for 75 to 85% of administered drug, are shown in Chart 1. Compounds that have not been identified as morphine metabolites but are probably also formed as human products include the O^3-glucuronide of O^6-acetylmorphine, the O^6-sulfate of morphine, the O^3-glucuronide and O^3-sulfate of normorphine (the expected conjugates), norcodeine, and unidentified conjugates of codeine. Since the deacetylation reaction occurs very rapidly, studies of the metabolism of diacetylmorphine after an initial period become equivalent to a study of morphine metabolism. The chief analytical problems arise from the relatively low concentration of drugs and their metabolites in urine and in blood.

The method chosen for hydrolysis of urinary and blood conjugates of morphine was adapted from techniques used in the study of human urinary steroids. The enzyme (Glusulase) contains both glucuronidases and sulfatases. The hydrolysis rates of glucuronides and sulfates with this enzyme mixture are influenced by steric effects, but both O^3- and O^6-conjugates of morphine are hydrolyzed relatively easily. Urinary rates may be slowed by inhibitors (107).

The method used for the extraction of morphine was based on studies of salt–solvent pair extraction of drugs (18). The fluid (urine or diluted 3–2 plasma or serum) was saturated with ammonium carbonate and extracted with ethyl acetate. Free morphine can be extracted under these conditions; the initial extraction process, however, yields a mixture that requires additional treatment. Morphine and its basic metabolites (normorphine, codeine) were returned to an aqueous phase by extraction of the organic phase with dilute hydrochloric acid. The reextraction of the aqueous solution was carried out with 3–1 chloroform–isopropanol after saturation with ammonium carbonate. This procedure provides a sample suitable for derivatization and instrumental analysis.

When enzymic hydrolysis of urine was employed, morphine and its basic metabolites were removed from aqueous solution by ion-exchange chromatography (AG 50W). After elution with hydrochloric acid (4*N*), the desired

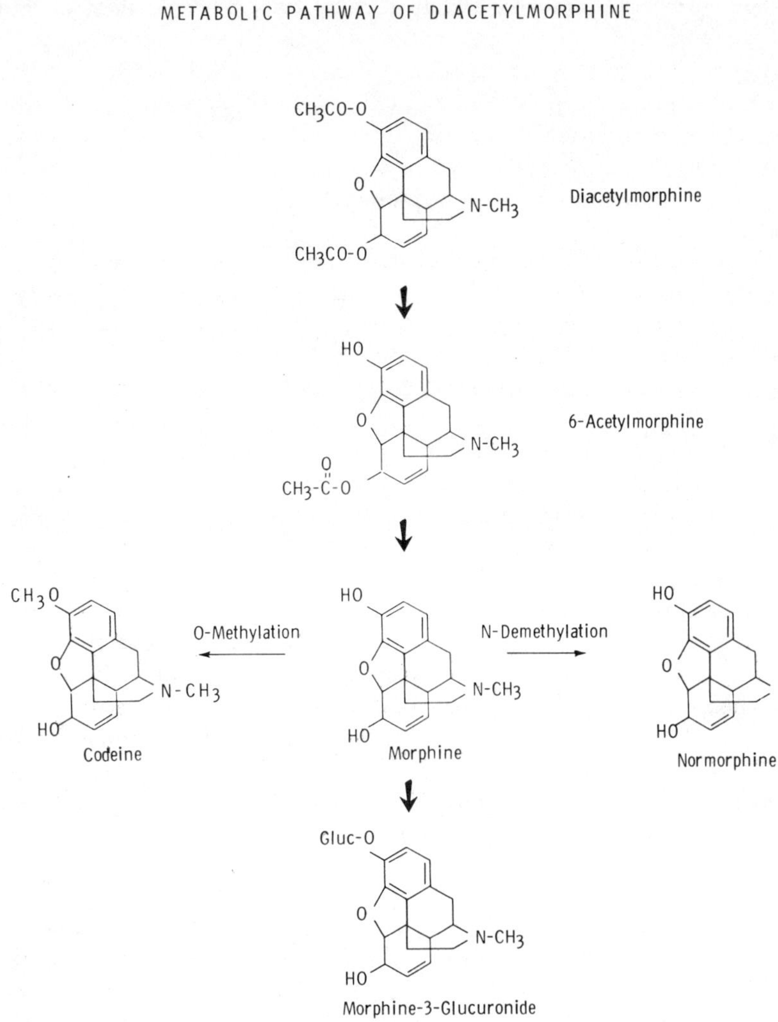

Chart 1. Metabolic pathway of diacetylmorphine.

compounds were extracted with ethyl acetate-ammonium carbonate.

The direct study of glucuronides by gas phase procedures is possible; the most satisfactory derivatives are the methylester trimethylsilyl ethers. Conjugates of morphine have not been studied in this way, but the analytical information that is usually needed is that resulting from an estimation of free and conjugated morphine. In recent studies, free and conjugated values were determined.

A. ANALYSIS OF URINE

EXPERIMENTAL. Ammonium carbonate (solid) was added to saturate 5.0 ml of urine, to which 1.5 μg of morphine-d_3 (NCD_3-morphine) had been added, and the aqueous solution was extracted twice with 5-ml portions of ethyl acetate. The combined organic extracts were dried with anhydrous sodium sulfate, and the solvent was evaporated with the aid of a nitrogen stream.

For the determination of morphine, the sample was converted to derivative form by treatment with bis-trimethylsilylacetamide (25 to 50 μl; 25 μl was used when morphine concentrations were low) at 60 °C for 20 min. Under these conditions morphine forms a ditrimethylsilyl derivative, and codeine forms a monotrimethylsilyl derivative. O^6-Acetylmorphine forms a monotrimethylsilyl derivative, and O^3,O^6-diacetylmorphine remains unchanged. For

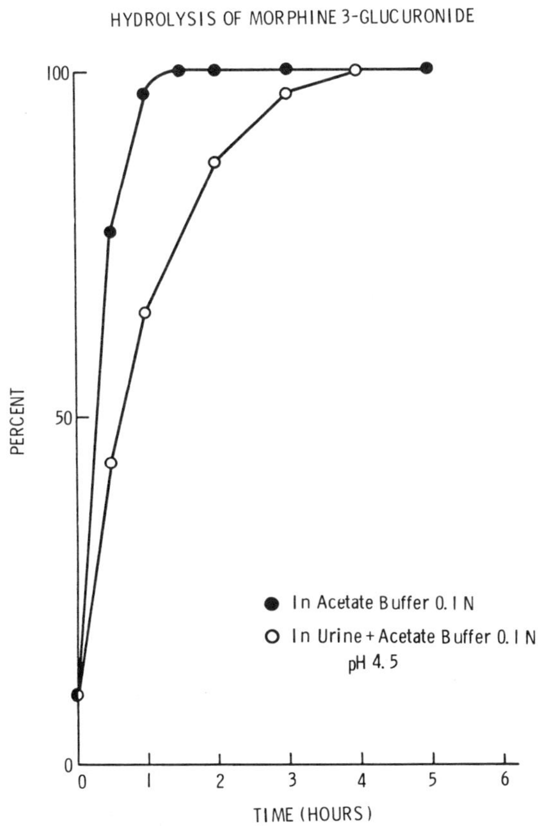

Figure 9. Hydrolysis rate of morphine-3-glucuronide in acetate buffer, 0.1*N*, pH 4.5, and in urine + acetate buffer, 0.1*N*, pH 4.5, with enzyme preparation Glusulase.

the determination of normorphine, 25 μl of *N*-trimethylsilylimidazole was used as the derivatizing agent under the same conditions; a ditrimethyl derivative was formed. (Conversion to tritrimethylsilyl derivative occurs with other silylating reagents such as BSA).

Under CI conditions with methane, the ions of interest for the quantification of morphine are at 430.2, 414.2, and 340.2 amu. These correspond to the ions MH^+, $(MH - 16)^+$, and $(MH - 90)^+$. The related ions for morphine-d_3 are 3.0 amu greater. Ratios of ion intensity values were used to calculate the morphine concentration in the sample; two pairs of ions were always monitored, to decrease the possibility of error due to unrecognized interference by other urinary components. The ions chosen are usually MH^+ and $(MH - R_2OH)^+$ from O^3-acetylmorphine, O^6-acetylmorphine, or codeine.

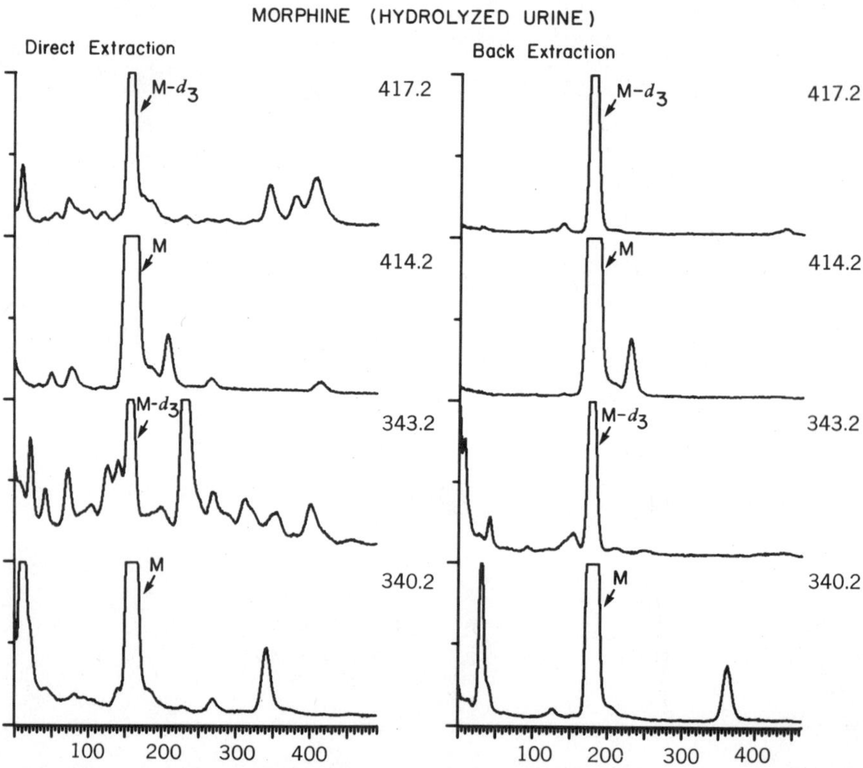

Figure 10. Selected ion monitoring charts for the analysis of morphine (M) in urine after enzymatic hydrolysis, using the TMS derivative and the CI mode (methane). The ion pairs at 340/343 and 414/417 were monitored. Left: analysis of a sample extracted directly from urine. Right: analysis of the same sample purified by back extraction. In both instances the internal standard was NCD_3-morphine (M-d_3).

Results. The determination of total morphine concentration in urine presents a number of difficulties. Yeh (107) found that the hydrolysis of the O^3-glucuronide of morphine was slow and dependent on the volume of urine. This effect may be attributable to unrecognized inhibitors, but a more likely source of difficulty lies in the presence of steroid glucuronides, which may act as competitive substrates. When a large excess of enzyme was used, the hydrolysis of conjugates was complete in several hours (Figure 9). Direct extraction, however, yielded a sample that was not suitable for analysis. Continued study of the problem indicated that steroids were responsible for the observed interferences.

Figures 10 and 11 show the nature of the problem, and two solutions. Ion

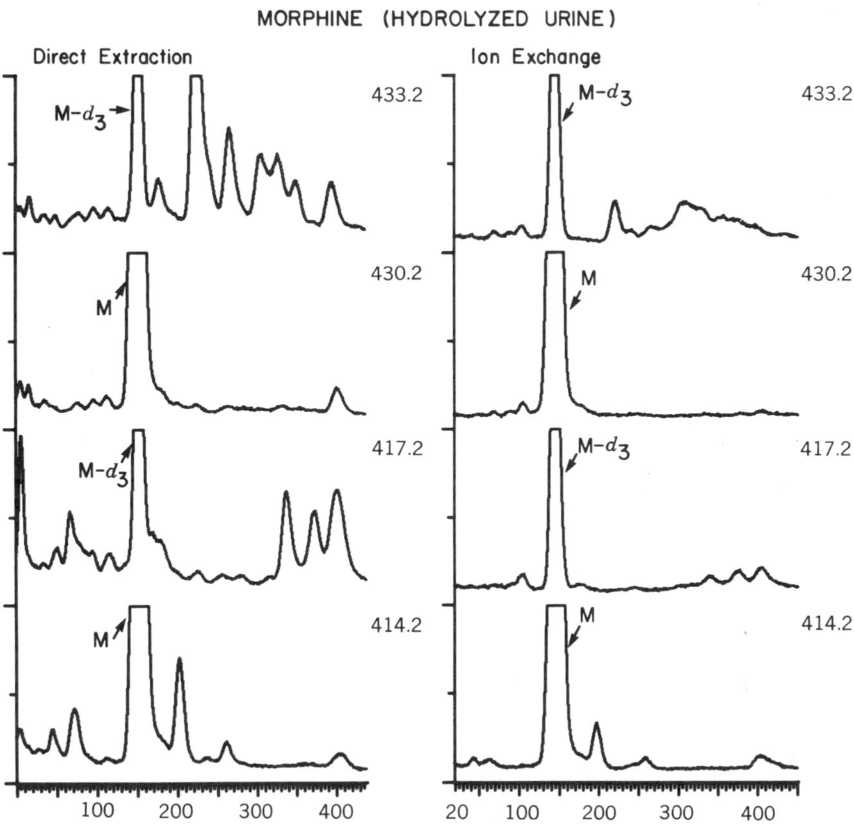

Figure 11. Selected ion monitoring charts for the analysis of morphine (M) in urine after enzymatic hydrolysis using the TMS derivative and the CI mode (methane). The ion pairs at 414/417 and 430/433 were monitored. Left: analysis of a sample extracted directly from urine. Right: analysis of the same sample purified by an ion-exchange procedure. In both instances the internal standard was NCD_3-morphine (M-d_3).

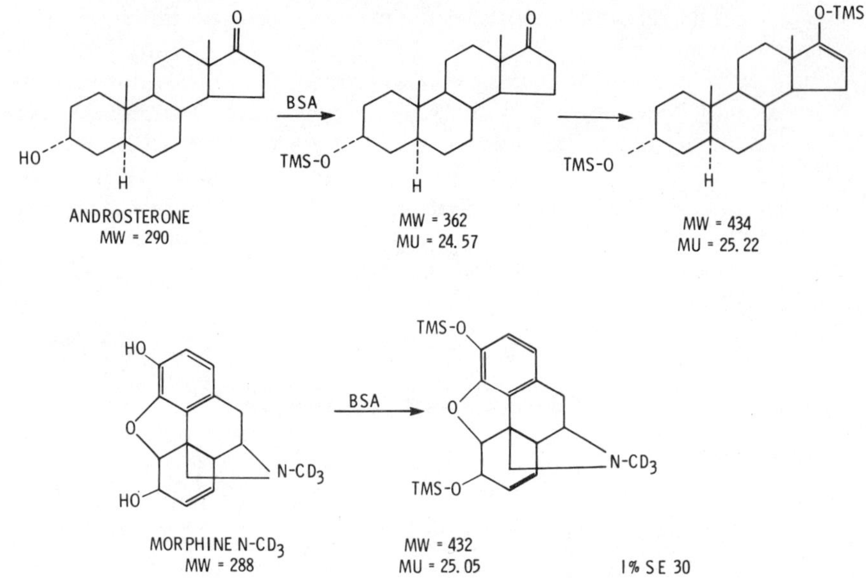

Chart 2. Trimethylsilylation of morphine NCD_3 and androsterone with bis-trimethylsilylacetamide (BSA). Addition of each trimethylsilyl group increases the molecular weight by 72 amu. Morphine, di-TMS, and O^3,O^{17}-ditrimethylsilyl androsterone are not separated on packed column (1% SE-30), as shown by the difference of their methylene unit (MU) value.

monitoring at 430.2, 414.2, and 340.2 amu for the ditrimethylsilyl ether of morphine, and at the corresponding masses for the derivative of morphine-d_3 (3 amu greater) showed a considerable amount of interference due to other compounds. Back extraction into an aqueous solution, followed by reextraction, gave suitable samples (Figure 10). The use of an ion-exchange column also gave good results. The analytical samples prepared in this way were free of interference from steroids (Figure 11).

Chart 2 shows the origin of the interference due to urinary steroids. Trimethylsilyl derivatives of androsterone, etiocholanolone, and dehydroepiandrosterone are eluted from nonpolar columns with nearly the same retention time as the trimethylsilyl derivative of morphine. These three steroids, under the conditions used for the derivatization of morphine, form both the expected O^3-trimethylsilyl derivatives and derivatives of the enol form of the steroids (O^3,O^{17}-ditrimethylsilyl derivatives), as indicated in Chart 2. The GC separation with an SE-30 column of the O^3-trimethylsilyl derivatives of androsterone and dehydroepiandrosterone, and of the ditrimethylsilyl derivative of morphine, is represented in Figure 12. This was obtained by selective ion monitoring under CI conditions. The O^3-trimethyl-

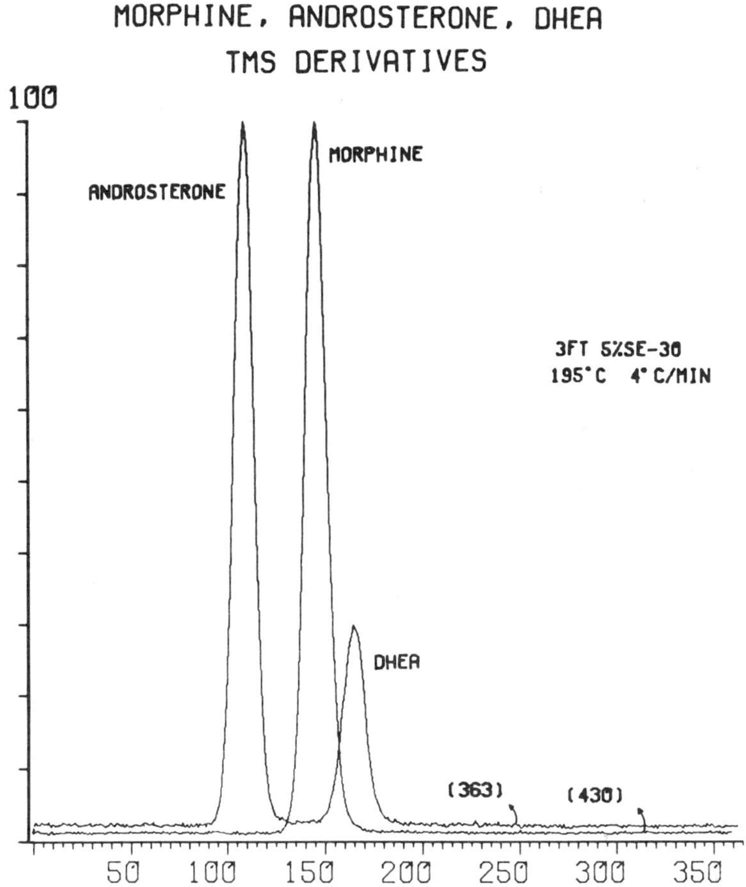

Figure 12. Selected ion monitoring chart showing the relative elution of di-TMS morphine [MH^+ = 430] and O^3-TMS androsterone [MH^+ = 363] and O^3-TMS dehydroepiandrosterone (DHEA) [MH^+ = 361]. The ion at 363 is arising from the [MH + 2] isotope of DHEA: 3 ft 5% SE-30 column, temperature programmed from 195°C at 4°C/min.

silyl derivatives of androsterone and dehydroepiondrosterone (DHEA:MH^+ = 363 and 361 amu, respectively) do not interfere with the analysis of morphine and morphine-d^3, as indicated in Figure 12. Enolization of the 17-ketone yields the O^3,O^{17}-ditrimethysilyl derivative with the increase of the MH^+ ion by 72 amu. The $(M - H)^+$ ion of the O^3,O^{17}-ditrimethylsilylether of androsterone, etiocholanolone, and the MH^+ ion the di-TMS dehydroepiandrosterone contribute to the MH^+ ion of the di-TMS derivative of morphine-d_3. Ikekawa et al. (24) have reported that the limit of sensitivity in the determination of morphine by gas chromatography was greatly reduced after acidic hydrolysis because of a "large amount of interfering substances." In the study of the disposition of codeine in man, using trifluoroacetyl derivatives, Ebbinghausen et al. (56) observed a large unknown peak eluted between codeine and normorphine. The ion at 386 amu observed by these authors is likely to arise from the O^3-trifluoroacetyl ketosteroids.

B. ANALYSIS OF SERUM

PREPARATION OF SAMPLE

First 1 ml of serum, to which 1.5 μg of morphine-d_3 had been added, was extracted twice with 5 ml of ethyl acetate. Hexane (1 ml) was added to the

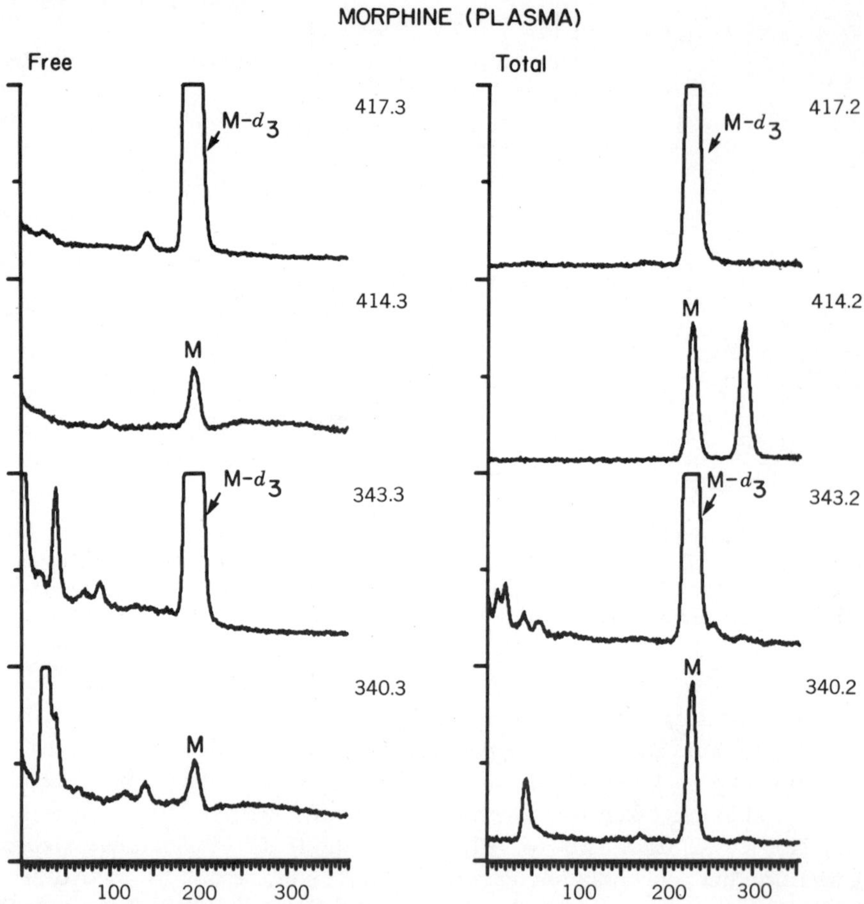

Figure 13. Selected ion monitoring charts. Left: analysis of a plasma sample for free morphine. (M) Right: analysis of a plasma sample for total morphine (M). The internal standard was NCD_3-morphine (M-d_3). The ions pairs at 340/343 and 414/417 were monitored in the CI mode (methane), using TMS derivatives.

combined ethyl acetate extracts, and the organic phase was washed twice with 1 ml of 0.1*N* hydrochloric acid. The aqueous phase containing morphine and other bases was then extracted with ethyl acetate or with a mixture of chloroform and isopropanol (3–1) after neutralization and saturation with solid ammonium carbonate. The organic solution, dried with anhydrous sodium sulfate, was evaporated with the aid of a nitrogen stream. The residue was treated with 25 to 50 μl of bistrimethylsilylacetamide at 60 °C for 20 min,. Trimethylsilyl derivatives of morphine, codeine, and normorphine were formed under these conditions.

In some experiments proteins were precipitated with tungstic or trichloroacetic acid before the extraction step. Lower recoveries were obtained because of protein binding. The most satisfactory extraction method was that described here. Figure 13 compares morphine analyses before and after enzymic hydrolysis. After morphine or diacetylmorphine ingestion, most of the morphine found in blood is in conjugated form.

2. Metabolism Studies

GC-MS techniques are widely used in drug metabolism studies. Detection and quantification of several metabolites are possible in a single experiment.

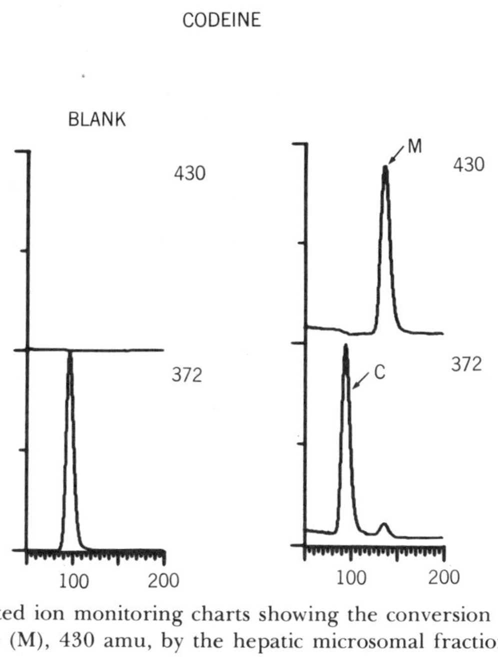

Figure 14. Selected ion monitoring charts showing the conversion of codeine (C), 372 amu, to morphine (M), 430 amu, by the hepatic microsomal fraction of male Sprague-Dawley rats. Left: analysis of the reaction mixture when heated microsomes were added. Right: analysis of the reaction mixture.

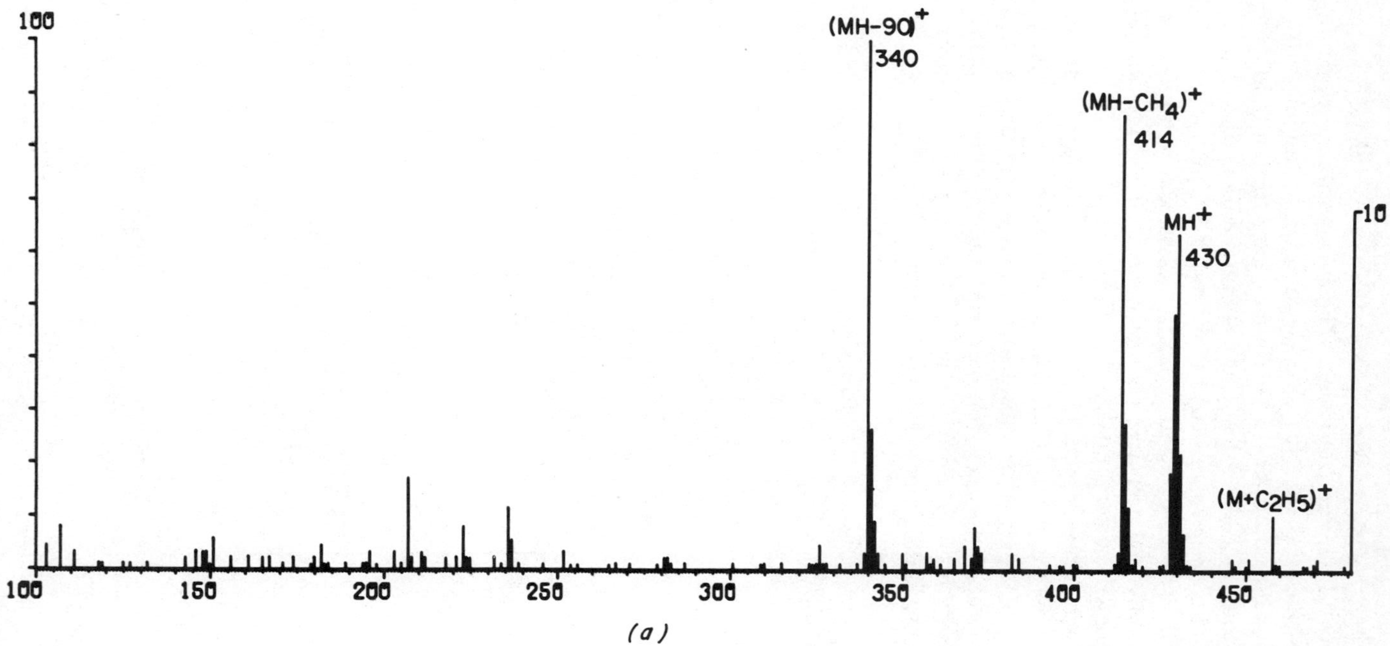

Figure 15*a* CI mass spectrum (methane) of the metabolite M in Figure 14.

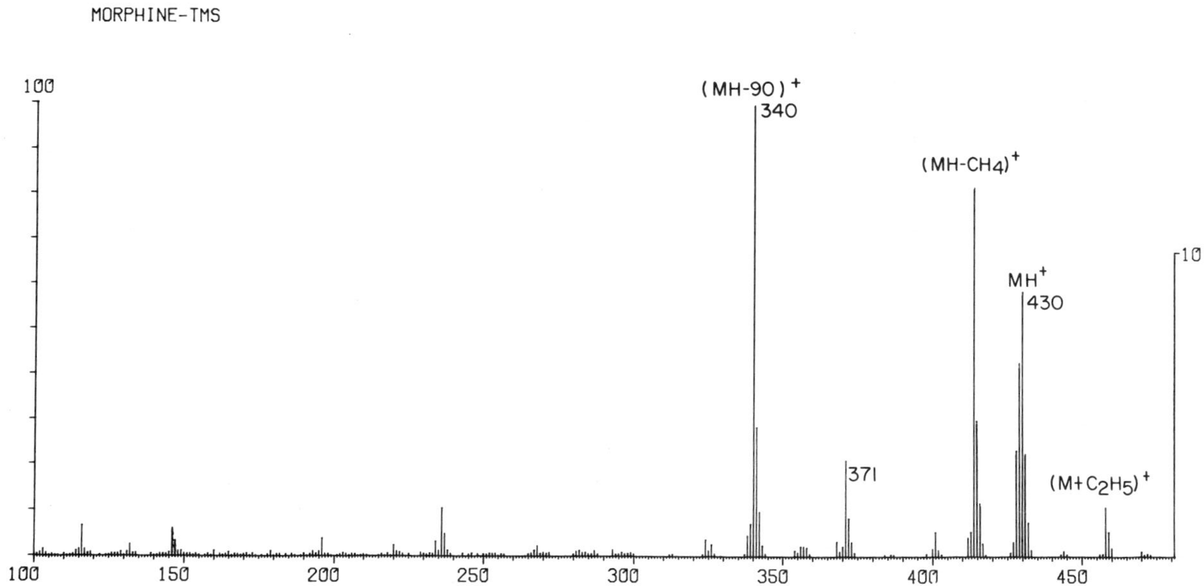

(b)

Figure 15*b*. CI mass spectrum (methane) of an authentic morphine sample as TMS derivative.

Metabolism studies of morphine and related compounds are numerous. *N*-Demethylation has been investigated under a wide range of conditions; the reaction was followed spectrophotometrically by measuring the liberation of formaldehyde (108). With this method, it is not possible to study in detail reactions with competitive substrates. With selected ion monitoring techniques, quantitative measurements at the nanogram level are feasible; the method is particularly suited for analyzing components of *in vitro* reactions with microsomal preparations starting with only 50 to 100 μg of substrate. In the morphine series *N*- and *O*-demethylation may occur, resulting in the formation of norcodeine or morphine. A hepatic microsomal fraction prepared from male Sprague-Dawley rats converted codeine into morphine (109). No trace of metabolites monitored at 430 amu (morphine or norcodeine TMS) was detected in the blank, which consisted of heated microsomes bearing the necessary cofactors. After 15 min of incubation, morphine was formed as shown by the trace at 430 amu (Figure 14). Morphine was identified by its retention time and mass spectrum (Figure 15).

3. Analysis of Legal and Illicit Preparations

It is sometimes necessary to analyze the content of legal or illicit preparations. A current need is the identification of narcotic drugs in street samples.

Furmamec (32) has reported the gas chromatographic quantification of the major alkaloids in gum opium. Numerous reports have been published dealing with the analysis of illicit diacetylmorphine samples (40, 110–113). The procedure is usually very simple and consists in extracting the sample directly with an organic solvent and injecting the extract. Using a computerized gas chromatographic–mass spectrometric method, Smith (114) identified morphine, codeine, thebaine, and papaverine in an opium extract. Identification of adulterants may give clues the sample origin. Miller (115) reported the determination of excipient sugar diluents in street samples. Quinine and/or procaine are very often incorporated in illicit heroin preparations. Jardine and Fenselau have shown the presence of methapyrilene. β-Chloromorphide, found by Herman and Kan (116) in a seized sample, may have been formed by the use of acetyl chloride in place of the usual acetic anhydride.

4. Analysis of Biological Fluids in Emergency Cases

Emergency cases due to drug overdose require immediate treatment. Rapid identification of drugs in the body fluid (blood, gastric content) is essential. Sensitivity of detection is not a problem, and there is no need for extensive sample concentration or purification. Extracts of gastric contents have been analyzed by computerized GC-MS systems in more than one hundred samples (118). The analysis time may be reduced by using a direct insertion probe, without separating the mixture by gas chromatography. Milne et al. (119) reported the use of chemical ionization mass spectrometry for the identification of dangerous drugs by this technique. In one case, where available evidence suggested that the patient had ingested Percodan, the MH^+ ion from 14-hydroxy-dihydrocodeinone (oxycodone) at *m/e* 316 was observed along with those of aspirin, phenacetin, caffeine, and amytal.

Acknowledgment

The authors gratefully acknowledge the encouragement and scientific expertise obtained from Dr. Evan Horning during the course of this investigation and his helpful discussions while preparing this manuscript. This work was supported by Contract DAMD 17-74-C-4052 of the U.S. Army Medical Research and Development Command.

References

1. H. L. Holmes, in *The Alkaloids*, Vol. 2, R. H. F. Manske and H. L. Holmes, Eds., Academic Press, New York, 1952, p. 1.
2. C. R. A. Wright, *J. Chem. Soc.*, *12*, 1031 (1874).
3. H. Dreser, *Ther. Monatsh.*, *12*, 509 (1898).
4. A. Morel-Lavallée, *Rev. Ther.*, *69* (1902).
5. A. H. Beckett and A. F. Casy, "The Testing and Development of Analgesic Drugs," in *Progress in Medicinal Chemistry*, Vol. 2, G. P. Ellis and G. B. West, Eds., Butterworths, London, 1962, p. 43.
6. K. Rice and A. E. Jacobson, *J. Med. Chem.*, *18*, 1033 (1975).
7. J. F. Taylor, "Methods of Chemical Analysis," in *Narcotic Drugs: Biochemical Pharmacology*, D. H. Clouet, Ed., Plenum Press, New York-London, 1971, p. 17.
8. S. J. Mulé, *J. Chromatogr.*, *55*, 255 (1971).
9. I. Jane and J. F. Taylor, *J. Chromatogr.*, *109*, 37 (1975).
10. R. J. Coumkis and B. Kaul, *J. Forensic Sci.*, *19*, 307 (1974).
11. M. G. Robinson, R. C. Howe, N. W. Ream, H. W. Siegel, and F. W. Hegge, *Clin. Pharmacol. Ther.*, *12*, 806 (1971).
12. S. J. Mulé, *J. Chromatogr. Sci.*, *12*, 245 (1974).
13. D. Sohn, J. Simon, M. A. Hanna, G. V. Ghali, R. A. Tolba, and V. Melkonian, *Anal. Chem.*, *45*, 1498 (1973).
14. J. Roboz, "Mass Spectrometry in Clinical Chemistry," in *Advances in Clinical Chemistry*, Vol. 17, O. Bodansky and A. L. Latner, Eds., Academic Press, New York, 1975, p. 109.
15. J. J. Kaufman, N. M. Semo, and W. S. Koski, *J. Med. Chem.*, *18*, 647 (1975).
16. J. S. Stas, *Bull. Acad. Roy. Med. Belg.*, *11*, 304 (1851).
17. S. K. Niyogi, *J. Forensic Med.*, *17*, (1970).
18. M. G. Horning, P. Gregory, J. Nowlin, M. Stafford, K. Lertratanangkoon, C. Butler, W. G. Stillwell, and R. M. Hill, *Clin. Chem.*, *20*, 282 (1974).
19. G. Schill and M. Marsh, *Svensk Farm. Tidsk.*, *67*, 385 (1963).
20. P. F. G. Boon and A. W. Mace, *J. Chromatogr.*, *41*, 105 (1969).
21. S. L. Tompsett, *Acta Pharmacol. Toxicol.*, *17*, 295 (1960).
22. J. H. Knox and J. Jurand, *J. Chromatogr.*, *82*, 398 (1973).

23. C.-Y. Wu, S. Siggia, T. Robinson, and R. Waskiewicz, *Anal. Chim. Acta*, *63*, 393 (1973).
24. N. Ikekawa, K. Takayama, E. Hosoya, and T. Oka, *Anal. Biochem.*, *28*, 156 (1969).
25. A. Stolman and C. P. Stewart, *Analyst*, *74*, 536 (1949).
26. D. Sohn, J. Simon, M. A. Hanna, and G. Ghali, *J. Chromatogr. Sci.*, *10*, 292 (1972).
27. P. A. F. Pranitis and A. Stolman, *J. Forensic Sci.*, *20*, 726 (1975).
28. E. Reid, *Analyst*, *101*, 1 (1976).
29. N. B. Eddy, H. M. Fales, E. Haahti, P. F. Highet, E. C. Horning, E. L. May, and W. C. Wildman, United Nations Secretariat, ST/SOA/Ser. K/114/Con. 1, October 1961.
30. E. Brochmann-Hanssen and A. B. Svendsen, *J. Pharm. Sci.*, *51*, 1095 (1962).
31. W. J. A. Vanden Heuvel, C. C. Sweeley, and E. C. Horning, *J. Am. Chem. Soc.*, *82*, 3481 (1960).
32. D. Furmamec, *J. Chromatogr.*, *89*, 76 (1974).
33. J. W. Fairbairn and K. Helliwell, *J. Pharm. Pharmacol.*, *27*, 217 (1975).
34. P. Liras, *J. Chromatogr.*, *106*, 238 (1975).
35. S. E. Hays, L. T. Grady, and A. V. Kruegel, *J. Pharm. Sci.*, *62*, 1509 (1973).
36. B. R. Rader and E. S. Aranda, *J. Pharm. Sci.*, *57*, 847 (1968).
37. H. J. Wesselman, *J. Pharm. Sci.*, *57*, 8 (1968).
38. E. B. Dechene, L. H. Booth, and M. J. Caughey, *J. Pharm. Pharmacol.*, *21*, 678 (1969).
39. A. S. Cury and D. A. Patterson, *J. Pharm. Pharmacol.*, *22*, 198 (1970).
40. A. Viala, F. Gouezo, J. Catalin, and J.-P. Cano, *J. Eur. Toxicol.* *5*, 375 (1971).
41. A. Hanson and J.-E. Lindgren, *Arch. Toxikol.*, *31*, 1 (1973).
42. E. Schmerzler, W. Yu, M. I. Hewitt, and I. J. Greenblatt, *J. Pharm. Sci.*, *55*, 155 (1966).
43. H. Yoshimura, M.-A. Mori, K. Oguri, and H. Tsukamoto, *Biochem. Pharmacol.*, *18*, 2353 (1970).
44. E. P. J. Van Der Slooten, H. J. Van Der Helm, and P. J. Geerlings, *J. Chromatogr.*, *60*, 131 (1971).
45. L. R. Goldbaum, P. Santingua, and A. M. Dominguez, *Clin. Toxicol.*, *5*, 369 (1972).
46. F. Medzihradsky and P. J. Dahlstrom, *Pharm. Res. Commun.*, *7*, 55 (1975).
47. G. N. Christopaulos, N. W. Chen, and A. J. Toman, *J. Chromatogr.*, *106*, 446 (1975).
48. M. W. Anders and G. S. Mannering, *Anal. Chem.*, *24*, 730 (1962).
49. S. J. Mulé, *Anal. Chem.*, *36*, 1907 (1964).
50. J. E. Wallace, J. D. Biggs, and K. Blum, *Clin. Chim. Acta*, *36*, 85 (1972).
51. L. H. Welsh, *J. Org. Chem.*, *19*, 1409 (1954).
52. C. I. Wright, *J. Pharmacol. Exp. Ther.*, *71*, 164 (1941).
53. J. E. Wallace, H. E. Hamilton, K. Blum, and C. Petty, *Anal. Chem.*, *46*, 14 (1974).
54. W. O. R. Ebbighausen, J. H. Mowat, P. Vestergaard, and N. S. Kline, *Adv. Biochem. Pharmacol.*, *7*, 135 (1973).
55. D. A. Smith and W. J. Cole, *J. Chromatogr.*, *105*, 377 (1975).
56. W. O. R. Ebbighausen, J. H. Mowat, and P. Vestergaard, *J. Pharm. Sci.*, *62*, 146 (1973).
57. G. R. Wilkinson and E. L. Way, *Biochem. Pharmacol.*, *18*, 1435 (1969).
58. J.-P. Thénot and E. C. Horning, *Anal. Lett.*, *5*, 801 (1972).
59. P. A. Clarke and R. L. Foltz, *Clin. Chem.*, *20*, 465 (1974).
60. D. E. Fry, P. D. Wills, and R. G. Twycross, *Clin. Chim. Acta*, *51*, 183 (1974).
61. F. Fish and W. D. C. Wilson, *J. Chromatogr.*, *40*, 164 (1969).
62. R. Truhaut, E. Esmailzadeh, J. Lebbe, J.-P. Lafarge, and N. P. Lich, *Ann. Biol. Chim.*, *32*, 429 (1974).
63. G. J. Digregorio and C. O'Brien, *J. Chromatogr.*, *101*, 424 (1974).
64. H. E. Sine, N. P. Kubasik, and J. Waytash, *Clin. Chem.*, *19*, 340 (1973).
65. S. Felby, H. Christensen, and A. Lund, *Forensic Sci.*, *3*, 77 (1974).
66. K. D. Parker, J. A. Wright, A. F. Halpern, and C. H. Hine, *Forensic Sci. Soc.*, *10*, 17 (1970).
67. J.-P. Thénot and E. C. Horning, unpublished results.

68. E. J. Corey and M. Chaikovsky, *J. Am. Chem. Soc.*, *84*, 866 (1962).
69. E. J. Corey and M. Chaikovsky, *J. Am. Chem. Soc.*, *87*, 1345 (1965).
70. S. I. Hakomori, *J. Biochem.*, *55*, 205 (1964).
71. K. D. Haegele, G. Holzer, W. Parr, S. H. Nakagawa, and D. M. Desiderio, *Biomed. Mass Spectrom.*, *1*, 175 (1974).
72. M. M. Abdel-Monem and P. S. Portoghese, *J. Med. Chem.*, *15*, 208 (1972).
73. T. A. Montzka, J. D. Matiskella, and R. A. Partyka, *Tetrahedron Lett.*, 1325 (1974).
74. K. C. Rice, *J. Org. Chem.*, *40*, 1850 (1975).
75. C. Elison, H. W. Elliott, M. Look, and H. Rapoport, *J. Med. Chem.*, *6*, 237 (1963).
76. J. Vessman, P. Hartvig, and M. Molander, *Anal. Lett.*, *6*, 699 (1973).
77. P. Hartvig and J. Vessman, *Acta Pharm. Sueica*, *11*, 115 (1974).
78. P. Hartvig and J. Vessman, *Anal. Lett.*, *7*, 223 (1974).
79. A. C. Moffat, *J. Chromatogr.*, *113*, 69 (1975).
80. B. Dahlström and L. Paalzow, *J. Pharm. Pharmacol.*, *27*, 172 (1975).
81. M. Riedmann, *Naturwissenschaften*, *59*, 306 (1972).
82. D. A. Smith and W. J. Cole, *J. Chromatogr.*, *105*, 377 (1975).
83. H. Nakata, Y. Hirata, A. Tatematsu, H. Tada, and Y. K. Sawar, *Tetrahedron Lett.*, 899 (1965).
84. H. Audier, M. Fetizon, D. Ginsburg, A. Mandelbaum, and T. Rull, *Tetrahedron Lett.*, *13* (1965).
85. A. Mandelbaum and D. Ginsburg, *Tetrahedron Lett.*, 2479 (1965).
86. D. M. S. Wheeler, T. H. Kinstle, and K. L. Rinehart, *J. Am. Chem. Soc.*, *89*, 4494 (1967).
87. K. D. Haegele, *Proceedings of the Twenty-third Annual Conference on Mass Spectrometry and Allied Topics*, Houston, Texas, May 25–30, 1975, p. 376.
88. M. S. B. Munson and F. H. Field, *J. Amer. Chem. Soc.*, *87*, 3294 (1965).
89. M. S. B. Munson and F. H. Field, *J. Am. Chem. Soc.*, *91*, 3413 (1969).
90. J.-P. Thénot and E. C. Horning, *Proceedings of the Twenty-third Annual Conference on Mass Spectrometry and Allied Topics*, Houston, Texas, May 25–30, 1975, p. 275.
91. G. W. A. Milne, H. M. Fales, and T. Axenrod, *Anal. Chem.*, *43*, 1815 (1971).
92. R. Saferstein and J.-M. Chao, *J. Assoc. Off. Anal. Chem.*, *56*, 1234 (1973).
93. I. Jardine and C. Fenselau, *J. Forensic Sci.*, *20*, 373 (1973).
94. I. Jardine and C. Fenselau, *Anal. Chem.*, *47*, 730 (1975).
95. E. C. Horning, M. G. Horning, D. I. Carroll, I. Dzidic, and R. N. Stillwell, *Anal. Chem.*, *45*, 936 (1973).
96. D. I. Carroll, I. Dzidic, R. N. Stillwell, M. G. Horning, and E. C. Horning, *Anal. Chem.*, *46*, 706 (1974).
97. E. C. Horning, M. G. Horning, D. I. Carroll, R. N. Stillwell, and I. Dzidic, *Life Sci.*, *13*, 1331 (1973).
98. I. Dzidic, D. I. Carroll, R. N. Stillwell, and E. C. Horning, *J. Am. Chem. Soc.*, *96*, 5258 (1974).
99. I. Dzidic, D. I. Carroll, R. N. Stillwell, and E. C. Horning, *Anal. Chem.*, *47*, 1309 (1975).
100. E. C. Horning, D. I. Carroll, I. Dzidic, K. D. Haegele, M. G. Horning, and R. N. Stillwell, *J. Chromatogr. Sci.*, *12*, 725 (1974).
101. E. C. Horning, D. I. Carroll, I. Dzidic, K. D. Haegele, M. G. Horning, and R. N. Stillwell, *J. Chromatogr.*, *99*, 13 (1974).
102. D. I. Carroll, I. Dzidic, C. Oertli, and E. C. Horning, *Proceedings of the Twenty-third Conference on Mass Spectrometry and Allied Topics*, Houston, Texas, May 25–30, 1975, p. 203.
103. R. A. Hites and K. Biemann, *Anal. Chem.*, *42*, 855 (1976).
104. C.-G. Hammar, B. Holmstedt, and R. Ryhage, *Anal. Biochem.*, *25*, 532 (1968).
105. F. C. Falkner, B. J. Sweetman, and J. T. Watson, *Appl. Spectrosc. Rev.*, *10*, 51 (1975).
106. U. Boerner, S. Abbott, and R. L. Roe, *Drug Metab.*, *4*, 39 (1975).

107. S. Y. Yeh, *J. Pharmacol. Exp. Ther.*, *192*, 201 (1975).
108. K. Milthers, *Acta Pharmacol. Toxicol.*, *18*, 199 (1961).
109. J.-P. Thénot, E. C. Horning, and M. G. Horning, in *Advances in Mass Spectrometry in Biochemistry and Medicine*, Vol. 2, A. Frigerio, Ed., Spectrum Publications, New York, p. 119.
110. A. S. Curry and D. A. Patterson, *J. Pharm. Pharmacol.*, *22*, 198 (1970).
111. P. DeZan and J. Fasanello, *J. Chromatogr. Sci.*, *10*, 333 (1972).
112. J. Moore and F. E. Bena, *Anal. Chem.*, *44*, 385 (1972).
113. A. Hanson and J.-E. Lindgren, *Arch. Toxikol.*, *31*, 1 (1973)
114. R. M. Smith, *J. Forensic Sci.*, *18*, 327 (1973).
115. M. D. Miller, *J. Forensic Sci.*, *17*, 150 (1972).
116. G. J. Herman and M.-N. N. Kan, *Biomed. Mass Spectrom.*, *1*, 350 (1974).
117. N. C. Law, V. Aandahl, H. M. Fales, and G. W. A. Milne, *Clin. Chim. Acta*, *32*, 221 (1971).
118. G. W. A. Milne, H. M. Fales, and T. Axenrod, *Anal. Chem.*, *43*, 1815 (1971).

Methods for Analysis of Connective-Tissue Macromolecules by Determination of Certain Constituents

NELLY BLUMENKRANTZ AND GUSTAV ASBOE-HANSEN, *University of Copenhagen Department of Dermatology (with Connective-Tissue Research Laboratories), Rigshospital, Copenhagen, Denmark*

I. INTRODUCTION

Collagen and related glycoproteins, elastin, and acid mucopolysaccharides (glycosaminoglycans) are the main macromolecular components of connective tissue. Since they participate in various pathological processes, we have been attempting for several years to develop techniques that, together with the existing approaches, could be helpful in the study of certain connective-tissue diseases. The study of the metabolites of the above-mentioned macromolecules in blood and in urine has been of some significance. Biopsies of affected and control areas have enabled us to study the biosynthesis and metabolism of collagen, elastin, and mucopolysaccharides by incubation with ^{14}C-tagged proline (Pro), lysine (Lys), and glucosamine.

The biosynthesis of collagen and related glycoproteins passes a series of sequential steps. For a review see ref. 1.

1. Assembly of protocollagen, a polypeptidic precursor rich in prolyl and lysyl residues, in which an extra noncollagenous peptide extension occurs at the C-terminal or at the N-terminal (1).

2. Hydroxylation of certain prolyl and lysyl residues to hydroxyproline (Hyp) and hydroxylysine (Hyl) by protocollagen proline hydroxylase (PPH) and protocollagen lysine hydroxylase (PLH), requiring the cofactors ascorbic acid, α-ketoglutarate, Fe^{2+} ions, and atmospheric O_2. The formation of Hyp is directly related to the twisting of collagen.

3. Glycosylation of certain hydroxylysyl residues. This step consists of two substeps: (*a*) galactosylation of certain hydroxylysyl residues by collagen galactosyl transferase in the presence of the cofactors Mn^{2+} and uridine

diphosphate galactose (UDPGal) and (*b*) glucosylation of certain galactosylhydroxylysyl residues by collagen glucosyl transferase in the presence of the cofactors Mn^{2+} and uridine diphosphate glucose (UDPGlu). At this step, the macromolecule is ready to be extruded to the extracellular compartment.

4. The extruded macromolecule is called procollagen. The action of the enzyme procollagen peptidase is needed to liberate the noncollagenous extra peptides from the procollagen molecule. The presence of the extra peptides gives procollagen its characteristic solubility that enables it to function as the "transport form of collagen" and prevents intracellular fibrillogenesis (1). Once the extra peptides are liberated, the macromolecule is converted into tropocollagen.

5. A new series of steps must be passed by the tropocollagen. One is the oxidative deamination of certain lysyl and hydroxylysyl residues by lysyl oxidase. The resulting aldehyde can react by formation of Schiff base or by aldol condensation with other original or oxidatively deaminated compounds. As a result, the inter- and intramolecular cross-linkages required to stabilize the collagen molecule are formed (1). The "milieu" is important for ensuring that the processes of fibrillogenesis (2) and transport take place. Acid mucopolysaccharides participate actively in both processes, especially depending on their type, (unsulfated, monosulfated, or polysulfated) and their degree of polymerization.

II. BIOSYNTHESIS OF COLLAGEN BY TISSUE EXPLANTS

Collagen biosynthesis can be studied in organ cultures of skin, cartilage, umbilical cord, and so on. The tissues are minced and preincubated at 37 °C for 1 hr under aerobic conditions in a medium of salts, buffer, and glucose (3). Depending on the length of the incubation, ampicillin (4) should be added. At the end of the preincubation time, 5 μC of (^{14}C)L-Pro or (^{14}C)L-Lys are added, and the incubation is allowed to proceed through 2 or 24 hr. In general, the 2-hr incubation period is used when studying collagen biosynthesis by embryonic tissues, whereas the 24-hr period is used for investigation of collagen biosynthesis by human mature tissues. After incubation, embryonic tissues are homogenized and dialyzed; human adult tissue, which is difficult to homogenize, is submitted to papain digestion, homogenization, and dialysis (5).

ASSAYS

(^{14}C)*Pro-Hyp*. Samples labeled with (^{14}C)Pro were hydrolyzed with 6*N* HCl; the HCl was evaporated, and the samples were analyzed for (^{14}C)Hyp according to the procedure of Blumenkrantz and Asboe-Hansen (6). There are two main variations of the procedure. One is used when (^{14}C)Pro also is to be counted; the other in case the investigator is interested only in measuring (^{14}C)Hyp. The procedures are based on the oxidation of (^{14}C)Pro or (^{14}C)Hyp by chloramine T and the extraction of the ^{14}C-labeled pyrrole derivatives with toluene, followed by counting in a liquid scintillation counter. The conditions for the oxidation of both imino acids are different; that is, Pro is oxidized at room temperature, and Hyp requires higher temperature for oxidation.

(^{14}C)*Hyl.* After dialysis, samples labeled with (^{14}C)Lys were divided into two parts. Part 1 was analyzed without previous hydrolysis, and part 2 was hydrolyzed under the conditions first indicated for the (^{14}C)Pro-labeled material. Both samples were analyzed according to the procedure of Blumenkrantz and Asboe-Hansen (7), based on the fact that unsubstituted hydroxy-amino acids are oxidized by periodate, giving formaldehyde and NH_4OH. The use of the two samples allows the detection of total and unglycosylated (^{14}C)Hyl, and, by subtraction of the latter from the former, of glycosylated (^{14}C)Hyl.

There may be decreased hydroxylation with normal incorporation of the precursor amino acids and with normal amounts of enzyme present if some of the necessary cofactors are absent.

ORGAN CULTURE

The experimental model, organ culture of chick embryo tibia, has been successfully used for the study of agents known to affect connective tissue (8–10).

1. Assays for ^{14}C-Hydroxyproline

Previous assays (11–14) either were complicated or did not fulfill the requirement of noninterference of (^{14}C)Pro. A modification of these procedures was therefore developed by the authors (6).

MATERIALS.

Sodium pyrophosphate (E. Merck A. G., Darmstadt). A 0.2*M* solution in water was adjusted to pH 8.0 with HCl. Solutions adjusted to pH 4.0, 6.0, and 10.0 were also prepared.

Tris buffer (E. Merck A. G., Darmstadt). A solution of Tris (1.0*M*) in distilled water was adjusted to pH 8.0 with HCl. Solutions adjusted to pH 4.0, 6.0, and 10.0 were also prepared.

Sodium acetate-citric acid buffer, pH 6.0. Fifty grams of citric acid, 15 ml of glacial acetic acid, 120 g of sodium acetate, and 34 g of sodium hydroxide were dissolved in water and brought to a volume of 1 liter. Toluene was added to avoid bacterial growth, and the buffer was kept at 4°C.

Chloramine T (E. Merck A. G., Darmstadt). *Solution I*: A 0.2*M* solution in water was prepared daily before use. *Solution II*: A 0.05*M* solution was prepared in a mixture of 2 parts of distilled water, 3 parts of methyl cellosolve, and 5 parts of sodium acetate-citric acid buffer, pH 6.

Silicic acid. Columns were prepared with silicic acid sized to eliminate small particles. Silicic acid, 1.5 g, was added to 5 ml of toluene, and the slurry was poured in a 10 × 300-mm glass column with a coarse sintered-glass disk at the bottom. Columns without sintered-glass disks but plugged with glass wool were also used.

Pyrrole (Fisher Scientific Co., Pittsburgh). A 0.1 μmole/ml standard solution in toluene was prepared. Pyrrole solutions of 0.02 and 0.04 μmole/5 ml toluene were used as standards for color recoveries.

Imino acids. L-(^{14}C)Pro uniformly labeled, more than 180 μCi/μmole (New England Nuclear Corp.); (^{14}C)*Hyp*, DL-(2 − ^{14}C)Hyp, 30 to 50 mCi/mmole (The Radiochemical Centre, Amersham). Dilutions of both ^{14}C-labeled imino acids were prepared and used for assay, either alone or in mixture.

L-*Pro* (E. Merck A. G., Darmstadt). A solution containing 1 mg per milliliter in distilled water. L-*Hyp* (E. Merck A. G., Darmstadt). A solution containing 2 mg per milliliter in distilled water.

Ehrlich's reagent (E. Merck A. G., Darmstadt). The reagent was prepared as indicated by Juva and Prockop (11).

Collagenase, purified (Worthington Biochemical Corp., Freehold, N.J.).

Phosphor solution. Six grams of 2,5-diphenyloxazole (PPO, Packard Instrument Co.) and 20 mg of 1,4-bis-(2-(4-methyl-5-phenyloxazolyl)) benzene (POPOP, Packard Instrument Co.) were dissolved in 1000 ml of toluene (E. Merck A. G., Darmstadt) and 600 ml of ethylene glycol monomethyl ether (methyl-cellosolve, E. Merck A. G., Darmstadt) was added.

Sodium thiosulfate (E. Merck A. G., Darmstadt). Solutions of sodium thiosulfate 0.05, 0.1, 0.72, 4, 5, and 7.2*N* in water.

Potassium thiocyanate (E. Merck A. G., Darmstadt). Aqueous solutions of potassium thiocyanate with respective normalities of 0.0025, 0.005, 0.0083, 0.0125, 0.0250, 0.05, 0.1, 0.72, 5.0, and 7.2 were prepared.

Procedures

A. PROCEDURE I [MODIFICATION OF JUVA AND PROCKOP'S METHOD (11)]

One milliliter of the solution of each carrier imino acid was added to 4 ml of a ^{14}C-labeled hydrolyzed and evaporated sample or a ^{14}C-labeled standard. The volumes were brought to 8 ml with distilled water, 6.0 ml of 0.2*M* pyrophosphate buffer was added, and the samples were oxidized with 1.0 ml of chloramine T solution I. After 20 min, 6.0 ml of 7.2*N* sodium thiosulfate was added. Four milliliters of Tris buffer was added, and the samples were saturated with NaCl or KCl. (Saturation with salt can also be performed before the oxidation step).

The tubes were immediately sealed and heated in boiling water for 25 min. The samples were cooled, and 12 ml of toluene was added. (Toluene can also be added before the boiling step.) The tubes were shaken in a Marius shaker (Utrecht) for 5 min and the samples centrifuged at approximately 600 rpm for 5 min or left for a few minutes until the aqueous and organic layers separated. Ten milliliters of the toluene phase was passed through a silicic acid column as indicated by Juva and Prockop (11). The remaining aqueous toluene phase was shaken for 5 min with 10 ml of toluene and, after separation of 10 ml of the organic phase, it was also passed through the same column. A third extraction was performed with 5 ml of toluene, and the shaking and passing through the column of 5 ml of the toluene phase was repeated. The final elution volume was 25 ml. Twenty milliliters of the 25-ml toluene elution was counted in a Beckman liquid scintillation counter after addition of 1 ml of phosphor solution.

To obtain the recovery, a colorimetric determination of Hyp was made on 0.1 ml of the 25-ml toluene elution. The sample was diluted to 5 ml with toluene, and 2 ml of Ehrlich's reagent was added. After 30 min, the absorbance of the chromogen was read at 560 nm against a reagent blank. The amount of pyrrole originating from the oxidation of Hyp was calculated by using pyrrole standards of 0.02 and 0.04μM concentration or by using a standard of Hyp of similar concentration. Taking the percentage color recovery into consideration, a factor was obtained which was multiplied by the dpm of the (^{14}C)Hyp pyrrole derivative. When uniformly labeled (^{14}C)Hyp was assayed, it was considered that only four out of five ^{14}C atoms are present in the pyrrole derivative. The fact that 20 ml of the 25-ml elution volume was counted also received consideration. The recoveries calculated by the color reaction and by the use of a (^{14}C)Hyp standard were compared. The same assay can be performed without addition of unlabeled imino acids, only using a (^{14}C)Hyp standard. Calculations were then related to the recovery of the ^{14}C-labeled pyrrole derivative of the (^{14}C)Hyp standard.

Alkaline or acid hydrolysis of the collagenase-digested peptides labeled with (^{14}C)Pro (15) was performed to study the possibility of using the former treatment, discussed earlier, when glycosylated (^{14}C)Hyp is to be determined. Assays were performed on collagenase-digested peptides obtained from 10-day-old chick embryo tibias labeled with (^{14}C)Pro. No difference in the (^{14}C)Hyp assay was noticed after hydrolysis. No glycosylation of (^{14}C)Hyp of chick embryo collagen can be assumed.

B. PROCEDURE IIA [MODIFICATION OF ROJKIND AND GONZÁLEZ'S METHOD (14)]

The evaporated sample was dissolved in 5 ml of buffer, centrifuged and the supernatant used for assay. Four milliliters of ^{14}C-labeled sample was mixed with 2 ml of chloramine T solution II. After 20 min, the oxidation was terminated by the addition of 1 ml of 4*N* sodium thiosulfate or an equal volume of 0.05*N* potassium thiocyanate, and mixing was performed in a Vortex mixer. (The same recovery was obtained with 1 ml of 0.72 to 7.2*N* sodium thiosulfate or with 1 to 7 ml of 2*N* sodium thiosulfate.) The solution was then saturated with NaCl or KCl. Seven milliliters of toluene was added, and the samples were mixed in a Vortex mixer. After the two layers had separated, 5 ml of the toluene phase was transferred to a counting vial. Three additional toluene extractions of 5 ml each were performed. Five milliliters of the toluene phase was removed from each extract, and the extracts were pooled to a total volume of 20 ml of extract; 1 ml of scintillation (phosphor) solution was added and the fluid counted as (^{14}C)Pro. The rest of the toluene phase was discarded. The aqueous phase containing (^{14}C)Hyp was then converted into the ^{14}C-labeled pyrrole derivative by boiling for 30 min in a water bath. After cooling, the samples were extracted with toluene four times, as indicated previously for (^{14}C)Pro, the only difference being that the extracts were passed through a silicic acid column as specified in procedure I. To 20 ml of the toluene extracts, 1 ml of scintillation solution was added and the fluid counted in a Beckman liquid scintillation counter.

Recoveries were calculated in relation to ^{14}C-labeled Pro and Hyp standards as well as to the total ^{14}C dpm added. They were counted on a 0.1-ml aliquot of each sample analyzed. Recoveries can also be calculated by a color assay as in procedure I, if carrier Hyp is added.

C. PROCEDURE IIB (MODIFICATION OF PROCEDURE IIA)

When (^{14}C)Pro recovery was not required, a simplified procedure was performed, omitting the toluene extractions after oxidation with chloramine T at room temperature. The samples were thus oxidized and, after stopping the oxidation by addition of 1 to 7 ml of 4*N* $Na_2S_2O_3$ and saturation with NaCl or KCl, they were boiled for 30 min. The samples were cooled, and extraction of the pyrrole derivative of (^{14}C)Hyp was performed as indicated in procedure IIA. The toluene extract was passed through a silicic acid column. To 20 ml of this extract, 1 ml of scintillation solution was added and the fluid counted.

The use of a Hyp standard instead of pyrrole or in addition to it, for calculation of recoveries based on colorimetric assay, is advantageous because it must be subjected to all steps of the procedure with the same reagents except the passing through the silicic acid column. The pyrrole standard gives an almost constant chromogen with Ehrlich's reagent, dependent only on the reagent. Since similar recoveries are obtained without the use of unlabeled Hyp, a standard of (^{14}C)Hyp can be used for calculations with the same advantages as in the colorimetric assay.

Procedures I and IIB are recommended when the purpose of the determination is to find (^{14}C)Hyp, expressing the result as percentage (^{14}C)Hyp of total ^{14}C, the latter determined on a small aliquot of the sample to be analyzed (10 to 100 μl). "Total ^{14}C" comprises (^{14}C)Pro and (^{14}C)Hyp.

Procedure IIA, which allows quantitative recovery of (^{14}C)Pro, is recommended when separate determinations of dpm (^{14}C)Pro and dpm (^{14}C)Hyp are necessary.

2. Assay for Protocollagen Proline Hydroxylase

A. PREPARATION OF (^{14}C)PRO-LABELED PROTOCOLLAGEN (SUBSTRATE)

Forty-eight 10-day old tibias were preincubated in 2.7 ml of modified Krebs glucose solution (3) for 1 hr at 37 °C under aerobic conditions. Next 0.3 ml of 10m*M* αα′-dipyridyl was added and the preincubation continued for 30 min more. Then 50 μCi (^{14}C)Pro in 0.5-ml medium was added, and the incubation was performed under the same aerobic conditions for 2 hr. At the end of the incubation period, the tissues and the medium were separated, and the medium was discarded. The tissues were homogenized in 6 ml of distilled water and centrifuged at 15,000*g* and 4 °C for 1 hr in a Servall centrifuge. The homogenate was dialyzed against 8 liters of 1*M* KCl-0.02*M* Tris (pH 7.6) overnight. After dialysis, the undialyzable, unhydroxylated material was boiled for 5 min and diluted with 1*M* KCl to obtain 60,000 cpm/ml. The sample was stored frozen in 20-ml ampoules. After defrosting, this material was used as substrate for hydroxylation.

B. PREPARATION OF ENZYME: RAT SKIN PREPARATION

First 1.5 g of rat skin was homogenized in a Virtis homogenizer in 25 ml of 0.01*M* KCl for 5 min at 45,000 rpm. Homogenization was performed while the glass was placed on crushed ice. Immediately after homogenization, the sample was centrifuged at 15,000*g* at 4 °C for 30 min in a Servall centrifuge. One ml of the supernatant served as source of the enzyme protocollagen proline hydroxylase (PPH).

C. PREPARATION OF COFACTORS

The following solutions must be prepared immediately before use.

1. Solution of $FeSO_4$: 111.2 mg of $FeSO_4$ is dissolved in 10 ml of water.
2. Solution of α-ketoglutarate-ascorbic acid: 29.2 mg of α-ketoglutarate and 140.8 mg of ascorbic acid are dissolved in 30 ml of water.

One-half milliliter of $FeSO_4$ was added to 30 ml of solution 2; then 10 ml of 0.5*M* Tris-HCl buffer, pH 7.8, was added at 25 °C. Each tube took 0.8 ml.

Procedure. The enzyme assay was performed directly on the labeled substrate in the ampoules (16): 0.8 ml of a mixture of the indicated cofactors was added to 1 ml of the defrosted substrate and 1 ml of crude enzyme preparation. Incubation was performed at 37 °C for 60 min under shaking. The incubation was stopped either (1) by the addition of an equal volume of concentrated HCl with immediate hydrolysis, or (2) by the addition of 15 ml of cold acetone and centrifugation at 2000 rpm for 20 min. The supernatant was eliminated by suction. The pellet was dried to eliminate the acetone, and 2 ml of 6*N* HCl was added. In both cases the hydrolysis was allowed to take place at 110 °C overnight. The HCl was eliminated under vacuum at 65 °C over pellets of NaOH. The evaporated sample was dissolved in 5 ml of buffer and centrifuged. A 0.1-ml aliquot of the supernatant was used to detect "total ^{14}C" and another aliquot, usually 4.5 ml, was assayed for (^{14}C)Hyp, according to procedure IIB of Blumenkrantz and Asboe-Hansen (6).

The percentage hydroxylation was calculated as dpm (^{14}C)Hyp × 100/ dpm total ^{14}C.

A blank was always run with substrate and cofactors, and water instead of enzyme. The value obtained was subtracted from that of the samples before calculation of percentage hydroxylation. In our experimental procedure for the assay of protocollagen proline hydroxylase, we found the bovine serum albumin, dithiothreitol (DTT), and catalase used by other authors to be unnecessary. Orr (17) reported on an inhibition of catalase by ascorbic acid. Considering that the latter was present as a cofactor in the hydroxylation system, it is easier to understand that catalase was unnecessary.

The alternative procedure of cold acetone precipitation before hydrochloric acid hydrolysis allows a reduction of the volume of acid used for hydrolysis. Consequently, the time required for the evaporation of HCl was reduced.

If evaporation of HCl is not possible, neutralization of the samples can be performed with comparable results. Previous authors recommend ultracentrifugation at 100,000*g* for the preparation of the substrate (18–24). Comparison of the substrates prepared under these conditions with those indicated by us did not show any difference. Therefore we used centrifugation in the Servall device.

3. Assay for ^{14}C-Labeled Hydroxylysine, Total and Glycosylated, in Collagen and Related Materials

The procedure is based on the periodate oxidation of the ϵ-amino group of Hyl. The liberated formaldehyde reacts with dimedon to form a formaldemethone complex, which is extracted with toluene and counted in a liquid scintillation counter. It is an indispensable condition for the periodate oxidation of this as well as other hydroxy amino acids that in this case, both ϵ-amino and Δ-hydroxy groups be unsubstituted. Thus by performing the assay on two aliquots [i.e., one hydrolyzed with acid (15) and the other nonhydrolyzed (7) or hydrolyzed with 2*N* KOH (15), which does not attack glycosidic linkages], we were able to distinguish between unglycosylated and glycosylated (^{14}C)Hyl. Therefore the assay was applicable to the study of the hydroxylation of lysine as well as the glycosylation of Hyl. (^{14}C)Hyl was assayed on dialyzed homogenates or on collagenase-digested peptides from the dialysate. For the analysis, the samples were divided into two aliquots. One was submitted to acid hydrolysis; the other was analyzed directly, without previous hydrolysis (7).

MATERIALS

Culture tubes, screw-capped with Teflon liners, 200 × 25 mm (Kimax).

Citrate-phosphate buffer. The buffer was prepared by mixing 154 ml of 0.15*M* citric acid (Fisher Scientific Co., Pittsburgh) with 346 ml of 0.3*M* dibasic sodium phosphate (J. T. Baker Chemical Co.). The pH of the final solution was about 6.4.

Carrier hydroxylysine solution. Hydroxylysine-HCl (mixture of DL and DL-allo, Sigma Chemical Corp., St. Louis, Mo.) was prepared at a concentration of 30 mg/ml.

Periodate solution. A 0.3*M* solution of sodium *m*-periodate (Fisher Scientific Co., Pittsburgh) in water was prepared and was stored in the dark in a brown bottle covered with aluminum foil for up to 2 weeks.

Dimedon solution. Dimedon (5,5-dimethyl-1,3-cyclohexanedione) was purchased from Mann Research Laboratories, and 2.8 g was dissolved in 50 ml of 50% ethanol and water. The solution was stable when stored in a brown bottle for several weeks.

Phosphor solution. Six grams of 2,5-diphenyloxazole (PPO, Packard Instrument Co.) and 20 mg of 1,4-bis-(2-(4-methyl-5-phenyloxazolyl))benzene (POPOP, Packard Instrument Co.) were dissolved in 1000 ml of toluene (Fisher Scientific Co., Pittsburgh), and 600 ml of ethylene glycol monomethyl ether (methyl cellosolve, Fisher Scientific Co.) was added.

^{14}C-Hydroxylysine standard. Twenty-eight cartilagenous tibias from 10-day-old chick embryos were incubated at 37 °C for 2 hr with 5 μ (^{14}C)Lys (uniformly labeled, New England Nuclear Corp.). The tissues were homogenized in 3.0 ml of distilled water; the homogenates were dialyzed, and collagenase-solubilized peptides were prepared. The peptides were hydrolyzed with HCl, and the (^{14}C)Hyl was isolated by chromatographing the hydrolysate on the short column of a Spinco amino acid analyzer (15). The fractions containing (^{14}C)Hyl were pooled, and the sample was desalted with a cation-exchange resin column (AG 50W8, 200 to 400 mesh, hydrogen form, Calbiochem) which was eluted with 1.5*N* NH_4OH.

METHODS

Stable solutions of the peptide-bound Hyl from collagen were prepared by treating (^{14}C)Lys labeled tissues with bacterial collagenase, and the peptides were isolated. To assay total (^{14}C)Hyl in the isolated peptides, the solutions of collagenase-solubilized peptides were hydrolyzed with HCl to give free amino acids, or they were hydrolyzed with perchloric acid under conditions that removed all the galactosyl or glucosyl-galactosyl substituents from the ^{14}C-Hyl. To assay nonglycosylated ^{14}C-Hyl in the isolated peptides, the collagenase-solubilized peptides either were assayed directly or were assayed after hydrolysis under alkaline conditions that split the peptide bonds but did not remove the *O*-glycosidic substituents.

a. Treatment of Tissue Homogenates with Bacterial Collagenase. From 15 to 30 mg wet weight of embryonic tissue was homogenized in 3 ml of cold water with a Teflon and glass homogenizer at about 4000 rpm for 1 to 5 min, and the homogenate was dialyzed exhaustively against distilled water, then against 1m*M* $CaCl_2$ and 50m*M* Tris-HCl, pH 7.2. The sample was then incubated with 200 μg of purified bacterial collagenase (Worthington Biochemical Corp., Freehold, N.J.) at 37 °C for 12 to 15 hr. An additional 100 μg of collagenase was added, and the incubation was continued for an additional 3 hr. The sample was heated at 60 °C for 15 min to coagulate undigested protein, and the insoluble proteins were removed by centrifugation at 15,000*g* for 20 min at 4 °C. The collagenase-solubilized peptides in the supernate contained more than 90% of the total protein-bound (^{14}C)-Hyl in homogenates of embryonic cartilage.

b. Hydrolysis with HCl for Assay of Total (^{14}C)Hyl. The samples were hydrolyzed in sealed ampoules for 18 to 20 hr with equal volumes of 12*N* HCl in a Temp-Block or in an oven at 110 °C. After hydrolysis, the HCl was evaporated under vacuum at 65 °C. The residue was then dissolved in a measured volume of distilled water and the assay for (^{14}C)Hyl performed.

c. Hydrolysis with Perchloric Acid for Assay of Total (^{14}C)Hyl. An equal volume of 4*N* perchloric acid was added to 0.3 to 2.0 ml of solution of collagenase-solubilized peptides, and the sample was sealed in a test tube under air. After hydrolysis at 110 °C for 2.5 hr, the tube was cooled. The sample was neutralized with 4*N* KOH, then with 0.1*N* KOH, until the pH was found to be about 6 with pH paper. The precipitate of potassium perchlorate was removed by centrifugation, and the precipitate was washed with a volume of water equal to the initial volume of collagenase-solubilized peptides (0.3 to 2.0 ml). The wash solution was added to the original supernate, and the combined sample was taken for assay of total (^{14}C)Hyl.

d. KOH Hydrolysis for Assay of Nonglycosylated (^{14}C)Hyl. An equal volume of 4*N* KOH was added to 0.3 to 2.0 ml of collagenase-solubilized peptides, and the solution was sealed in alkali-resistant tubes (Corning 7280) under N_2. The sample was hydrolyzed at 105 °C for 18 to 20 hr and cooled. The hydrolysate was neutralized with 4*N*, then 0.01*N* perchloric acid, to a pH of about 6. The precipitate was removed and washed as described previously, and the wash solution was added to the original supernate. The combined sample was used for the assay of unglycosylated (^{14}C)Hyl.

e. Assay of Unhydrolyzed, Nonglycosylated (^{14}C)Hyl. Similar results were obtained for unglycosylated (^{14}C)Hyl when the assay was performed on an unhydrolyzed aliquot (7).

Procedure. Samples for the assay of (^{14}C) hydroxylysine were adjusted in such a way that they contained 1000 to 25,000 dpm (^{14}C)Hyl. (^{14}C)Hyl accounted for 1 to 20% of the total ^{14}C in the sample used. The sample, in a volume of 0.5 to 8.0 ml, was placed in a screw-capped culture tube.

a. Preliminary Extraction of Interfering Radioactive Materials. From 10 to 12 ml of 0.3*M* citrate-phosphate buffer, pH 6.4, was added, and the solution was mixed. About 15 ml of toluene was added and the tube sealed with a Teflon-lined screw cap. The tube was placed in a Marius shaker (Utrecht) and vigorously shaken for 15 min. The tube was centrifuged briefly and the toluene phase discarded. These extractions were repeated, usually three times, until the toluene extract counts reached the background values.

b. Periodate Oxidation and Extraction of ^{14}C-Formaldemethone Complex. First 0.1 ml of the carrier hydroxylysine and 2 ml of 0.3*M* sodium-*m*-periodate were added, and the solution was stirred. Then 3.0 ml of the dimedon solution was added and the tubes stirred briefly in a Vortex mixer. Exactly 15.0 ml of toluene was added and the tube sealed with a Teflon-lined cap. The tubes were shaken in a Marius shaker for 15 to 30 min. Thereafter they were centrifuged briefly at low speed to separate the aqueous and toluene phases. Exactly 10.0 ml of the toluene phase was placed in a glass counting vial, and 10 ml of the phosphor solution was added. The sample was mixed and counted in a liquid scintillation counter.

c. Calculations. The dpm (^{14}C)Hyl in the sample was calculated by the formula:

$$\text{dpm } (^{14}\text{C})\text{Hyl} = (\text{observed cpm} - \text{background cpm}) \times 6 \times \frac{100}{75} \times \frac{15}{10} \times \frac{100}{E},$$

where the factor 6 corrects for the fact that only the ϵ-carbon of uniformly labeled (^{14}C)Hyl is recovered as ^{14}C-formaldehyde in the assay, 100/75 corrects for the 75% recovery of ϵ-carbon, 15/10 corrects for the aliquot of toluene extract placed in the counting vial, and E is the efficiency of the counting system.

4. Assay for Protocollagen Lysine Hydroxylase

The above-mentioned assay for (^{14}C)Hyl has been used for the detection of protocollagen lysine hydroxylase (22). A (^{14}C)Lys-labeled protocollagen was used as substrate for PLH in the presence of the required cofactors (i.e., α-ketoglutarate, ascorbic acid, Fe^{2+}, and atmospheric oxygen). The (^{14}C)Hyl generated was detected by periodate oxidation and toluene extraction of the formaldemethone complex (7).

Since the syntheses of Hyl and Hyp are accompanied by conversion of α-ketoglutarate to CO_2 and succinate, Rhoads and Udenfriend (23) and Krane et al. (24) used sodium ($5 - {}^{14}$C)α-ketoglutarate among the cofactors and measured the hydroxylation by the liberation of (^{14}C)CO_2.

Although both the formation of the ^{14}C-formaldemethone complex and the liberation of (^{14}C)CO_2 are part of the same process, the assay of the former is more specific and more sensitive than the assay with (^{14}C)α-ketoglutarate.

The procedure was described by Blumenkrantz and Prockop (15) and modified by Blumenkrantz and Asboe-Hansen (7). The modification refers to the detection of unglycosylated (^{14}C)Hyl by periodate oxidation of an unhydrolyzed sample instead of the KOH hydrolysis as indicated in ref. 15.

5. Glycosylation of Hydroxylysine in Collagen

A. ASSAYS FOR COLLAGEN GALACTOSYL AND GLUCOSYL TRANSFERASE

a. Substrate for Collagen Galactosyl Transferase. The substrate for collagen galactosyl transferase is unglycosylated collagen. The preparation of the substrate is based on the fact that 10-day-old chick embryo tibias incubated with deoxyglucose can synthesize a fully hydroxylated macromolecule (viz. Hyp and Hyl), although the Hyl is not glycosylated.

Fifty-four tibias (64.2 mg dry weight) of 10-day-old chick embryos were incubated for 12 hr with deoxyglucose ($10^{-2}M$) in a simple medium consisting of buffer phosphate, salts, and inosine ($10^{-2}M$) as source of energy. The tibias were homogenized in 2.5 ml of distilled water after several washes with distilled water.

b. Substrates for Collagen Glucosyl Transferase

1. Samples of neutral salt-soluble collagen from rat skin were hydrolyzed for 30 min at 100 °C with $1N$ HCl. The hydrolysate was neutralized with $1N$ NaOH and dialyzed against distilled water for 16 hr. The product was lyophilized and stored at 0 °C over $CaCl_2$.

2. Boiled homogenates of 8- and 9-day-old chick embryo tibias were assayed as substrates for both enzymes. Enzymes for the assays were prepared with twelve 8-day-old chick embryos (4.70 g wet weight) and twelve 9-day-old chick embryos (6.45 g wet weight) homogenized in 10 times their volume of $0.1M$ KC1 in 0.1% Triton X-100 at 4 °C. Wet weights represent

decapitated embryos. Extracts of these homogenates were then prepared by stirring for 16 hr at 4 °C. The suspension was centrifuged at 20,000*g* for 1 hr at 3 °C. The supernatant solution was used as crude enzyme. Aliquots of the crude enzyme were used for the assay of both transferases. They were assayed according to our modification of the method of Bosmann and Eylar (26).

c. Enzyme assay

Collagen Galactosyl Transferase. The assay of this enzyme requires the following materials.

1. 50 μl crude enzyme.
2. 50 μl substrate (unglycosylated).
3. 10 μl 0.25*M* $MnCl_2$ (freshly prepared).
4. 10 μl UDPGal-^{14}C (40.000 cpm per 1.4×10^{10} mole).
5. 10 μl 0.1% Triton X-100.
6. 20m*M* sodium acetate buffer pH 5.8 to a final volume of 0.54 ml.

Collagen Glucosyl Transferase. Material as just indicated, except that UDPGlu-^{14}C is used instead of UDPGal-^{14}C.

Under the conditions of the assays, extracts of 8- and 9-day-old chick embryos containing 710 μg of protein served as substrate.

Incubation was performed at 37 °C for 3 hr. The incubation was stopped by precipitating the protein with 0.1 ml of 1% phosphotungstic acid in 0.5*M* HCl. The protein-bound ^{14}C-labeled precipitate was retained in a Millipore filter and counted in a liquid scintillation counter after addition of 20 ml of scintillation solution.

The 8-day-old chick embryo tibia homogenate was a good substrate for collagen galactosyl transferase. No collagen glucosyl transferase activity was found in the crude enzyme from 8-day-old chick embryos. Homogenates of 9-, 10-, and 12-day-old chick embryos showed decreased or no capacity to be further glucosylated.

III. ANALYSIS OF COLLAGEN AND ITS METABOLITES

Determination of the two amino acids characteristic of collagen (i.e., Hyp and Hyl), and the ratio Hyp to Hyl, can give information on the structure of the collagen molecule. Quantitative estimation of both amino acids can be performed on one sample of hydrolyzed collagen, on collagen-containing tissue, or on biological fluids containing collagen metabolites (e.g., urine and serum).

ANALYSIS OF COLLAGEN

A few years ago, most authors defined the collagen molecule solely by determination of Hyp. However under certain conditions only Hyp or only

Hyl levels are altered in tissues or biological materials (9,27–30). This indicates an abnormal buildup or an abnormal degradation of collagen. Thus the parallel determination of Hyp and Hyl in tissues and biological fluids is essential. The amino acid analyzer can perform complete amino acid analysis, but the minimum time required for Hyp and Hyl determination by chromatographic separation exceeds 2.5 hr per sample (31).

The specific assays for Hyp and Hyl have been adapted to the autoanalyzer. Sixty determinations of Hyp and 50 of Hyl can be performed in an hour (32,33).

1. An Assay for Hydroxyproline and Proline on One Sample and a Simplified Method for Hydroxyproline

A simplified assay for quantitative determination of Hyp has been reported by the present authors (34). The assay has the advantage of reducing unnecessary steps and increasing the recovery yield as compared to available methods based on the same principle (i.e., oxidation of Hyp with chloramine T to a pyrrole derivative). The pyrrole derivative is extracted with toluene and gives a chromogen with λ max at 565 nm with Ehrlich's reagent.

Pro and Hyp can be detected in the same sample. The mixture is oxidized with chloramine T at room temperature. The product of oxidation of Pro, which is assumed to be Δ-pyrroline-carboxylic acid, is extracted with toluene and reoxidized with sodium periodate. The product of periodate oxidation produces a chromogen with λ max 565 nm with Ehrlich's reagent.

Although more sensitive, the previously reported method of Troll and Linsley (35) for the determination of Pro is less specific. The need to oxidize Pro further with sodium periodate to give a chromogen with the Ehrlich reagent excludes Pro as a substance interfering with the color reaction of Hyp. This was further confirmed by our simplified assay of mixtures of known contents of the two imino acids as well as of hydrolyzed samples of urine. It showed results that were not significantly different from those obtained with the more complicated assay of Prockop and Udenfriend (36) or the modification of Kivirikko et al. (37).

A. MANUAL ASSAYS

MATERIALS

Culture tubes, screw-capped with Teflon liners (Kimax) 123 × 15 mm.

Standard Hyp (E. Merck A. G., Darmstadt): L-Hyp solutions of concentrations ranging from 1 to 6 μg/ml.

Standard Pro (E. Merck A. G., Darmstadt): L-Pro solutions of 10, 20, and 50 μg/ml.

BUFFER SOLUTIONS

1. *Borate buffer pH 8.7.* 1*M* buffer solution; 61.84 g of boric acid and 225 g of KCl were mixed with about 800 ml of distilled water. The pH was adjusted to 8.7 with KOH and the volume supplemented with distilled water to 1000 ml; 1–5, 1–10, and 1–100 dilutions of the buffer were also used (0.2, 0.1, and 0.01*M*).
2. *Alanine solution* (0.112*M*). Alanine (10 g) was dissolved in about 90 ml of distilled water. The pH was adjusted to 8.7 and the volume adjusted with distilled water to 1000 ml. In addition, 1–5, 1–10, and 1–100 dilutions (0.0224, 0.0112, 0.00112*M*) of the alanine solution were used together with 1–5, 1–10, and 1–100 dilutions of the borate buffer.
3. *Borate-alanine buffer.* First 100 ml of the borate buffer and 50 ml of the alanine solution were mixed. The resulting solution was also used in a 1–10 concentration (0.066*M* in terms of boric acid).
4. *Pyrophosphate buffer.* A 0.2*M* sodium pyrophosphate buffer, pH 8.0.
5. *Tris buffer.* 1*M*, pH 8.0.
6. *Citric acid buffer.* After 50 g of citric acid $1H_2O$ was mixed with 12 ml of acetic acid (96%), 120 g of sodium acetate $3H_2O$, and 34 g of NaOH, the mixture was dissolved in distilled water to a final volume of 1000 ml.

Chloramine T solution (E. Merck A. G., Darmstadt). A 0.2*M* chloramine T solution was prepared. The solvents tested were (1) H_2O, (2) methyl cellosolve, and (3) ethyl cellosolve. Sodium *m*-periodate (E. Merck A. G. Darmstadt). A 0.3*M* solution in distilled water. Sodium thiosulfate (E. Merck A. G., Darmstadt). A 3.6*M* solution in distilled water. Potassium chloride (E. Merck A. G., Darmstadt). *n*-Propanol (E. Merck A. G., Darmstadt). Toluene (E. Merck A. G., Darmstadt). Isobutanol (E. Merck A. G., Darmstadt). *p*-Dimethylaminobenzaldehyde (PDMAB; Ehrlich reagent analytical grade, E. Merck A. G., Darmstadt). Concentrated sulfuric acid was slowly added to 200 ml of absolute ethanol in a beaker and the mixture cooled; 120 g of PDMAB was added to 200 ml of absolute ethanol in another beaker, and this solution was also cooled. The acid-ethanol mixture was then added to the second beaker in an ice bath. This reagent was used for the Hyp assay (36).

Ehrlich reagent for Pro assay (PDMAB, analytical grade, E. Merck A. G., Darmstadt): Fifteen milliliters of isobutanol was added to 4 g of PDMAB in a beaker after which 4.5 ml of perchloric acid was added. The solution was stored for several weeks at 4 °C in a dark bottle.

METHODS

PREPARATION OF SAMPLES

Samples of urine, skin, cartilage, or purified collagen were hydrolyzed overnight at 118 °C in 6*N* HCl. The hydrolyzed samples were evaporated at 65 °C under vacuum.

a. Assay for Proline and Hydroxyproline on One Sample. Evaporated samples were dissolved in distilled water or directly in the borate-alanine buffer. To 1 ml of the sample containing 0.5 to 10 μg of Hyp, 1.5 ml of the borate-alanine buffer (1–10) and 0.6 ml of chloramine T were added. The oxidation proceeded for 20 min. Then 2 ml of thiosulfate was added, and the samples were mixed. At this moment, or before the addition of chloramine T, NaCl or KCl was added to saturation. The time of oxidation was kept constant for all samples. Next, 3 ml of toluene was added, and the tubes were tightly capped and shaken in a Marius shaker (Utrecht) for 15 min. The tubes were centrifuged at 1000 rpm for 10 min, and the organic phase was separated from the aqueous phase.

To determine Pro, 1.25 ml of the toluene extract was mixed with 1.25 ml of isobutanol and 0.5 ml of *n*-propanol; then 0.5 ml of 0.3*M* $NaIO_4$ and 6 ml

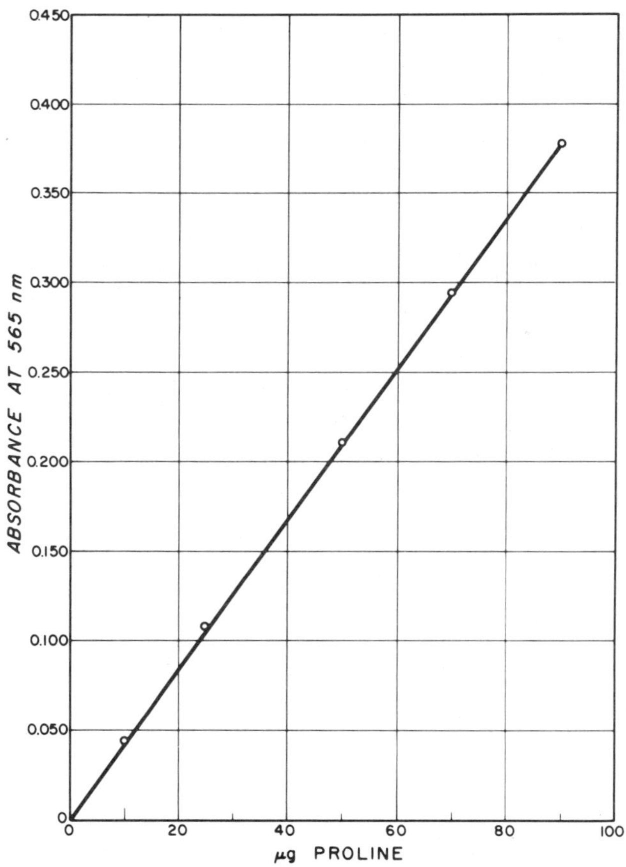

Figure 1. Colorimetric assay of proline. The chromogen produced by the product of oxidation of proline after addition of Ehrlich's reagent follows Lambert and Beer's law.

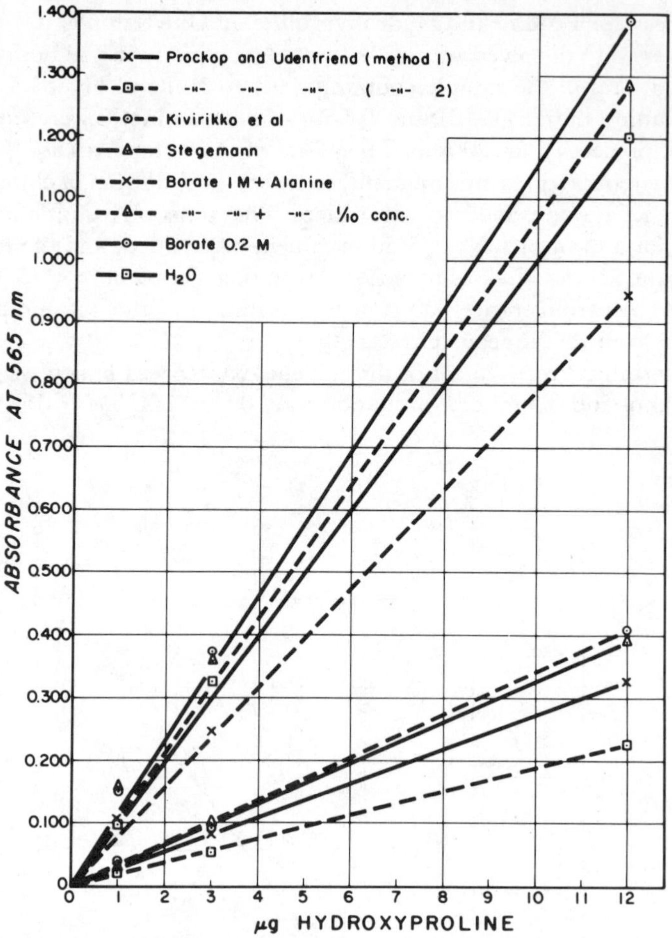

Figure 2. Comparison of the simplified assay for hydroxyproline with other previously reported methods.

of phosphate-citrate buffer, pH 7.0, were added (38). The samples were shaken in a Marius shaker (Utrecht) for 30 min, the tubes being protected from the light. The tubes were centrifuged, and 2.0 ml of the organic phase was added to 0.5 ml of PDMAB in $HClO_4$ and isobutanol. The chromogen was allowed to develop for 15 min, then read in a Beckman DU spectrophotometer, at 565 nm, against a blank of reagents. Standards containing 10, 25, and 50 μg of Pro were run (Figure 1).

Hydroxyproline determination was performed according to Kivirikko et al. (37) on the remaining aqueous phase reduced to a microscale, though no neutralization was performed and toluene was added to the aqueous phase before the sample was boiled. Three milliliters of toluene was added to the aqueous phase and, after boiling, cooling, shaking, and centrifuging, 0.4 ml of Ehrlich's reagent in sulfuric acid-ethanol was added to 1 ml of the organic phase. The color reaction was allowed to develop for 30 min, and absorbances were read at 565 nm (Figure 2).

b. Simplified Assay for Hydroxyproline. Evaporated samples were dissolved and the oxidation and reduction steps carried out as just indicated. At this time, or before, chloramine T, NaCl, or KCl was added to saturation. The buffer used was borate-alanine, 1–10 dilution. No extraction of the oxidation product obtained at room temperature was performed. Then the toluene was added and the tubes placed in a boiling water bath for 20 min. After cooling, the tubes were shaken and centrifuged, and 1 ml of the organic phase was treated and read as specified previously.

2. Quick Assay for Hydroxyproline (Manual)

A rapid assay for Hyp originated from the finding that this substance can be oxidized to a pyrrole derivative by periodic acid and various periodates in strong acid medium. This pyrrole derivative gives a chromogen with λ max at 565 nm with Ehrlich's reagent. Neither Pro nor Hyl is oxidized by periodate in acid medium. The assay is quick, specific, and reproducible (38).

MATERIALS

Culture tubes, screw-capped with Teflon liners (Kimax), 120 × 15 mm.

Periodic acid solution. A $1M$ solution of periodic acid (Fisher Scientific Co., Pittsburgh) in $9M$ H_3PO_4 (E. Merck A. G., Darmstadt) was prepared and stored in brown bottles for up to 6 months in the dark.

Citrate-phosphate buffer. The buffer was prepared by mixing 154 ml of $0.15M$ citric acid (Fisher Scientific Co., Pittsburgh) with 346 ml $0.6M$ dibasic sodium phosphate (J. T. Baker Chemical Co.). The final pH of the buffer was 7.0.

Extraction solution. The extraction solution was prepared by mixing 250 ml of toluene with 250 ml of isobutanol and 100 ml of *n*-propanol.

Standard hydroxyproline solution. L-Hyp was prepared at a concentration of 10 μg/ml.

Standard proline solution. L-Pro was prepared at two different concentrations, 50 and 200 μg/ml.

p-Dimethylaminobenzaldehyde (*Ehrlich reagent*) (analytical grade, E. Merck A. G., Darmstadt). Fifteen milliliters of isobutanol was added to 4 g of *p*-dimethylaminobenzaldehyde in a beaker, followed by 4.5 ml of perchloric acid (70%, E. Merck A. G., Darmstadt). The solution could be stored for several weeks at 4 °C in a dark bottle.

METHOD

Assay Procedure. Samples were hydrolyzed overnight in 6*N* HCl in sealed tubes at 118 °C. The hydrolysates were evaporated to dryness *in vacuo* at 65 °C. Evaporated samples were dissolved in distilled water.

One milliliter of the sample (containing 2.5 to 50 μg of Hyp) was added to 6 ml of buffer and 0.5 ml of 1*M* periodic acid in 9*M* H_3PO_4, and the solution carefully mixed. Then 2.0 ml of the extraction solution was added, and the tubes were stirred in a Vortex mixer. The tubes were placed in a test tube rack, and the rack was covered with aluminum foil and shaken on an Equipoise horizontal shaking machine for 30 min. The tubes were then centrifuged at low speed for 10 min to separate the aqueous and organic phases. A semimicro or micro adaptation of the assay procedure was performed when a low content of Hyp was suspected. The only differences from the above-mentioned procedure were the volumes of the samples and the reagents added. For the semimicro assay, 3 ml of buffer, 0.5 ml of periodic acid, and 1.75 ml of the extraction mixture were added to 0.5 ml of the sample. For the microassay, 1.5 ml of buffer, 0.5 ml of periodic acid, and 1.2 ml of extraction mixture were added to 0.25 ml of the sample.

To 600 μl of the organic phase, 150 μl of Ehrlich reagent was added. The mixture was stirred vigorously. The color was allowed to develop for 15 min at room temperature, and the absorbance was read at 565 nm on a Beckman DU spectrophotometer. The color was stable for 4 hr.

3. Automated Procedure for Quantitative Determination of Hydroxyproline in Biological Materials

The method was developed on the basis of Stegemann's manual assay (39,40) for Hyp. As compared to an automated assay previously described by Grant (41), the procedure for the quantitative determination of Hyp has the advantages of a simpler flow pattern and a higher sensitivity. The procedure allows 60 determinations of Hyp per hour. It is used for determination of Hyp in hydrolysates of urine, blood, skin, and other tissues (32).

MATERIALS

Citrate-phosphate buffer pH 6.0 was prepared as follows:

1. A 0.005*M* solution of citric acid.
2. A 0.01*M* solution of dibasic sodium phosphate.

For this buffer 17.9 ml of solution 1 was mixed with 32.1 ml of solution 2 and the total volume diluted with distilled water to 100 ml.

Chloramine T Solution. A 0.0177*M* solution in distilled water was prepared. When kept in the refrigerator in a dark, tightly capped bottle, the solution could be used for up to 15 days.

Ehrlich's reagent. Seven grams of *p*-dimethylaminobenzaldehyde was dissolved in 15 ml of 70% perchloric acid and the solution adjusted to 100 ml with isopropyl alcohol.

Sodium hydroxyde solution. A 6*N* NaOH solution in distilled water.

Standard solutions. Solutions of Hyp containing 2, 4, and 6 μg/ml buffer were prepared as indicated below. (See "Preparation of Samples.") Solutions of higher and lower concentrations were also used to determine minimum and maximum sensitivity. Solutions of proline (Pro) 12, 18, and 24 μg per milliliter of buffer were also assayed.

PREPARATION OF SAMPLES

First 24-hr urines were collected from individuals on a collagen-free diet. Filtered urine (1 ml) was added to an equal volume of 12*N* HCl, then hydrolyzed at 110 °C for 16 hr or at 130 °C for 3 hr. Two different procedures were used after hydrolysis. (1) Evaporation of the HCl under vacuum at 65 °C followed by suspension in 2 ml of buffer; 0.5 ml of the suspension was diluted to 10 ml with buffer. (2) Neutralization of 0.5 ml of the suspension with 0.5 ml of 6*N* NaOH followed by addition of 9 ml of buffer. In both cases the diluted samples were centrifuged at 2000 rpm for 10 min, and the supernatant was used for Hyp determination. Standards were submitted to a similar procedure (i.e., heating with 6*N* HCl followed by evaporation or neutralization of the acid and further dilution with the buffer). The Hyp standards with a final concentration of 2, 4, and 6 μg/ml were used in every run.

For the determination of Hyp in other biological materials (e.g., tibias of chick embryos and skin), a similar procedure was followed.

Autoanalyzer. The instrument used for the analysis was the Autoanalyzer produced by Technicon Instruments Corp., Chauncey, New York.

Procedure. The flow diagram for the quantitative automated assay of Hyp appears in Figure 3. Alternating cups of sample and water are placed in the sampler. The samples are fed into the machine at the rate of 60-2/1 samples per hour; that is, of the 60 sec corresponding to each sample, 40 sec are used for delivery of the sample and 20 sec for the water. Accordingly, the actual volume of the samples delivered is two-thirds of that indicated in the flow diagram (i.e., 0.53 ml). Acidflex tubing is used for the intake and delivery of Ehrlich's reagent and for the waste flow, whereas in all the other steps clear Tygon tubing is used.

The reaction is carried out as follows. Chloramine T is added to the sample and the substances flow together through the stream to be mixed in a 14-turn single coil. The oxidation is performed while the combined flow passes a time-delay coil of 10 turns (made from an original $23\frac{1}{2}$-turn coil). After the oxidation, Ehrlich's reagent is added and the flow mixed in another single coil. The current is heated at 75 °C in a heating bath with a delay coil of $28\frac{1}{2}$

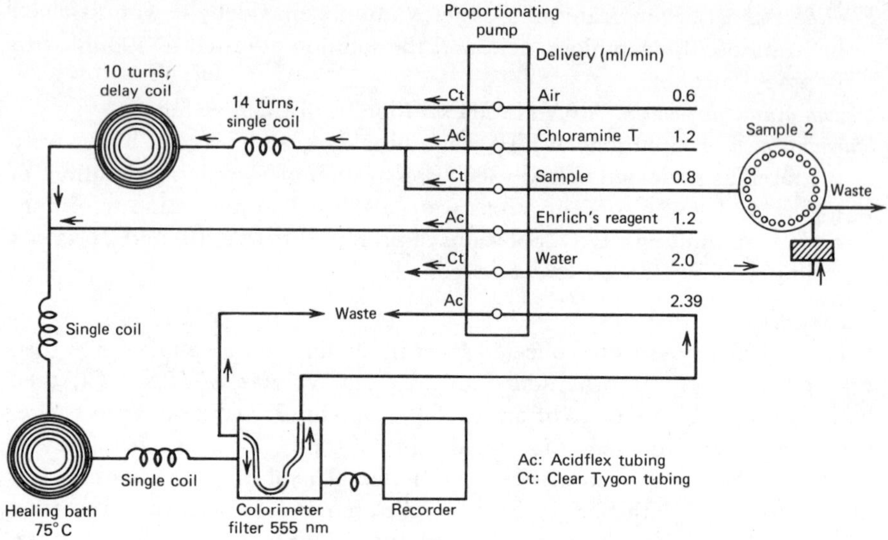

Figure 3. Flow diagram for automated assay for hydroxyproline.

turns, then passed through a third single coil. The color produced is measured with a 555-nm interference filter in the colorimeter attached to the machine. The colorimeter is fitted to the machine with a 555-mm flow cell. Transmission is registered in the recorder connected to the colorimeter. One standard is run for every 10 samples. When low concentrations of Hyp are measured, a Goerz-Servogor S-RE-543 recorder is used. Transmission is converted into absorbance by using the ruler belonging to the machine or by using the Technicon semilogarithmic paper directly.

When the run is finished, the tubing used for the intake of Ehrlich's reagent is washed with isopropanol, and the rest of the system is washed with 1*N* HCl.

4. Quantitative Determination of Hydroxylysine in Biological Materials

Blumenkrantz and Prockop (42) have developed a specific assay for Hyl based on periodate oxidation of Hyl to glutamic semialdehyde and the conversion of the latter into its cyclic form, Δ′-pyrroline-5-carboxylic acid, which, on further oxidation, gives a chromogen with Ehrlich's reagent. Interference of Pro was noticed and prevented by a preparative separation by thin-layer chromatography (42). The authors have reported (43) a modification based on the finding that by using highly diluted periodate for the oxidation of Hyl, the interference from Pro can be avoided. This modification makes the assay simpler, quicker, and more specific. The sensitivity of the original procedure was increased to make it applicable at the micro scale. The manual assay described by Blumenkrantz and Asboe-Hansen (43) has been adapted to the autoanalyzer. The extraction of the Δ′-pyrroline-5-carboxylic acid derivative with organic solvents as reported in the manual assay is not used in the automated procedure (33).

The automation of the quantitative manual assay adds to its sensitivity and specificity the advantages of rapidity and higher reproducibility. The Hyl content of 50 samples can be assayed in an hour by this procedure. Another procedure used at present for the determination of hydroxylysine is chromatography followed by the ninhydrin reaction, as performed by the amino acid analyzer (31). By this procedure, the determination of Hyl in one sample requires at least 2.5 hr. The ninhydrin reaction, which is not specific, does not show a possible interference of other ninhydrin positive substances eluting at the Hyl chromatographic position.

A. MANUAL ASSAY

MATERIALS

Standard hydroxylysine solution. Δ-Hydroxylysine HCl (mixture of DL and DL-allo; Sigma Chemical Corp., St. Louis, Mo.) was prepared at a concentration of 1.0 mg per milliliter. Subsequent dilutions were prepared from this solution.

Standard proline solution. (E. Merck A. G., Darmstadt) was prepared at a concentration of 1.0 mg per milliliter.

Periodate solutions.

1. Solutions containing 0.3, 0.1, 0.05, 0.025, 0.01, 0.0025, and 0.0015*M* sodium *m*-periodate (Fisher Scientific Co., Pittsburgh) in water were prepared; they could be stored up to 4 months in a dark bottle covered with aluminum foil.
2. Periodic acid solution (E. Merck A. G., Darmstadt) of similar concentrations as indicated for sodium *m*-periodate were prepared and stored.
3. Potassium *m*-periodate (J. T. Baker Chemical Co.) solutions from 0.1 to 0.0015*M* in water were prepared and stored as just indicated.
4. Lithium periodate (Alfa Inorganics, Inc.) solutions from 0.05 to 0.0015*M* in water were prepared and stored as indicated for the other periodates.

Culture tubes, screw-capped with Teflon liners (Kimax) 120 × 15 mm. The citrate-phosphate buffer, extraction solution, and *p*-dimethylaminobenzaldehyde (PDMAB) were prepared as indicated by Blumenkrantz and Asboe-Hansen (38).

Assay Procedure. Protein samples were hydrolyzed overnight in 6*N* HCl in sealed tubes at 118 °C, and the hydrolysates were evaporated to dryness *in* *vacuo* at 65 °C. Samples were washed twice with distilled water and reevaporated to dryness. Samples were dissolved in 0.001*N* NaOH.

Semimacro Method. To 1 ml of the sample containing 0.5 to 10.0 μg of hydroxylysine, 6 ml of the buffer, 0.5 ml of 0.0015*M* periodic acid, and 2 ml of the extraction solution were added.

Macro Method. To 1 ml of the sample containing 5 to 15 μg of hydroxylysine, 6 ml of buffer and 0.5 ml of 0.0015*M* periodic acid were added. The solution was stirred. Then 3 ml of the extraction solution was added, and the tubes were stirred in a Vortex mixer.

Semimicro Method. To 0.5 ml of the sample containing 1 to 4 μg of hydroxylysine were added 3 ml of the buffer, 0.25 ml of 0.0015*M* periodic acid, and 1.75 ml of the extraction solution.

Micro Method. To 0.25 ml of the sample containing 0.25 to 2 μg of hydroxylysine, 1.5 ml of the buffer, 0.2 ml of 0.0015*M* periodic acid, and 1.2 ml of the extraction solution were added.

After mixing, the tubes covered with aluminum foil were shaken for 15 min in a Marius shaker (Utrecht). The tubes were then centrifuged at low speed for 10 min to separate the organic phase. Exactly 600 μl of the organic phase was separated and placed in a test tube. Thereafter, 150 μl of Ehrlich's reagent was added. The tubes were stirred vigorously, and color was allowed to develop for 15 min at room temperature. The absorbance was read at 565 nm on a Beckman spectrophotometer (model DU-2).

B. AUTOMATED ASSAY

REAGENTS

Citrate-phosphate buffer. The buffer was prepared by mixing 154 ml of 0.15*M* citric acid (Fisher Scientific Co., Pittsburgh) with 346 ml of 0.6*M* dibasic sodium phosphate (J. T. Baker Chemical Co.). The pH of the final solution was 7.0.

Dilute citrate-phosphate buffer. Citrate-phosphate buffer, pH 7.0, was diluted 1–3 with distilled water.

Standard hydroxylysine solution. Hydroxylysine-HCl (a mixture of DL and DL-allo; Sigma Chemical Corp., St Louis, Mo.) was prepared at a concentration of 1–10 μg/ml in phosphate citrate buffer.

L-*Pro and* L-*Hyp* were products of E. Merck A. G., Darmstadt. Solutions containing 100 μg/ml of the imino acids were prepared.

Wetting agent solution (Brij 30 to 35%; Technicon Chemicals, S.A., Belgium). A solution containing 1 ml of Brij per 1 water was prepared.

Sodium m-periodate (E. Merck A. G., Darmstadt). A 0.0015*M* solution in the diluted citrate-phosphate buffer.

Ehrlich's reagent. Four g of *p*-dimethylaminobenzaldehyde (PDMAB) analytical reagent was dissolved in 15 ml of perchloric acid, and 15 ml of isopropanol was added. The reagent was prepared daily. The drugs used for the preparation of is reagent were products of E. Merck A. G., Darmstadt.

Autoanalyzer. An autoanalyzer (Technicon Instruments Corp., Chauncey, N.Y.) was used for the procedure.

PREPARATION OF SAMPLES

Pure Collagen. For the determination of total Hyl, a 5-mg sample was hydrolyzed with 6*N* HCl at 110 °C for 16 hr. The acid was then evaporated under vacuum at 65 °C and the evaporated sample diluted with the citrate-phosphate buffer to a final Hyl concentration of 1 to 5 μg/ml (Table 1). The sample was now ready for analysis after centrifugation at 1800 rpm for 15 min.

Urine Samples. Twenty-four-hour urines were collected under standard conditions from individuals on a collagen-free diet. One milliliter of urine was precipitated directly in ampoules with 5 ml of acetone and kept on ice for 30 min. The ampoules were centrifuged at 1800 rpm for 15 min. The supernatant was separated by vacuum suction, the precipitate was washed with 5 ml of acetone, then centrifuged, and the supernatant discarded. The process of washing with acetone and centrifugation was repeated twice. After drying in air, the samples were submitted to acid hydrolysis, as indicated earlier. Thus hydrolysis was performed in the same ampoules as used for precipitation. The analyses were carried out after dissolving the evaporated sample in the citrate-phosphate buffer.

Normal Human Skin. Skin biopsies were dried and defatted by treatment with acetone (3 changes daily for 3 days), with acetone-ether, 1–1 (3 changes on one day), and with ether (1 change). The ether was evaporated, and the samples were placed in a stainless steel vacuum desiccator (Nikortanks, cat. no. 800) to constant weight (3 days). The dry weight of the samples was determined.

The material was hydrolyzed with 2 ml of 6*N* HCl. After hydrolysis, the samples were centrifuged at 1800 rpm for 10 min. Two aliquots were taken from the supernate; that is, one of 0.1 ml and another of 1.5 ml, to be used for determination of Hyp and Hyl, respectively. Evaporation of HCl of both aliquots was performed as indicated previously. The dried samples were diluted to 10 and 3 or 6 ml with the correspondent buffers and used for analysis with the automated assays for Hyp and for Hyl.

Standards were treated with 6*N* HCl and submitted to the same steps as the samples, (i.e., heating, evaporation, and dilution in buffer). The amounts of sample used for Hyl assay on different biological materials with the automated method are given in Table I.

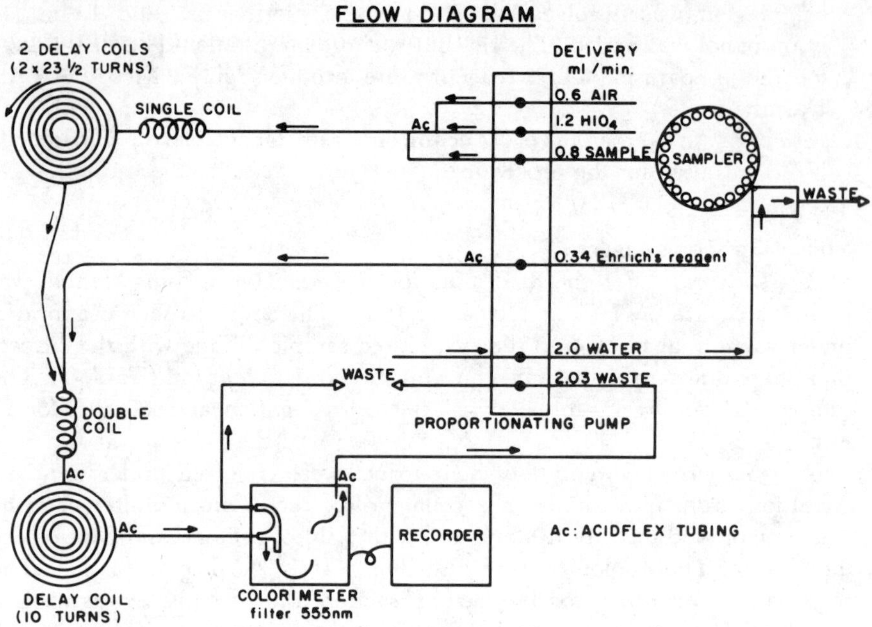

Figure 4. Flow diagram for automated assay for hydroxylysine.

Procedure. The amounts of air, sample, and reagents delivered through the corresponding tubing were expressed as milliliters per minute (Figure 4). Since the speed was 50 samples per hour (50-2/1) and 0.8 ml of the sample was delivered per minute, 40 sec is used for the sample and 20 sec for water. The actual volume of the sample and standard used for the reaction was 0.64 ml. To avoid air in the system, 0.70 ml was the minimum volume per cup that could be measured. Alternating cups of sample and water containing Brij were placed in the sampler. One standard was run per 10 samples.

TABLE I
Amount of Different Biological Materials Used for the Automated Assay for Hydroxylysine

Sample	Range of sample hydrolyzed	Amount of buffer added after evaporation for analysis (ml)
Urine (acetone precipitated)	1 ml	4
Collagen (purified)	1 mg	2
Skin (dry, defatted)	0.5–4 mg	3
	5–10 mg	6

The sample was combined with the periodate solution and mixed through a single coil. The oxidation was performed while the stream passed through two $23\frac{1}{2}$-turn delay coils. The delay coils were protected from the light with aluminum foil. Ehrlich's reagent was then added to the stream, which was mixed while passing through a double coil, and the color was allowed to develop during the passage through a 10-turn delay coil protected with aluminum foil. The color was measured with a 555-nm interference filter in the colorimeter attached to the machine. Absorbances were registered by the machine's recorder, equipped with the corresponding absorbance-registering paper. When solutions containing less than 1 μg of Hyl per 0.64-ml sample were assayed, a Servogor S recorder was used. During the run, the system was protected from the light by thick black plastic foil. All the tubing of the system was Acidflex except for the tubes transporting air and sample, which were clear Tygon. When the run was finished, the tubing used for the intake of the Ehrlich reagent was washed with *n*-propanol, and the rest of the system was washed with 1*N* HCl.

5. Color Assays for PPH or PLH Activity

Assays for protocollagen proline or protocollagen lysine hydroxylase activity on a series of synthetic peptides have been performed by using the assays for hydroxyproline or hydroxylysine, respectively (44,45). The synthetic peptides are hydroxylated by PPH or PLH in the presence of α-ketoglutarate, ascorbic acid, ferrous ions, and atmospheric O_2. Hyp or Hyl (final products of the hydroxylation) can be detected by the colorimetric assays described.

6. Hydroxyproline and Hydroxylysine in Urine

Most of the studies on collagen metabolism performed up to now have focused on the excretion of hydroxyproline. In the past, the lack of an easy, specific, sensitive, and reproducible assay for hydroxylysine hampered the determination of this amino acid in parallel with Hyp. Most previous authors (47) have taken into consideration the total Hyp and, in some cases, the free Hyp (48). It has been shown that large Hyp-containing peptides originate from degradation of newly synthesized collagen and that 5 to 20% of the urinary Hyp is nondialyzable (48). An increased urinary excretion of Hyp associated with growth consists of peptides derived from soluble collagen. By precipitation of 1 ml of urine with 5, 9, and 19 ml of acetone, we isolated different fractions containing Hyp (46). The fraction precipitated with 5 ml of acetone was approximately 20% of the total Hyp. This fraction increased in normal adolescents and in some pathological conditions. It is increased *inter alia* in active phases of scleroderma and follows the degree of activity and progression.

Procedure. A collagen-free diet was given to hospitalized patients (95% of the total material). A 24-hr urine was collected one day before the diet and another on the day of the diet. In addition, urines of some nonhospitalized controls (5%) were collected. The latter subjects had an unrestricted food intake. Samples were grouped according to age (decades), sex, and diet. The effect of the diet was analyzed by comparing the parameters studied on the two 24-hr samples collected from each individual. Analyses of Hyl and Hyp were performed according to the methods of Blumenkrantz and Asboe-Hansen (32,33).

PREPARATION OF URINE SAMPLES FOR ANALYSIS

Total Hydroxyproline. Filtered urine (1 ml) was hydrolyzed in 6*N* HCl at 110 °C for 16 hr. After hydrolysis, HCl was evaporated under vacuum at 65 °C. Evaporated samples were dissolved in 2 ml of phosphate-citrate buffer, and 0.5 ml of the solution was diluted with 9.5 ml of the same buffer. The samples were centrifuged and the supernatants used for the assay of "total Hyp."

Hydroxyproline Containing Peptides. These substances were precipitated with different amounts of acetone and, according to the proportions added, called fraction 1 + 5, 1 + 9, and 1 + 19. Urine (1 ml) was added to 5, 9, and 19 ml of acetone in glass ampoules of 10 and 25 ml. The samples were left at 4 °C for 30 min, then centrifuged at 3000 rpm for 15 min. The supernatant was aspirated by vacuum. The small amount of acetone remaining in the pellet was allowed to evaporate, and samples were hydrolyzed, evaporated, and dissolved as described for total Hyp.

Hydroxylysine. Filtered urine (1 ml) was precipitated with 5 ml of acetone in a 10-ml ampoule. The samples were refrigerated at 4 °C for 30 min, then centrifuged at 3000 rpm for 10 min. The supernatant was aspirated by vacuum and the pellet hydrolyzed with 6*N* HCl at 110 °C for 16 hr. Hydrolysis may be performed in 2 to 6*N* HCl under these conditions. After hydrolysis, HCl was evaporated under vacuum at 65 °C and the dried sample was dissolved in 4 ml of the correspondent phosphate-citrate buffer, pH 7.0 (33). The samples were centrifuged in the same ampoules in which these steps were performed, and the supernatant was used for Hyl determination.

A. EFFECT OF DIALYSIS OF URINE

Urine samples were dialyzed against running tap water for 15, 30, and 120 min, and 1 ml of the samples was precipitated with 5, 9, and 19 ml of acetone. After hydrolysis and evaporation, the samples were assayed for Hyp and Hyl. Dissolved 1 + 5, 1 + 9, and 1 + 19 fractions were submitted to dialysis after they had been precipitated. Dialysis was performed against running tap water for 15, 30, and 120 min. The undialyzable material was then hydrolyzed and analyzed for Hyl and Hyp. The effect of 120-min dialysis on the determination of free glucosyl-galactosyl hydroxylysine (Hyl-Gal-Glu) was tested. Collagenase digestion of the precipitated fractions and reprecipitation with different volumes of acetone, as well as dialysis of the collagenase-digested peptides, were performed. Both procedures were followed by hydrolysis and Hyl and Hyp determinations.

B. CALCULATIONS

The results were expressed in milligrams or millimoles per 24 hr for Hyp and Hyl in relation to age, expressed as decades of life. The patients were grouped as control (not suffering from systemic disease) and generalized scleroderma in various degrees of activity. The percentage Hyp content in the 1 + 5, 1 + 9, and 1 + 19 fractions in relation to total Hyp was calculated. The ratios of total Hyp to total Hyl as well as the molar ratio of total Hyp to Hyp in the 1 + 5 precipitates were determined.

C. SCLERODERMA

The results of the analyses of Hyp and Hyl on urines of scleroderma patients and controls were remarkable. There was no significant difference in the contents of total Hyp, the 1 + 5 Hyp-containing fraction, and Hyl between the urines of control individuals in their first, third, fourth, fifth, sixth, and seventh decades. In the second decade, there was an increased urinary output of collagen metabolites, as represented by 1 + 5 Hyp and Hyl. Analyses of the fractions containing Hyp in six normal teenagers showed that the only fraction that was really increased was the 1 + 5 fraction. No sex differences were observed in the parameters studied (i.e., Hyp and Hyl) in the different decade groups.

Patients with generalized scleroderma in an active phase showed increased urinary excretion of 1 + 5 Hyp and Hyl. There was an increase in the 1 + 5 fraction that closely correlated with the degree of activity of the generalized scleroderma. This increase was most clearly demonstrated by the percentage of 1 + 5 Hyp in relation to total Hyp. The ratio of total Hyp to Hyl showed a slight decrease in active generalized scleroderma.

The ratio of Hyp to Hyl in the 1 + 5 fraction was more constant than that of total Hyp to Hyl. Urinary excretion of Hyl followed the same pattern as that of 1 + 5 Hyp. No statistically significant differences were observed in relation to sex or diet in the patients with generalized scleroderma, active or nonactive, belonging to the second decade.

Our studies show that "total hydroxyproline" represents only an average sum of the fractions, which themselves do not change in a parallel way. In fact, values of total hydroxyproline within the normal range may be found with an increased or decreased 1 + 5 fraction. In active phases of generalized scleroderma, an increase of collagen metabolism appeared. This was particularly evident from the variation in the percentage of Hyp in the 1 + 5 fraction.

The increase of the 1 + 5 fraction in the active phases of generalized scleroderma probably represents active biosynthesis and metabolism of collagen. According to Krane et al. (49), the high molecular weight peptides represent metabolites of the most recently synthesized collagen. It is evident that the 1 + 5 fraction corresponds to the high molecular weight peptides, especially if one considers that after 2 hr of dialysis, the Hyp content of the fractions precipitated with 9 and 19 ml of acetone become equal to that of the 1 + 5 fraction.

It is worth noting that Hyl is precipitated *in toto* by addition of 5 volumes of acetone to urine, whereas normally only approximately 20% of the total Hyp is present in that fraction. The collagenase sensitivity of the precipitated fractions confirms their collagen origin. Although in man most of the urinary Hyp is present in the form of low molecular weight peptides that are readily dialyzable, a small fraction occurs in the form of higher molecular weight material (4500 to 10,000). The amino acid composition of the latter fraction has shown that not only Hyp is present, but also Hyl (49).

Effect of Dialysis. After 120 min of dialysis, the Hyp content of the 1 + 9 and 1 + 19 became equal to that of the 1 + 5 fraction. There was no difference in the Hyl content of these three fractions, and it was similar to that of the nondialyzed 1 + 5 precipitate. With shorter dialysis time, fractions 1 + 9 and 1 + 19 showed decreasing Hyp contents until equality with the 1 + 5 fraction was reached. Standard Hyl-Gal-Glu dialyzed completely after 120 min.

The fact that Glu-Gal-Hyl standard in urine is dialyzable, whereas the Hyl content of the 1 + 5 fraction is virtually nondialyzable indicates that free Glu-Gal-Hyl or Gal-Hyl per se cannot be taken as representatives of the Hyl excreted in urine.

D. GLYCOSYLATED HYDROXYLYSINE IN URINE

Urine was precipitated with 4 volumes of acetone (1 ml of urine plus 4 ml of acetone). The precipitate was suspended in $0.01N$ NaOH and two aliquots were separated. One was oxidized with periodate, and H_2O was added to the other instead of periodate. Both samples were dialyzed and submitted to acid hydrolysis, and the acid was allowed to evaporate under vacuum at 65 °C. Then both samples were suspended in buffer and submitted to the reaction for Hyl, which was described by Blumenkrantz and Asboe-Hansen (33). In the sample oxidized without previous hydrolysis, the unglycosylated ϵ-amino group of Hyl was eliminated, thus avoiding the possibility of a posthydrolysis reaction by these oxidized residues. The hydroxylysine residues were protected only if glycosylated. The other sample allowed the detection of total Hyl. The difference between total and glycosylated Hyl gave the fraction of unglycosylated Hyl. The percentage or degree of glycosylation can be calculated from the results obtained.

7. Determination of Hydroxyproline- and Hydroxylysine-Containing Fractions in Serum and Plasma (50)

Keiser et al. (51) described the presence of a hydroxyproline—containing plasma protein and suggested its possible implications in medicine (e.g., in excessive fibrosis of some connective-tissue disorders). "Leakage" from cells that are rapidly synthesizing collagen was suggested as a source of this serum protein. The clinical interest in the determination of this protein in relation to aging and disease and its unknown relationship to the C_{1q} fraction of complement, also containing Hyp, has been hampered by the tardiness of the procedures available (51,52). Most techniques aim at avoiding an interfering unknown substance, present in serum. After oxidation with chloramine T at room temperature, one toluene extraction, boiling, extraction of the pyrrole derivative of Hyp with toluene, and reaction with Ehrlich's reagent, two absorption peaks are seen,—namely, the expected one at 560 nm, and another at 450 nm. Pure Hyp submitted to the same procedure gives only one chromogen with an absorption peak at 560 nm. Proline reacts with Ehrlich's reagent after oxidation with chloramine T at room temperature and gives a chromogen with a peak of λ max at 450 nm.

This finding, together with the fact that one toluene extraction as used in previous procedures (11,29) is insufficient to eliminate this oxidation product, prompted us to test the elimination by repeated toluene extractions. Although successful extraction was obtained by this procedure, we found that, in the presence of chloramine T, the prolongation of the boiling time from 20 min to 3 hr produced a reduction of the Pro oxidation product without affecting that of Hyp. Only one toluene extraction is necessary in the latter case. Both procedures are simple and can be used in clinical chemistry laboratories. They give values similar to other methods previously reported, that were more complicated. At the same time, information on the nature of the above-mentioned "interfering material" (36) was obtained.

Isotope dilution or chromatographic separation followed by colorimetric assay for Hyp, as used by Keiser et al. (51) and Le Roy et al. (52), and our two procedures showed values of 0.60 to 0.85 mg of Hyp per 100 ml of serum. In these methods, serum was precipitated with ethanol, trichloroacetic acid, and other media.

By dialyzing 50 ml of serum, Tomaszewski (53) isolated a euglobulin fraction that contained Hyp. We precipitated the euglobulin fraction in 2 ml of serum or plasma by dilution with distilled water. The precipitate was centrifuged and analyzed for Hyp and Hyl. Values ranging between 0.300 and 0.450 μg of Hyp, similar to those reported by Tomaszewski, were found. The procedure requires small volumes of blood.

MATERIALS

Two samples of blood were drawn. One was obtained the day before, the other the day after the patient was submitted to a collagen-free diet. No anticoagulant was used, and, after clotting, the serum was separated from the clot by centrifugation. Samples were kept frozen up to the time they were analyzed. Samples of plasma were also analyzed for the Hyp-containing euglobulin.

METHODS

PREPARATION OF SAMPLES

Total Hyp-Bound Protein in Serum. One milliliter of serum was precipitated with 4 ml of absolute ethanol and kept in an ice bath for 1 hr. The procedure from the precipitation to hydrolysis was performed in a 10-ml ampoule. The supernatant was separated from the precipitate by centrifugation at 3000 rpm for 15 min. The supernatant was discarded and the precipitate washed twice with 95% ethanol. The alcohol washing was separated from the pellet by centrifugation and discarded. The alcohol remaining in the pellet was allowed to evaporate. The pellet was hydrolyzed in 3 ml of 6*N* HCl at 110 °C for 16 hr. Then the acid was evaporated under vacuum at 65 °C, and the sample was suspended in 3 ml of water; 1 ml was used for the assay of Hyp.

The procedure of Blumenkrantz and Asboe-Hansen (34) was used, modified only as follows: (1) Seven toluene extractions were performed after the oxidation with chloramine T at room temperature (34), or (2) the samples were boiled for 3 hr instead of 20 min after the oxidation with chloramine T at room temperature and addition of $Na_2S_2O_3$. In the latter procedure, which is less laborious than the former, the samples were extracted only once with toluene after boiling, and Ehrlich's reagent was added to the toluene extract (34). The chromogen that now developed was read at 450 and 560 nm in a Beckman DU spectrophotometer. At this point, 800 mg of Pro did not give any interference at 450 nm by either procedure.

The euglobulin, which contains Hyp, can be assayed by either procedure. By our adaptation to the autoanalyzer, this latter fraction has also been analyzed for Hyp and Hyl as mentioned in this chapter. Values obtained for plasma were as follows: Hyp, 338 ± 10 μg per 100 ml of plasma, and for Hyl, 134 ± 7 μg per 100 ml of plasma.

IV. ACID MUCOPOLYSACCHARIDES

1. Quantitative Determination of Uronic Acids with *m*-Hydroxydiphenyl

Uronic acids heated in concentrated sulfuric acid-tetraborate at 100 °C for 5 min develop a chromogen with λ max at 520 nm after treatment with *m*-hydroxydiphenyl. The classical methods for the detection of uronic acid, using carbazole or orcinol reagents, have the disadvantages of interference either by pentoses (54) or by hexoses (55,56) (Figure 5).

A procedure for the determination of uronic acid containing acid mucopolysaccharides (*syn.* glycosaminoglycans) in urine by precipitation of the anionic macromolecules with cetyltrimethylammonium bromide and determination of the uronic acid content with the *m*-hydroxydiphenyl reagent has been reported by Blumenkrantz and Asboe-Hansen (57). Compared with previously reported methods (54–56), the new procedure is quicker, more specific, and more sensitive; in addition, it uses considerably reduced volumes of urine in comparison with Teller et al. (58).

The *m*-hydroxydiphenyl method for uronic acid has also been automated (59).

For years investigators have needed a method for total hexosamine having equal intensities of chromogen for equal concentrations of glucosamine and

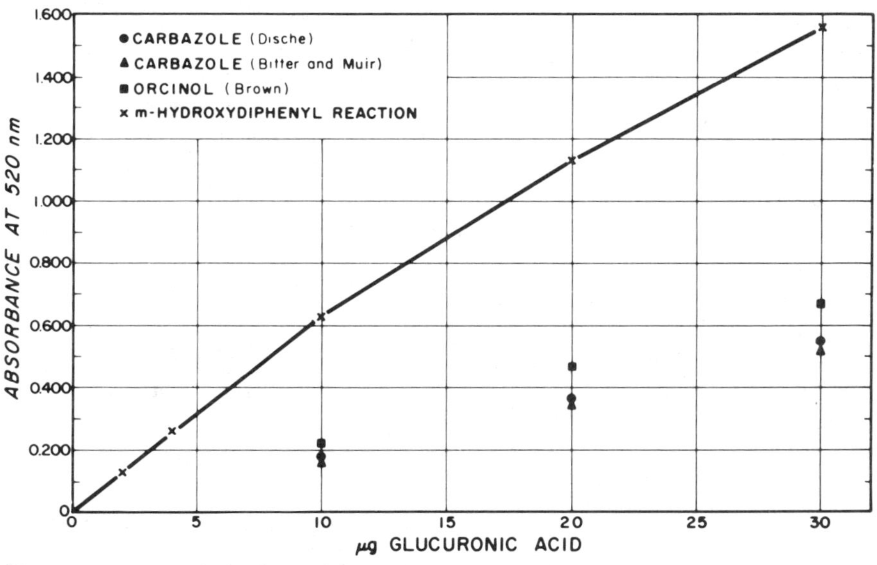

Figure 5. Comparison of *m*-hydroxydiphenyl reaction for uronic acids with other procedures currently used.

galactosamine or their mixtures, as well as a differential assay for only one of them (i.e., galactosamine). New sensitive and reproducible procedures for these purposes were worked out (60).

A. URONIC ACID: MANUAL PROCEDURE (57)

MATERIALS

Glucuronic acid, galacturonic acid, iduronic acid, mannuronolactone, and chondroitin sulfates A and B used were gifts from Dr. Martin B. Mathews, University of Chicago.

Hyaluronic acid, chondroitin sulfate C, and heparin were kindly donated by Prof. Karl Meyer, Yeshiva University, New York.

Glucose and L-arabinose were obtained from E. Merck A. G., Darmstadt; glucosamine and galactosamine were products of Sigma Chemical Corp., St. Louis, Missouri; concentrated sulfuric acid, specific gravity 1.84 (analytical grade) E. Merck A. G., Darmstadt; sodium tetraborate (p.a.), Riedel-de-Haën A. G., Seelze, Hannover; *m*-hydroxydiphenyl (analytical grade) was obtained from K & K Laboratories, Inc., Plainview, New York; *p*-hydroxydiphenyl (p.a.), Koch-Light Laboratories Ltd., Colnbrook, Buckshire, England; *o*-hydroxydiphenyl (analytical grade), Eastman Organic Chemicals, Rochester, New York; cetyltrimethylammonium bromide (reagent grade) was obtained from British Drug Houses (BDH).

REAGENTS

m-Hydroxydiphenyl solution. A 0.15% solution of *m*-hydroxydiphenyl in 0.5% NaOH. The reagent solution could be kept in the refrigerator covered with aluminum foil for more than a month.

A solution of o-hydroxydiphenyl was prepared in a similar way. *p*-Hydroxydiphenyl was dissolved in 10% NaOH, and the solution was diluted with distilled water to a final concentration of 1.5% in 0.5% NaOH. The solution was then diluted 1–10 with 0.5% NaOH and used as reagent.

H_2SO_4-sodium tetraborate solution. A 0.0125*M* solution of tetraborate in concentrated sulfuric acid.

Cetyltrimethylammonium bromide solution. A 5% solution of cetyltrimethylammonium bromide in distilled water.

METHOD

To 0.2 ml of the sample containing from 0.5 to 20 μg of uronic acids, 1.2 ml of sulfuric acid-tetraborate was added. The tubes were refrigerated in crushed ice. The mixture was shaken in a Vortex mixer and the tubes heated in a water bath at 100 °C for 5 min. After cooling in a water-ice bath, 20 μl of the *m*-hydroxydiphenyl reagent was added. The tubes were shaken and,

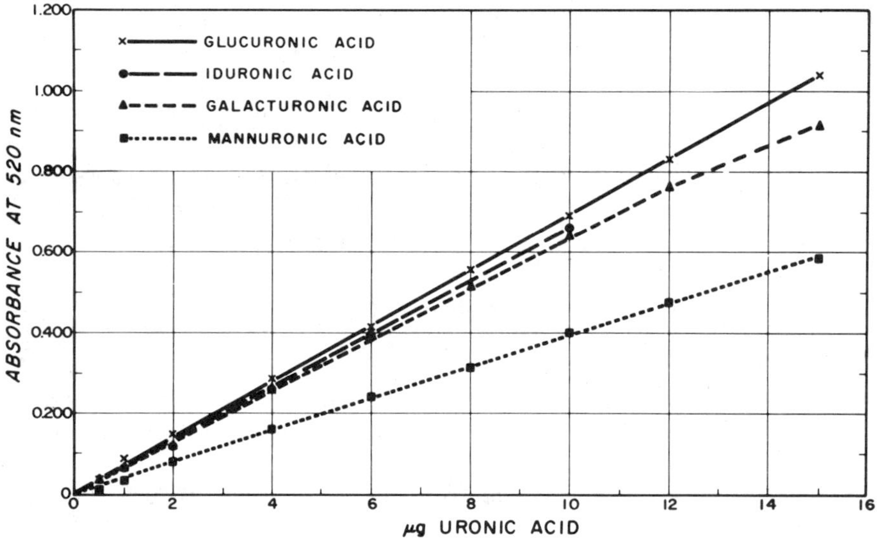

Figure 6. Reaction of different uronic acids with *m*-hydroxydiphenyl (abscissa: micrograms of uronic acid per 0.2-ml sample).

within 5 min, absorbance measurements made at 520 nm in a Beckman DU spectrophotometer (Figure 6).

Since carbohydrates produce a pinkish chromogen with sulfuric acid-tetraborate at 100 °C, a blank sample was run without addition of the reagent, which was replaced by 20 μl of 0.5% NaOH. The absorbance of the blank sample was subtracted from the total absorbance.

a. Determination of Acid Mucopolysaccharides in Urine with *m*-Hydroxydiphenyl. Urinary acid mucopolysaccharides were precipitated with cetyltrimethylammonium bromide according to Teller et al. (58). However a modification of this method was introduced by using 3 or 6 ml of urine instead of 15 or 30 ml as stated by Teller. When the specific gravity of the urine was 1.020 or higher, 3 ml of urine and an equal amount of water were added, then precipitated with 0.2 ml of the quaternary ammonium compound. In case of lower specific gravity, 6 ml of urine was used. Before the addition of the precipitating agent, the urine was brought to pH 5.0 by 2*N* HCl with the use of indicator paper. After the addition of cetyltrimethylammonium bromide solution (0.2 ml), the sample was cooled in a water-ice bath for 60 min. The precipitate was centrifuged at 3000 rpm for 20 min. The supernatant was discarded and the pellet washed twice in 96% ethanol saturated with NaCl. The final precipitate was dissolved in 1 ml of distilled water. The determination of uronic acids by the *m*-hydroxydiphenyl reaction required 100 to 200 microliters of this solution.

B. URONIC ACID: AUTOMATED PROCEDURE

MATERIALS

An autoanalyzer (Technicon Instruments Corp., Chauncey, N.Y.) was used for the quantitative assay.

Glucuronic acid, galacturonic acid, mannuronolactone, and chondroitin sulfates were gifts from Dr. Martin B. Mathews (University of Chicago).

Hyaluronic acid and heparin were kindly supplied by Prof. Karl Meyer (Yeshiva University, New York).

D-Glucose and D-fructose were products of E. Merck A. G., Darmstadt, and L-fucose, D-xylose, and L-arabinose were purchased from Sigma Chemical Corp., St. Louis, Missouri.

m-Hydroxydiphenyl (analytical grade) was obtained from K & K Laboratories Inc., Plainview, New York, and from Eastman Organic Chemicals, Rochester, New York.

Cetyltrimethylammonium bromide (reagent grade) was a product of BDH.

REAGENTS

Glucuronic acid standard. Solutions containing 5, 10, and 20 μg per milliliter of water were used as standards. Solutions more or less concentrated than the standards were also prepared to determine the degree of sensitivity.

Galacturonic acid was used at the concentrations of 5, 10, and 20 μg per milliliter of water.

Mannuronolactone solutions contained 10, 20, and 40 μg in 1 ml of water.

Acid mucopolysaccharide solutions contained 20, 50, 75, 100, and 200 μg in 1 ml of water.

Sodium hydroxide solution contained 0.5 g of NaOH per 100 ml of water.

m-Hydroxydiphenyl solution, 0.15% in 0.5% NaOH. The reagent solution in a bottle covered with aluminum foil keeps for months when stored in a refrigerator.

H_2SO_4-sodium tetraborate solution. Sodium tetraborate was dissolved in concentrated sulfuric acid to give a final concentration of 0.0125*M*.

Cetyltrimethylammonium bromide solution, 5% in water.

METHODS

The flow diagram is given in Figure 7. Minimum amounts of 210-μl samples can be used in the standard cups, whereas with conical cups, even 200 μl is sufficient. The samples are fed into the machine at the rate of 60-2/1 samples per hour. This means that the delivery is one sample per minute. However 40 sec is required for the sample and 20 sec for the water. Each sample cup is followed by 3 cups of distilled water to ensure satisfactory washing of the system between samples. This procedure allows 15 quantitative assays to be performed per hour.

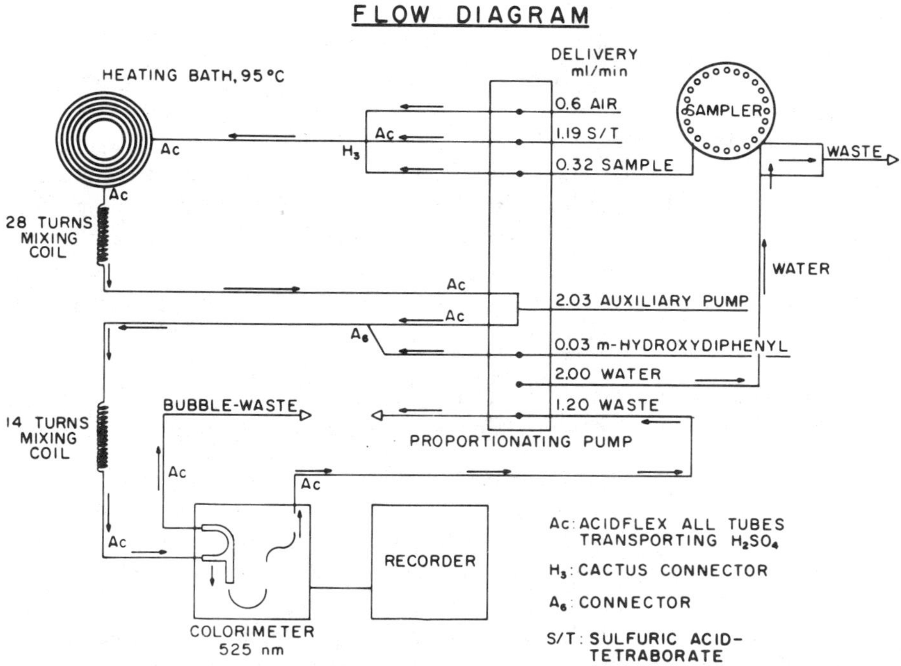

Figure 7. Flow diagram for automated assay for uronic acid.

As indicated in the flow diagram, Acidflex tubing is used in all steps of the system, except for the air, sample, reagent, and water intakes, where clear Tygon tubing is used. Tubing clamps with a screw are employed in all connections of the system where the stream containing sulfuric acid is flowing. For safety purposes a lid of Plexiglas covers the proportionating pump. The samples are mixed with sulfuric acid-tetraborate, and the combined flow is heated to 95 °C in a delay coil of $28\frac{1}{2}$ turns in a bath. The current is then passed through a double coil to mix it well, and, after addition of 30 μl of *m*-hydroxydiphenyl reagent, the stream is passed through a single coil. The color developed is read at 525 nm in the colorimeter of the autoanalyzer. The colorimeter is fitted with a 15-mm tubular flow cell. The extinctions are registered by the recorder attached to the machine, or, when low concentrations of uronic acids are measured (< 0.20 μg/0.2 ml, i.e., 1 μg/ml), a Servogor S-RE-543 is used. The extinctions are converted into absorbances by using the machine's ruler or recorded on the absorbance registering paper.

If nonpurified material is analyzed, a parallel run is performed by adding 30 μl of 0.5% NaOH instead of the reagent *m*-hydroxydiphenyl. The absorbances observed under these conditions are due to the impurities. The difference between the absorbances obtained with and without *m*-hydroxydiphenyl correspond quantitatively to the uronic acid content of the samples analyzed.

For the determination of acid mucopolysaccharides in tissue and biological fluids, samples are prepared as indicated in the manual assay of Blumenkrantz and Asboe-Hansen (57) and uronic acids assayed by the automated procedure reported earlier (59). Parallel determinations have been performed by manual assay. The results are expressed as values for uronic acids contained in the anionic macromolecules precipitated by cetyltrimethylammonium bromide (Table II).

2. Manual and Automated Assay for Total Hexosamine and a Manual Differential Assay for Glucosamine and Galactosamine

Current procedures for the quantitative determination of glucosamine and galactosamine, with the unique exception of the total hexosamine assay of Ludowieg and Benmaman (61), do not give equal chromogens for same concentrations of either hexosamine. Thus it is difficult to quantitate them, or even to detect the relative proportion of each one in a mixture, since it is impossible to decide what standard should be used. A simple procedure for the detection of total hexosamine, manual and automated, with equal sensitivity for glucosamine and galactosamine has been elaborated in this laboratory (60). A quick and sensitive procedure for the differential determination of glucosamine or galactosamine in acid mucopolysaccharides has also been elaborated (60). Glucosamine can be determined by noting the difference between total hexosamine and galactosamine, the latter detected by differential assay.

Previously, Ludowieg and Benmaman (61) described an assay for total hexosamine and a differential assay for glucosamine and galactosamine. Both

TABLE II
24 hr Yield of Urinary Acid Mucopolysaccharides, Expressed as Uronic Acid Values

	Uronic acid[a] (mg)	
	m-Hydroxydiphenyl method	
No. Samples	Manual	Automated
22	4.39 ± 0.29[b]	4.13 ± 0.29

[a] Measured as glucuronic acid.
[b] Mean values and standard error of the mean.

procedures suggested by these authors are less sensitive and more complicated than the procedures discussed below. Balazs et al. (62) reported on an automated assay for hexosamine. With their assay, a higher intensity chromogen was obtained with glucosamine than with galactosamine. This inconvenience is overcome with the automated assay of Blumenkrantz and Asboe-Hansen (60,63).

MATERIALS

p-Dimethylaminobenzaldehyde (p.a.), *acetylacetone*, *trisodium phosphate*, and *hydrochloric acid* were products of E. Merck A. G., Darmstadt.

2-Propanol (p.a.) was purchased from Riedel-de-Haën A. G, Seelze, Hannover.

Potassium tetraborate was a product of Fisher Scientific Co., Pittsburgh.

REAGENTS

1N trisodium phosphate. An aqueous solution of the indicated normality was prepared.

0.5N potassium tetraborate solution in water.

Trisodium phosphate-potassium tetraborate solution. A mixture of 98 ml of the former plus 2 ml of the latter solution was used as solvent of acetylacetone when total hexosamine was assayed.

Acetylacetone I. 3.5% (v/v) in sodium phosphate-potassium tetraborate solution was used in the total hexosamine assay.

Acetylacetone II. 3.5% (v/v) in 0.5*N* potassium tetraborate was used in the differential assay of glucosamine and galactosamine.

Ehrlich's reagent. 3.2 g of *p*-dimethylaminobenzaldehyde was dissolved in 30 ml of 12*N* HCl and diluted with 210 ml of 2-propanol.

PREPARATION OF SAMPLES

Samples of acid mucopolysaccharides (AMP) containing from 6 μg of AMP were analyzed after hydrolysis, which could be performed from 30 min at 120 °C to 12 to 14 hr at 100 °C. After hydrolysis, the hydrochloric acid was evaporated under vacuum (over NaOH pellets) at 65 °C. In all cases, standard glucosamine and galactosamine were submitted to the same conditions as the samples. After evaporation, AMP samples and standards were suspended in distilled water (at this point the pH was neutral). When samples containing a high concentration of proteins together with AMP were hydrolyzed and evaporated as described, the pH was still acid. Therefore, the evaporated samples were dissolved in 0.01*N* NaOH.

A. MANUAL ASSAY

a. Total Hexosamine. First 0.8 ml of sample or standard was added to 0.6 ml of 3.5% acetylacetone (I). The mixture was heated for 30 min at 100

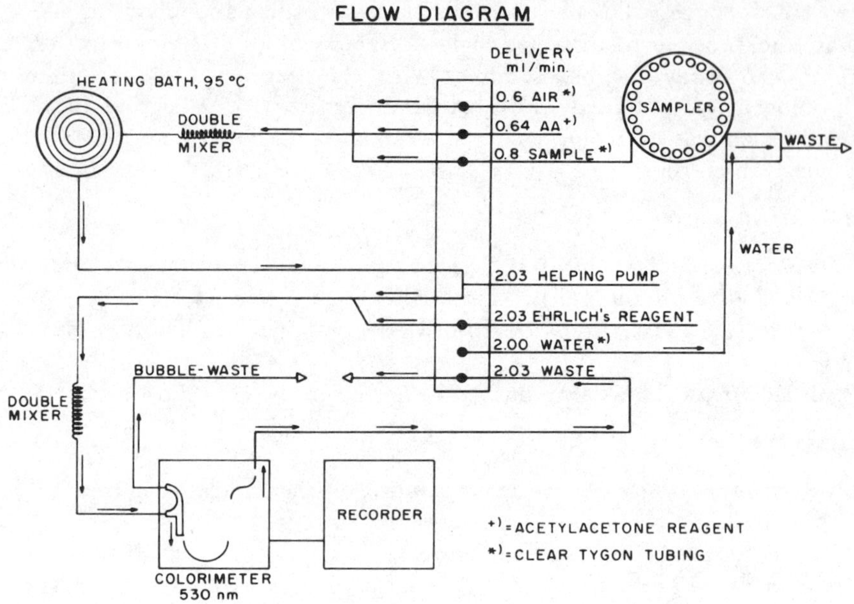

Figure 8. Flow diagram for automated assay for hexosamine.

°C; 2.0 ml of Ehrlich's reagent was added and, after 5 min, the solution was read at 535 nm in a Beckman spectrophotometer.

b. Differential Assay for Glucosamine and Galactosamine. A 0.8-ml sample or standard was mixed with 0.6 ml of the 3.5% acetylacetone solution in 0.5*N* potassium tetraborate (II). The mixture was left for 2 hr in crushed ice; heated to 50 °C for 15 min, 2 ml of Ehrlich reagent was added, samples allowed to stand at room temperature for 30 min, and read at 530 nm in a Beckman spectrophotometer.

B. AUTOMATED ASSAY: TOTAL HEXOSAMINE

Analyses were performed in an auto-analyzer (Technicon Instruments Corp., Chauncey, N.Y.). The reagents were the same as those given for the manual assay for total hexosamine. The amounts of samples and reagents delivered through the corresponding tubing are shown in the flow diagram (Figure 8). The procedure can be run at a speed of 60 determinations per hour. Air, sample, and water were delivered through clear Tygon tubing. All the other tubing was Acidflex.

3. Micromethod for Fractionation of Acid Mucopolysaccharides

The following procedure allows fractionation of unsulfated, monosulfated, and polysulfated acid mucopolysaccharides. The method is based on the formation of a complex between the anionic AMPs and the cationic detergent cetylpyridinium chloride (CPC), as well as differing sensitivities of the complexes to various concentrations of HCl (64). The complex of nonsulfated AMP-CPC is split by relatively low HCl concentration, whereas that of mono- and polysulfated AMP-CPC requires higher HCl concentrations to be split.

A new procedure using protamine sulfate instead of CPC has now been elaborated (65). Our procedure is easier than those previously reported (66,67). The use of chondroitinases or of testicular hyaluronidase made the procedure more specific. When these enzymes were used, their activity on the AMPs was evaluated by the lack of precipitability of the resulting low molecular weight oligosaccharides by CPC. They were eluted instead of being precipitated with the precipitating agent. The eluted fractions were assayed for uronic acid with the *m*-hydroxydiphenyl reaction (57), for neutral sugars with the anthrone reaction (68), and for hexosamines (60).

MATERIALS

Acid mucopolysaccharides. Hyaluronic acid (HA), under and oversulfated chondroitin sulfate (CS), chondroitin sulfates A, B, and C (CSA, CSB, CSC; *syn.* chondroitin-4-sulfate, dermatan sulfate, chondroitin-6-sulfate), heparitin monosulfate (HMS; *syn.* heparan sulfate), keratosulfate (KS; *syn.* keratan sulfate), and heparin (Hep) were gifts from Drs. Martin B. Mathews and John Cifonelli, University of Chicago.

Glucuronic acid was a gift from Dr. Mathews; *galactose* p.a. was a product of E. Merck A. G., Darmstadt.

Hyaluronidase (Leo, Hälsingborg, Sweden) was prepared from bovine testes. The activity was 15.000 to 20.000 U/mg.

Chondroitinases. Chondroitinase A, B, C. (Ch-ase A, B, C) chondroitinase A, C (Ch-ase A, C), and chondrosulfatase 4 and 6 were purchased from Miles-Seravac Ltd, Moneyrow Green Holyport, Maidenhead, Berkshire, England. They were products of Sukagaku, Kogyo Co., Ltd, Tokyo, Japan.

Microcolumns. Disposable plastic micropipettes (Dansk Laboratorieudstyr, Copenhagen, Denmark).

Celite 545, 30-80 mesh (Bie & Berntsen, Copenhagen).

Papain (Worthington Biochemical Corp., Freehold, N.J.), a twice-crystallized suspension in 0.05*M* sodium acetate pH 4.5, 39 mg/ml. Activity 15.8 units/mg.

EDTA. Ethylene diamine tetraacetate disodium salt p.a. (E. Merck A. G., Darmstadt).

Cysteine HCl. (E. Merck A. G., Darmstadt).

m-Hydroxydiphenyl (analytical grade) was obtained from K & K Laboratories, Inc., Plainview, New York, or Eastman Kodak Co., Rochester, New York.

Anthrone p.a. was a product of E. Merck A. G., Darmstadt.

Sulfuric acid. Concentrated sulfuric acid, specific gravity 1.84; analytical grade (E. Merck A. G., Darmstadt).

Sodium tetraborate p.a. (Riedel de Haën A. G., Seelze, Hannover).

Sodium hydroxide p.a. (E. Merck A. G., Darmstadt).

Hydrochloric acid, analytical grade, specific gravity 1.19. (E. Merck A. G., Darmstadt).

Cetylpyridinium chloride p.a. (E. Merck A. G., Darmstadt).

REAGENTS

Eluents. The solutions of HCl in distilled water were adjusted to normalities of 0.005; 0.5; 2.0; 3.0 and 6.0.

10% CPC in distilled water.

0.02N NaCl in a 10% aqueous solution of CPC.

A *1% CPC* solution was prepared by dilution of the 10% aqueous solution.

A *0.5% NaOH* in water.

m-Hydroxydiphenyl, a 0.15% solution in 0.5% NaOH. The reagent was kept in the refrigerator covered by aluminum foil.

H_2SO_4-sodium tetraborate, a 0.0125*M* solution of tetraborate in concentrated sulfuric acid.

Standard glucuronic acid. Aqueous solutions containing 1, 2, and 4 μg of glucuronic acid per 0.2 ml of water.

Standard galactose. A solution of galactose containing 12.5 μg per 0.25 ml of water.

Preparation of celite column. Celite was suspended in 10% CPC, and 1 ml of the suspension was applied to a disposable 3-cm plastic column with glass wool at the bottom. The column packed with 1 to 3 cm of celite was eluted with 0.02*N* NaCl in 10% CPC to ensure saturation with the quaternary ammonium compound. The columns were prepared immediately before use. Different lengths of the celite columns were used according to the AMP content of the sample submitted to fractionation (i.e., 1-cm column was used for an AMP content up to 20 μg; 2 cm for 20 to 40 μg, and 3 cm for $>$ 40 μg AMP per 50 μl aqueous solution of the sample).

Tris-HCl buffer, pH 7.3. Stock solutions of Tris 0.2*M* and HCl 0.2*M* were prepared; 50 ml of the former was mixed with 41.4 ml of the latter, and the solution was diluted with water to a total of 200 ml. A 1–100 dilution of the buffer in water was used in the assay with chondroitinase A, C.

Tris-HCl buffer, pH 8. The same stock solutions as for the pH 7.3 buffer were used, and 50 ml of the 0.2*M* solution of Tris was added to 26.8 ml of 0.2*M* HCl. The solution was then diluted with water up to 200 ml. A 1–100 dilution of the buffer in water was used in the assay with chondroitinase A, B, C.

Protamine sulfate (Sigma Chemical Corp., St. Louis, Mo.). A 1% aqueous solution.

METHOD

The tissues to be analyzed were defatted with acetone, acetone-ether, and finally ether. The tissues were stored in a stainless steel vacuum desiccator (Nikortanks, cat. no. 800) until constant weight was attained. The dry defatted tissue weight was used as a reference value. For extraction of AMP, the dry defatted tissues were submitted to papain digestion in a medium of distilled water brought to pH 6.0 with sodium hydroxide and containing 0.005*M* cysteine-HCl and 0.005*M* disodium versenate (EDTA). Papain was added in the amount of 50 μg per milligram of dry defatted tissue, and the tissues were incubated overnight at 58 °C. If incomplete solubilization was noted the next day, papain was added again and incubation was continued for an additional 12 hr. The samples were then dialyzed overnight against distilled water.

Next 10 to 30 μl of the papain-digested material containing 2.5 to 10 μg of uronic acid or 10 to 30 μg of pure individual AMP or mixture was added to the column. The column was eluted with 0.02*N* NaCl in 10% CPC, 0.005*N* HCl, 0.5*N* HCl, 2.0*N* HCl, 3.0*N* HCl, 4.0*N*, 5.0*N* and 6.0*N* HCl; 1 ml of each eluent was added in fractions of 0.5 ml. The eluted material was collected and analyzed for uronic acid, neutral sugar and hexosamines (Tables III to VI).

Uronic acid was determined according to the *m*-hydroxydiphenyl method (57) on 0.2 ml of the elution collected in every tube. Neutral sugars were assayed on 0.25 ml of the eluted material. The anthrone reaction, as described by Dreywood (68) was used for the detection of KS, a unique AMP that contains galactose instead of uronic acid. Determinations of total hexosamine as well as glucosamine and galactosamine were performed on evaporated hydrolyzed aliquots (60). Hydrolysis was performed at 120 °C for 30 min or at 100 °C for 12 to 14 hr. If total hexosamine and galactosamine are demonstrated, the relative proportion of the galactosamine- and glucosamine-containing AMPs can be calculated. Aliquots of the samples were submitted directly to the fractionation procedure or to fractionation after being treated for 15 min with chondroitinase A, C at different temperatures (1.5, 4, 22, and 37 °C). The digestion mixture contained 0.0025 to 0.005 unit of the enzyme and 20 μg of AMP (or samples containing between 2.5 and 10 μg of uronic acid) in a final volume of 50 μl. After the enzymatic digestion, the samples were passed through a celite column.

a. Effect of Testicular Hyaluronidase Digestion on the Elution Pattern of AMP. The effect of preliminary testicular hyaluronidase digestion was investigated using pure samples of hyaluronidase-sensitive and hyaluronidase-resistant AMP as well as mixtures of different proportions of both. The samples were dissolved in distilled water and brought to pH 7, which is suitable for the enzyme. Incubation was performed at 37 °C for 30 min, 1, 2, or 12 hr. The samples were then submitted to the fractionation procedure.

TABLE III
Fractionation Diagram

Eluents	Direct fractionation[a]		Fractionation after digestion with Ch-ase A, C[a] (15 min)					
			1.5°C		4°C		22–37°C	
	UA	NS	UA	NS	UA	NS	UA	NS
0.02*N* NaCl in 10% CPC		KS	HA (CSA, CSC)	KS	HA (CSA, CSC)	KS	HA, CSA, CSC (CSB)	KS
0.005*N* HCl								
0.5*N* HCl	HA							
1.0*N* HCl	UCS							
2.0*N* HCl	CSA, CSC, HMS, UCS, OCS	KS	CSA, CSC, HMS	KS	CSA, CSC, HMS	KS	HMS	KS
3.0*N* HCl	CSA, CSB, CSC, HMS, OCS	KS	CSA, CSB, CSC, HMS	KS	CSA, CSB, CSC, HMS	KS	CSB, HMS	KS
4.0*N* HCl	CSB, HMS, OCS		CSB, HMS		CSB, HMS		CSB, HMS	
5.0*N* HCl	Hep		Hep		Hep		Hep	
6.0*N* HCl	Hep	KS	Hep	KS	Hep	KS	Hep	KS

[a] UA = uronic acid, NS = neutral sugar, UCS = undersulfated chondroitin sulfate, OCS = oversulfated chondroitin sulfate, parentheses indicate trace amounts.

TABLE IV

Elution Pattern of Different AMPs, Indicating Percentage Recoveries[a] as UA (UA-containing AMP) or galactose (KS)

Eluents	HA	CS A	CS B	CS C	CS Under-sulfated	CS Over-sulfated	HMS	Hep	KS
0.02*N* NaCl in 10% CPC									28 ± 2
0.005*N* HCl									
0.5*N* HCl	97 ± 4								
1.0*N* HCl					59 ± 2				
2.0*N* HCl		59 ± 2		46 ± 1	21 ± 2	4 ± 1	52 ± 2		35 ± 2
3.0*N* HCl		41 ± 2	18 ± 2	53 ± 2	19 ± 1	59 ± 2	32 ± 1		21 ± 2
4.0*N* HCl			80 ± 3			37 ± 2	14 ± 1		
5.0*N* HCl								28 ± 2	
6.0*N* HCl								60 ± 2	15 ± 2

[a] Each value represents an average of six determinations. The added amount was considered to be 100%.

TABLE V

Fractionation Diagram for Direct Fractionation[a]

Eluents	Analysis: UA	NS	Total hexosamine	Differential assay (galactosamine)
0.02*N* NaCl in 10% CPC		KS	+	−
0.005*N* HCl				
0.5*N* HCl	HA		+	−
1.0*N* HCl	UCS		+	+
2.0*N* HCl	CSA, CSC, HMS	KS	+	CSA, CSC + HMS − KS −
3.0*N* HCl	CSA, CSB, CSC, HMS	KS	+	CSA, CSB, CSC + HMS − KS −
4.0*N* HCl	CSB, HMS		+	CSB + HMS −
5.0*N* HCl	Hep		+	Hep −
6.0*N* HCl	Hep	KS	+	Hep − KS −

[a] The combination of this diagram with the use of hyaluronidase and/or chondroitinase A, C disentangles the overlapping observed with 2 and 3*N* HCl.

TABLE VI
Fractionation Diagram for Conditions After Digestion of Sample with Chondroitinase AC, 15 min at 22 to 37° C

	Analysis[a]			
Eluents	UA	NS	Total hexosamine	Differential assay (galactosamine)
0.02*N* NaCl in 10% CPC	HA, CSA, CSC, (CSB)	KS	+	HA− KS− CSA, CSB, CSC+
0.005*N* HCl				
0.5*N* HCl				
1.0*N* HCl				
2.0*N* HCl	HMS	KS	+	−
3.0*N* HCl	CSB, HMS	KS	+	CSB+ HMS− KS−
4.0*N* HCl	CSB, HMS		+	CSB+ HMS−
5.0*N* HCl	Hep		+	−
6.0*N* HCl	Hep	KS	+	−

[a] UA = uronic acid, NS neutral sugar; parentheses indicate trace amounts.

Comparison was made with the procedures of Svejcar and Robertson (66) and Thunell et al. (67).

Determination of uronic acids was performed according to Bitter and Muir (56) when these procedures were used. Samples of hyaluronidase-sensitive AMPs, submitted to the action of the enzyme for different lengths of time, were also fractionated by our procedure and with the eluting solutions used by Svejcar and Robertson (66) and Thunell et al. (67), to determine the influence of the degree of polymerization on the method.

b. Effects of Undersulfation and Oversulfation of Chondroitin Sulfate on the Elution Pattern. To evaluate the importance of the degree of sulfation for the complex formation with CPC, samples of undersulfated and oversulfated chondroitin sulfates were passed through the celite column and eluted with the previously mentioned eluting solutions.

c. Effect of Chondroitinase A, C on the Elution Pattern of AMP, Pure or in Mixture. Twenty micrograms of AMP was submitted to the action of 0.025 to 0.0050 unit of chondroitinase A, C per 40 μl of water. The incubation was performed at different temperatures (1.5, 4, 22, and 37 °C) for different lengths of time in the presence of sodium acetate and/or Tris buffer, or without these substances (69). The samples were submitted to the micromethod for fractionation of AMP already described, using 50 μl for the fractionation. The recovery yield and the fractionation pattern were studied under the conditions mentioned. The effect of different pH values in the presence or absence of sodium acetate and/or Tris buffer on the assay was also tested. Complete degradation was considered to have occurred when the material reacting as uronic acid was only present in the first eluent (namely, 10% CPC in 0.02*N* NaCl), which indicates that the material cannot be precipitated by the quaternary ammonium compound solution.

d. Effects of Chondroitinase A, B, C on the Elution Pattern of AMP. Chondroitinase A, B, C, 0.020 unit per 40 μl in the presence of 40 μl of Tris-HCl buffer, pH 8, and 80 μl of H_2O were added to 80 μg of AMP in 40 μl water at 37 °C for 10 min. The elution pattern was compared with the same mixture having the same volume of enzyme, which was previously inactivated by heating at 100 °C, and also with the absence of the enzyme. Fifty microliters of the final volume of 200 μl was used for the fractionation. The shifting of the elution to a less concentrated HCl eluent was a measure of degradation.

e. Effects of Chondrosulfatases 4 and 6. Pure samples of AMP or mixtures were submitted to the action of chondrosulfatase-4 or chondrosulfatase-6. These enzymes were added to polymerized AMP as well as to material previously degraded.

f. Comparison of Different Procedures for Determination of Keratosulfate in a Mixture of AMPs. Determination of neutral sugars (68) was performed on aliquots of the different elution substances when samples of pure KS or KS in mixtures with other AMPs were submitted to the fractionation procedures of Svejcar and Robertson (66), Thunell et al. (67), and the procedure reported herein.

g. Fractionation of a Mixture of HA and CSB Previously Submitted to Enzymatic Digestion with Testicular Hyaluronidase. First 5 μg of HA was mixed with 5, 10, and 15 μg of CSB, and 1.25 μg of testicular hyaluronidase per 5 μg of AMP was added. The substrate and enzyme were dissolved in distilled water and were incubated at 37 °C for 30 min. Then they were passed through a 1-cm celite column previously equilibrated with 10% CPC in 0.02*N* NaCl, and eluted as described previously. Samples of HA and CSB of the same concentration as indicated for the mixtures were run separately.

h. Fractionation of a Mixture of HA, CSA, and CSC Previously Submitted to Enzymatic Digestion with Testicular Hyaluronidase. Mixtures of HA, CSA, and CSC were submitted to enzymatic digestion with testicular hyaluronidase. The relative proportions of HA, CSA, and CSC were as indicated for HA and CSB. The incubation conditions were also similar. Separate samples of HA, CSA, and CSC were run simultaneously.

i. Percentage Distribution of AMPs in Tissues of a 19-week-old Human Embryo. Papain-digested material from umbilical cord, articular and epiphyseal cartilage (tibia), and skin, prepared as already described, was submitted to fractionation. Samples containing 2 to 5 μg of uronic acid per 50 μl were passed through a 1-cm celite column. The eluents were, as usual, 0.5 ml of the solutions previously mentioned.

RESULTS

Highly reproducible results were obtained for the fractionation of AMPs with the micro method presented. The use of protamine sulfate instead of CPC as precipitating agent leads to even better separation of KS (65).

Elution Pattern of AMP. The complex of HA-CPC was split by 0.5*N* HCl. Consequently, HA was eluted with that solution. Undersulfated chondroitin sulfate was eluted with 1, 2, and 3*N* HCl, while normally sulfated CSA and CSC were eluted with 2 and 3*N* HCl. CSB was eluted with 3 and 4*N* HCl, oversulfated chondroitin sulfate with 2, 3, and 4*N* HCl. HMS was eluted by 2, 3, and 4*N* HCl, and Hep with 5 and 6*N* HCl. KS fractions were eluted with 10% CPC in 0.02*N* NaCl, and 2, 3, and 6*N* HCl, as shown by the anthrone reaction. The recovery of AMP was 100 ± 9% (Table III and IV).

Evaluation of Chondroitinases and Chondrosulfatase 4 and 6. The enzyme Ch-ase A, C has shown more substrate specificity for HA than for CSC as evidenced by the presence of UA containing material in the first eluent (i.e., 10% CPC in 0.02*N* NaCl). There is reduction of the chromogen obtained after the enzymatic action of Ch-ase A, C; that is, when complete depolymerization is obtained, the chromogen eluted with 10% CPC in 0.02*N* NaCl is 66 ± 2% for HA and 44 ± 1% for CSC as substrate, respectively.

Although, according to the information sheet of Miles-Seravac Ltd. (Moneyrow Green Holyport, Maidenhead, Berkshire, England) it is desirable to keep the solution of Ch-ase A, C in the refrigerator, no effect on its depolymerizing activity was observed after keeping this enzyme sample deep-frozen. A comparison was made of the HA digestion by two aliquots of the enzyme, one kept at 4 °C, the other thawed after having been kept frozen at − 20 °C for 3 days. Incubations were performed at 1.5, 4, 22, and 37 °C for 15 min. The depolymerizations were similar with both enzyme preparations, whether kept deep-frozen or at 4 °C.

When samples of HA, CSA, CSB, or CSC were submitted to the influence of Ch-ase A, B, C under the conditions indicated by Yamagata et al. (69) (i.e., 10 min incubation at 37 °C in the presence of 0.005 unit of the enzyme), little effect was observed, as indicated by the shift to elution with an eluent of lower concentration. Only when incubation was performed for 16 hr at 37 °C, was a consistent depolymerization of the order of 90% for CSC and 20% for HA observed. The need of Tris buffer, pH 8.0, and sodium acetate for the enzymic action to proceed was confirmed insofar as no depolymerization was observed in water or in buffer alone. The enzyme was not active in solution beyond 1 week. The recovery, as calculated by the uronic acid content of AMP samples incubated overnight, with or without Ch-ase A, B, C, was 100 ± 2%.

In our system, chondrosulfatases 4 and 6 were shown to be nonactive on the polymers CSA, CSB, and CSC. When the samples were first submitted to the action of Ch-ase A, C for 10 min at 37 °C and afterward to the chondrosulfatases 4 or 6, the only compound remaining almost fully polymerized was CSB.

Effect of Tris Buffer and Sodium Acetate on the Activity of Ch-ase A,C. No difference in the percentage of degraded HA, CSA, and CSC was observed, regardless of whether the digestion mixture contained sodium acetate and/or Tris buffer or the enzyme was just in aqueous medium (Table III). Comparisons were made in relation to temperature (1.5, 4, 22, and 37 °C) and to time of incubation in the presence or absence of buffer and sodium acetate. The degradation increased with increasing temperature or time of incubation. No differences were noted when samples were incubated at the same temperature or for the same length of time in the presence or absence of Tris buffer, sodium acetate, or both.

Effect of AMP Resistant to Ch-ase A, C on the Degradation of CSA and CSC. When mixtures of CSA or CSC with HMS or KS were submitted to degradation with Ch-ase A, C in the presence or absence of Tris buffer and sodium acetate (64), no inhibition of the enzymatic degradation of either CSA or CSC was observed.

Effect of pH on the Enzymic Activity of Ch-ase A,C. No differences in the sensitivity of HA, undersulfated chondroitin sulfate, and CSC were observed when assaying the effect of pH from 6.0 to 8.0.

Effect of a Hyaluronidase-Resistant AMP on the Degradation of HA. Samples of HA and CSB run separately or as mixtures without enzyme were eluted with 0.5*N* and 3 to 4*N* HCl, respectively. The recoveries obtained were 90 ± 5%.

The presence of a hyaluronidase-resistant AMP (CSB) inhibited the degradation of HA by the enzyme. The elution pattern showed that after incubation with the enzyme for 30 min, HA was eluted with 0.5*N* HCl while the CSB of the mixture was eluted with 3 and 4*N* HCl. When mixtures of HA and CSC were treated with hyaluronidase and fractionated, most of the HA and CSC were degraded as shown by uronic acid-containing material being eluted with all the eluents from 10% CPC in 0.02 NaCl to 2*N* HCl, the uronic acid material eluted with the latter being minimum. The recoveries obtained were 92 ± 4%. Undepolymerized CSC was eluted with 2 to 3*N* HCl only (Table III).

Analysis of AMPs from Human Embryonic Tissues. A predominance of HA was found in umbilical cord and skin, whereas chondroitin sulfates predominated in cartilage.

DISCUSSION

The method presented offers the following advantages over the most widely used procedures for fractionation of AMPs. (1) The precipitation and fractionation is performed on a standard disposable column, (2) the results are highly reproducible, and (3) HCl is a most convenient eluent, especially because it allows further hydrolysis in case determination of hexosamines is also desired. The concentration of HCl in the eluents may be changed to one that is normally desirable for hydrolysis, and the samples may be hydrolyzed without previous treatment. In other procedures in common use for fractionation of micro quantities of AMP [e.g., the method of Thunell et al. (67)], the analysis for hexosamine must be preceded by alcoholic precipitation of the sodium and magnesium salts.

The elution of an individual AMP with more than one eluent may express polydispersity and/or different degrees of sulfation. This is shown by our experiments of time-course degradation with testicular hyaluronidase and fractionation of a purified undersulfated chondroitin sulfate. As demonstrated by Meyer et al. (70), chondroitin sulfates A and C from different sources possess considerable polydispersity and, at least in some tissues, varying degrees of sulfation.

Comparison of the elution patterns of hyaluronidase-sensitive AMPs digested for different lengths of time shows that these AMPs are eluted differently in relation to the degradation process. Since this observation is valid when using the eluents of Thunell et al. (67) and Svejcar and Robertson (66) as well as ours, it is almost impossible to assess the type of AMP eluted by the eluent and determination of hexuronic acid only. The same is true of the other means of determining the type of AMP by a certain eluent and the presence of glucosamine or galactosamine, particularly if one considers the overlapping of some AMPs eluted by more than one of the eluents.

The nonhomogeneous pattern of AMP elution after testicular hyaluronidase digestion may be explained by the mechanism of action of testicular hyaluronidase, an enzyme that exhibits both hydrolytic and transglycosidase action (71,72). The main product of testicular hyaluronidase digestion of HA, CSA, and CSC is a tetrasaccharide, although a series of oligosaccharides, ranging from di- to octosaccharides, has in fact been isolated. The transglycosidation occurs when both the higher polymers and the oligosaccharides are submitted to the action of hyaluronidase. The lack of retention of the UA-containing material by the column indicates depolymerization. The end product of Ch-ase A, C degradation is an unsaturated disaccharide (69); thus we may conclude that the disaccharide cannot form an insoluble complex with the quaternary ammonium compound.

Our results are in accordance with those of Linker et al. (73) in that, after complete degradation (all material eluted with $0.02N$ NaCl in 10% CPC), a reduction in the UA chromogen in the order of 66 ± 2% for HAand 44 ± 1% for CS was found. The depolymerization of a small proportion of CSB by Chase A, C may be explained by the presence of a hybrid region containing glucuronic instead of iduronic acid. Considering the inhibitory effect of CSB on the digestion of HA by testicular hyaluronidase, it is worth noting that the use of this particular enzyme in separating mixtures of hyaluronidase-sensitive from hyaluronidase-resistant AMPs may involve serious pitfalls.

4. Detection of Chondroitinase A, C and Hyaluronidase Activity

Chondroitinase A, C and hyaluronidase action on hyaluronic acid and chondroitin sulfates A and C can be determined with our micro method for fractionation of AMPs (64). Since the depolymerized substrates cannot form insoluble complexes with CPC, they are not retained by a celite column but are eluted instead of being precipitated with the quaternary ammonium compound. The presence or absence of the enzymes in the material assayed is indicated if uronic acid and hexosamines are found in the material eluting with CPC after incubation with the above-mentioned enzymes, while the substances are absent in an aliquot submitted to the same incubation conditions without the enzymes. In case of overlapping, the use of the *m*-hydroxydiphenyl reaction for UA, and the total and differential assays for hexosamines, help resolve a mixture of acid mucopolysaccharides.

References

1. P. M. Gallop and M. A. Paz, *Physiol. Rev.*, *55*, 418 (1975).
2. B. P. Toole and D. A. Lowther, *Arch. Biochem.*, *128*, 567 (1967).
3. D. J. Prockop and K. Juva, *Proc. Nat. Acad. Sci.* (*U.S.*), *53*, 661 (1965).
4. J. Uitto, *Biochim. Biophys. Acta*, *201*, 438 (1970).
5. N. Blumenkrantz, L. Danielsen, and G. Asboe-Hansen, *Acta Dermatol.-Venereol.*, *53*, 429 (1973).
6. N. Blumenkrantz and G. Asboe-Hansen, *Anal. Biochem.*, *66*, 330 (1975).
7. N. Blumenkrantz and G. Asboe-Hansen, *Mech. Ageing Dev.*, *1*, 445 (1972–1973).
8. N. Blumenkrantz and G. Asboe-Hansen, *Acta Dermatol.-Venereol.*, *53*, 94 (1973).
9. N. Blumenkrantz and G. Asboe-Hansen, *In Vitro*, *8*, 342 (1973).
10. N. Blumenkrantz and G. Asboe-Hansen, *Acta Dermatol.-Venereol.*, *54*, 35 (1974).
11. K. Juva and D. J. Prockop, *Anal. Biochem.*, *15*, 77 (1966).
12. B. Peterkofsky and D. J. Prockop, *Anal. Biochem.*, *4*, 400 (1962).
13. D. J. Prockop, S. Udenfriend, and S. Lindstedt, *J. Biol. Chem.*, *236*, 1395 (1961).
14. M. Rojkind and E. González, *Anal. Biochem.*, *57*, 1 (1974).
15. N. Blumenkrantz and D. J. Prockop, *Anal. Biochem.*, *30*, 377 (1969).
16. N. Blumenkrantz and G. Asboe-Hansen, *Clin. Biochem.*, *9*, 256 (1976).
17. C. W. M. Orr, *Biochem. Biophys. Res. Commun.*, *23*, 854 (1966).
18. K. Juva and D. J. Prockop, *J. Biol. Chem.*, *244*, 6486 (1969).
19. B. Peterkofsky and R. De Blasio, *Anal. Biochem.*, *16*, 384 (1966).
20. O. R. Harwood, M. E. Grant, and D. S. Jackson, *Biochem. J.*, *144*, 123 (1974).

21. J. Ristelli and K. I. Kivirikko, *Biochem. J.*, *144*, 115 (1974).
22. E. Weinstein, N. Blumenkrantz, and D. J. Prockop, *Biochim. Biophys. Acta*, *191*, 747 (1969).
23. R. E. Rhoads and S. Udenfriend, *Proc. Nat. Acad. Sci.* (*U.S.*), *60*, 1473 (1968).
24. S. M. Krane, S. R. Pinnel, and R. W. Erbe, *Proc. Nat. Acad. Sci.* (*U.S.*), *69*, 2899 (1972).
25. N. Blumenkrantz and D. J. Prockop, *Biochim. Biophys. Acta*, *208*, 461 (1970).
26. H. B. Bosmann and E. H. Eylar, *Biochem. Biophys. Res. Commun.*, *30*, 89 (1968).
27. M. L. Efron, E. M. Bixby, and C. V. Pryles, *N. Engl. J. Med.*, *272*, 1299 (1965).
28. B. P. Toole, A. H. Kang, R. L. Trelstad, and J. Gross, *Biochem. J.*, *127*, 715 (1972).
29. J. D. Cipera and A. G. Charian, *Calcif. Tissue Res.*, *3*, 30 (1969).
30. S. R. Pinnel, S. M. Krane, J. E. Kenzora, and M. J. Glimcher, *N. Engl. J. Med.*, *286*, 1013 (1972).
31. R. Askenasi, *J. Lab. Clin. Med.*, *83*, 673 (1974).
32. N. Blumenkrantz and G. Asboe-Hansen, *Clin. Biochem.*, *7*, 251 (1974).
33. N. Blumenkrantz and G. Asboe-Hansen, *Clin. Biochem.*, *8*, 177 (1975).
34. N. Blumenkrantz and G. Asboe-Hansen, *Anal. Biochem.*, *63*, 331 (1975).
35. W. Troll and J. Linsley, *J. Biol. Chem.*, *215*, 655 (1955).
36. D. J. Prockop and S. Udenfriend, *Anal. Biochem.*, *1*, 288 (1960).
37. K. I. Kivirikko, O. Laitinen, and D. J. Prockop, *Anal. Biochem.*, *19*, 249 (1967).
38. N. Blumenkrantz and G. Asboe-Hansen, *Anal. Biochem.*, *55*, 288 (1973).
39. H. Stegemann, *Z. Physiol. Chem.*, *311*, 41 (1958).
40. H. Stegemann and K. Stalder, *Clin. Chim. Acta*, *18*, 267 (1967).
41. R. A. Grant, *J. Clin. Pathol.*, *17*, 685 (1964).
42. N. Blumenkrantz and D. J. Prockop, *Anal. Biochem.*, *39*, 59 (1971).
43. N. Blumenkrantz and G. Asboe-Hansen, *Anal. Biochem.*, *56*, 10 (1973).
44. R. E. Rhoads and S. Udenfriend, *Arch. Biochem.*, *133*, 108 (1969).
45. K. I. Kivirikko, R. Shudo, S. Sakakibara, and D. J. Prockop, *Biochemistry*, *12*, 4966 (1973).
46. N. Blumenkrantz and G. Asboe-Hansen, *IRCS Med. Sci.*, *4*, 23 (1976).
47. H. E. Jasin, C. W. Fink, W. Wise, and M. Ziff, *J. Clin. Invest.*, *41*, 1928 (1962).
48. M. Ziff, A. Kibrick, E. Dresner, and H. J. Gribetz, *J. Clin. Invest.*, *35*, 579 (1956).
49. S. M. Krane, A. J. Muñoz; and E. D. Harris, *J. Clin. Invest.*, *49*, 716 (1970).
50. N. Blumenkrantz and G. Asboe-Hansen, in press.
51. H. Keiser, E. C. LeRoy, S. Udenfriend, and A. Sjoerdsma, *Science*, *142*, 1678 (1963).
52. E. C. LeRoy, A. Kaplan, S. Udenfriend, and A. Sjoerdsma, *J. Biol. Chem.*, *239*, 3350 (1964).
53. J. Tomaszewski, *Acta Biochim. Pol.*, *18*, 277 (1971).
54. A. H. Brown, *Arch. Biochem.*, *11*, 269 (1946).
55. Z. Dische, *J. Biol. Chem.*, *167*, 189 (1947).
56. T. Bitter and H. M. Muir, *Anal. Biochem.*, *4*, 339 (1962).
57. N. Blumenkrantz and G. Asboe-Hansen, *Anal. Biochem.*, *54*, 484 (1973).
58. W. M. Teller, E. C. Burke, J. W. Rosevear, and L. B. F. McKenzie, *J. Lab. Clin. Med.*, *59*, 95 (1962).
59. N. Blumenkrantz and G. Asboe-Hansen, *Biochem. Med.*, *11*, 60 (1974).
60. N. Blumenkrantz and G. Asboe-Hansen, *Clin. Biochem.*, *9*, 269 (1976).
61. J. Ludowieg and J. D. Benmaman, *Anal. Biochem.*, *19*, 80 (1967).
62. E. A. Balazs, K. O. Berntsen, J. Karossa, and D. A. Swann, *Anal. Biochem.*, *12*, 559 (1965).
63. N. Blumenkrantz and G. Asboe-Hansen, in press.
64. N. Blumenkrantz and G. Asboe-Hansen, *Clin. Biochem.*, *9*, 9 (1976).
65. N. Blumenkrantz and G. Asboe-Hansen, in press.
66. J. Svejcar and W. B. Robertson, *Anal. Biochem.*, *18*, 333 (1967).
67. S. Thunell, C. A. Antonopoulos, and S. Gardell, *J. Atheroscler. Res.*, *7*, 283 (1967).
68. R. Dreywood, *Ind. Ing. Chem.* (*Anal. Ed.*), *18*, 499 (1946).

69. T. Yamagata, H. Saito, O. Habuchi, and S. Susuki, *J. Biol. Chem.*, *243*, 1523 (1968).
70. K. Meyer, E. Davidson, A. Linker, and P. Hoffman, *Biochim. Biophys. Acta*, *21*, 506 (1956).
71. B. Weissmann, *J. Biol. Chem.*, *216*, 783 (1955).
72. P. Hoffman, K. Meyer, and A. Linker, *J. Biol. Chem.*, *219*, 653 (1956).
73. A. Linker, P. Hoffman, K. Meyer, P. Sampson, and E. D. Korn, *J. Biol. Chem.*, *235*, 3061 (1960).

Determination of Fluorine in Biological Materials

P. VENKATESWARLU, *Department of Biochemistry, the Medical School, University of Minnesota, Minneapolis, Minnesota, 55455.*

I. INTRODUCTION

1. Past Background and Present Significance

At about the turn of this century, a particular hypoplastic condition of the tooth called "mottled enamel" held the interest of several investigators in the field of dental research. In 1926, when attention was eventually focused on the mineral composition of water in search of the factor responsible for the causation of mottled enamel, Frank Hanan, chemist at the Filtration Plant in Toronto, Ontario, Canada, observed, "But when we consider fluorine, all is at present shrouded in obscurity. The detection and estimation of small doses of fluorine are tedious and troublesome and quite outside the province of the ordinary waterworks chemist ." He believed that a deficiency of fluorine caused mottled enamel. However Hanan's comments received very little attention, presumably because contemporary techniques were inadequate for the routine determination of fluorine in water.

Five years later, at the Aluminum Company of America, during a purely routine spectrographic examination of waters for aluminum, Churchill and his assistant Petry happened to observe the calcium fluoride band in a water sample from an area where mottled enamel was endemic. Fortunately the water sample contained sufficient calcium to yield the broad calcium fluoride absorption band, which made possible the identification of fluoride in water. It was soon found that the severity of mottled enamel was directly proportional to the fluoride content of water. At about the same time, the etiology of mottled enamel was clearly traced to fluorine through animal experiments as well, done independently by Smith in America and by Velu in Africa. An interesting finding emerged from these epidemiological studies: the incidence of dental carries among children was less in communities where mottled enamel was prevalent than in those whose populations did not have mottled enamel. These findings eventually led to the fluoridation of public water supplies as a mass caries-control measure (1,2).

Fluorine exhibits diverse physiological properties. Though beneficial in small doses, it is harmful in large amounts. In some severe cases, chronic fluorine intoxication, involving extensive damage to the skeleton (skeletal fluorosis) has been reported among men and animals. This condition results from protracted ingestion of waters naturally high in fluorine (endemic fluorosis), continuous exposure to fluoride dusts and gases in cryolite mines or industries processing fluorides, or consumption of forage contaminated with fluoride dusts or gases in the vicinity of such industries (industrial fluorosis).

Plants such as gladiola are extremely sensitive to increased levels of fluorine in the atmosphere. Others, like the tea plant, have a prediliction for the acquisition of fluorine from the soil. Most plants store the acquired fluoride as inorganic fluoride, but some transform the inorganic fluoride to toxic fluoroorganic compounds like fluoroacetic acid, which was responsible for the mysterious deaths some years ago of several cattle grazing on "gifblarr," or *Dichapetalum cymosum* (chaillelia cymosa) in South Africa (3).

Organic fluorine compounds find applications as pesticides, herbicides, rodent poisons, and so on. There has been considerable interest in the clinical applications of fluorine compounds (4,5), such as the use of inorganic fluoride in the treatment of osteoporosis and organic fluorine compounds in anesthesia, in cancer therapy, and possibly as blood substituents (for oxygen transport) in the foreseeable future. Attention is now being directed to the fate and the biological effects of several thousand tons of fluorocarbons in the form of aerosol propellants, being continually released every year into the atmosphere.

Very recently, Allmann and Benac (6) have reported that administration of low doses of fluoride to rats raised on an experimental low-fluoride diet has resulted in an increase in tissue synthesis and urinary excretion of cyclic AMP.

The success of investigations aimed at elucidating the mechanism of action of fluorine compounds, their biological effects, the "environmental cycle," and metabolic pathways of fluorine depends, among other things, on the availability of reliable methods for the determination of fluorine in soil, water, atmospheric samples, plants, foods, animal tissues, and body fluids. The last 10 years have witnessed the development of several new approaches to the determination of fluorine, which in turn have led to newer concepts in the biochemistry of the element. Based on a new diffusion technique, Taves

(7) has reported, for the first time, the existence of two fractions of fluorine in blood serum. While exploring the application of a new adsorption technique (not involving diffusion) for the determination of fluorine in serum, Venkateswarlu et al. (8) also encountered two fractions of fluorine in serum.

2. An Overview of Fluoride Methodology

One of the earliest colorimetric reagents reported for the fluoride ion is zirconium-alizarin sulfonate. In the early 1930s this reagent was employed for the direct determination of fluorine in water samples. The susceptibility of the method to interfering ions present in natural waters was not unknown. The most significant development at this stage was the classical perchloric acid-sulfuric acid distillation procedure for the separation of fluoride from interfering ions, described in 1932 independently by Willard and Winter in the United States (9) and by Tananajev in Russia (10). Furthermore, this isolation technique made it possible to determine fluorine in a variety of materials other than water. It marked the beginning of an era of fluoride determination in biological materials. Although the Willard-Winter technique was instrumental in many important advances in the biochemistry of fluorine, a need was always felt for a less tedious method for the isolation of fluoride, particularly when the analysis of a large number of samples was required. In 1954 Singer and Armstrong (11) described a novel method for the isolation of fluoride by diffusion as hydrogen fluoride. The main feature of the method is its relative simplicity, which permits setting up of as many as 20 to 30 diffusions per day by a single analyst. The diffusates are analyzed colorimetrically on the following day.

There had been some rather unexpected and interesting developments in the colorimetric methods for the determination of fluorine. Until 1959 all the colorimetric methods were "indirect"; they were based on the "color fading" effect of fluoride on colored metal-dye complexes. Belcher et al. were engaged in the syntheses of *NN*-di(carboxy methyl)aminoethyl derivatives of some hydroxyanthraquinones for use as indicators in complexometric titrations, especially the ethylenediaminetetraacetate titration of Pb^{2+}, Zn^{2+}, Co^{2+}, and In^{3+} in acid solution (12). It turned out that one of the indicators, alizarin complexan, also developed colors with Th^{4+} and Ce^{3+}. Presumably during research on the effects of various interfering anions on the foregoing color reactions, these investigators found that on adding fluoride to the red complex of cerium with alizarin complexan, instead of the expected decrease in the color, the red color altogether changed to lilac blue! This pleasant surprise was indeed the beginning of the development of "direct" colorimetric methods for fluoride (13). These direct methods are superior to the indirect methods, particularly with respect to tolerance toward interfering ions.

The most significant recent advance in methods for fluoride determination is the development of the fluoride ion electrode by Frant and Ross in 1966 (14). The electrode can be used to determine fluoride ion activity over a wide range, down to $1 \mu M F^-$ in the presence of several other ions, including even the most notorious of the interfering ions encountered in the colorimetric and titrimetric methods. Recently the assembling of a hanging drop fluoride electrode, which is an adaptation of the original fluoride electrode of Frant and Ross, has been described by Venkateswarlu (15) for the measurement of fluoride in sample volumes as small as 5 μl and containing no more than 2.5 pmole of fluoride. Although the fluoride electrode can be employed directly with most samples, some analytical situations may call for separation and concentration of fluoride ions, which now can be performed readily by adsorption (8,16) and reverse extraction (15) techniques. Reverse extraction offers a new approach with potential advantages over the conventional distillation and diffusion techniques for the separation of fluoride from inorganic and organic materials. The advantages are simplicity of procedure, extremely low fluoride blank, and a very high degree of concentration of fluoride, hitherto unachievable. Fluoride so isolated can be determined with colorimetric and fluorometric reagents or with the fluoride electrode.

Another extremely promising method is the gas chromatographic method originally developed by Bock and Semmler in 1967 (17). Other recent methods that are useful additions to the present repertory of fluoride analytical methods are based on mass spectrometry, polarography, and radioactivation analysis.

It would be presumptuous to favor exclusively any particular method of analysis as the best for the determination of fluorine in the microgram and submicrogram range. It is fortunate that we have more than one sound method. Laboratories that can afford to do so may switch frequently to more recent and supersensitive methods, but other laboratories with modest means and for other reasons may prefer to continue with the time-honored methods, which are indeed adequate for their needs. The success of a laboratory engaged in fluoride analysis, as in any other microanalysis, is largely dependent on the experience of the analyst and on his comprehension of the merits and limitations of the different available methods. Furthermore, competence in more than one approach to the determination of fluorine in any given biological material is helpful in successfully tackling several analytical problems that develop from time to time. Thus both the earlier methods that remain applicable and the recent approaches are discussed in this chapter.

Some of the earlier major reviews on the subject of analytical chemistry of fluorine have been written by Busch et al. (18), McKenna (19), Elving et al. (20), Horton (21), Macdonald (22,23), and Nikolaev et al. (24).

To provide a quick overview of the methodology, some of the available alternative approaches to the determination of fluorine in biological materials are presented in Table I.

TABLE I
Alternate Approaches to the Determination of Fluorine in Biological Materials

Pretreatment of samples	Separation and concentration of fluoride	Final measurement of fluoride
None	Distillation (SiF_4)	Titrimetry
Unmasking the fluoride ion	Distillation (H_2SiF_6)	Spectrophotometry
Open ashing	Diffusion (HF)	Fluorimetry
Fusion	Diffusion (TMFS[a])	Electroanalytical methods:
Preliminary distillation	Pyrohydrolysis (HF)	Spontaneous electrolysis
Confined combustion:	Pyrolysis (NH_4F)	Polarography
Oxygen flask	Solvent extraction:	Potentiometry (fluoride electrode)
Oxygen bomb	Complexes of Sb, Ta, La, etc.	Gas chromatography
Tubular furnace	Fluorosilanes	Mass Spectrometry
Oxyhydrogen flame	Separation as flouride ion	Radioanalytical methods:
Digestion with acid	Ion exchange	Isotope dilution
Digestion with alkali	Chromatography	Isotope interaction
Fusion with alkali metals	Electrodialysis	Isotope displacement
Reduction with alkali metals in organic solvents	Adsorption	Radioactivation
	Reverse extraction	Catalytic methods
		Enzymatic methods

[a] Trimethylfluorosilane.

II. PRETREATMENT OF SAMPLES

The procedures for pretreatment of samples for fluorine analysis depend on the nature of the sample and the final method employed for fluoride measurement. In the case of most potable water samples, which normally do not contain significant amounts of interfering ions, no preliminary preparation of the samples is required for fluoride measurement by colorimetric and titrimetric procedures or with the fluoride electrode.

This section describes the circumstances and the types of pretreatment of samples for fluoride analysis.

1. Unmasking the Fluoride Ion

Natural waters contain several types of cation (e.g., those of Ca, Mg, Al, and Fe), which could complex fluoride ions to different degrees. The existing equilibrium between fluoride ions and the fluoride complexes in a given water sample is essentially unaltered during fluoride measurement with the fluoride electrode; only the free fluoride ion concentration is measured. To release the fluoride from the metal complexes and make possible the determination of total fluoride, the water samples are treated with such metal-chelating agents as citrate, tartaric acid, EDTA, and 1,2-diaminocyclohexane tetraacetic acid. However this pretreatment procedure is inapplicable in the analysis of natural waters with colorimetric reagents.

2. Open Ashing

Ashing of biological materials serves two important functions: (1) it destroys all organic matter, which otherwise would partly distill or diffuse over (particularly volatile fatty acids) and interfere with the spectrophotometric methods for fluoride, and (2) it makes covalently bound organic fluorine available for isolation and subsequent determination. Furthermore, distillation of unashed biological material with perchloric acid could be hazardous because of the risk of explosion.

Since fluoride is lost because of volatilization at the 500–600 °C ashing temperature range, "fixatives" are employed. The usual fixatives chosen to prevent such a loss are calcium phosphate, and oxides and acetates of calcium and magnesium. The bicarbonates and chlorides of calcium and magnesium also have been used.

The level of fluoride contamination of commercially available fixatives sometimes precludes their use in the fluorine analysis of biological materials like body fluids and soft tissues, which are normally very low in fluoride. Methods are now available for the laboratory preparation of fixatives that are extremely low in fluorine. These methods are described in Section VI.1.B.

Ordinarily, approximately 20 g of a dry material or 50 to 100 g of a liquid sample are ashed with 1 g of magnesium oxide or 5 ml of 60% magnesium acetate or 2.5 ml of 30% suspension of calcium oxide in water. The use of a solution of acetate of calcium or magnesium, rather than a solid fixative like magnesium oxide, ensures a thorough mixing of the fixative with the sample. Unlike the corresponding oxides, bicarbonates, or chlorides, however, the acetates provide additional organic matter for combustion, thereby prolonging the time required for ashing. In the case of incomplete ashing, as indicated by the presence of considerable amounts of carbonaceous matter in the ash, the material is turned over with a clean spatula and reashed. Ikert (25) claims that the use of a mixture of chromium and calcium acetates results in speedier ashing and better recovery of fluorine. Gregorowicz et al.

(26) mix the ground sample with calcium hydroxide and ammonium nitrate before incineration. Brüning and Quast (27) consider that ashing of organs is best carried out with copper oxide.

If smaller samples are ashed, proportionately less fixative may be employed. The recommended amount of a fixative relative to the size of the sample is rather arbitrary and is best determined by the investigator to ensure good recoveries of fluoride added to test samples. Singer and Armstrong employ 75 mg of magnesium oxide to ash 5 to 10 ml of serum (28). They have also used 10 to 50 mg of calcium phosphate for ashing 5 ml of serum with a retention of 97 to 102% of radioactive fluoride added to serum (29). Taves (7) ashes 10-ml serum samples with 5-ml fixative solution (1 *M* magnesium chloride and 1m*M* sodium carbonate), with a recovery of 94 to 99% of added radioactive fluoride. However it is important to recognize that the foregoing recoveries do not necessarily guarantee the safety or success of the open ashing procedure relative to the covalently bound fraction of fluorine in the sample. The added (radioactive) fluoride that does not exchange with organic fluorine could be quantitatively retained by the fixative even before the release of the organic fluorine, which may or may not be quantitatively retained. Perhaps a better way of gaining insight into this problem is to study the recovery of fluorine following addition of representative fluoroorganic compounds to biological materials prior to open ashing. There is currently a paucity of information in this regard. Kakabadse et al. (30) have been able to recover only 21.7 to 71.3% of fluorine present in selected organic fluorine compounds, following ashing at 500 °C for 24 hr with magnesium oxide, or magnesium succinate plus lithium hydroxide.

Although the ashing of bones and other materials adequately high in fluorine content may be carried out in porcelain crucibles, ashing of body fluids and soft tissues low in fluoride is performed in platinum dishes or crucibles. Evaporation of solutions containing fluorine in porcelain and silica dishes has been shown to result in incomplete recovery of fluorine, presumably because of adsorption of fluoride ions by the dishes (31). If the fluoride content of the sample is low, the fluoride blank resulting from the use of such dishes may be significant.

3. Fusion

According to Willard and Winter (9), materials that cannot be decomposed by perchloric acid must be fused with sodium carbonate before isolation of fluoride by distillation. Ashing of biological materials with certain fixatives results in the formation of refractory compounds of fluoride from which fluoride is not quantitatively recovered under the normal conditions of distillation, as reported by Rowley et al. (32), Remmert and Parks (33), and Hardin et al. (34). These authors recommend fusion of the ash with sodium hydroxide or potassium hydroxide before distillation. Oelschläger (35) also recommends that samples high in silica be submitted to fusion with caustic soda to decompose the silicon oxyfluorides formed during ashing in the muffle furnace. Fusion of soil and rock samples is always necessary when total fluorine determinations are to be made.

4. Preliminary Distillation

A preliminary distillation with sulfuric acid is required for the analysis of fluorine in certain biological materials like bone meal and fatty materials, which cannot be thoroughly mixed with the fixative preparatory to ashing, and also of fluorine in refractory fluorine compounds present naturally in the test samples or formed as a consequence of ashing. The temperature of preliminary distillation is normally maintained at 135 to 145 °C, but it is elevated to 170 °C to facilitate rapid release of fluoride from refractory compounds. The distillate is concentrated by evaporation and ashed with a fixative before carrying out a perchloric acid distillation for final isolation of fluorine. Or alternatively, the time-consuming evaporation of 150 ml of distillate and ashing may be avoided by adsorbing the fluoride in the distillate onto magnesium oxide, centrifuging, and submitting the MgO-F sediment to final perchloric acid distillation, as reported by Venkateswarlu and Narayanarao (36).

Churchill et al. (37) have encountered rather high fluorine values following conventional Willard-Winter distillation of some samples of food; the high values have been traced to phosphates that partly distilled over and behaved like fluoride in the analytical procedure. To avoid such contamination of distillates, these authors propose a preliminary distillation from sulfuric acid. Brunisholz and Michod (38) have designed a special bubble column trap to prevent carryover of phosphoric acid during steam distillation; in fact, they use phosphoric acid instead of perchloric acid for distillation of samples and claim that double distillation is unnecessary. Smith and Gardner (39), who have observed loss of fluorine (believed to be as volatile iron fluoride) during open ashing of whole blood samples, and Venkateswarlu and Narayanarao (36), who also have encountered loss of fluorine during ashing of tea samples despite the use of fixatives, resort to preliminary distillation to avoid ashing of biological samples.

5. Confined Combustion

Open ashing procedures for the determination of submicrogram amounts of total fluorine in biological materials like serum and soft tissues have two drawbacks: (*a*) variable loss of fluorine during ashing, and (*b*) variable degree of fluoride contamination that invalidates the blank correction. However the open ashing procedures seem to succeed perhaps mainly because of balancing errors due to loss of fluorine on the one hand and contamination on the other (40,41). These problems are not encountered when the samples are submitted to "closed ashing" or "confined combustion," as in the following procedures.

A. COMBUSTION IN THE OXYGEN FLASK

Ever since Schöniger (42) successfully demonstrated the usefulness of the oxygen flask technique for the decomposition of organic compounds in the elementary analysis of halogens, this simple and elegant technique has drawn the attention of innumerable workers. A very valuable review of the oxygen flask technique has been written by Macdonald (43). The technique basically involves burning the sample (folded in a strip of ashless filter paper) in an atmosphere of oxygen confined in a flask, in which a few milliliters of water or an alkaline solution is placed to trap the gaseous products of combustion. The contents of the flask together with the rinsings are analyzed for the required ion. This technique is not suitable for analysis of highly volatile compounds without some special adaptations.

Schöniger (44) has employed this technique for the analysis of fluorine in organic compounds. Small reported losses of fluoride have been attributed to interaction of fluoride with boron in the borosilicate glass by Johnson and Leonard (45). To avoid such losses, the borosilicate flasks have been replaced by quartz flasks (46,47) and also by strong polypropylene flasks (48). With the latter flasks, the attack of heat on the walls is negligible if the flask is rotated during the combustion of the sample.

Although the oxygen flask technique is satisfactory for compounds containing a few C-F covalent bonds, Fernandopulle and Macdonald (48) consider that the decomposition of the C-F bonds tends to be incomplete in compounds containing in excess of 40% F. They find that the practice of adding to the sample such auxiliary combustants as sucrose, sodium peroxide, and potassium nitrate or chlorate, before ignition of the compounds, though recommended by others, is not really helpful in achieving quantitative decomposition of thermally stable fluorocompounds. The authors recommend the Kirsten hot flask method, in which the sample is ignited in a special oxygen-filled flask at 1000 °C. Samples as small as 10 to 50 μg have been analyzed for fluoride using smaller oxygen flasks (48,49).

Although the oxygen flask technique has been employed mostly for elemental analysis of fluorine in organic compounds since its inception about 20 years ago, only recently has it been used to determine fluorine in biological materials such as samples of vegetation (50), rodent baits impregnated with fluoroacetate (51), and "spots" of fluorinated compounds separated on paper chromatograms (52).

The fluoride in the washings from the oxygen flask can be determined directly by a variety of colorimetric, fluorimetric, and titrimetric methods when no interferring ions are present. When sulfate and phosphate are present they should be separated by distillation, diffusion, or ion exchange (48,53). Such separation steps are not required if the fluoride in the washings is determined with the fluoride electrode, after the addition of an appropriate complexing-buffer solution (CDTA-buffer). The procedure is simple and rapid; 12 to 15 samples of vegetation have been analyzed in a day by Levaggi et al. (50). The limit of determination of fluorine by this procedure is 0.2 μg F per 50 mg (i.e., 4 ppm).

Venkateswarlu has investigated the use of the oxygen flask technique for the determination of fluorine in biological materials such as serum and has found it to be inadequate because of the relatively large blank (0.16 μg F) and the limitation of the maximum sample size to 50 mg. A 50-mg lyophilizate of serum with a total fluorine concentration of 0.1 ppm and 90 mg solids per milliliter would contain only 0.055 μg F, which is considerably less than the fluoride blank itself.

The Parr peroxide bomb has been employed for combustion of samples of vegetation by Mavrodineanu and Gwirtsman (54) and other biological materials including whole blood by Waldo and Zipf (55) preparatory to Willard-Winter distillation. This overall technique might have been satisfactory for the analysis of samples of vegetation (containing $>$ 6 ppm F); it is undoubtedly inadequate for determining the low levels of fluoride in blood, particularly when samples of whole blood no larger than 1 g are employed for combustion.

B. COMBUSTION IN THE OXYGEN BOMB

The use of the Parr oxygen bomb technique for "confined combustion" of serum preparatory to total fluorine determination has been reported very recently by Venkateswarlu (41). The oxygen bomb permits combustion of a serum lyophilizate as large as 1 g. In the case of the same serum sample that could not be analyzed by the oxygen flask method as mentioned before, a 1-g lyophilizate would contain as much as 1.1 μg F compared to a blank value of only 0.05 μg F, a feature that ensures reliable determination of fluorine.

The fluoride in the bomb washings is determined with the fluoride ion electrode. The problem of excessive dilution of the submicrogram amounts of fluorine by the large volume of wash water to levels below the limits of sensitivity of the fluoride electrode is circumvented by taking recourse to the reverse extraction technique (15). The preliminary extraction of fluoride, following combustion of the sample, is carried out in the bomb itself, avoiding the need for laborious scrubbing. The reverse extraction is performed outside the bomb with a severalfold gain in the concentration of fluoride to be measured with the electrode. If the fluoride content of the sample is satisfactorily high and no interfering cations are present in the final washings, the fluoride therein could be determined directly with the fluoride electrode, after the addition of an appropriate buffer. If the fluoride content is low and a variety of unknown concentrations of fluoride-complexing cations are expected in the washings, however, the reverse extraction procedure has decided advantages.

The percentage of recovery in the bomb washings of ^{18}F added to 2 ml of bovine serum and submitted to combustion with a l-g pellet of filter pulp, was 95 ± 1.2 ($n = 4$). The mean recoveries of 0.2 μg F fired along with pellets of benzoic acid, bovine albumin, and lyophilized and nonlyophilized bovine sera, ranged from 94 to 103%. The recovery, by reverse extraction, of ^{18}F added to an aliquot of bomb washings of products of combustion of rat liver and muscle was 97%. The recovery of 0.2 μg F placed in the bomb by carrying out the preliminary extraction in the bomb itself and the reverse extraction outside the bomb was 97 ± 1.5%($n = 20$). Total fluorine values for animal and human sera obtained with the oxygen bomb technique were found to be slightly higher than those obtained by open ashing (41).

The concept of "ionic" and "nonionic" fluorine fractions in serum originally emerged from the observations that the values of concentration of fluorine in sera, determined after open ashing of the samples, are higher than those obtained by procedures that do not involve ashing and do measure ionic fluoride (8,56). Any uncertainties with regard to such a concept, arising from the possibility of contamination of the samples with extraneous fluoride during open ashing, are now satisfactorily excluded in the light of similar higher fluorine values also confirmed with the oxygen bomb technique, in which the chances for contamination with extraneous fluorine are altogether precluded. Although in general the nonionic fluorine values (1.0 to 3.4μM F) are greater than the ionic fluoride values (0.6 to 1.4μM F), there seems to be no clear-cut relationship between the ionic and nonionic fluorine fractions in any given serum sample.

C. COMBUSTION IN THE TUBULAR FURNACE

Among the earlier investigators to perceive the problems of loss of fluorine and contamination with extraneous fluoride during open ashing of biological materials were McClendon and Foster (57), who designed a tubular platinum furnace for confined combustion of biological materials. Recoveries of 1.0 to 10.0 μg F added to 5-g diet samples containing 1.5 to 1.9 μg F are 98 to 120%.

Pyrohydrolytic techniques by means of which combustion of biological materials is carried out while simultaneously the fluorine is isolated as HF are discussed in Section III.5.

D. COMBUSTION IN THE OXYHYDROGEN FLAME

Wickbold (58) has developed an oxyhydrogen flame technique to achieve complete decomposition of thermally stable organic compounds with high fluorine content. The compound is vaporized in a stream of oxygen and caused to mix and burn with a stream of hydrogen at a temperature of about 2000 °C. The exhaust gases are cooled and passed through sodium hydroxide to retain the fluoride.

The oxyhydrogen flame combustion method is applicable to a wide variety of fluoro compounds and is more precise than the oxygen flask method. However the latter method is satisfactory and economical for routine analysis of organic compounds. The use of the oxyhydrogen flame method in elementary fluorine analysis of organic compounds has been reported by Sweetser (59), Martin and Floret (60), Martin et al. (61), and Ehrenberger (62). Apparently, however, this method has not been explored for the determination of fluorine in biological materials.

6. Digestion with Acid

Preliminary distillation of biological materials with sulfuric acid and the purpose of this procedure were mentioned before. Quantitative cleavage of the covalent C-F bond in metabolites of methoxyfluorane has been found to occur under the acidic conditions of diffusion (63) and extraction (15). The covalently bound fluorine, thus released under acidic conditions, is referred to as "acid-labile fluorine." Considering that fluorine is the most electronegative element forming the strongest covalent bond with carbon, acid-labile fluorine should occur only rarely, which indeed is the case. The C-F covalent bond in fluoroacetic acid is stable even in the presence of hot concentrated sulfuric acid.

7. Digestion with Alkali

The acid-stable C-F covalent bonds may be cleaved under alkaline conditions. The cleavage of the C-F bond in the toxic principle "potassium cyomonate" from "gifblarr," *Dichapetalum cymosum* (Hook) Engl., was originally achieved by refluxing the preparation with 30% potassium hydroxide; fluoride and glycollic acid were obtained, leading to the identification of the toxic principle as fluoroacetate (3). Rogers and Wilson (64) employ alkaline digestion for cleaving fluorine from labile trifluoromethyl compounds. The fluride so released is measured with the fluoride electrode.

Although some C-F bonds are cleaved by acid digestion and some by alkaline digestion, it is important to realize that the C-F covalent bonds in most organic compounds are extremely strong, and in some cases their complete cleavage requires more drastic measures, such as combustion at elevated temperatures, fusion with alkali metals, and reduction with alkali metals in organic solvents. In fact, the other halogens in halofluoro compounds can be cleaved without rupturing the C-F bond (65).

8. Fusion with Alkali Metals

The procedure originally described by Elving and Ligett (66) for the decomposition of organic fluorine compounds by fusion with alkali metals has been further developed by other workers (43) to meet the problems of analysis of the more stable fluorine compounds that are being increasingly produced. The method involves fusion of the compound with metallic sodium (or, better, potassium for more stable compounds) in a specially designed nickel bomb heated to 600 to 900 °C for 1 to 2 hr. The washings and the rinsings from the bomb are analyzed for fluoride by appropriate methods as described by Belcher and Macdonald (67), Belcher et al. (68), Banks et al. (69), and Schröder and Waurick (70). This method has not yet been employed for decomposition of biological materials for fluorine analysis.

9. Reduction with Alkali Metals

Stepanow (71) described a rather simple method of cleaving halogens from organic compounds by reduction with alkali metals; the compound is refluxed with an alkali metal in a suitable organic solvent. The method has been modified by others (see refs. 20,43), but it has not been extensively used because of the availability of the simpler oxygen flask technique. Vaughan and Nieuwland (72) have cleaved halogens from organic compounds with metallic sodium in a liquid ammonia medium. Strahm (73) has proposed the use of lithium in propylamine. Little (74) has employed Strahm's lithium reduction method to determine organic fluorine in tooth enamel.

A more promising approach to the cleaving of the covalent fluorine bonds is by the use of the sodium-biphenyl reagent proposed by Benton and Hamill (75). The reagent is prepared in a dry inert atmosphere of nitrogen and preserved in tightly sealed bottles at low temperature in a refrigerator. Details of preparing the reagent have been published by Ligett (76), Chambers et al. (77), and Pecherer et al. (78). The reagent is also commercially available. Wilson and Marczewski (79) have recently described a nitrogen chamber and a reagent-handling system that make dispensing the caustic and rather readily oxidizable reagent a little easier and more efficient. They have determined fluorine in petroleum products and petroleum process catalysts. Clark et al. (80) have employed the biphenyl reagent in conjunction with the fluoride electrode for the determination of residual organic fluorine compounds in the livers of rats to which the compounds had been administered. The use of the biphenyl reagent in the determination of total fluorine in blood serum has been explored by this writer. There are several formidable problems; the amount of fluoride involved is very low compared to the relatively high fluoride blank due to the reagent itself.

III. METHODS FOR SEPARATION AND CONCENTRATION OF FLUORINE

Since most of the methods for the determination of fluorine are susceptible to varying degrees to interference, and since the chemical compositions of different samples are so diverse, a preliminary separation of fluorine is desirable in most cases and mandatory in others.

1. Distillation as Silicon Tetrafluoride

Some of the earliest attempts to separate fluorine from insoluble fluorides were based on volatilization of fluorine as silicon tetrafluoride (81–83). The material is decomposed by sulfuric acid in the presence of silica in a perfectly dry apparatus. Armstrong (84) has designed a highly simplified and efficient apparatus for the recovery of fluorine as silicon tetrafluoride. However this procedure has not been much used because of the availability of another simpler and more practical procedure published in the same year. In 1933 Willard and Winter (9) announced their distillation procedure for separation of fluoride as hydrofluosilicic acid. The technique met with instant success because of its relative convenience: it does away with the need to maintain the whole system dry—a prerequisite for isolation of fluorine as silicon tetrafluoride.

Nevertheless, even as recently as in 1975, isolation of fluoride by volatilization as silicon tetrafluoride was preferred for rapid alkalimetric titration of fluorine in specific situations (85). A U-tube containing zinc shot is included in the line to trap hydrogen chloride before the absorption of silicon tetrafluoride.

2. Distillation as Hydrofluosilicic Acid

A. MACRODISTILLATION

The Willard-Winter distillation method (9) involves decomposing the test material with sulfuric acid or perchloric acid with a continuous drip feed of water into the distillation flask maintained at 135 to 140 °C. The continuous and controlled admission of water results in a steady generation of steam with which hydrofluosilicic acid distills over. The distillate containing fluoride, separated from interfering ions, is analyzed by an appropriate method.

In the following 20 years and more, there have been several modifications and improvements of the Willard-Winter method, although the original approach has remained essentially the same. The improvements relate to using steam instead of dripping water into the flask for smoother distillation (86), redesigning the flask to prevent superheating of contents and consequent bumping (87,88), improvising special spray traps to prevent contamination of the distillate (57), employing constant boiling liquids in a jacket around the distillation flask or installing various other control devices to maintain the distillation temperature rigidly at 136 °C (86,89–93), and designing multiple stills and semiautomation (94,95) to permit rapid analysis of a large number of samples. Eberius and Kowalski (96) employed air instead of steam during the distillation of fluoride from ores and slags. A recycling apparatus for isolation of fluoride by distillation was designed by Pietzka and Ehrlich (97).

Distillation for an hour may be required to collect 100 to 150 ml of the distillate to obtain quantitative recoveries of fluoride in the test sample. Boric acid and gelatinous silica retard distillation of fluoride and collection of larger volumes of distillate is required. To counteract the retarding effect of aluminum and iron on the distillation of fluorine, phosphoric acid and mixtures of phosphoric acid with sulfuric acid have been employed for distillation (38,98–101). Under such circumstances, aluminum and iron form complexes with phosphoric acid rather than with fluoride, which is therefore easily available for distillation. At times the selection of the acid is based on other analytical considerations. Soft-tissue samples ashed with fixatives other than calcium salts may be distilled with the less expensive sulfuric acid. When calcium fixatives are employed, insoluble calcium sulfate is produced in the distillation flask and may cause annoying bumping of the contents. In such cases it is better to use perchloric acid instead. However it is preferable to avoid perchloric acid while distilling plant ash because potassium perchlorate is insoluble and may cause bumping, which may be a problem particularly when small microdistillation stills are employed.

Quantitative distillation of fluorine from some refractory materials (e.g., certain minerals, rocks, soil, and glass) cannot be accomplished at 135 to 140 °C. A higher temperature of distillation is required. Although sulfuric and phosphoric acids can be used for distillation at elevated temperatures (180 °C and above), perchloric acid cannot be put to this use because of the risk of explosion. During distillation at elevated temperatures, interfering ions do distill over and interfere with colorimetric and titrimetric procedures. The distillate is concentrated and submitted to a following perchloric acid distillation. To prevent the distillation of chloride, silver sulfate or silver perchlorate is added to the distillation flask (31,102). A silver spiral has been used to retain hydrogen sulfide and free chlorine (103). A rapid distillation procedure applicable to hot-cell preparations has been described by Kubota (104). Seel (105) has proposed the use of superheated steam for rapid distillation of fluoride from sulfuric acid medium. Reynolds and Hill (106) report that separation of fluoride from pyritic substances and organic matter is improved by distilling the samples in the presence of excess of permanganate.

Andersson and Gelin (107) believe that incomplete distillation of fluorine could be due to the presence of metals such as aluminum, uranium, and plutonium, which may be introduced into the separation by impure chemicals or by dissolution from the glass walls of the distillation apparatus. They report that recoveries of 99.5% fluorine are achieved using purified chemicals and a distillation apparatus made of fused silica. Interferences from uranium and plutonium in low valence states are eliminated by oxidation to U(VI) and Pu(VI).

Deutschman et al. (108) recommend the use of all-quartz apparatus for distillation as a means of keeping the blank value low.

B. MICRODISTILLATION

Besides the tediousness of operation, another main drawback of the conventional Willard-Winter distillation process is that the fluoride is collected in a large volume of the distillate. If the fluoride concentration is low, the distillate needs concentration by evaporation, which may introduce some contamination. Furthermore, the fluoride blank with the macrostills ranges from 1 to 3 μg F. In the analysis of samples low in fluoride, it is essential to collect the fluoride in a smaller volume of the distillate by a suitable procedure with an acceptable low fluoride blank. To this end several types of microstill have been designed by different workers to meet specific needs (28,109–114).

In most cases, the microstills are scaled-down units of the macrostill. Some authors have used steam with or without a stream of inert gas such as nitrogen or compressed air, but Kubota (104) utilizes the vapor from the distilling solution to transport fluoride; the still head (Kjeldahl spray trap) is heated to 170 °C and the vapor containing the fluoride (HF - H_2SiF_6) is condensed (20 ml).

In one important application of these microstills in biological investigations, Singer and Armstrong (28) employed their microdistillation procedure for the determination of fluoride in serum; the analyses were performed on 5- to 10-ml samples of sera, following ashing with fluoride-low magnesium oxide. The microstills were successfully employed by this writer for the determination of fluorine in bones of rats raised on a special low-fluoride diet. The fluorine concentration in the skeleton of newborn pups was 1 ppm of ash and the absolute amount of fluoride involved per analysis was as low as 0.1 μg (115).

3. Diffusion as Hydrogen Fluoride

Taking advantage of the introduction of polyethylene plasticware for routine laboratory use, Singer and Armstrong (11) developed a technique for isolation of fluorine by a procedure based on diffusion of hydrogen fluoride. The method involves treating the test sample with perchloric acid in a stoppered polyethylene bottle in which a strip of polyethylene with a film of sodium hydroxide is placed. The bottle is maintained at 50 °C overnight in an oven. The hydrogen fluoride that is evolved is trapped by the sodium hydroxide on the strip; the sodium hydroxide is dissolved from the strip and analyzed for fluoride.

This simple procedure of high research potential did not appear in scientific literature until five years after its first publication in 1954. The original authors themselves turned to the development of a microdistillation technique for isolation of fluoride that is just as tedious as the Willard-Winter macrodistillation. Of course unlike macrodistillation, the micro process has the definite advantage that the fluoride can be isolated in a smaller volume, but this is the case with the diffusion method as well.

Stegemann and Jung (116) carried out the diffusion in the hollow space of polyethylene stoppers fitted with specially devised closures and determined fluorine in biological tissues and fluids. Hall (117) employed 50-ml polyethylene bottles for diffusion and used a filter paper strip impregnated with sodium hydroxide to trap the fluoride. These bottles are convenient for use in the adsorption-diffusion technique; the fluoride is adsorbed onto calcium phosphate and sedimented by centrifuging the bottles; the supernatants are discarded, and the diffusions are carried out in the same bottles (16). Buttner (118) improved on Hall's diffusion bottle and carried out determination of 5 to 30 μg F in calcified tissues. Bäumler and Glinz (119) improved upon the Stegemann-Jung diffusion vessels, and later Bäumler (120) described a much improved and a very practical diffusion vessel. The sample and the acid are placed at the bottom of the vessel; a disk of filter paper impregnated with sodium hydroxide is positioned in the upper recess. The wide mouth of the diffusion vessel allows a platinum crucible with the ash of the biological sample to be directly inserted in the diffusion vessel, a feature that minimizes the problem of quantitative transfer of ash. Furthermore, the tight fit of the stopper does not require the cumbersome application of silicone grease, which is necessary for carrying out diffusion in the Conway-type polypropylene cells used recently by many workers (121–124). A diffusion apparatus containing mechanically sealable diffusion chambers, and not requiring the use of sealants, has been designed (125). Curiously, it was the practice of using the silicone grease as a sealant for diffusion vessels that led to the development of a more rapid diffusion method, carried out at room temperature rather than at 50 °C (see Section III.4).

Several investigators reported satisfactory recoveries of fluoride by diffusion over the temperature range 50–70 °C and diffusion periods of 13 to 48 hr (116,118,122,126–128). Wharton (127) believes that for the range of 1–3 μg F, diffusion for 16 hr at 60 °C is necessary and shaking the solutions does not speed diffusion. Frere (129) carried out successful diffusions at room temperature (24 hr). Greenland (130) recovered 1 to 10 μg F by 4-hr diffusion at 60 °C, employing sulfuric acid for diffusion. The diffusion of fluoride is much more easily achieved from aqueous solutions than from the test samples, and it is advisible to determine the efficiency of diffusion of metabolized fluorine and fluoride added to the biological samples. According to Stuart, organic material does not interfere with the recovery of fluorine except in the case of human serum (131).

The same substances that interfere with the separation of fluorine by the Willard-Winter distillation would undoubtedly retard diffusion of fluoride for the same reasons, and even more so because of the lower temperature in diffusion, which is less favorable for the decomposition of rather stable fluoride complexes. For example, although magnesium salts do not interfere with the release of fluoride by distillation at 135 °C, they retard release of fluoride by diffusion at 60 °C if present in excessive amounts. If relatively large amounts of magnesium compounds are employed for fixation of fluorine during ashing of biological materials, the diffusion could be incomplete. About 10 to 20 mg of magnesium oxide may be employed for ashing. Ideally, for ashing any given type of material, one must determine individually the minimum but safe amount of magnesium oxide that ensures quantitative fixation as well as complete release of fluoride by diffusion.

Weinstein et al. (132) have reported an automated microdistillation procedure for fluoride in which the sample is conveyed through a heated helix along with acid, and the hydrogen fluoride is removed under vacuum, condensed, fed into a colorimetric manifold, mixed with buffered lanthanum-alizarin complexan reagent, and measured spectrophotometrically.

Mandl et al. (133) have described a compact microdiffusion unit in which the sample mixed with acid is flushed under vacuum through a Teflon coil maintained in a heated oil bath. As a result of the use of this compact unit in the place of the digestor module employed in their previous automated procedures, the authors have been able to reduce the sample size from 7 to 3 ml and to perform 12 to 20 analyses per hour, and as many as 30 analyses of samples that are not too greatly different in fluoride content.

4. Diffusion as Trimethylfluorosilane

When substitutes were used for silicone grease sealant in the Singer-Armstrong diffusion procedure, Taves (134) encountered very low recoveries of fluoride, despite all precautions against leakage of gaseous hydrogen fluoride. The recoveries were in fact as poor as those obtained when no sealant was used at all. This led him to explore the possibility of the presence in the silicone grease of some volatile fraction that participated in a still unknown way in the quantitative diffusion of fluoride from acidified solutions. In a sense, all this seems to have been preordained. Innumerable successful diffusions have been carried out in several laboratories by many workers, including this author, who used sealants other than silicone grease or employed diffusion bottles that do not require the application of any sealant at all. Nevertheless Taves was unique in encountering the circumstances that led him to discover that the silicone grease does indeed contain a volatile fraction that can participate in the efficient diffusion of fluoride.

Based on such a finding, Taves (56) described a new procedure for rapid diffusion of fluoride at room temperature. It involves direct introduction of a silicone, hexamethyldisiloxane (HMDS), into the acid employed for diffusion. A recovery of about 98% of radioactive fluoride added to bone, urine, and serum samples was achieved by diffusion at room temperature within 1 to 2 hr, which is about one-tenth the time required by the previous diffusion procedures. HMDS is presumed to enhance the rate of diffusion by the formation of trimethylfluorosilane (TMFS), which is highly volatile (bp 16.4 °C) and hydrophobic, thus readily escapes from the aqueous acidic solution. The alkaline absorbing solution decomposes TMFS, trapping fluoride and releasing trimethylsilanol (TMS). Two molecules of TMS recondense to form the parent HMDS, which returns to the acidic solution to repeat the cycle of events, resulting in a rapid diffusion of fluoride. Taves calculated that under the conditions of his technique, 80 nmole of HMDS brought about the diffusions of 45,000 nmole of fluoride from bone in 1 hr.

At about the same time that Taves in the United States was developing his technique for rapid diffusion of fluoride as trimethylfluorosilane, some remarkable investigations of the quantitative extraction of fluoride as fluorosilanes into immiscible organic solvents were being pursued in Prof. Rudolph Bock's laboratory in Germany. This work has opened innumerable approaches to the determination of fluorine, including some of the most sensitive methods now available (gas chromatography, reverse extraction, hanging drop fluoride electrode).

5. Pyrohydrolysis as Hydrogen Fluoride

The principle of separation of fluorine as hydrogen fluoride by pyrohydrolysis is based on the original observation of Fremy in 1856. When water vapor was passed over fluorspar in a platinum tube heated to redness, hydrogen fluoride was obtained in the condensate. In 1926 Prideaux and Roper (135) employed the pyrohydrolytic technique for the analysis of fluorides of zirconium. Shortly thereafter, however, interest in this technique was revived by the need to find rapid and reliable analytical methods for the determination of fluoride and other halides in connection with the production of uranium fluoride and thorium fluoride in the Manhattan Project (1942–1945). Detailed publications on this technique appeared about 10 years later.

Warf et al. (136) described macro and semimicro procedures, requiring no more than 15 to 20 min, for the pyrohydrolytic separation of fluoride from metal fluorides heated at 1000 °C in a platinum apparatus. The hydrogen fluoride is titrated and has been found to be equivalent to the weight of the residual metal oxides in the platinum tube. The pyrohydrolysis of the light metal fluorides, which is rather sluggish, is hastened by mixing accelerators such as uranium oxide with the metal fluorides before pyrohydrolysis. Banks et al. (137) find chromium(III) oxide and uranium oxide (U_3O_8) to be the best among the several accelerators tested for hastening pyrohydrolysis.

Susano et al. (138) described a less expensive apparatus constructed entirely from nickel and stainless steel, rather than platinum and quartz. The apparatus produced 5000 determinations before the reactor tube had to be replaced. Surak et al. (139) and Gahler and Porter (140) also designed nickel apparatus for pyrohydrolysis. Employing a stream of moist oxygen, instead of steam alone, Powell and Menis (141) were able to achieve quantitative pyrohydrolysis of several inorganic fluorides in a silica apparatus at 800 °C, which is lower than the temperatures required in the previous procedures. They discovered that tungstic oxide is a better accelerator than uranium oxide. The method is applicable to measurement of 10 μg F, in conjunction with a spectrophotometric method for fluoride. Yamamura et al. (142) and Rao and Subramanian (143) find that moist air is as good as moist oxygen for pyrohydrolysis. Berns and van der Zwaan (144) use nitrogen supersaturated with steam for pyrohydrolysis of minerals (e.g., fluorspar) in a ceramic reaction tube. Using the Wickbold oxyhydrogen flame for heating the sample holder, Hoel et al. (145) were able to complete the pyrohydrolysis in a period as brief as 15 min.

Prior to the publication of details of the pyrohydrolytic separation of fluoride in the 1950s, Schumb and Radimer (146) reported decomposition of fluorocarbons by passing the volatilized fluorocarbons with a mixture of nitrogen and oxygen through a platinum tube heated at 1100 °C; the hydrogen fluoride collected in the condensate was determined with peroxidized titanium.

Pyrohydrolytic separation of fluorine has some advantages over the conventional distillation and diffusion techniques for the separation of fluorine from materials of interest in biological investigations. The technique is rapid, requiring only 10 to 20 min to isolate fluorine. The problem of difficult-to-decompose refractory compounds of fluorine (encountered in the distillation procedure) does not arise in the pyrohydrolytic separation of fluorine. Furthermore, the fluoride is collected in a smaller volume of the absorbant solution rather than in a large volume of the distillate, which is an advantage in determining small amounts of fluorine. However the technique has one drawback. When elements such as sulfur and phosphorus are present in the sample, they are carried over along with fluorine and interfere with colorimetric and titrimetric methods of analysis of fluoride; thus it becomes necessary to remove these elements by ion-exchange resins or other suitable methods. Such separation is not required if fluoride is determined with the fluoride electrode.

Van Gogh (147) has described a pyrohydrolitic method for the decomposition of biological materials (urine, tea, fossilized bone) and organofluorine compounds. Kakabadse et al. (30) have described a similar technique for the determination of fluoride in tea, cocoa, bacteria, tobacco, and ground human teeth. Recently Kirsten and Shah (148) have described the pyrolytic decomposition of rock phosphate, opal glass, and organic compounds, using a split-type furnace. The sample containing 5 to 17 μg F and placed in a platinum boat is treated with H_3PO_4 (3 μl) or in some instances with a flux prepared from NaH_2PO_4 in least amount of water and H_3PO_4 and heated to 900 °C in a current of oxygen. Before absorption in water, the gaseous products are led through another chamber at 800 °C supplied with hydrogen. Phosphate and sulfate are reduced to phosphine and hydrogen sulfide, neither of which interferes with the Ce-alizarin complexan colorimetric method for fluoride. More recently, Kirsten (149) has modified further the foregoing technique to permit the determination of fluoride down to 0.2 μg F in samples containing up to 1 g of inorganic or organic materials. The samples are decomposed with tungsten trioxide and phosphoric acid. To avoid the risk of explosion when large samples are burned, the samples are pyrolized under nitrogen at first. The technique has been employed for the determination of fluorine in oxalic acid, sucrose, green fodder pills, sewage waste, manure, pine needles, and organic fluorine compounds.

Curry and Mellon (150) have carried out pyrohydrolysis of organic fluorine compounds in a quartz tube packed with quartz chips and platinum gauze. Fluoride is absorbed as silicon tetrafluoride (not as HF), hydrolyzed, and determined in terms of silicon by conversion of silicon to molybdosilicic acid and reduction to heteropoly-blue.

Gamsjäger and Schwarz-Bergkamf (151) reported the isolation of fluoride from insoluble water matrices such as calcium fluoride and aluminum fluoride as chromyl fluoride by heating samples with a mixture of sulfuric and chromic acids.

6. Pyrolysis as Ammonium Fluoride

Volodina et al. (152) developed a procedure for the pyrolytic separation of fluorine as ammonium fluoride. They mixed 5 to 12 mg of the sample (glasses and silicates) with 200 to 250 mg of ammonium sulfate and heated the mixture in a silica tube at 800 to 850 °C in a stream of ammonia for about 10 min, followed by nitrogen. The condensate of ammonium fluoride collected on the colder part of the tube was dissolved in water and analyzed for fluoride.

7. Solvent Extraction of Fluoride

Solvent extraction of fluoride from aqueous solution has some promising applications. If the extraction is selective, it provides a simple means for the separation of fluoride from other interferences. It could also provide a means for concentrating fluoride in a small volume of solution, which can be advantageous in subsequent analytical steps.

A. EXTRACTION AS COMPLEXES OF ANTIMONY

The first report on solvent extraction of fluoride for separation of fluoride from interfering ions like sulfate is due to Moffett et al. (153). It is based on the solubility differences of tetraphenylstibonium salts in water (sulfate, > 60g%; fluoride, 0.06 g%) and in carbon tetrachloride (sulfate, 0.16%; fluoride, 10.3%). Tetraphenylstibonium sulfate reagent is added to a solution containing fluoride and sulfate, then extracted with carbon tetrachloride. Fluoride recoveries (0 to 0.2 nmole F) based on the weight of tetraphenyl stibonium fluoride after evaporating the carbon tetrachloride ranged from 97 to 98%. Iron(III) and aluminum(III) retard fluoride extraction. Chloride and bromide, which are extractable as tetraphenylstibonium salts, tend to give higher results. However 99.2% recovery of fluoride was obtained by removing chloride by silver nitrate preciptation before extraction. Aulagnier et al. (154) have achieved 95% extraction of $0.1 \mu M$ F^- as $(C_6H_5)_4SbF$ into carbon tetrachloride from an aqueous solution at pH 6.0. Chermette et al. (155) have also employed triphenylantimony dichloride for extracting fluoride from aqeous solutions; 1,2-diaminocyclohexanetetracetic acid is added to the aqueous phase to release fluoride for extraction from complexes bearing aluminum, zirconium, and other elements.

B. EXTRACTION AS COMPLEXES OF TANTALUM

Triphenylmethane dyes (methyl violet and malachite green) interact with fluoride complexes of tantalum to form triple complexes of tantalum that have been selectively extracted from aqueous solution into solvents such as benzene by Kakita and Goto (156). Based on these findings, Dahl (157) developed an extraction-spectrophotometric method, involving the tantalum-fluoride-malachite green complex, for the determination of fluoride in organic fluorine compounds, following their combustion in the Parr oxygen bomb. This extraction procedure has not yet been employed in the fluorine analysis of biological materials. The presence of cations that form stronger complexes with fluoride than tantalum results in low fluorine recovery. It would therefore be necessary to isolate fluorine free from such interferences, which are often encountered in the analysis of biological materials.

Tantalum fluoride complex has been extracted from a strong acid solution into 2,6-dimethyl-4-heptanone. This extraction is the basis of a radiometric method for analysis of fluoride (158).

C. EXTRACTION AS COMPLEXES OF LANTHANUM

A blue triple complex is formed when the lanthanum chelate of alizarin complexan is added to a solution containing fluoride. This is the basis of the spectrophotometric method for the determination of fluoride described by Belcher and others. The methods would be more sensitive if the triple complex could be extracted more or less selectively from the unreacted lanthanum chelate, which also contributes a significant fraction of the measured optical density. This has been accomplished by several workers and is discussed later (Section IV.2.B).

D. EXTRACTION AS COMPLEXES OF OTHER METALS

Kletenick and Bykhovskaya (159) find that in the presence of certain quadrivalent metals (e.g., Zr, Th, Ti, and Sn), fluoride is extractable from an aqueous solution into a solution of bis-(2-ethylhexyl)phosphate in immisible organic solvents. Preferably fluoride is extracted from an acid solution with zirconium into a solution of bis-(2-ethylhexyl)phosphate in hexane.

E. EXTRACTION AS FLUOROSILANES

Halide ions react with silanols to form the corresponding silanes. Except for fluorosilanes, the silanes readily hydrolyze in water. Thus the stable fluorosilanes, which are soluble in organic solvents, can be extracted from the aqueous phase. The extraction is rather slow, but it progresses rapidly in solutions of high acidity.

$$R_3SiCl + H_2O \rightarrow R_3SiOH + HCl$$

$$R_3SiOH + H^+ + F^- \rightarrow R_3SiF + H_2O$$

where R = alkyl or aryl radical

Bock and Semmler (17) have described detailed studies on the extraction of fluoride as 14 different fluorosilanes by 25 different solvents and the analyses of the fluoride so isolated by a variety of physical and chemical methods. They finally recommend extraction of fluoride as triethylfluorosilane and its determination by gas chromatography or mass spectrometry.

8. Separation as Fluoride Ions

A. SEPARATION BY ELECTRODIALYSIS

Bazille (160) has proposed electrodialysis for extracting fluoride from organs for toxicological analysis of fluorides. Fabre (161) employs electrodialysis to separate fluoride ions from blood and soft tissues. The electrodialysis is run for 15 hr; the contents of the anode compartment are withdrawn and replaced with water; the electrodialysis is continued for another 3 hr. The contents of the anode compartment are added to the previous lot, neutralized with sodium hydroxide, evaporated, and distilled by the Willard-Winter technique.

B. SEPARATION BY ION EXCHANGE

Ion-exchange resins have been used for separating fluoride ions from other interfering anions as well as for removing fluoride-complexing cations of iron, aluminum, and so on, from the sample solutions. Anion-exchange resins are used in the former case and cation exchange resins in the latter case.

a. Anion-Exchange Resins. The most important interfering anions are the polyvalent ions phosphate, arsenate, sulfate, and so on, which are adsorbed more strongly on anion-exchange resins than are the univalent ions (fluoride, chloride, nitrate, etc.). Furthermore, among the univalent ions, fluoride is less strongly adsorbed than the others. These differences are the basis of procedures for ready separation and elution of fluoride from interfering anions using anion-exchange resin columns.

Since a given ion-exchange resin column has a definite capacity, the total ionic strength of the test solution has a bearing on the retention of fluoride. To avoid "leaks" of fluoride, larger column lengths may be needed and should be determined by the investigator.

Several anion-exchange resins and appropriate eluents have been employed by different workers for the separation of fluoride from different anions: Amberlite IRA-400 for separation from carbonate, sulfate, and phosphate (Funasaka et al., 162); Dowex I-X10 resin for separation from phosphate (Zipkin et al., 163); DeAcidite FF for separation for phosphate (Newman, 164); Dowex I-X8 for separation from phosphate, chloride, and sulfate (Nielsen, 165; Nickl and von Braunmuehl, 166); Dowex I-X4 for separation from other halides and thiocyanate (Zalveskaya and Starobinets, 167); AV-17 resin for separation from bromide (Eristavi et al., 168); and A-25 anionite for separation from other halides and thiocyanate (Staribinets and Zakhartseva, 169).

b. Cation-Exchange Resins. Cation-exchange resins have been employed occasionally to remove interfering cations from water samples; more frequently they are used to remove the high mineral content resulting from sodium peroxide combustion of organic fluorine compounds for fluorine determination or from fusion of insoluble and refractory mineral and soil samples before fluorine analysis.

The likelihood of retention of some fluoride as positively charged complexes with the adsorbed polyvalent cations of aluminum, titanium, thorium, plutonium, zirconium, iron, and so on, should not be overlooked. Coursier and Saulnier (170) have observed unsatisfactory separation of fluoride in the presence of aluminum with acidic solutions, presumably due to retention of some fluoride by aluminum-mediated bonding. They achieve better separation of fluoride from aluminum by adsorbing the latter as anions at alkaline pH on an anion-exchange resin and eluting aluminate with 0.2*M* sodium hydroxide and fluoride with 1*M* sodium hydroxide solution. Newman (164) converts metal ions to their anionic ethylenedinitrolotetraacetate complexes and uses anion-exchange resin to retain the complexes; the fluoride is eluted with ammonium chloride solution, pH 9.2. In such situations, the silanol extraction and reverse extraction techniques have a definite advantage over those involving cation-exchange resins. Radioactive fluoride has been quantitatively recovered from solutions of salts of aluminum, cerium, thorium, and zirconium by reverse extraction techniques (171).

C. SEPARATION BY PAPER CHROMATOGRAPHY AND OTHER METHODS

Separation of fluoride from other anions, mostly other halide ions, has been achieved by paper chromatography (Broomhead and Gibson, 172; Gera and Tomkowiak, 173), by thin-layer chromatography using silica gel (Seiler and Kaffenburger, 174; Gagliardi and Pokorny, 175), by cellulose (Bark et al., 176), on Sephadex C-15 (Deguchi, 177) and by electrophoresis on anion-exchange paper (Tokutomi and Kamiya, 178). Separation of fluoride from phosphate on Whatman 3MM filter paper strips has been reported by Zipkin et al. (163).

D. SEPARATION BY ADSORPTION

a. Adsorption on Magnesium Oxide. The application of adsorption techniques in the microdetermination of fluorine had rather fortuitous beginnings. Samples of water from endemic areas of fluorosis in India were found to contain 0.5 to 18.0 ppm F according to a conventional colorimetric method; when analyzed 6 months later by a rapid titrimetric method just developed in the laboratory, however, considerably smaller amounts of fluorine (0.0 to 3 6 ppm F) were reported. The later analyses were purely incidental to the need to evaluate the new method and to the availability of the water samples from a previous analysis by different means. The water samples were stored in glass bottles of various types, and careful examination revealed a film of fine sediment in most of the bottles. The sediments were found to contain calcium, magnesium, carbonate, chloride, and significant amounts of fluoride. It was surmised that during the long period of storage of samples, the soluble bicarbonates of calcium and magnesium present in the water samples slowly decomposed, yielding insoluble carbonates of calcium and magnesium that adsorbed the fluoride from the water samples. Laboratory preparations of calcium and magnesium carbonates were tested and found to adsorb fluoride. As a means of reactivating the "spent" magnesium carbonate, it was heated in a furnace at 600 °C and the product turned out to be even more efficient in adsorbing fluoride than the parent magnesium carbonate. In this reactivation process, the magnesium carbonate was inadvertently converted to magnesium oxide. Light magnesium oxide was found to be superior to heavy magnesium oxide. It is possible to adsorb as much as 10 μg F per milligram of magnesium oxide. The method involves boiling the solution containing fluoride with magnesium oxide; most of the fluoride is adsorbed onto magnesium oxide almost instantaneously, and continuing the boiling for 5 min is more than adequate for quantitative adsorption.

Details of the factors governing adsorption of fluoride on magnesium oxide and the application of the findings to the determination of fluorine in water and some biological materials by procedure involving distillation, but not requiring evaporation and ashing, were reported by Venkateswarlu and Narayanarao (36,179–182). At about the same time, Singer and Armstrong (11) introduced the diffusion technique for isolation of fluoride, and this writer joined Armstrong's laboratory in 1955. We attempted to combine the two procedures to develop an adsorption-diffusion technique that does not require ashing; such an approach would have been used in the determination of fluorine in biological materials, but the efforts were unsuccessful. Fluoride adsorbed on magnesium oxide was not available for quantitative diffusion under the experimental conditions. The difficulties encountered in these investigations were circumvented as described in the following section.

b. Adsorption on Calcium Phosphate. Since the Singer-Armstrong diffusion technique (11) is applicable to the analysis of fluoride in bone and tooth minerals, which are essentially "calcium phosphate," an adsorption-diffusion technique for the determination of fluorine in biological materials seemed feasible, if only fluoride in such samples could be quantitatively adsorbed onto calcium phosphate. With these objectives in view, factors governing the adsorption of fluoride on calcium phosphate were investigated. On the basis of these observations, procedures have been devised by Venkateswarlu and Sita (16) for the quantitative adsorption on calcium phosphate of fluoride present in natural waters, salt solutions, urine, and plant extracts. The calcium phosphate with the adsorbed fluoride is then sedimented by centrifugation and subjected to microdiffusion. The fluoride in the diffusate is analyzed by a suitable spectrophotometric procedure. Recoveries of 2 μg F added to the above-mentioned test samples are 97 to 100%.

Determination of fluoride contamination in nitrates by direct diffusion or distillation technique, is conjunction with a colorimetric procedure, would not be very practical because the nitrous fumes evolved exert an oxidizing effect on the dyes employed in colorimetric analysis. The problem of analysis of bromides and iodides is also complicated by the liberation of HBr and HI, which tend to dissociate during diffusion or distillation. Such salts can be conveniently handled by the calcium phosphate adsorption-diffusion technique.

When the Singer-Armstrong diffusion colorimetric method proposed for animal tissues (183) is employed for analysis of plant materials, false high results have been obtained (184). The advantage of the calcium phosphate adsorption-diffusion colorimetric technique is that the supernatant, which contains the large bulk of interfering substances, is discarded. Furthermore, it is possible to concentrate traces of fluoride from a large volume of the sample low in fluoride, without recourse to time-consuming evaporation in platinum dishes, a feature that permits more reliable determination of fluoride. The procedure precludes the need for ashing of the biological samples to eliminate the interfering substances, thus averting the hazards of loss of fluoride due to volatilization and contamination with extraneous fluoride. Furthermore, the fluoride adsorbed on calcium phosphate can be determined directly with the fluoride ion electrode after dissolving in a suitable acid-buffer solution, and the diffusion step can be altogether omitted.

The calcium phosphate adsorption technique has been employed by Venkateswarlu et al. (8) to determine fluoride in a wide variety of biological fluids—serum, saliva, urine, milk, solutions of albumin fraction V, extracts of muscle, and extracts of tea. The recovery of 0.5 μg F added to 5-ml volumes of these samples is 94 to 103%; the recovery of 0.05 to 0.2 μg F added to bovine serum is 93 to 103%. The recovery of ^{18}F added to serum, saliva, milk, and urine is 96 to 99%. The fluoride adsorbed onto calcium phosphate can be determined by diffusion colorimetric methods, directly with the fluorideelectrode, or indirectly by an adsorption-restoration technique. The application of this technique to the determination of serum fluoride has led to the concept of "ionic" and "nonionic" covalent fluorine fractions in serum. This is discussed in detail later.

c. Adsorption on Other Compounds. Vialatte et al. (185) find that 74% of fluoride is adsorbed onto antimony pentoxide by agitation for 1 hr in 11N hydrochloric acid. Fedoroff and Debove (186) have carried out radiochemical separation of fluoride by adsorption of fluoride at 100 °C on lead strontium hydroxylapatite.

E. SEPARATION BY REVERSE EXTRACTION TECHNIQUES

The extraction of fluoride as a fluorosilane for gas chromatographic measurement of fluoride was mentioned earlier. Another inseparable chemical process that occurs concomitantly with and as a consequence of such extraction is the separation of fluoride from all other inorganic ions (cations as well as anions) that might be present in the original solution and could interfere in some cases with the fluoride determination. To take advantage of such a separation of fluoride from interferences and to render the extraction technique amenable to the determination of fluorine by spectrophotometric, fluorimetric, and fluoride electrode methods, which are simpler than gas chromatography or mass spectrometry, the reverse extraction technique has been conceived. This would be possible if the fluoride, after extraction as fluorosilane into an immiscible organic solvent, could be quantitatively reconverted into the fluoride ion in an aqueous phase. This reconversion has been accomplished by back extracting the fluorosilane with an alkaline aqueous solution, whereby the fluorosilane is decomposed and the fluoride ions are released into the aqueous phase (Figure 1).

Fresen et al. (187) have extracted fluoride as trimethylfluorosilane (bp 16.4 °C) with benzene (bp 80 °C) for gas chromatographic determination of fluoride. When the reverse extraction technique is carried out promptly after the extraction, good recoveries (99.8%) are obtained employing trimethylchlorosilane in benzene for extracting 50 nmole of fluoride. Studies involving ^{18}F, however, have revealed that trimethylfluorosilane partly escapes from the organic phase on standing. To avoid errors due to such losses of fluorine or due to excessive reduction in volume of the organic phase through solvent evaporation, particularly while handling a large number of samples, it has

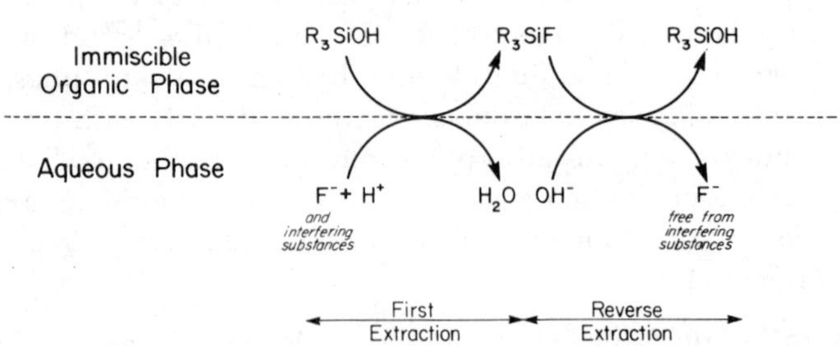

Figure 1. Schematic representation of the principles of the reverse extraction technique for the isolation of fluoride; R stands for acyl- or aryl-groups.

been found advantageous to extract fluoride as diphenyldifluorosilane (bp 157 °C) with toluene (bp 110 °C). The recovery of 10 nmole of fluoride from aqueous solutions is 99.8 ± 1.39 (S.E.)%, (n = 17) (15).

a. Reverse Extraction–Fluoride Electrode Macro Procedure. Four milliliters of extractant (2.5m*M* diphenylsilanediol, DPSD, in toluene—i.e., 530 mg of DPSD per liter) are layered on the top of 2 ml of the sample (serum) in a 15-ml polypropylene centrifuge tube. Ice-cold concentrated perchloric acid (1 ml) is added and the tube stoppered promptly. The contents are vigorously shaken for 1 hr on a laboratory shaker, which serves to transfer the fluoride in the aqueous phase as diphenyldifluorosilane into the organic phase. The tubes are centrifuged for 10 min at 3000 rpm. Protein, if present in the sample, is partly precipitated at the bottom of the tube and partly at the interphase of the aqueous and nonaqueous layers. Three milliliters of the extract (top layer) containing the fluorosilane is carefully withdrawn and transferred to another 15-ml polypropylene tube containing 50 μl of 2.5*N* sodium hydroxide solution. The tube is stoppered and shaken for 1 hr, and the fluoride is back extracted from the organic phase into the 50-μl sodium hydroxide solution. The tube is centrifuged for 5 min at 2000 rpm. The organic layer is removed by suction. The residual benzene is removed by placing the tube in a continually evacuated desiccator for 5 min. The air admitted into the desiccator for breaking the vacuum is drawn through scrubbers containing 5*N* thorium nitrate and 5*N* sodium hydroxide solutions. The back extract is neutralized with 50 μl of 2.5*N* hydrochloric acid and buffered with 500 μl of 0.05*M* histidine buffer, pH 6. 1. The contents are transferred to a 10-ml polystyrene tube, a small magnetic stirring bar is dropped in, and the fluoride electrode is introduced. Contact with the reference electrode, immersed in a few milliliters of saturated potassium chloride solution, is made through an agar bridge (4% agar in 0.2*M* sodium chloride and 0.05*M* histidine buffer).

The fluoride standards (0.5, 1.0, 1.5, 2.0, and 2.5 nmole F/600 μl) are prepared by adding 100 μl each of the sodium fluoride stock solutions (5, 10, 15, 20 and 25μ*M* F) to a mixture of 50 μl of 2.5*N* sodium hydroxide, 50 μl of 2.5*N* hydrochloric acid, and 400 μl of 0.0625*M* histidine buffer solutions. For recording the potentials of the electrode with the standards and the samples, a response time of 30 min is followed.

In the foregoing procedure, the histidine buffer may be replaced with 0.05*M* acetic acid buffer, pH 4.7. The 1-hr shaking period on a laboratory shaker for preliminary and reverse extractions is prescribed for handling a large number of samples (10 to 100 or more). If only a few samples are to be analyzed, both the preliminary extraction and the reverse extraction can be rapidly performed in a matter of 2 to 4 min by mixing the solutions vigorously on a vortexing machine (188).

The net gain in the concentration of fluoride prior to measurement with the electrode is two-to threefold compared to that in the original sample. The gain in fluoride concentration is further enhanced to thirty- to fiftyfold if the reverse extraction is carried out with concentrated ammonia as described in the next subsection.

Recoveries of 0.53 nmole of fluoride added per milliliter to bovine, rabbit, and human sera ranged from 99 to 101%. Recoveries of ^{18}F incorporated *in vivo* into rat body fluids and soft tissues (plasma, kidney, muscle, liver) and of ^{18}F added to samples of serum, urine, extracts of pea plant, and to washings of products of combustion of oxygen bomb of rat liver and muscle, were all satisfactory, 96 to 101%. (The only low recovery, 88%, was in the case of unashed rat liver.)

The fluorine content of samples of sera determined by the calcium phosphate adsorption technique (8), by measurement of ultrafiltrates with the fluoride electrode (189) and by the reverse extraction technique, just described, were the same, suggesting that the silanol-extractable fluoride occurs as ionic fluoride in normal sera. However, the extraction is carried out at a very low pH (about 20% perchloric acid), and if the sample contains any acid-labile organic fluorine compounds, the reverse extraction technique will measure the total of ionic fluoride and the acid-labile fluorine. Overnight acid diffusion of serum samples (63) to determine "acid-labile" fluorine can be replaced by the more rapid silanol extraction. The acidified sample (20% perchloric acid) contained in a 15-ml polypropylene centrifuge tube and preferably covered with a layer of silanol extractant is placed in the water bath at 70 °C for a predetermined time for complete cleaving of the acid-labile C-F covalent bonds. Following this, the preliminary extraction and the reverse extraction are carried out as usual.

b. Reverse Extraction–Fluoride Electrode Micro Procedure. The micro procedure is essentially the same as in the macro procedure. Both the preliminary extraction and the reverse extraction are carried out in 4-ml stoppered polypropylene tubes. The performance of the preliminary extraction involves 1 ml of sample, 1 ml of extractant, and 0.5 ml of concentrated perchloric acid. The reverse extraction is carried out by shaking 800 μl of the toluene extract with 100 μl of concentrated ammonium hydroxide. The neutralization of the back extract with hydrochloric acid as carried out in the macro procedure and the consequent formation of salt (which to some extent affects the fluoride activity measurement with the electrode), are avoided by removing the excess ammonia by placing the tubes containing the back extracts in a continually evacuated desiccator for 1 hr. As a consequence, the volume of the back extract is also reduced to 30 to 40 μl, with a twenty- to thirtyfold gain in the concentration of fluorides. Five microliters of 0.25*M* histidine buffer, pH 6.1 (or the acetate buffer), is added to a 20-μl concentrated back extract. The buffered back extract (5 μl) is applied to the hanging drop fluoride electrode, which is assembled as described later (Section IV.4.D, Figure 2).

The response of the fluoride electrode diminishes markedly below $10^{-6}M$ fluoride concentration. The reverse extraction technique coupled with the hanging drop fluoride electrode provides a means of isolation and concentration of fluoride from samples, permitting the investigator to attain fluoride concentrations greater than $10^{-6}M$, thus ensuring reliable measurements with the electrode.

If the evacuation of the desiccator is too strong or unduly prolonged, the volume of the ammonium hydroxide back extract may diminish to below 5 μl, or the back extract may become totally dry. Under these circumstances the recoveries of fluoride tend to fall, presumably because of the onset of sublimation of ammonium fluoride. Such losses can be prevented by adding to the ammonium hydroxide back extract trivial amounts of fixatives, 20 μl of $10^{-3}M$ sodium hydroxide solution, or calcium chloride solution. Appropriate modifications are made in the standards to compensate for such an addition (188).

The reverse extraction and hanging drop electrode techniques have been employed to determine fluoride in pooled microtome sections of rat epiphyseal plates (tibia). The samples weigh 102 to 335 μg and contain 0.03 to 1.13 nmole F; 50-mg samples of rat tail tendon containing as low as 0.08 nmole (1.6 nmole F/g) have been analyzed by the same procedures. It would not be possible to determine fluoride in such small amounts by the standard diffusion technique (183), which has a blank of 0.0 to 4.2 nmole of fluoride, or by the modified rapid diffusion procedure (56) with a blank ranging from 0.00 to 0.52 nmole F. The reverse extraction–hanging drop fluoride electrode approach has the advantage of having a low and less variable fluoride blank 0.2 ± 0.013 (S.E.) nmole F, a feature that has made possible the measurement of nanomole quantities of fluoride.

c. Reverse Extraction–Spectrophotometric Procedure. The extraction and reverse extraction steps are the same as in the reverse extraction macro procedure except that the extractant employed is 60 mg% trimethylchlorosilane (TMCS) in toluene. After the back extract is neutralized with hydrochloric acid, the volume is built to 3 ml with distilled water; 2 ml of cerium-alizarin complexan reagent are added, and after 1 hr optical density (620 nm) is determined on a Beckman model B spectrophotometer, using cells with a 5-cm optical path. The range of the fluoride standards is from 0.0 to 1.0 μg F. The cerium-alizarin complexan reagent is prepared by mixing 20 ml of $0.0005M$ alizarin complexan (stored at pH 5), 4 ml of $1.0M$ acetate buffer, pH 4.3, 20 ml of $0.0005M$ cerous nitrate, 16 ml of distilled water, and 40 ml of acetone, in that order (119).

The fluoride contents of selected samples determined by the foregoing reverse extraction (and the conventional diffusion technique) were as follows: low-F bone ash, 0.0032% (0.0031%); high-F bone ash, 0.400% (0.401%); two urine samples, 0.84 and 1.1 ppm (0.88 and 0.95 ppm). Such good agreement between the results for bone by both the methods clearly establishes the efficiency of the reverse extraction procedure in the separation of fluoride from the maximum amount of phosphate, the most notorious interfering ion encountered in any spectrophotometric analysis.

Fluoride so extracted from serum is not amenable to spectrophotometric determination, however, because of the low levels of fluoride and also because of the carryover, from the serum to the back extract, of traces of substances that interfere with the colorimetric reagent.

d. Reverse Extraction–Fluorimetric Procedure. The extraction and reverse extraction steps are the same as in the foregoing procedure. After neutralization of the back extract, 4 ml of thorium-morin reagent is added and the volume is built to 5 ml. The contents are transferred to 5-ml disposable polystyrene tubes for fluorimetric measurement with a Turner fluorimeter, model 110 (primary filter, 420 nm; secondary filter, 510 nm). The range of the fluoride standards is 0.00–0.10 μg F. The fluorimetric reagent is prepared according to Taves (190); 0.5 ml of thorium-morin stock reagent (0.001*M* morin, 0.0005*M* thorium nitrate in 95% ethanol) and 25 ml of acid-alcohol mixture (100 ml 95% ethyl alcohol and 2 ml 11.4*M* perchloric acid diluted to 1 liter with water) are made up to 100 ml with distilled water.

When extractions are carried out on 25% perchloric acid supernatants of serum and finally analyzed with the fluorimetric procedure, the results obtained are distinctly higher than those found with the fluoride electrode, according to the reverse extraction macro procedure, indicating the presence of interfering substances in the back extract. Two aliquots of the serum supernatant are extracted separately, one with the usual extractant (i.e., toluene containing trimethylcholorosilane) and the other with plain toluene (without TMCS). These extracts are submitted to the usual reverse extraction and analyzed with the fluorimetric procedure. The difference between the two values obtained with the two aliquots mentioned represents the fluoride value corrected for interferences. The values (0.011 and 0.018 ppm) so obtained for two samples of sera have been found to be in agreement with the values (0.010 and 0.017 ppm, respectively) obtained with the reverse extraction macro procedure involving the use of fluoride electrode. This procedure suggests a way for spectrophotometric and fluoremetric analysis of fluoride following reverse extraction of fluoride, when extractable interfering substances are encountered (191).

e. Adsorption–Reverse Extraction–Fluorimetric Procedure. A fifth procedure has been designed to eliminate the need for employing two aliquots of the sample, as required in the reverse extraction–fluorimetry procedure. First 2 ml of the sample (serum), diluted with 5 ml of distilled water, is shaken vigorously with 10 mg of fluoride-low calcium phosphate in a 15-ml stoppered polypropylene centrifuge tube. The contents are centrifuged for 15 min at 3000 rpm. The supernatant containing the interfering substances is removed by suction. The fluoride adsorbed onto the calcium phosphate, after washing with 2 ml of redistilled water, is isolated by reverse extraction and determined fluorimetrically as described in the foregoing procedure. The results of fluoride content of two samples of bovine sera—0.015 ± 0.0005 ppm (21) and 0.021 ± 0.0005 ppm (12)—obtained with the fluorimetric method are slightly higher than those obtained with the fluoride electrode—0.013 ± 0.005 ppm (11) and 0.017 ± 0.0003 ppm (4), respectively. Although the results are not too far divergent, they do suggest the adherence of small traces of interfering substances in serum onto calcium phosphate and their subsequent carryover into the back extracts (191).

Chermette et al. (192) extract fluoride from an aqueous solution at pH 4 to 6.5 (phosphate buffer) into triphenylantimony dichloride solution in carbon tetrachloride. Addition of 1,2-diaminocyclohexanetetraacetic acid is recommended to mask fluoride complexing ions of aluminum, zirconium, and so on. The fluoride is back extracted into dilute ammonium hydroxide solution and determined with the fluoride ion electrode. The method is reported to permit measurement of fluoride down to 0.4 ppb. Chermette et al. (193) later described a spectrophotometric procedure for the determination of fluoride following the solvent extraction technique. Kletenik and Bykhovskaya (159) have also reported a method for spectrophotometric measurement of fluoride following a preliminary extraction of fluoride with Zr-bis-(2-ethylhexyl) phosphate in hexane. Reports of the application of these methods for the microdetermination of fluoride in biological fluids and tissues are awaited.

IV. METHODS FOR THE DETERMINATION OF FLUORINE

1. Titrimetric Methods

The first practical volumetric method for the determination of fluorine in biological materials, following its isolation by distillation, was reported by Willard and Winter (9). Fluoride in the distillate is titrated with thorium nitrate solution in a 50% ethyl alcoholic medium using zirconium-alizarin sulfonate lake as the indicator. Since the zirconium in the lake itself reacts with the fluoride ions and since the zirconium is superfluous in the presence of thorium ions, Armstrong (194) modified the procedure by using alizarin sulfonate alone as the indicator in the place of its complex with zirconium. The titration is carried out in a totally aqueous medium, which gives better results because of more exact equivalence of thorium nitrate solution for fluorine through the range 0.5–10.0 μg F. These findings have been confirmed by Rowley and Churchill (195). Further improvements relate to the control of pH (2.7 to 3.5) during titration using, for example, monochloroacetate buffer (196).

Dahle et al. (197) have proposed a back-titration procedure involving thorium nitrate for the determination of fluorine in 30–350 μg range. This approach involves two titrations, compensates for salt effects, and minimizes errors due to variations in the thorium-fluoride ratio at the completion of titration. Williams (198) has conducted a critical study of thorium nitrate titration of fluoride in the microgram range and has developed a simplified technique, comprising (*a*) the use of an acidified standard thorium nitrate solution of convenient strength with a constant fluorine equivalence, (*b*) dissociation control by the addition of sodium chloride as a means of reducing interferences, and (*c*) a single titration against a permanent color standard in the place of two separate titrations required in the back-titration procedure; the method is more rapid and has a greater tolerance for interfering ions. This procedure, as adapted by Smith and Gardner (199), has yielded excellent results in the hands of this author for the determination of fluorine in distillates of biological materials as well as for routine rapid direct determination of fluorine in potable waters, requiring no more than 5 to 10 min per sample (200).

Several different indicators have been investigated from time to time for the thorium nitrate titration of fluoride, although none has totally replaced the time-honored indicator, alizarin sulfonate. Willard and Horton (201) have tested 200 dyestuffs and concluded that purpurin sulfonate is a shade better than alizarin sulfonate. Some of these indicator dyes have also been proposed as the chromophores in metal-dye complexes used in spectrophotometric determination of fluorine and tabulated in a review by Horton (21).

To eliminate subjective errors of visual detection of the end point during titration of fluoride, many electroanalytical instrumental approaches have been devised (Section IV.4.A). It is not difficult, however, to train oneself to detect the end point visually.

Some of the indicators proposed for thorium nitrate titration, in addition to alizarin sulfonate, are xylenol orange (202), sodium-2-(*p*-sulfophenylazo)-1,8-dihydroxy naphthalene-3,6-disulfonate (SPADNS) (203), methyl thymol blue (204–206), solochrome brilliant blue S (207), chrome azurol S (208), gallocyanine (209), and arsenazo I (210). To improve the sharpness of the end point, indigocaramine (211) or methylene blue (212) may be added to the alizarin sulfonate indicator during the thorium nitrate titration of fluoride.

Most of the earlier methods using aluminum chloride titrant are not applicable to the determination of fluoride in the microgram range. The most sensitive of them, involving titration of fluoride with aluminum chloride in the presence of eriochromecyanin R indicator, is suitable for more than 5 mg of fluoride (213). Arsenazo I (214) and morin (215) have been suggested as alternate indicators.

There have been very few reports on the use of zirconium salts in the volumetric determination of fluoride because of the inherent problem of formation of varied soluble and insoluble complexes (21). Lubyanskaya (216) has reported measurement of 10 to 100 μg F by zirconium titration in a weak acidic solution, using alizarin sulfonate as the indicator.

Cianflone and Braddock (217) have reported rapid determination of fluoride in such complexes as BF_4^-, FeF_6^{3-}, CrF_6^{2-}, TiF_6^{3-} and SiF_6^{2-} by procedures not involving special separation steps. Except for SiF_6^{2-}, which is decomposed by alkali before it is analyzed, the test solution is made strongly acidic and the fluoride determined by any of the common volumetric reagents. Everson and Ramirez (218) determine fluoride by thermometric titration with thorium, cerium, aluminum, and calcium salt solutions in the presence of moderate to considerable amounts of sulfate, borate, phosphate, or silicate. Awad et al. (219) have reported a thiosulfate titrimetric method for fluoride based on an amplification reaction; fluoride displaces iodate from clacium iodate, and the displaced iodate liberates iodine from potassium iodide. An indirect complexometric micromethod involving cerium, EDTA, and magnesium for the determination of fluoride has been reported by Singh and Narasaraju (220). A similiar indirect method involving samarium and EDTA has been reported by Combs and Grove (221) for determination of fluorides, fluoborates, and fluosilicates; mercaptoic acid is used to mask tin. Hg(II)-diphenylcarbazide complex, benzopurpurin 4B, and dichlorofluorescein have been proposed as adsorption indicators for microtitration of fluoride with cerium (222). Reviews of titrimetric methods for fluoride are available (223,224).

2. Spectrophotometric Methods

A. "INDIRECT" SPECTROPHOTOMETRIC METHODS

Metal ions such as Zr^{4+}, Th^{4+}, Al^{3+}, and Fe^{3+} form intensely colored lakes with certain dyes (alizarin sulfonate, eriochromecyanin R, etc.). Because fluoride ions form stronger complexes with the metal ions than do the dyes, the intensity of the color due to the metal-dye complex is diminished in the presence of fluoride ions, proportionate to the concentration of the latter. This property forms the basis of "indirect" spectrophotometric methods for the determination of fluoride.

Valach (225) cautions against labeling the above-mentioned methods involving interactions among fluoride (F), metal (M), and dyes (D) as "indirect" methods on the basis of decrease in the visible color intensity of the metal dye (MD) in the presence of fluorine. He discusses evidence for the existence of fluoride-metal-dye (FMD) ternary complexes that may have different colors as well as the same color as D or MD, and the interested reader should consult the original paper.

a. Zirconium-Dye Complexes. Whereas thorium methods predominate in the volumetric determinations of fluoride, the spectrophotometric determinations feature zirconium approaches.

The zirconium colorimetric methods for fluoride are based on the original color reaction reported by De Boer (226) in which fluoride bleaches the color of zirconium-alizarin sulfonate complex. Elvove (227) was among the first to describe a colorimetric method for the determination of fluoride in water. The color reaction is susceptible to interferences particularly by sulfate and phosphate among the anions and by aluminum, iron, thorium, and others, among the cations. To overcome the interference due to sulfate and chloride ions normally present in water, Sanchis (228) further modified the technique by adding measured amounts of hydrochloric acid and sulfuric acid to both the standards and the test samples, in amounts totally overshadowing the normal levels of chloride and sulfate in the water samples. Furthermore, the high acidity serves to minimize the effect of interfering cations. The method also has the advantage of possible use for analysis of Willard-Winter distillates, employing perchloric acid as well as sulfuric acid for the distillation. Sanchis method has undergone several modifications. Lamar and Seegmiller (229) point out that replacing the hydrochloric acid in the Sanchis reagent with an equivalent amount of sulfuric acid, reduces the sulfate interference even further. By changing the zirconium-alizarin-sulfonic acid ratio in the Sanchis reagent to 1–6, Boonstra (230) reports a sixfold increase in the sensitivity: 0.5 to 1.0 μg F/ml is recovered quantitatively in the presence of added phosphate, varying from 0.5 to 4.0 μg P_2O_5/ml. The sensitivity of the method is 0.05 μg F/ml. Bumsted and Wells (230a) have modified Sanchis reagent to develop a photometric method sensitive to 0.1 μg/ml. Megregian and Maier (231) also have modified the Sanchis method to improve its sensitivity and have adapted it for spectrophotometric measurement of fluoride in the range 0–3 ppm F in water, with a sensitivity of 0.1 ppm. The advantage of this method is that the reaction time of 1 hr at room temperature is adequate compared to 2 hr in the method by Bumsted and Wells and 4 hr in Boonstra's method, which moreover requires heating.

More recent zirconium methods for fluoride involve the use of zirconium complexes with dyes other than alizarin-sulfonic acid. Most popular among them is the rapid zirconium-eriochromecyanin R method introduced by Megregian (232), which is capable of measuring fluoride in the range 0.00 to 1.4 μg/ml, with a standard deviation of ±0.0163 μg/ml, within only 5 min. The method is applicable to the analysis of distillates and of potable waters, which normally do not contain excessive interfering ions. Sulfate is the most serious interfering ion. Rather than removing sulfate prior to analysis, the author prefers to make a rapid determination of sulfate and apply a correction factor for fluoride results. By scaling down proportionately the amounts of reagents, the author considers it possible to determine fluoride as little as 0.25 μg F in a 5.0-ml sample. Singer and Armstrong (28) have modified the Megregian method for the determination of fluoride in the 0–3 μg F/20 ml range. Oelschläger (233) has reported an improvement in sensitivity by reducing the concentration of zirconium.

Yet another zirconium method that has received considerable attention is the one due to Bellack and Schouboe (234). The method, which involves the use of the chelate Zr-SPADNS, is applicable to the determination of fluoride in water samples in the 0.00–1.40 ppm range, with an accuracy of 0.02 ppm. The reaction is immediate and the readings can be taken immediately or at any subsequent time. As in the Zr-eriochromecyanin R method, phosphates and aluminum interfere here also. A reaction time of 2 hr tolerates 3 ppm Al and 4 hr up to 30.0 ppm. Perhaps, as in the Megregian method, the time required to overcome aluminum interference could be reduced also in the Bellack-Schouboe approach if aluminum ions were converted to aluminate by treating the sample with a few drops of 1.0*N* sodium hydroxide solution before adding the colorimetric reagent. The Zr-SPADNS method has greater tolerance for sulfate than the Zr-eriochromecyanin R method. A comparatively broad range of the spectrum over which the sensitivity to fluoride could be determined makes the Zr-SPADNS method readily adaptable for use with the less expensive photoelectric colorimeters fitted with glass filters.

Other dyes proposed for zirconium reagents for colorimetric determination of fluoride are 4-dimethyl aminoazobenzene-4-arsonic acid (235,236), stilbazo (237), pyrocatechol violet (238,329), 2,3,7-hydroxy-9-phenylfluorone (240), xylenol orange (241,242), arsenazo (243,244), methyl thymol blue (242), hematoxylin (245,246), purpurin (247), and solochrome cyanin R (248).

b. Thorium-Dye Complexes. Talvitie (249) is one of the first workers to use thorium-alizarin-sulfonic acid complex for visual colorimetric determination of fluoride in waters. Icken and Blank (250), Nommik (251), Lothe (252), and others adapted thorium-alizarin-sulfonic acid method for the spectrophotometric determination of fluoride. As in the case of zirconium-dye complexes, in the thorium-dye methods such conditions as pH, ratio of thorium to dye, temperature, and reaction time, are just as critical and important for reliable results. The method developed by Nommick measures fluoride in the 0–7 μg/50 ml range, with an average deviation of 0.05 μg F. These methods are also susceptible to more or less the same interferences that afflict the zirconium methods.

Revinson and Harley (253) have reported a thorium-chromeazurol S spectrophotometric method for determining fluoride in the 5–90 μg/50 ml range, with relative accuracy of 10%; by reducing the volumes and using 5-cm optical cells, as little as 0.1 μg F may be determined. High salt concentrations are tolerated, according to the authors.

Other dyes used in association with thorium for determining fluoride are methyl thymol blue (254), [2-(1,8-dihydroxy-3,6-disulfo-2-naphthylazo)] phenoxyacetic acid (PAC) (255), and arsenazo (256,257). Moizhes and Stolyarova (257) claim a sensitivity of 0.01 to 0.02 μg F/ml for the thorium-arsenazo complex method.

Lambert (258) has described a method in which the intensity of the color measured is proportional to the fluoride content. An insoluble thorium-amaranth complex supported on a filter paper exchanges dye molecules for fluoride ions in a given solution. The amaranth color is measured. Yasuda and Lambert (259) have reported a similar "positive" color method in which fluoride releases proportionate amounts of alizarin-sulfonic acid from cellulose-supported thorium-alizarin-sulfonic acid reagent. Liddell (260) has given results of a visual comparison method for determining 0 to 0.6 μg F/ml using a specially prepared suspension of a lake of thorium-4-(*o*-arsenophenylazo)-*N*-(1-naphthyl)-ethylenediamine. Unlike the case of the "indirect" methods, the brown colloidal dispersion becomes bright red within a few minutes of adding the fluoride solution. Color differences produced by less than 0.1 μg F/ml can be readily recognized. Liddell (261) has described yet another method in which fluoride ions alter the bluish-violet color of thorium-chromotrope 2B dispersion (in ethanol containing ethyl cellulose) to the intense bright red of the original dye. Interesting as they seem, these methods apparently lack precision. Mehra and Lambert (262) determine fluoride with a solid ion association reagent consisting of buffalo black (I) and brilliant green (II) supported on an activated alumina column. When the test solution is passed through the column, II reacts with fluoride, releasing an equivalent amount of I in the ionic form, which can be measured after destroying II with sodium bisulfite. The limit of detection is 1 μg F. Recently Hinze and Humphrey (263) have described a photometric method for the determination of fluoride (0 to 15 ppm) following reaction of fluoride with $Th(IO_3)_4$, reduction of the displaced iodate to iodine, and measurement of the starch-iodine color.

Ballczo (264) described a filter paper spot test for semiquantitative determination of 5 to 5000 μg F in 5 to 20 μl of solution, based on a precipitation exchange principle. It involves the reaction between fluoride ions and LaTl $[Fe(CN)_6]$, followed by spraying the spot with a reagent mixture containing bathophenanthroline, glycerol, Me_2CO, and MeOH, and heating. The intensity of the color of the spot increases proportionately with the concentration of fluoride. Spot tests such as this one may be useful in the development of paper densitometric methods for microdetermination of traces of fluoride following separation and concentration into a few microliters-volume by appropriate techniques (e.g., reverse extraction).

c. **Aluminum-Dye Complexes.** Of all the dyes used with aluminum for colorimetric determination of fluoride, eriochromecyanin R seems to be the best. The molecular absorbancy of aluminum complex of eriochrome-cyanin R is about 3, 4, and 7 times the absorbancies of aluminum complexes with aluminon, hematin, and chromazurol S, respectively (265).

Thrun (266) has described an aluminum-eriochromecyanin R complex method for the field determination of fluorine in potable waters. Based on a comprehensive study of the interaction of fluoride with aluminum-eriochrom-ecyanin R complex, MacNulty et al. (265) have reported a method for analysis of 0 to 20 μg F/50 ml. The reaction mixture is heated at 60 °C for 15 min to hasten the reaction, which would require 16 hr for completion at room temperature. However the instability of the blanks at elevated temperatures has posed a problem in microanalysis of fluoride in the 0–2.4 μg F/15 ml range. Venkateswarlu et al. (267) prescribe adding aluminum reagent to the test solution followed by the eriochrome reagent; the color is allowed to develop at room temperature and is read after 1 hr.

The fluoride analytical methods employing aluminum have greater tolerance for such commonly occurring interfering ions as sulfate and phosphate than do those employing zirconium and thorium. For example, in the preceding method 300 μg of phosphate produces a positive error of 1 μg F, and in a comparable micro method involving Zr-eriochromecyanin R (28) as little 25 μg of phosphate yields the same magnitude of error. It may be recalled that in routine analysis of water samples by the Zr-eriochromecyanin R method, Megregian applies a correction factor for the sulfate content of water; such a correction is not required for most samples analyzed by the aluminum-eriochromecyanin R method.

Hematoxylin has been employed as a dye for determination of fluoride with aluminum salts (268,269). MacNulty and Hunter (270) concluded that the aluminum complex is not formed with hematoxylin, but with hematin, its oxidation product; this finding explains the necessity for aging the reagent for analysis. Other dyes proposed in aluminum methods for fluoride are a mixture of aluminon and ferron (271), eriochrome red B, arsenazo I (272), xylenol orange (273), and chromeazurol S (274). Pyrogallol red (275) has been used for the determination of fluoride between 0 and 25 μg F/10 ml. Readings are taken 15 min after adding the reagents. The error is reported to be 6% at 10 μg F level. It has been reported that Mg^{2+}, Ca^{2+}, Ba^{2+}, Sr^{2+}, Zn^{2+}, Co^{2+}, Ni^{2+}, Mn^{2+}, Cd^{2+}, Hg^{2+}, Cr^{6+}, Pb^{2+}, Br^-, I^-, Cl^-, SO_4^{2-}, and $H_2PO_4^-$ do not interfere; Cu^{2+} and Fe^{3+} interfere. This method seems to merit further investigation.

d. Iron Complexes. Fluoride has also been determined by several colored complexes of iron, such as those with acetyl acetone (276), ferron (277), salicylic acid (278–280), 5-sulfosalicylic acid (281), thiocyanate (282–284), 5-phenyl salicylic acid and resacetophenone (285), and more recently with *N,N′*-bis(*β*-hydroxy-propyl)-*o*-phenylenediamine (286) and 1-(2-carboxy-4-sulfonatophenyl)-3 hydroxy-3-phenyltriazen [2-(3-hydroxy-3-phenyltriazeno)-5-sulfobenzoic acid] (287).

e. Titanium Complexes. Determination of fluoride with titanium solution, reported near the turn of the century (288,289), is one of the oldest methods available. It is based on the fading by fluoride of the yellow-orange color of titanium chloride solution treated with hydrogen perioxide. This method has been somewhat ignored by many workers because of inadequate sensitivity. Interest has been revived in 1948 by Monnier et al. (290), who have drawn attention to the possibility of applying this approach in the presence of moderate amounts of phosphate, which interfere readily in all other colorimetric methods for fluoride. When an excessive amount of phosphate is present, Monier et al. (290) suggest taking an additional reading after adding an excess of aluminum to the solution to complex all fluoride, thereby nullifying the effect of fluoride on the titanium yellow color. The difference between the two readings is a measure of the fluoride concentration in the sample. Ascorbic acid has been proposed in the place of hydrogen peroxide for the titanium method for fluoride (291,292). This eliminates the inconvenience of having to handle the rather corrosive hydrogen peroxide and also the uncertainties of the method due to lack of total stability of the reagent.

Babko and Khodulina (293) have reported a color reaction for fluoride with titanium-chromotropic acid reagent, which is 100 times more sensitive than the peroxide method. This method ought to be investigated, particularly in relation to phosphate interference and its adaptation to colorimetric determination of fluoride in natural waters with high mineral content, in soil extracts, and in industrial and sewage waters. Recently Alonso (294) employed this method for the determination of fluoride in rocks, following fusion.

f. Miscellaneous Methods.

Chloranilate Methods. Bertolacini and Barney (295) have demonstrated the displacement by fluoride of equivalent amounts of chloranilic acid from insoluble strontium chloranilate suspended in 50% isopropanolic medium; chloranilic acid in the supermatant, after centrifugation, can be measured by ultraviolet spectrophotometry (absorption 332 nm); the limit of detection is 5 ppm F. More sensitive methods involve the use of lanthanum chloranilate (296,297) or thorium chloranilate (46,298,299). West et al. (300) have confirmed the claims of Hensley and Barney (298) and have further reported that in an aqueous medium at pH 4.7, chloranilate and fluoride ions yield 1 chloranilate ion in solution for each 2 fluoride ions, and thorium difluorochloranilate is retained on the solid phase. In 1–1 methyl cellosolve-water medium, the colored thorium difluorochloranilate is released into the solution phase, further enhancing the sensitivity of the method. Phosphate and molybdate, which interfere, are removed by anion-exchange resins. Cationic interferences are removed by adsorption on Dowex 8 resin or Amberlite IR-120.

Molybdenum Blue Method. Curry and Mellon (150) describe a method in which fluorine in the sample is isolated as silicon tetrafluoride, which is then hydrolyzed, converted into silicomolybdic acid and thence, by reduction, to molybdenum heteropoly-blue and measured spectrophotometrically. Fluorine in organic compounds has been measured with molybdenum blue method (301). Earlier, a procedure based on the very same principles has been reported for the determination of fluorides and gaseous hydrofluoric acid (302).

Other Methods. Jean (303) proposes determination of fluoride based on extinction measurements (240 to 245 nm) of tetraphenyl antimony fluoride after extraction into hexane-chloroform (8–5) mixture. Some of the complexes (precipitates) of other ions with tetraphenyl antimony are not extractable, but some are extractable and can cause interference. Such interfering ions need to be removed before the extraction of fluoride. Complexes of scandium with catechol violet (304) and with methyl thymol blue (305) have been proposed for the determination of fluoride in 0–40 μg range.

B. "DIRECT" SPECTROPHOTOMETRIC METHODS: THE ALIZARIN FLUORINE BLUE COMPLEXES

The observation by Belcher et al. (13) that a blue color is formed when fluoride is added to a red complex of cerium(III) with 1-2-dihydroxyanthranquinone-3-methylamine-*N*-*N*-diacetic acid (also known as alizarin complexan, alizarin complexone, alizarin fluorine blue) has marked a significant development in fluorine analysis. Based on this reaction, the authors have described a method for the determination of fluorine. The work of Belcher and West (306) suggests that the blue color involves the formation of a ternary complex between fluoride and cerium-alizarin complexan.

A search has been made for other metals that could react with alizarin complexan and fluoride ions to form colored complexes. Scandium and yttrium and all the lanthons share this property (307–309). Among them, however, only lanthanum and praseodymium have been found to be promising for fluoride determination; lanthanum has some advantages over cerium.

Belcher and West (306) have reported a detailed study on interferences and on the optimal pH for the cerium method for fluoride that is more or less applicable to the lanthanum method also. These procedures have greater tolerance to interfering ions than the zirconium and thorium methods. Cations such as aluminum and iron need to be removed. Moderate amounts of calcium, barium, and magnesium do not interfere. Greenhalgh and Riley (309) have found that 20 μg of fluoride could be determined accurately in the presence of about 1 mg of bromide, 2 mg of perchlorate, 5 mg of silicate, or 25 mg of sulfate. Yamamura et al. (310) have observed that 1–1 ratio between Ce(III) and alizarin complexan provides a higher tolerance toward common ions that do not complex cerium, and toward many cations. A 1–2.5 ratio improves tolerance toward complexing anions such as sulfate and phosphate. However in the determination of a few micrograms of fluorine, or submicrogram amounts, one needs to ascertain the extent of interference due to the anticipated concentrations of other ions. The tolerance for phosphate is low and is of the order of 5 or 10 μg of phosphorus in the determination of 10 or 25 μg F (48,309).

Addition of acetone improves the stability and the sensitivity of the lanthanum method (308,309). Dimethylformamide has been proposed in the place of acetone (311). Yamamura et al. (310) found that addition of acetone or acetonitrile in the cerium method improves its sensitivity. Hanocq and Molle (312) have reported that with the use of an aqueous 25% v/v solution of dimethylsulfoxide, the speed of the reaction could be greatly enhanced and the maximum sensitivity obtained (625 nm). The same authors (313) note that the addition of methyl sulfoxide enhances the absorption of both the cerium and lanthanum complexes with fluoride and alizarin complexan at 283 nm, and at this wavelength the extinction of the La complex is 34% greater than that of the Ce complex. The interference due to nitrate, carbonate, and magnesium ions is greater at 283 nm than that at 625 nm.

The sensitivity of the lanthanum method has been further improved by the use of solvent extraction techniques for the separation of the ternary metal-dye-fluoride complex from the metal-dye complex, which also contributes significantly to the measured absorbance of unextracted reaction mixtures.

Johnson and Leonard (314) have obtained a fivefold improvement in sensitivity by extracting the ternary complex with tribenzylamine-pentanol-secondary butanol mixture. Hall (315) has observed that boric acid-buffered reagent produces a more intense blue color in the lanthanum method than the acetate-buffered reagent. Succinate-buffered reagent enhances the speed of reaction to give greater sensitivity and also provides lower blank values. He prescribes extraction of succinate-buffered lanthanum-fluoride-alizarin complexan into isobutanol containing hydroxylamine hydrochloride. Davies and Foreman (316) report an eightfold gain in sensitivity by extracting the blue complex into 10% *n*-propylamine in butanol. In a semiautomated method, Manly et al. (317) employ isoamyl alcohol containing hydroxylamine hydrochloride for the extraction of the blue complex. Hirano et al. (318) use isoamyl alcohol containing 5 vol% *N-N*-diethylaniline for extraction of the blue complex. Iyer et al. (319) extract the complex with benzyl alcohol, after adding solid sodium chloride and methanol to the reaction mixture in succinate buffer, pH 4.6. Micro and submicro procedures for the determination of fluoride following extraction also of Ce-F-alizarin complexan with triethylamine-pentanol (1–19) have been reported by Haarsma and Agterdenbos (320).

Hall (315) mentions the possibility of back extracting the dye from the 4-ml volume of isobutanol into a smaller volume of alkali to make the lanthanum method even more sensitive. Venkateswarlu (321) has been able to back extract the dye into 100 μl of sodium hydroxide and measure the optical density in a 50-μl microcell. The gain in sensitivity is thirty- to fortyfold.

In connection with fluorine analysis of "hot" cell products with the foregoing methods, the effect of radiation from ^{60}Co on the stability of the alizarin complexan dye has been investigated. The dye lost nearly 1% of absorbance per 100 rads absorbed. This is not considered to be of any consequence in the analytical method in relation to the levels of radiation encountered during analysis (104).

A color reaction has been reported in which fluoride develops a positive color with a complex of iron; the yellow color of cocotheline-iron(II) complex is transformed to violet on reaction with fluoride ions (151). In the present form, the sensitivity of the reaction is not adequate for quantitative work.

Wilson and Cooke (242) have reported sensitive spot tests for detection-of as little as 3 ng F in 30 μl of test solution (F, 0.1 ppm). These are based on the formation of positive colors, blue and red, on addition of methyl thymol blue and xylenol orange to acidic solutions of zirconium in the presence of traces of fluorine. Excessive fluoride ($>$ 50 ppm) bleaches the positive colors. The color reactions are reported to be 50 times more sensitive than the cerium-alizarin complexan reaction with fluoride ions. Macejunas (322) has used the positive color formation with zirconium-xylenol orange for determination of fluoride in waters; sulfate is removed by precipitation as barium sulfate.

Methods for colorimetric determination of fluoride have been reviewed by Moraru (323).

3. Fluorimetric Methods

Fluoride is measured in terms of the quenching of fluorescence when it is added to fluorescent complexes of organic reagents with metals.

Willard and Horton (324) have reported two methods for fluorimetric determination of fluoride. In one, a standard solution of aluminum is added to the test fluoride solution, the unreacted aluminum is extracted with chloroform containing oxine, and the fluorescence (520 to 570 nm) is measured. In the other method, the decrease in fluorescence of aluminum-morin complex in 50% alcohol at pH 4.9 on reaction with the fluoride in the unknown solution is determined. The method is sensitive down to 0.2 μg F in a final reaction volume of 100 ml. The authors recommend isolation of fluoride by distillation or by pyrohydrolysis before using the fluorimetric method.

Bouman (325) has adapted the Willard-Horton fluorimetric method for the determination of fluorine in as little as 0.25 ml of water samples. The greatest decrease in fluorescense is obtained when the analysis is carried out at pH 4.4 with the aluminum-morin complex in 10% ethyl alcohol.

Powell and Saylor (326) have described fluorimetric methods based on quenching by fluoride of fluorescence of aluminum complexes with dihydroxyazo dyes, eriochrome red B and superchrome garnet Y. The waiting periods for the two methods are 3 and 1 hr, respectively. These methods are sensitive down to 0.2 μg F/50 ml.

Guyon et al. (327) have described a fluorimetric method for traces of fluoride, 0.002 to 0.01 ppm, using zirconium-flavonal complex. Most of the 42 ions tested in several thousandfold excess have not interfered. Aluminum, molybdate, citrate, tartrate, and oxalate ions have interfered even at tenfold excess. Unlike in the zirconium-eriochromecyanin R spectrophotometric method, sulfate ions do not interfere in this method; the zirconium reagent is prepared in sulfuric acid medium. Taves (190) has reported a rapid fluorimetric method for measurement of fluoride in the range 0–10 nmole F (i.e., 0.00 to 0.19 μg F/4 ml using thorium-morin reagent). No waiting period is required after the reagent has been added. Taves has employed this method for the determination of serum fluoride after separation by a diffusion procedure (56). Schenk and Dilloway (328) describe a method based on the quenching by fluoride of the flurescence of aluminum-1-(2-pyridylazo)-2-

naphthol for the determination of nanomolar to micromolar fluoride. A thousand times as much phosphate as fluoride causes only a 25% relative error; as much sulfate causes an error that is relatively very small.

Har and West (329) have reported a "direct" fluorimetric method for the determination of fluoride sensitive to 0.01 ppm F. It is based on the formation of a fluorescent ternary complex between zirconium, calcein blue, and fluoride ions. The fluorescence is measured at 410 nm (excitation at 350 nm).

4. Electroanalytical Methods

A. EARLY ELECTROANALYTICAL METHODS

In the early phases of development of fluoride determinations, electroanalytical methods have served mainly as an adjunct to the standard titrimetric procedures for fluoride. They have been devised primarily to eliminate the subjective errors in visual detection of the end points with the indicators. Subsequently there have been rather noteworthy developments in both the indirect and the direct electrometric methods for fluoride analysis. The most sensitive are some of the polarographic procedures and the potentiometric measurement of fluoride with the ion-specific electrode.

A conductometric method for the determination of fluoride with aluminum chloride, in the range 0.012 to 20.0 mg F, has been described by Harms and Jander (330). Ions of Cl, NO_3, SO_4, and SiO_2 do not interfere. Israel et al. (331) have described a conductometric end point detection of thorium nitrate titration of fluoride (0.3 to 5 mg F/10 ml); the end point detection for the 0.1-mg level is less certain. High-frequency oscillometry is also employed for the end point determination of 0.03 to 0.8 mg F/10 ml with thorium and lanthanum nitrate solutions (332–335). Titrations of fluoride with solutions of uranium salts and with aluminum salts have been performed using potentiometric methods (336,337).

Amperometric procedures have been devised for the titration of fluoride with thorium and lanthanum salts (338), with lead salts (339) and, for the determination of 0.2 to 6.0 ppm F in water, with aluminum salts (340–344). An amperometric procedure described by Castor and Saylor (345) involves titration of fluoride with aluminum-superchrome garnet Y complex. Johannesson (346) has reported an amperometric method for thorium nitrate titration of microgram and submicrogram amounts of fluoride, using a rotating aluminum electrode and a stationary platinum electrode. Harris (347) has published an account of titration of fluoride with thorium using a rotating palladium electrode. The method is sensitive to 0.2 ppm F. Liplavk (348) has described procedures for amperometric titration of several elements, including fluorine.

B. DIRECT ELECTROANALYTICAL METHODS

Low and Pryde (349) have described a procedure for rapid determination of fluoride (0.2 to 60 ppm F) with a simple concentration cell using the Fe^{2+}/Fe^{2+} - Pt electrode. Measurement of current from spontaneous electrolysis in an electrochemical cell, Al/CH_3COOH (0.2*M*)/Pt, following the addition of fluoride, is the basis of a simple and sensitive method developed by Baker and Morrison (350) for the determination of fluoride in the microgram range. The upsurge in current is caused by the interaction of fluoride with the aluminum electrode. A saturated solution of benzoic acid gives a much smaller blank current than acetic acid and is to be preferred for measuring less than 5 μg F. The method has been suggested for adaptation in portable detection units in field work. Baker (351) has employed this method for determining hydrofluoric acid in nitric acid. Howard and Weber (352) have designed an automatic analyzer, based on spontaneous electroylsis, for monitoring atmospheric fluoride levels. Megregian (353) has described the use of a zirconium-platinum electrode pair for the measurement of 2.5 to 20 ppm F in a 0.1*M* perchloric acid medium. A British patent (354) covers an instrument based on this principle for monitoring atmospheric samples. The solution or gas is introduced into a cell containing 1 to 5*N* HNO_3, provided with platinum and zirconium electrodes, and maintained at between − 1 and + 1 V with respect to a standard calomel electrode. A known concentration of F 0.1*M* is initially provided to obtain a stable and fixed current. Increase in the current following the introduction of fluoride-containing gases or solution is measured; as low as 0.05 ppm F can be determined. Turner (355) has employed a single-crystal *n*-type silicon electrode in a similar procedure for nitric acid solutions.

Changes in the conductivity of boric acid solutions following addition of fluoride is the basis of a method described by Manning and White (356); the method is sensitive down to 0.1 μg F.

C. POLAROGRAPHIC METHODS

Haul and Griess (357) have reported a method for the polarographic determination of fluoride ions as lead chlorofluoride. MacNulty et al. (358) have described a means of determining fluoride in microgram quantities of fluoride, featuring the addition of an excess of aluminum to the fluoride solution and the polarographic determination of the unreacted aluminum. Later the same authors experimented with one of the most sensitive methods for fluoride determination based on the depression by fluoride of the polarographic step due to the reduction of the aluminum-solochrome violet R. S. complex. The step depression is linearly related to the fluoride concentration down to 0.001 μg/ml. A very sensitive cathode ray polarograph is required (114). Shoemaker (359) has described a polarographic method, using rotating platinum microelectrodes, in which the iron fluoride complex is reduced irreversibly at potentials more negative than the reduction of ferric ion. Microgram amounts of fluoride can be determined by this method. The procedure described by Iwase (360) is based on the change, due to fluoride (10 to 150 ppm), in the limiting current of Cd^{2+} in a mixture of Cd-EDTA, (Cd^{2-}), Zr(IV), and NH_4SCN. A method reported by Bond and O'Donnell (361) utilizes the shift of the U(V) to U(III) half-wave potential due to fluoride (0.05 to 20 ppm). The method has been suggested by the authors for fluorine analysis of potable waters.

Zittel and Florence (362) described a voltammetric method based on measurement of the anodic reaction of the alizarin red S released by fluoride from zirconium-alizarin red S complex. Since the measurement is carried out at anodic potentials, oxygen does not interfere and deaeration is unnecessary. The method is sensitive to $1.0\mu M$ F^-. Berge and Bruegmann (363) have also reported measuring 0.5 to $5.0\mu M$ F^- by polarographic determination of alizarin red S released by fluoride from zirconium-alizarin red S complex.

D. THE FLUORIDE ION ELECTRODE

In 1966 Frant and Ross (14) developed an electrode-type sensor that gives a Nernstian reponse over a wide range of fluoride ion activity. The construction of the electrode is similar to that of the glass pH electrode; the membrane material is a section of single crystals of fluorides of rare earths (lanthanum, neodymium, or praseodymium, etc.). These crystals have high electrical conductance because there are mobile fluoride ions in the crystal lattice. The conductivity is further enhanced by doping the membrane with a divalent ion (e.g., europium). The electrode can be used for direct measurement of the fluoride ion activity using calibration curves prepared with appropriate standard fluoride solutions or as an end point indicator in the titration of fluoride with suitable fluoride complexing reagents.

The electrode is essentially specific to fluoride ions even in the presence of several other ions that interfere in other methods of fluoride analysis. Hydroxyl ions interfere with the electrode response to fluoride, but only when the hydroxyl concentration is more than one-tenth the concentration of the fluoride ion. The problem is easily overcome by neutralizing alkaline solutions and measuring low fluoride concentrations ($1\mu M F^-$) in the pH range 4.5–6.0 using appropriate buffers. Srinivasan and Rechnitz (364) have shown that the electrode responds to fluoride ion activity in highly acidic media. Mesmer (365), Warner (366), and others have demonstrated that the electrode is sensitive to submicromolar concentrations of fluoride and has an extremely high tolerance toward chloride ions, thus permiting studies on fluoride complexes in chloride media.

For all practical purposes, the electrode responds similarly toward fluoride ions and fluosilicate ions (367), although SiF_6^{2-} may give a slightly less response than fluoride (368).

Recently some observations on nonspecific and nonionic responses of the fluoride ion electrode have been reported. Zentner (369) finds that the fluoride ion electrode responds to 1-fluoro-2,4-dinitrobenzene (FDNB) in buffered and unbuffered solutions, although no fluoride ions are present. In the presence of $10^{-4}M F^-$, a tenfold change in the concentration of FDNB gives rise to a change in potential of − 35 mV, as compared to a change in potential of − 59.2 mV with a similar change in the concentration of the fluoride ions. The analogous chloro compound, 1-chloro-2,4-dinitrobenzene, has no effect on the chloride (or fluoride) electrode response. Zentner believes that the fluoride atom imparts an appreciable dipole moment to the nonionized FDNB, and the strongly polar FDNB molecule may be adsorbed at the crystal surface of the electrode. The electrode has been found to respond to calcium ions above millimolar concentration. Berndt and Stearns (370) have observed response of the fluoride electrode to $(NH_4)_2PO_3F$ solution buffered at pH 5.9, corresponding to 16% of the amount of fluorine in the compound; no fluoride impurity was detected by other means and hydrolysis under the conditions was considered highly unlikely. Venkateswarlu (171), who noted that 1.5% of fluorine present in a sample of fluoroacetate responded to the fluoride electrode, attributed this property to the possible presence of fluoride impurity in the sample, rather than to any nonspecific response of the electrode. Further work is indicated in this area. The foregoing observations serve as a caution in the use of the electrode, but in no way should they deter anyone from choosing this equipment.

The use of the fluoride electrode as the end point indicator in the titration of fluoride has been investgated by many workers. Lingane (371,372) finds that La^{3+} is the most suitable among the titrants (La^{3+}, Ca^{2+}, Th^{4+}) because it gives the largest rate of potential change at the quivalence point. Stirring helps titrations of solutions with pF greater than 4.5. Harzdorf (373) has investigated titration of fluoride in the range 0.05mM to 0.1M and considers that the fluoride titrations do not correspond to exact stoichiometry and that an empirical factor is required. According to Bazzelle (374), 60% 2-propanol is the optimal medium for titration of small amounts of fluoride with La^{3+} solution. Heckel and Marsh (375) describe titrimetric approaches to the determination of subnanomole amounts of fluoride in polar nonaqueous solvents containing 5% water. The electrode response is slow in nonaqueous solutions, and the titration procedure is hastened by the use of the time response graph paper to determine by extrapolation the electrode potential at infinite time. Anfalt et al. (376) find evidence for the formation of mixed fluoride-acetate complexes when fluoride is titrated in acetate buffer, and they recommend titration of fluoride in unbuffered solutions. The optimal conditions for potentiometric titration of fluoride with lanthanum and thorium in unbuffered solutions have been investigated by Eriksson and Johansson (377). Fluoride ion electrode has also been used in coulometric titration of fluoride by Muto and Nozaki (378). Evans et al. (379) find that the carboxylate ions do not affect the normal operation of the fluoride electrode but could interefere in potentiometric titration of fluoride with lanthanum nitrate solution because of the formation lanthanum acetate complexes. In direct measurement of fluoride ions with the electrode, using calibration curves prepared with fluoride standards containing the same amount of acetate as the test solutions, the presence of acetate causes no problems. Elimination of interferences from fluoride-complexing cations is described in Section V.2.

A standard (or known) addition technique with the fluoride electrode has been developed to minimize error due to differences in such factors as the ionic strength or the ionic composition between the unknowns and the standards, that might alter the activity coefficients or the junction potentials. It permits measurement of fluoride concentrations in samples with widely differing ionic strengths and compositions without the use of calibration curves or adjustment of the ionic strengths of the unknowns. Essentially, the procedure involves determining the difference in potentials (ΔE) of the test sample before and after addition of a known amount of fluoride. A very small volume of the fluoride solution of a suitably high concentration is added, to prevent the alteration of the ionic strength of the sample solution. If C_o represents the original concentration of the fluoride in the sample, C_a the altered concentration following the addition of fluoride, ΔE the change in the potential, and S the electrode slope or Nernst factor (theoretically 59.2 at 25 °C), the fluoride concentration of the sample is calculated by

$$C_o = \frac{C_a}{\text{antilog}\,(\Delta E/S) - 1}$$

It is also assumed that the complexing factors are present in such an excess that the same proportion of fluoride is complexed after the addition of fluoride as in the original test solution. Therefore this approach is not suitable when the level of the complexing agents in the sample is inadequate or too low for complexing the added fluoride in the same ratio as in the original sample. Yet this procedure may be employed by initially totally decomplexing the fluoride ions from ions of aluminum, iron, thorium, and other metals, by precomplexing the latter with such metal-complexing agents as $5M$ phosphoric acid. Under these circumstances, an abundance of a new species of fluoride-complexing cation (H^+) is provided in the medium. Bauman

(380) was among the first to investigate this approach for fluoride analysis. Since under the circumstances just described the pH is rather low and a significant fraction of total fluoride exists as HF to which the fluoride electrode does not respond, the initial level of total fluoride must be high enough to provide a concentration of free fluoride ions to which the electrode can respond. Between 10^{-3} and $10^{-5}M$F, the relative error is 10% and the relative standard deviation less than 5%. Lower concentrations of total fluoride have been determined by this approach under less acidic conditions, pH 5.5, and by using cyclohexanediaminetetraacetic acid (381) for precomplexing the interfering cations.

The standard addition technique has been employed for determining fluoride in ill-characterized natural waters (381), calcium halophosphates (382), phosphate rock (383), silicates (384), extracts of feed stuffs (385), seawater (386), and blood plasma or serum (387).

For a more accurate measurement of fluoride, Liberti and Mascini (388) propose using Gran's plot to locate the equivalence point. Small volumes of fluoride solution (concentrated and having the same ionic strength as the test solution if possible) are successively added to the test solution, and the potential is determined after each such addition. Fluoride concentrations in the range 10^{-5} to $10^{-6}M$ have been calculated successfully. One of the requirements for success of these techniques is a linear relationship between the response of the electrode and the logarithm of the fluoride concentrations over which the measurements are made. This method has been employed by Liberti and Mascini for the determination of fluoride in mineral and waste waters, by Selig for the determination of fluoride in 0.08–3.00 ppm range (389), and by Schick for determining fluoride in several household products (390).

Based on an observation that the permissible limit of aluminum concentration increases markedly with decreasing fluoride concentration, Shiraishi et al. (391) have proposed a successive dilution technique for the complete demasking of fluoride from its complexing agents (e.g., aluminum). Successive 1–10 dilutions are carried out with 0.5*M* citrate buffer, pH 6, until two successive results of fluoride analysis agree. This approach has been employed to determine fluoride in samples of glass, phosphate rock, and well water.

The "analate additions" technique described by Durst (392) is the inverse of the standard addition technique. It is considered useful in the case of small sample volumes that cannot be analyzed by the standard addition method without prior dilution.

The minimum sample size required for fluoride measurement with the electrode and the Orion microsample dish is 0.3 ml. Durst and Taylor (393) have modified the original Orion fluoride ion electrode for use with sample volumes less than 1 ml. They have shown a usable response in the range 10^{-1} to $5 \times 10^{-6}M$ for fluoride in 50-μl sample volumes. Later Durst (394) has described a linear null point potentiometric procedure that is suitable for determining 190 to 0.38 ng F in 10-μl samples. This procedure calls for the use of two fluoride ion electrodes rather than one fluoride electrode and one standard calomel electrode. Durst (395) has also fabricated a micro-fluoride electrode, which in conjunction with a microcapillary reference electrode is capable of measuring fluoride in 2-μl samples. Further miniaturization along similar lines could eventually permit intracellular fluoride determination. Orion Research, Inc., have introduced a combination fluoride electrode (model 96-09) which has been also recommended for measurement of fluoride in 10-μl sample volumes, using a confined spot test paper. In the latter situation, however, the results are influenced by the leaks of potassium chloride solution from the combination electrode, particularly when the fluoride concentration is low and long periods of response time must be allowed. Nevertheless the combination electrode may be convenient in other situations when the sample volume is large and relatively shorter response times are adequate. Errors due to progressive dilution of microvolumes of samples with the saturated potassium chloride solution leaking from the reference electrode or from the combination fluoride electrode, and errors caused by concentration of the microsample by evaporation, are obviated by using the hanging drop fluoride electrode assembly described by Venkateswarlu (15).

The hanging drop fluoride electrode assembly (Figure 2) is self-explanatory except for the following points. The agar bridge contains 0.4% agar in 0.05*M* histidine buffer, pH 6.1. The 5-μl sample applied to the crystal of the electrode occasionally tends to drift, and as a result contact with the agar bridge is broken. To overcome this problem the following improvisation is made. A small square of Parafilm is stretched taut over the crystal of the fluoride electrode and tightly secured by adhesion. A small hole is made in the Parafilm with a sharp plastic probe, exposing a small surface of the lanthanum fluoride crystal. The end of the electrode is then pressed evenly against a pad of filter papers covered with a strip of glazed weighing paper to eliminate all crevices between the crystal and the Parafilm into which the sample otherwise could be drawn by capillary action. After assembling the hanging drop fluoride electrode, the agar bridge is secured in position with a strip of Parafilm wrapped around the nozzle, the sleeve, and the bridge. A completely closed system, ensuring against the evaporation of the 5-μl sample, is thus achieved. Between reading samples, the electrode is wiped gently with a cotton sponge, without disturbing the Parafilm cover, and rinsed with 5 μl of the "blank" solution and with a 5-μl aliquot of the following sample. Also, the droplet of the previous sample adhering to the tip of the agar bridge is removed by wiping with a cotton swab. Reading times with the electrode are kept the same for all the standards and the unknowns. For weak fluoride solutions (1μ*M*), a maximum of 30 min has been employed.

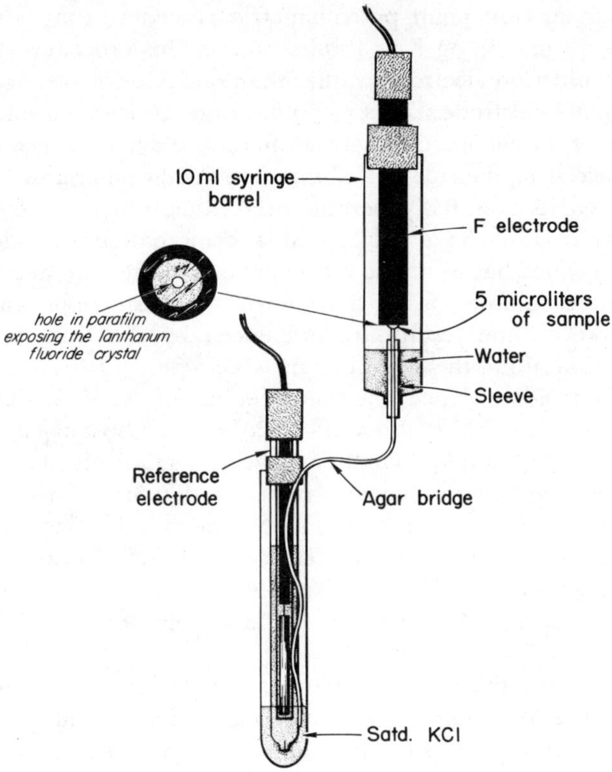

Figure 2. Hanging drop fluoride electrode assembly.

Substitutes for the single-crystal lanthanum fluoride membrane are being investigated. Materova et al. (396) have reported a precipitation and anion-exchange membrane electrode reversible with respect to fluoride. It is prepared by mixing powdered calcium fluoride with a solution of poly(methoxyphenylsiloxane) resin on PTFE plates. This membrane is sensitive to about $0.05M$ F^-. Bausovo et al. (397) have prepared a polycrystalline membrane by hot moulding of powdered LaF_3 drifted with 0.1 to $0.15M$ concentration of EuF_2 and pressing into disks. The membrane responds to $10\mu M$ to $0.1M$ F^- in aqueous solutions and down to 0.1 to $1\mu M$ F^- in a mixed solvent (80% aqueous acetone). Freiser et al. (398) have described an electrode consisting of a conductive wire of platinum, copper, or graphite bearing a film of anion-exchange material in a polymer matrix. The ion-exchange material may contain any potentially measurable ion and may be incorporated in the polymer structure. Electrodes are conditioned by immersion in a $0.1M$ solution of the ion for a few minutes. These electrodes are considered to be sturdier and less expensive than the conventional solid-membrane and liquid-membrane electrodes. Hirata and Ayuzawa (399) have reported the preparation of lanthanum fluoride ceramic membrane for use as selective electrode for fluoride ions. The response of the electrode at 25 °C is a 59-mV change per tenfold change in fluoride concentration from $10\mu M$ to $0.1M$ and 56 mV in the 1–$10\mu M$ range. The authors claim that the response in the latter range is better and more rapid than with the single-crystal electrode and requires 40 to 60 sec response time at $10^{-5}M$ fluoride.

5. Radiometric Methods

Radiometric methods provide yet another approach to the determination of fluorine. In some cases the difficulties, inaccuracies, and uncertainties of chemical analysis of fluorine can be resolved by resorting to analysis by radiometric methods. It is also possible to carry out nondestructive analysis by radioactivation techniques. However such techniques require special facilities available only at nuclear centers and advanced laboratories.

A. ISOTOPE DILUTION

A sensitive isotope dilution method using ^{18}F to determine fluoride in the 0.01–10 ppm range in water and tooth enamel has been reported by Fremlin et al. (400). In a method described by Van der Mark and Das (401), the test solution is spiked with ^{18}F and the fraction of ^{18}F acquired by chemisorption onto glass beads is determined. A rectilinear relationship exists between the reciprocal of the count rate and the fluoride concentration. The lower limit of detection is 0.03 μg F/ml. The method has been applied to the determination of fluorine in water and urine samples. Das et al. (402) have reported yet another radioisotope dilution method for fluoride determination in urine and water; it is based on substoichiometric extraction of ^{18}F with trimethylchlorosilane in benzene. The inverse value of the fraction of the ^{18}F extracted is found to be a quadratic function of the concentration of fluoride. A substoichometric isotope dilution method following extraction of fluoride into chloroform with tetraphenylstibonium sulfate reagent has been described (403). The use of this procedure calls for separation of fluoride by appropriate means, when interfering ions such as chloride are present.

B. INTERACTION WITH AN ISOTOPE

Some methods depend on the reaction of fluoride with a radioactive isotope of a metal, followed by measurement of the reacted or unreacted fraction of the isotope. Onstott and Ellis (404) have titrated 20 to 30 mg F with samarium ion containing europium carrier-tracer. The end point is determined by measuring the excess of the titrant. The titration can be carried out on a microscale down to 35 μg F with less than 2% error. The complexation of hafnium by fluoride is the basis of a radiometric procedure reported by Siegel (406). With increasing amounts of fluoride in the aqueous phase (0.25*M* sulfuric acid, $10^{-5}M$ hafnium, ^{181}Hf), progressively less ^{181}Hf is extracted into *n*-hexane containing 0.026*M* trioctylphosphine; a 10 μg F change in the aqueous solution causes 28 to 48% of hafnium to be retained in the aqueous phase.

C. ISOTOPE DISPLACEMENT

A spot of $(ZrO)_3(PO_4)_2$, labeled with ^{32}P, is prepared on a paper strip and chromatographed with appropriate solvents following the application of 2 μl of the fluoride solution. The activity of the ^{32}P displaced from the original spot is linear with the fluoride content above 4 μg (407). The chief drawback of this method is the need to concentrate fluoride without introducing contamination. Carmichael and Whitley (408) have described a radio-release method for the determination of 20 to 100 μg F in aqueous solution, which is passed through a polyethylene column packed with $Zr(^{75}SeO_3)_2$ – PTFE. Phosphate, iron, and aluminum interfere. This method has been suggested for use in routine continuous monitoring of fluoride in waters.

D. RADIOACTIVATION ANALYSIS

By appropriate irradiation techniques, fluorine present in a test sample is converted into a radioactive isotope or isotopes of fluorine or of some other element, whose radioactivity is determined by suitable techniques. The activity so ascertained is a measure of the fluorine content of the sample. In actual analytical situations interferences may develop because of the formation of the same radioactive isotope by different nuclear reactions, from fluorine as well as from some other element present in the sample or by the production of isotopes with overlapping energy spectra of their radiations. These problems are resolved to some extent by taking advantage of the differences in threshold activation energies for fluorine and the interfering elements, by employing an irradiation beam with the optimal and intermediate energy and also by using high-resolution spectrometers for isolating and measuring specific peaks of the resulting radiation. These techniques permit rapid, nondestructive, and simultaneous analyses of fluorine and other elements as well, and they are less dependent on the composition of the sample. Some analyses, however, call for radiochemical separation of fluorine.

a. Neutron Activation Analysis (NAA). NAA had been carried out almost exclusively with fast thermal neutrons in nuclear reactor centers until about 1960, when 14-MeV neutron generators became commercially available. These generators provide considerably lower neutron flux and reduced cross-sectional area in the 15-MeV region than did the nuclear reactors. Consequently there is a loss in sensitivity; nevertheless, the neutron generators are an extremely convenient source of neutrons for activation analysis.

Neutron activation of fluorine results in the following nuclear reactions:

$$^{19}F(n,\gamma)\,^{20}F \quad (E_\gamma\ 1.63\ \text{MeV}, t_{1/2}\ 10.7\ \text{sec}) \quad [1]$$
$$^{19}F(n,2n)\,^{18}F \quad (E_\gamma\ 0.51\ \text{MeV}, t_{1/2}\ 109\ \text{min}) \quad [2]$$
$$^{19}F(n,p)\,^{19}O \quad (E_\gamma\ 0.2\ \text{MeV},\ 1.37\ \text{MeV}, t_{1/2}\ 29\ \text{sec}) \quad [3]$$
$$^{19}F(n,\alpha)\,^{16}N \quad (E_\gamma\ 5.7\ \text{MeV}, t_{1/2}\ 7.2\ \text{sec}) \quad [4]$$

where ^{18}F is a positron emitter.

Fluorine determination may be made by discerning any one of the foregoing reactions, subject of course to such practical considerations as analytical interferences and availability of the measuring devices.

In measurements based on [1], sodium, which is present in biological materials, interferes because of the reaction $^{23}Na(n,\alpha)\,^{20}F$. Atchinson and Beamer (409) recommend correction based on sodium content of the sample. Bakes and Jeffery (410) point out that [2] is subject to interference from other positron emitters formed from other elements and that [3] requires an expensive high-resolution multichannel γ-spectrometer for isolating 0.2- and 1.37-MeV activities. They prefer quantitation of fluorine by [4], which requires a single-channel spectrometer. Oxygen interference due to the reaction $^{16}O(n,p)\,^{16}N$ is avoided by using neutrons of intermediate energy; threshold values for activation of fluorine and oxygen are 2 and 10 MeV, respectively. Results of analyses by neutron activation of several mill products containing fluoride showed close agreement with those obtained by pyrohydrolysis. On the other hand, Garrec and Fourcy (411) find it most convenient to quantitate fluorine by [3]; they employ 14-MeV neutrons and γ-ray spectrometry. For activation analysis of fluoride without interference from oxygen, Wilkniss (412) proposes the use of photoneutrons, generated by bombarding a tantalum target with an electron beam from a linear accelerator.

Andersen et al. (413) have used [2] for the determination of F in a sample of cereal (43 ± 6ppm) with Cockcroft-Walton 14-MeV neutron generator. Interference from other positron emitters, notably ^{64}Cu, has necessitated radiochemical separation of ^{18}F during activation analysis of the same sample with reactor neutrons.

Reitief et al. (414) have failed to determine fluorine in human teeth by thermal neutron activation. Because of the higher Cl content of enamel and dentine, they have encountered interference of 1643-keV photopeak of ^{38}Cl over the 1630-keV photopeak of ^{20}F.

Henkelmann et al. (415) have developed a very rapid method, requiring about 1 min total time for pneumatic transfer of sample to and from the reactor, irradiation, and assay by [1]; the detection limit is about 1 μg F. The limits of detection of F by pulsed neutron activation analysis system described by Naughton and Jester (416) are 0.23 μg F by [4] and 0.19 μg F by [1]. Wing and Wahlgren (417) use fast neutrons from ^{241}Am - ^{242}Cm - Be source and [4] for measuring 0.4 mg F/10-g sample; the source is produced by 4-month irradiation of l-g slugs of Am_2O_3 mixed with beryllium metal (1–50) with a thermal neutron flux of 8.5×10^{13} n cm $^{-2}$ s^{-1} Lauff et al. (418) have used ^{252}Cf as the source of neutrons for NAA of fluorine. Automatic systems of analysis of neutron activation for on-line measurement have been developed (419).

b. Photonuclear Activation Analysis (PNAA). Bremsstrahlung is produced by absorbing a beam of electrons from a linear accelerator in a water-cooled tungsten target. By adjusting the electron energy, the photon energy (Bremsstrahlung energy) can be regulated. PNAA of fluorine is based on the reaction $^{19}F(\gamma, n)\,^{18}F$; 0.1 μg F can be determined using irradiation with 15-MeV (E_{max}) bremsstrahlung. Levels of 1.1 ± 0.1 to 22.2 ± 0.5 ppm F in samples of vegetable oils have been determined. Interferences due to Na and Cl are avoided by employing 12-MeV (E_{max}) photoactivation. As a result, however, the limit of detection (10 μg F) is affected. The method has been employed to determine F in a tea sample (F, 150 ± 6ppm) and is considered to be more accurate and more precise than analysis by means of instrumental reactor NAA. Despite greater sensitivity, reactor NAA has the disadvantage of simultaneous activation of several major sample constituents. Because of the comparatively slight interference, 12-MeV PNAA is considered superior to NAA. For example, fluorine determination of teeth could be performed by PNAA, but not by NAA. A comparative study of PNAA and NAA of various elements including fluorine has been made by Andersen et al. (413). With advanced instrumentation, it would seem possible to increase the sensitivity of PNAA of fluorine. Hislop (420) emphasizes the advantages of γ-photon activation analysis.

Brune et al. (421) report a detection limit of 3 μg F by PNAA. According to Engelman and Scherle (422), amounts as low as 0.1 μg F can be detected. Wilkniss (423) places the limit of detection at 0.05 μg F. PNAA has been employed for the determination of F in seawater (424), urine (425), pharmaceuticals, fluorite, and vanadium concentrates (426).

PNAA of fluorine may also be based on the reaction $^{19}F(\gamma, 2p)\,^{17}N$ and measurement of delayed neutrons emitted by ^{17}N (427).

c. Proton Activation Analysis. The method of determining fluorine by proton activation is based on the reaction $^{19}F(p, p'\gamma)\,^{19}F$. The fluorine-containing sample is irradiated by a beam of energetic 4.75-MeV protons from a cyclotron. Both the 110- and 197-keV levels in ^{19}F are excited by inelastic scattering, and they decay via γ-emissions to the ground state. The γ-radiation is measured with a 20 - cm^3 Ge(Li) detector of 1.8-keV resolution at the ^{19}F 197-keV line. The entire sample preparation, involving essentially fixing 500 μg of the finely pulverized powder of the sample onto a thin Formvar foil, requires no more than 20 min. Rudolph et al. (428) have employed this technique for the determination of fluorine in coal and tooth samples. At the present stage of development of this technique, the limit of detection is 10 ng F in a 500-μg aliquot of the sample (i.e., 20 ppm). Coote et al. (429) have reported a sensitivity of 10 to 50 ppm F by inelastic scattering of protons.

Bewers and Flack (430) determined fluorine by prompt γ-radiation from proton bombardment, based in the reaction $^{19}F(p,\alpha\gamma)\,^{16}O$. Determination in liquids are possible in the range 20 ppm to 10% F in aqueous solutions and in gases at a relatively high level (5 to 100%) and in rock samples; the only interferences found are aluminum and lithium. Demortier (431) has described determination of fluorine in samples of biological materials by prompt nuclear reactions, $^{19}F(p,p'\gamma)\,^{19}F$ and $^{19}F(p,\alpha\gamma)\,^{16}O$. Based on the latter reaction, Anttila and Keinonen (432) have reported a method sensitive to amounts as low as 0.3 pg F. The technique avoids the need for depositing thin layers of sample on a carbon foil and permits the measurement of fluorine content both at the surface and within the sample.

Lee et al. (433) report determination of fluorine in LiF, Teflon, and PbF_2 by ^3He activation, based on the reaction $^{19}F(^3He, 2p)\,^{20}F$, and measurement of 1.63-MeV γ-ray of 10.8 sec ^{20}F. The detection limit is about 0.2 $\mu g/cm^2$, corresponding to 2 ppm in the matrix 100 mg/cm^2 thick.

Preparation for comparison standards, of stable stoichiometric fluorine compounds that are anhydrous and do not absorb detectable moisture during a 90-min exposure to atmosphere has been reported. Such compounds are especially prepared by horizontal zone refining or sublimation in HF atmosphere (434).

Radiometric methods for the determination of fluorine have been reviewed by Foreman (435).

6. Catalytic Methods

Lambert (436) has reported a quantitative method for fluoride based on its accelerating effect on the reaction between iodine ion and the hydrolysis products of ceric sulfate. The speed of the reaction is measured by timing the appearance of the blue linear starch-triiodide (I_3^-) ion complex. The method is sensitive to 0.3 to 0.9 ppm F in water. However it is susceptible to several interfering substances present in potable waters.

Armstrong and Singer (437) have described a method for fluoride determination based on the marked catalytic effect of fluoride on the cerous-ceric electron-exchange reaction. The catalytic effect of fluoride is about 900 times that of sulfate and 270 times that of phosphate. By making constant the effect of interfering ions, the authors have determined fluoride content of bone ash. The exchange is carried out at a strictly controlled temperature (−14.2 °C) using labeled cerous ions; ceric ions labeled as a result of adding fluoride are extracted and counted. The method is rather involved and has not been adapted for routine fluorine analysis.

Cabello-Tomas and West (438) have reported a kinetochromic spectrophotometric method for the determination of 0.5 to 5 μg F, based on the catalytic effect of fluoride ions on the reaction of zirconium with xylenol orange. The color reaches maximum intensity within 90 min and remains stable for 1 hr; the molecular extinction coefficient is 2×10^5 for F^-. The reaction is not suitable for determining large amounts of fluoride because it is substoichiometric. Hems et al. (439) described a similar method for the determination of fluoride, based on the catalytic effect of fluoride on the slow reaction between methyl thymol blue and zirconium(IV) in aqueous solution. The method is applicable over the 0.5–4.75 μg range. The method is less susceptible to cationic interferences but more susceptible to anionic interferences than is the alizarin complexan method involving lanthanum or cerium.

Mori (440) has reported a spectrophotometric method for zirconium as well as for fluoride with gallein. Fluoride catalyzes the formation of a water-soluble reddish violet complex of Zr(IV) with gallein, particularly in the presence of cetylpyridinium chloride.

An indirect kinetic method for the determination of fluorine in the parts per billion range has been developed by Klockow et al. (441). It is based on the inhibitory effect of fluoride (by complexing Zr) on zirconium-catalyzed reaction between perborate and iodide. The reaction rates are monitored by an automatic potentiostatic unit. The method is claimed to be sensitive to 19 ng F/50 ml (0.38 ppb).

7. Enzymatic Methods

The enzymatic approach to the determination of fluorine has one major attraction. Because of the specificity of enzyme reactions, enzymatic methods may be less susceptible to the multitudinous interfering substances present in the biological materials from which fluoride is required to be isolated for measurement with any colorimetric method. Several of the enzymes known to be inhibited or activated by fluoride have been investigated in this connection. Some of the most sensitive and promising methods are presented here, although none have been adapted for routine work beyond the laboratories in which they were developed. This is perhaps because of several uncertainties and difficulties in the preparation and preservation of enzymes and because of the availability of other methods that can be readily adapted for routine analysis.

Stetter (442) described a micro method for fluoride sensitive to 0.02 μg F/ml. It is based on the inhibitory effect of fluoride on potato phosphatases. Titanium, aluminum, beryllium, and molybdate ions interfere. The effect of the Ti and Al is eliminated by adding excess of these ions to the enzyme. The applicability of the method to the determination of fluoride in water, plant, and animal materials has been described.

Linde (443) has reported an enzymatic method for the determination of 0.1 to 0.5 μg of fluoride per milliliter of body fluids; it is based on the inhibitory effect of fluoride on the hydrolysis of ethyl-*n*-butyrate by hog liver lipase. McGaughey and Stowell (444) have modified Linde's method to permit the estimation of a few nanograms of fluoride. They also describe a procedure for preparing an acetone powder of stable hog liver esterase. The enzymatic activity is measured by titration with sodium hydroxide solution of the liberated butyric acid. They have been able to determine fluoride in solutions of small samples of tooth enamel and dentin without the need for removing phosphate, a remarkable accomplishment before the advent of the fluoride ion electrode. Further improvements in their method have been reported subsequently (445,446).

Geike (447) recently described a sensitive method for thin-layer chromatographic and enzymatic detection of fluoride with phosphatase. The limit of detection is 19 ng.

8. Gas Chromatographic Methods

As mentioned earlier, Bock and Semmler (17) were the first to propose gas chromatography for the determination of fluoride following its extraction as triethylfluorosilane into xylene. The silicone oil Sterchemol gas chromatography column is operated at 75 °C, and nitrogen carrier gas is used. With a hydrogen flame ionization detector, the detection limit is 0.1 μg F/ml. The method has been employed to determine fluorine in samples of water and soluble and insoluble phosphates (the latter after fusion). Based on the foregoing work, Fresen et al. (187) have reported a gas chromatographic procedure for the determination of fluorine in biological materials—bone, saliva, serum, and urine. These authors prefer extraction of fluoride as trimethylfluorosilane. It is possible to determine 0.01 to 10 μg F/ml with a relatively small standard deviation. Bock and Strecker (448) suggest the use of tartaric adic to overcome the interference of aluminum during extraction of fluoride as triethylfluorosilane.

Cropper and Putnam (449) have employed the procedure just given for determining fluorine in dental creams. Russel (450) has adapted the gas chromatographic method for determining fluoride in urine, blood, and soft tissues, following destruction of the organic matter in the latter samples by prior digestion with hot hydrochloric acid. The author suggests the use of this method as a rapid means to the detection of acute fluoride poisoning. The high levels of fluoride in body fluids and tissue homogenates encountered in acute fluoride poisoning can be readily monitored now with the fluoride ion electrode without the need for acid digestion and extraction.

Groff (451) has described a gas chromatographic method for determining fluorine in bone. In a Teflonized reaction chamber, HF is released from the bone sample by the action of sulfuric acid. The HF is caused to react with SiO_2 to form SiF_4, which is dried and directed over the Kel-F 40 activated phase of 80-100 mesh by a helium carrier at 25 ml/min. MacGee and Allen (452) have described a novel ion-exchange alkylation-GSC technique for determining halides including fluoride. It is based on thermal degradation of the tetraalkylammonium halides formed by passage of the halide solution through a column of Dowex 50W-X4 in tetraalkylammonium form; fluoride is ideally determined in the form of the tetramethyl salt. Gutsche et al. (453) have devised a fluoride-specific gas chromatographic method. The sample from the gas chromatograph and calcium vapor in a stream of argon are fed into an oxyacetylene microburner; the intensity of the CaF emission band at 529.9 nm is a measure of the F content of the sample. Spears and Hackerman (454) have reported gas chromatographic techniques for analysis of gas mixtures containing HF, N-F compounds, and elemental fluorine. Pitak (455) has demonstrated successful gas chromatographic separation of corrosive inorganic volatile fluorine compounds (F_2, MoF_6, UF_6, and SbF_5). Modified with a fluorinated polymer, Al_2O_3 provides an efficient packing for absorption columns. A similar technique has been described by Pappas and Million (456).

Gas chromatographic methods have been reported for the determination of halothane in blood (457), fluoroacetate in the gastrointestinal contents of dog, coyote, and bird (458), and perfluorocarbons in blood and tissue homogenates (459).

9. Mass Spectrometric Methods

Bock and Semmler (17), who proposed gas chromatography for the determination of fluorine following extraction as triethylfluorosilane, described in the same publication a mass spectrometric procedure for measuring fluorine. The sensitivity is about 50 μg F/ml, and a linear calibration curve is obtainable over the range 100 to 600 μg F/ml benzene solvent.

Hardwick and Martin (460) have reported a pilot study using mass spectrometry for the estimation of the trace element content of dental tissues. The presence of 100 to 1000 ppm F in enamel and dentin was confirmed by a doubly charged ion, and 1 to 10 ppm F was found in ashed plaque (2% by wt of the unashed sample). The authors hope to improve the accuracy of the method by employing prepared matrix electrodes of known composition and by using a microdensitometer for comparing the line image on the photographic plate. Swift (461) employs spark source mass spectrometry to determine the majority of elements in dental tissues in a single analysis. Gregory (462) has recently achieved a high degree of resolution in the mass spectroscopic analysis of ashed blood samples. With this enhanced resolution it might be possible to use mass spectrometry to determine fluorine, magnesium, aluminum, silicon, and other elements in blood samples.

10. Other Physical Methods

Spectrographic methods for fluoride depend on measuring the intensity of the calcium fluoride band (5291 Å). McNally et al. (463) describe a specially designed hollow cathode source applicable to spectrographic analysis for halogens and gases. The detection limit is 0.01 μg F and the greatest concentration sensitivity is 1 ppm. Chakraborty et al. (464) report emission spectroscopic estimation of fluorine in rock phosphates. To mitigate the effect of the volatilization difference between F and Ca and to promote the ready formation of calcium fluoride, they provide an abundance of calcium by mixing the sample with $CaCO_3$. To overcome the same problem of difference in the volatility of Ca and F during the excitation step, Schoenfeld (465) recommends the Stallwood jet for burning the samples. The spectrographic methods reported by Czakow (466) and by Ioffe and Leonova (467) are capable of determining fluoride down to 1 ppm; the latter investigators measure the BeF_2 band (3009 Å). Berezin et al. (468) have explored determination of fluorine in the vacuum region of the spectrum (F, 954.82 Å) by using a hollow cathode; the sensitivity is 10 ppm F. Karyakin et al. (469) have reported the use of pulsed vacuum excitation in spectrographic determination of halogens.

Bond and O'Donnell (470) have adapted atomic absorption spectrometry to the determination of fluorine over the 0.2–20 μg/ml range. Their technique is based on the depression by fluoride of the absorption of magnesium in the air-coal gas flame. Alternately the enhancement by fluoride of the zirconium absorption in the nitrous oxide-acetylene flame may be employed for fluorine determination. The latter method, although less sensitive (range 5–200 μg F/ml), is less susceptible to interferences than the former modality. In the presence of phosphate, which interferes with the zirconium method, the authors advocate measurement of enhancement of titanium absorption for fluorine analysis in the 40–400 μg/ml range. These atomic absorption methods are no doubt considerably more rapid than the spectrophotometric methods requiring isolation of fluoride. However the levels of fluoride measurable in this way are rather high, and nowadays smaller amounts of fluoride can be rapidly determined with the fluoride electrode without the need to remove the interfering ions.

It is not possible to determine low atomic number elements like fluorine by conventional X-ray spectrographic methods. Luke (471) has reported an indirect method for the determination of amounts as low as 1 μg F by an X-ray procedure; fluoride is precipitated as lanthanum fluoride in a 75% ethanol-water mixture and the heavy element La is measured.

Loch (472) has shown that the intensity of a desired fluorescene line can be enhanced or interference can be reduced by the use of specific gases in flow proportional counters. Use of argon and methane (9–1) has improved the detection limit of fluorine. Jecht (473) has reviewed the determination of fluorine by X-ray fluorescence analysis.

A halide meter (474) for determining halogens in gaseous mixtures is based on the analysis of the spectra produced by the spark from the device. It was believed that the meter operated on the Cu or Cu-halide excitation principle. It is now realized that it is the nitrogen spectrum that becomes more intense in the presence of the halide and is recorded by the photodetector. A stainless steel and platinum electrode system with an ultraviolet transmitting filter were found to enhance the sensitivity fourfold. Freon 12 and SF_6 have been measured with this meter (475).

Nuclear magnetic resonance (NMR) techniques have been employed mainly on structural and fundamental studies on fluorine in relation to such specific materials as fluorite (476), calcified tissues (477–479), and fluorapatite (480). The NMR line of F nuclei in mineral fluorapatite is a poorly resolved triplet. An infrared absorption method for detection and evaluation of gases, including HF, has been reported (481). Laser radiation absorption has been used to detect pollutants (e.g., HF) emitted from burning plastics (482). Electron probe microanalysis of fluoride has been reported by some investigators (483–485). X-Ray emission spectroscopy has been suggested for obtaining structural information with regard to carbon, nitrogen, oxygen, fluorine; and other elements (486).

V. ASPECTS RELEVANT TO FLUORINE ANALYSIS OF SPECIFIC MATERIALS

1. Soil

Fluorine in water, which is the main natural source of supply of fluorine to man, animals, and plants, is derived by the solvent action of water on the rocks and soils of the earth's crust. According to Fleischer (487), fluorine is seventeenth in the order of abundance of elements in the earth's crust (F 0.032%). There are two main problems associated with fluorine analysis of samples of soil, rock, and minerals. Most of the samples are not readily soluble in water or mineral acids. Moreover, they have a very diverse mineral composition, and some of the chemical constituents hinder quantitative separation of fluorine. These circumstances call for special procedures for sample preparation before the fluorine is isolated for analysis.

MacIntire et al. (488) have observed that the conventional practice of calcining soil samples at high temperatures with calcium peroxide results in poor fluorine recoveries. Direct distillation from sulfuric acid at 165 °C of unignited soil samples, followed by the usual redistillation from perchloric acid at 135 °C, is satisfactory for analysis of fluorine-enriched soil samples. Volatilization of fluorine during the calcining step is believed to be the cause of poor recoveries, rather than the conversion of fluorine into undecomposable refractory form in the calcines. For improved separation of fluorine from pyritic sulfur and organic matter, Reynolds and Hill (106) recommend distillation in the presence of excess permanganate.

Von Fellenberg (88) has suggested that values of acid-decomposable fluorine of the soil would be more interesting than acid-undecomposable fluorine, since it is to be expected that the former can be more easily taken up by the plants, when both types of fluorine are present in the soil. For this purpose, he proposes 1- to 2-hr extraction of soil with 1*N* perchloric acid at room temperature, with occasional shaking and analyzing the acid extract for fluorine following distillation. Venkateswarlu (489), however, has found that not all acid-decomposable fluorine is obtained in a single extraction. Fluorine is again and again obtained on successive extractions. It is therefore considered better to determine all acid-distillable fluorine in the soil. About 100 mg of finely ground soil sample is submitted to Willard-Winter distillation, and 20- to 50-ml collections of the distillate are analyzed for fluorine by the rapid thorium nitrate titration (198) while the distillation is in progress. The distillation is continued until two successive 50-ml collections of the distillate are free from fluorine. The distillation time is reduced by employing smaller samples, which are finely pulverized. Ke and Regier (490) have recently determined fluorine in soils with the fluoride electrode, following nonfusion distillation. Fluorine content of powdered rock samples, as determined by diffusion colorimetric methods, has been found to vary with the mesh size of the sieves through which the samples were passed; the higher the mesh, the higher the fluoride (491). It must be pointed out that none of these methods measures total fluorine content of the soil.

The first step in the analysis of total fluorine in soil and similar samples is alkaline fusion to achieve thorough decomposition of the sample and to facilitate complete solubilization of fluoride. The next step is usually removing aluminum and silicate ions, which interfere with the isolation of fluorine by distillation or diffusion. Shell and Craig (98) find that a single direct distillation of a sample after fusion with sodium carbonate indicates only the approximate quantity of fluorine in the soil. Quantitative results are obtained by again fusing the residue from the distillation flask, redistilling, and combining the results. Similar recommendations were made earlier by Von Fellenberg (88). Shell and Craig (98) employ sodium carbonate-zinc oxide fusion for better decomposition. They find no loss of fluoride during fusion of soil samples; on the other hand, they point out that incomplete fusion is a larger factor in fluoride "loss" than is volatilization. Fusion with zinc oxide, however, results in the formation of insoluble zinc salts that hinder the dissolution of soluble salts and possibly also retard thorough decomposition. These problems are much less significant with the low-temperature sodium peroxide fusion described by Chu and Schafer (492). Finely pulverized sample, mixed with 9 to 10 times the sample weight of sodium peroxide is fused in a platinum crucible at 500 to 540 °C for 1 hr in a muffle furnace. The melt is cooled and dissolved in water. The completeness of dissolution indicates the thoroughness of the decomposition. Silicate and aluminate ions are then precipitated with zinc sulfate solution; the filtrate and washings are analyzed for fluoride following distillation. Based on recovery studies, Hall (493) concludes that potassium hydroxide is better than sodium hydroxide for the fusion of soil samples.

The presence of both calcium and phosphate in the samples, as in rock phosphate, prevents the quantitative separation of fluoride following fusion and precipitation of silicate and aluminate ions with zinc salts (98). This is probably due to the formation of an insoluble calcium phosphate matrix, which could trap fluoride. Samples such as rock phosphate are best analyzed by direct distillation without recourse to fusion. Mineral samples that are readily decomposed by acid (e.g., cryolite) also do not require fusion. It is perhaps advantageous to employ a perchloric acid-phosphoric acid mixture for distilling all soil and mineral samples, whether directly or following fusion. Phosphoric acid has been shown by Grimaldi et al. (99) to facilitate quantitative separation of fluorine by distillation in the presence of fluoride-complexing cations naturally present in soils and rocks (e.g., aluminum). Seel et al. (494) resort to phosphoric acid distillation at 220 °C for separation of fluoride from zirconium. Rechenberg (495) has recently reported rapid fusion of samples with lithium tetraborate in a gold-platinum (1–19) crucible at 1000, 1100, or 1200 °C for 15, 10, or 5 min, respectively, before distillation. However it was noted earlier that high concentrations of borate in the sample could lead to low fluoride results following distillation (98), because of the formation of nonionized fluoborate, which also distills over. Quantitative results are obtained by fusing the evaporated distillate for 2 min with a Fischer burner at medium heat and completing the analysis. Low concentrations of borate in the distillation mixture do not seem to favor the formation of nonionized fluoborate.

Following fusion of the samples, fluorine may be separated from the fusion melts also by diffusion or ion exchange, after which it may be determined by a suitable method. Hall and Walsh (496) remove interfering ions from the extracts of fusion melts by digesting the extracts with ammonium carbonate and filtering; fluoride in the filtrate is determined spectrophotometrically with the lanthanum-alizarin complexan reagent, without recourse to further separation of fluoride. The fluoride in the fusion extracts has also been determined directly with the fluoride ion electrode, after adding an appropriate reagent (e.g., sodium citrate) to complex the interfering cations (497). Fluoride in phosphate rocks that are sufficiently soluble in mineral acid has been determined directly with the fluoride electrode, without fusion (498). Fluoride in phosphate rock has also been calculated on the basis of the difference in sodium hydroxide titration values of a nitric acid solution of the sample before and after precipitation of fluoride with 4 to 5 g of potassium chloride (499). Taneeva and Freze (500) propose decomposing phosphate rock by means of ion-exchange resins: 0.5 to 1.0 g of the sample is stirred for 30 min under water with a mixture of AN-1 (OH^- form), particle size more than 1.2 mm and KU-1 (H^+ form), particle size less than 1.2 mm. The dried resin mixture is passed through a 1.2-mm ssieve. Fluoride is eluted with 0.1*N* sodium hydroxide and analyzed.

Based on X-ray diffraction examination of fusion melts and chemical analysis, the following decomposition procedures have been recommended: Na_2CO_3 fusion and $H_2SO_4 - H_3PO_4$ distillation for aluminum fluoride, NaOH fusion and distillation for glass, $H_2SO_4 - H_3PO_4$ distillation for cryolite (no fusion), $HClO_4$ distillation for fluorite and sodium hexafluorosilicate (no fusion), and simple dissolution in dilute HCl for phosphate rock (501).

Newman (502) has determined fluorine in aluminosilicates, without preliminary fusion, by a pyrohydrolysis procedure. Recoveries from sodium fluoride and fluorspar (98.6 and 98.9%) by pyrophydrolysis are better than the 95% recoveries obtained by Willard-Winter distillation. Pyrohydrolytic determination of fluorine in rocks has also been reported more recently by Clements et al. (503) and Shiraishi et al. (504). Fluorine analyses of minerals have been carried out by radioactivation and spectrographic methods, as mentioned before.

2. Water

In public health programs involving fluoridation of communal water supplies, fluorine analysis of water is carried out to ensure optimal levels of fluorine in the drinking water. In studies on endemic fluorosis, the water analysis serves to correlate the manifestation of fluorosis with the degree of ingestion of fluoride from drinking water. In the context of the latter, the following points are worthy of attention. The fluorine content of natural waters is likely to vary with the season of the year, generally being low in the rainy season and high in summer, particularly in the case of well waters. It is therefore necessary to collect samples on a year-round basis, preferably every month. An assessment of fluoride ingestion from drinking water during the different seasons would be of value, particularly because larger amounts of water are consumed in summer months when the fluorine content of well waters in some areas is also high. Collection and storage of water samples in glass bottles, particularly of soft glass, could vitiate the results through the adsorption of fluoride on the glass surface or dissolution of traces of fluoride from the fresh glass surfaces. It is best to collect water samples in polyethylene bottles and to perform the analyses as soon as possible. It is inadvisable to filter a water sample through filter papers, since most varieties of analytical grade "ashless" filter papers are washed with hydrochloric acid and hydrofluoric acid solutions. If the water sample is turbid and some precipitation is suspected, the sample may be clarified by the addition of a few drops of dilute hydrochloric acid. If this fails, the water sample may be centrifuged and the supernatant and the sediment separately analyzed for fluoride. Particulate fluorides suspended in water require more careful attention than they have received so far, particularly in studies on endemic fluorosis.

Potable waters seldom contain significant amounts of interfering ions, and without isolating the fluoride, they may be directly analyzed for fluorine by several colorimetric methods (228,232,234,249,266,268,505). Potable water samples can also be analyzed rapidly (within 5 min) by direct titration with thorium nitrate solution after flooding the sample with sodium and chloride ions (200). An aluminum-morin-based fluorimetric method requires only 0.25 ml of the water sample (325). All these methods have different limits of tolerance for the various interfering ions, and if the interferences exceed these limits, the fluoride must be separated before analysis.

In some methods, such as the thorium-alizarin sulfonate colorimetric method of Talvitie (249), the presence of interfering phosphate and aluminum ions can be recognized; in others this may not be possible. Interference can be suspected when there is reasonable ground to believe that the water is highly mineralized, when results computed by analysis employing aliquots of different volumes of the water sample do not tally, or when the results do not conform to the clinical picture. Ideally each laboratory should develop its own "limit tests" for the interfering ions with respect to the colorimetric method adapted in the laboratory and rapidly screen the water samples for interferences before fluoride analysis.

Whereas distillation and diffusion procedures might be followed for separating fluorine, other simpler procedures have been suggested for removing or masking the effect of interfering ions present in water samples. Sulfate is removed by precipitation with barium chloride (322,506,507), with benzidine hydrochloride (508), and with ion-exchange resins (509). Most of the interfering substances in public waters (e.g., aluminum, carbonates, hydroxides, iron, manganese, phosphates, and sulfates) can be removed by precipitation with cadmium and mercury salt solutions at pH 8, using sufficient boric acid-sodium hydroxide buffer (510). Thioglycolic acid has been added to water samples to convert the interfering ferric Fe, manganic Mn, and free chlorine to noninterfering ferrous Fe, manganous Mn, and chloride ions (511). Thiosulfate addition and ultraviolet irradiation have been proposed to convert free chlorine to chloride and hydrogen peroxide addition for reduction of MnO_2 (512). Sodium thiosulfate has been shown to be effective in overcoming ozone interference in the fluoride analysis of ozonized potable waters (513).

Li and Huang (514) remove cations by H-type cation exchanger and determine fluorine with Th-dimethylphenol orange. Zolotavin and Kazakova (515) use a cation exchanger KU-2 to separate fluoride from iron and aluminum in natural and waste waters. Kelso et al. (516) absorb fluoride on the anion-exchange resin, Dowex 2X-8, elute with beryllium acetate, and analyze with the SPADNS reagent. If aluminum in the water sample is more than 0.5 ppm, they add ammonium-EDTA solution and sodium hydroxide to both the standards and samples before passing through the ion-exchange column. This procedure permits the analysis of 20 to 30 samples per day. Fujinuma et al. (517) have overcome the interference of as much as 250 μg Al^{3+} in the determination of 5 to 50 μg F by coprecipitation of Al^{3+} with added Fe^{3+}. Light et al. (518) achieve separation of fluoride from interferences by ion-exchange concentration; fluoride is eluted with O.1*N* potassium chloride and determined titrimetrically with thorium nitrate. The authors claim better precision by the ion-exchange concentration-titration method (S.D. 0.013 ppm) than by direct fluoride electrode measurement (S.D. 0.047 ppm). Palin (519) had described a colorimetric procedure using a prepared blank in which the fluoride is rendered totally "inactive" by adding excess aluminum and removing the unreacted excess of the added aluminum by an ion-exchange resin. Shapiro and Kolesnikova (520) report a method in which the fluoride ion is isolated from the water sample by adsorption on magnesium hydroxide and determined with iron-protocatechin-aldehyde and the thorium-alizarin methods. It is difficult to judge how well this method would succeed on water samples containing excessive amounts of interfering ions, which might also be partly adsorbed on the magnesium hydroxide. A magnesium oxide adsorption method for concentrating traces of fluoride from water containing excessive amounts of such interfering ions as phosphate has been described earlier by Venkateswarlu and Narayanarao (181); however the adsorbed fluoride is isolated by distillation before titration with thorium nitrate. Shiraishi et al. (521) remove aluminum by extraction as oxine into chloroform. Meyling and Meyling (522) describe a modified zirconium-alizarin method for direct determination of fluorine in waters. Alizarin is extracted with pentanol to eliminate the effect of variations in color from different samples. Interference due to phosphate is overcome by adding an equivalent amount of phosphate to the standards or by precipitating phosphate by adding ferric chloride and sodium acetate solutions and boiling the water samples.

Dirks and Cox (509) point out that removal of sulfate by ion exchange is required for fluoride determination with the zirconium-eriochromecyanin reagent, but not with the cerium-alizarin complexan reagent. Quentin and Rosopulo (523) isolate small amounts of fluoride from alkaline evaporation residues of water samples (1 liter containing less than 30 μg F) by rapid distillation with superheated steam in the presence of sulfuric acid (105). Hey and Jenkins (524) resort to diffusion for analysis of waste waters, river samples; and sewage effluents. Shaw (525) suggests simple chemical treatment for removal of turbidity and color by adsorption with calcium sulfate and decolorizing carbon before direct colorimetric analysis of fluoride in lysimeter leachings and soil extracts. For monitoring fluoride levels in public water supplies, Morrow and Henry (526) describe a continuous amperometric method featuring a cell equipped with a rotating electrode of 99.99% pure aluminum and an aluminum alloy counterelectrode (93% Al). The precision between 0.2 and 2 ppm F in the water is 2.2%. Interference by Cu(III) or by chlorine or oxygen is prevented by adding EDTA or sodium thiosulfate. Interferences due to phosphate, iron, and aluminum can be compensated for.

Thompson and Taylor (527) have determined fluoride in seawater by titration with zirconium nitrate-alizarin reagent. To overcome interferences due to Cl^- and SO_4^{2-}, equivalent amounts of these ions are added to the comparison standards. The same reaction has been used in the spectrophotometric analysis of seawater by Anselm and Robinson (528). Grasshoff (529) and Chan and Riley (530) employ lanthanum-alizarin complexan reagent of Greenhalgh et al. (309) for direct analysis of fluorine in seawater using the Technicon Autoanalyzer. Little or no interference by major ions in the seawater has been encountered. Kletsch and Richards (531) use the cerium-alizarin complexan reagent of Yamamura et al. (310) for analysis of seawater. Wilkniss and Linnenbom (424) have reported a proton activation analysis method; the limit of detection is 0.002 ppm F when a 5-g sample of seawater is used. The reproducibility of the results is less than that in the spectrophotometric methods.

The fluoride ion electrode of Frant and Ross (14) is extremely well suited for direct measurement of fluoride in natural waters over a very wide range of fluoride concentrations. Whereas in the spectrophotometric procedures the interfering ions may represent both the cations and the anions, in the fluoride electrode methods the interference is almost exclusively from the cations, as long as the measurements are carried out at the right pH, which should not be alkaline. Since the fluoride electrode responds only to the fluoride ion activity and not to the fraction of fluoride complexed by aluminum, iron, magnesium, and calcium ions present in natural waters, the fluoride values obtained with the fluoride electrode will be somewhat lower than those obtained with the distillation and spectrophotometric methods (532,533). However total fluorine concentration in waters can be determined with the fluoride electrode by adding to the water sample and the standards a total ionic strength adjustment buffer (TISAB) containing citrate ions, which can complex the metal ions, thereby releasing the fluoride from its metal complexes. McFarren et al. (534) report that the fluoride electrode procedure involving addition of citrate is simpler and gives somewhat better precision and accuracy than the spectrophotometric procedure applied to fluoride isolated by ion-exchange concentration and elution with beryllium acetate (516); acceptable results, however, are obtained with the latter procedure as well.

Ke and Regier (535) recommend preliminary extraction of interfering polyvalent cations as their 8-hydroxyquinoline complexes into a butyl cellosolve-chloroform layer, before measurements are made with the fluoride electrode. This method should be useful in the analysis of polluted waters containing high levels of calcium, magnesium, iron, and aluminum ions. According to Duff and Stuart (536), 25 ppm iron, zinc, nickel, cobalt, manganese, chromium, and copper is tolerated when fluoride measurements are carried out using citrate and triethanolamine buffer; aluminum, however, interferes seriously. Palmer (537) recommends the use of ammonium sulfosalicylate for masking aluminum. Bock and Strecker (448) have investigated the problem of interferences and masking them by appropriate measures.

The fluoride electrode technique has been employed by Dougherty and Johnson (538) and Takemoto et al. (539) for fluorine analysis of seawater. Anfalt and Jagner (386) have reported a standard addition titration method for determining fluoride in seawater with the fluoride electrode. They have developed equations for computer treatment of the titration data.

3. Atmospheric Samples

Samples of ambient air and stack effluents contain fluorides in different physical and chemical forms that vary in their chemical reactivity and in their potential toxicity to plants and animals. It is a useful practice to determine separately gaseous fluorides (HF, H_2SiF_6, SiF_4, etc,) and particulate fluorides (cryolite, CaF_2, Na_2SiF_6, NaF, FeF_3). The former, which can be readily absorbed from the atmosphere, are more toxic to plants.

Semiquantitative results are obtained by exposing to air filter paper disks or strips wetted with various reagents that react with atmospheric fluorides. These methods may be fairly indicative of levels of HF in working chambers or green-houses, but in actual field work they are subject to errors due to conditions of exposure, wind, weather, and even the levels of F concentration

(540). Methods involving aspiration of known volumes of air into fluorine-absorbing or -retaining devices give quantitative results.

There are both wet and dry sampling procedures for collecting atmospheric fluorides. Scrubbers and impingers have been designed to increase the rate of dissolution of gaseous pollutants. For absorbing fluorides, Mavrodineanu and Coe (541) employ a specially designed 500-ml flask, from which the fluoride is later distilled. The efficiency of absorption of atmospheric fluorides by gas washing bottles fitted with porous glass disks is reported to be increased to 10 times that of the impinger bottles by the addition of a few drops of butyl alcohol as a foam-forming agent (542). Bailey et al. (543) follow an aspiration rate of 15 liters/min in 100 ml of water. Two serially connected Richter absorbers containing 6 ml of distilled water are considered adequate for collecting HF from air by aspiration at the rate of 10 liters of air per minute (544). Marshall and Wood (545) absorb the fluoride (HF) by aspirating the ambient air directly into an acidic solution of zirconium reagent for fluoride and measure the decrease in absorbance due to fluoride. Since Al^{3+}, phosphate, and other ions interfere with this procedure, it is not recommended for testing industrial atmospheres polluted with fluorine dusts (cryolite, apatite, rock phosphate, etc.). West et al. (546) have employed a sulfuric acid prescrubber between the sample source and the absorption solution to achieve an effective dissociation and release of fluoride from its complexes with aluminum, iron, zinc, and cobalt. The fluorine analysis is completed with the lanthanum-alizarin complexan-fluorine reaction. Perhaps the use of phosphoric acid in the prescrubber would be even better. Chaikin et al. (547) pass the sample of air containing particulate F through a platinum tube at 1000 °C; fluorine, by reaction with the moisture normally present in the air, is converted to HF (pyrohydrolysis).

Nielsen and Dangerfield (548) employ an anion-exchange resin for the isolation of fluoride from interfering cations. In a procedure described by Paez and Vasquez (549), the air sample is passed through an ion-exchange resin and the absorbed fluoride is eluted with sodium chloride solution. The resin needs preconditioning to avoid irreversible retention of fluoride. Suvorova et al. (550) use silica gel wetted with 3% solution of triethanolamine in a U tube to absorb fluorides from atmospheric samples. The absorbed fluoride is determined by the thorium-thoron method.

As a means of separating gaseous fluorides from particulate fluorides, the air or stack gas sample is usually passed through a tube with a reactive surface such as sodium carbonate film, then through an appropriate filter to collect the particulate matter. A variety of filters have been employed. Pack et al. (551) have described the use of glass fiber filters, which give insignificantly lower results for gaseous fluorides and slightly higher values for particulate fluorides than the impingers. The filters with the particulate matter retained thereon are subjected to distillation. Fusion of the filter disk may be required when very nonreactive refractory minerals of dusts are collected. The average fluoride blank is 9.4 μg, and enough fluoride should be collected to offset the fluoride blank and obtain reliable results. Killick (552) reports that glass fiber filters have rather high fluoride content and silver membrane filters are less efficient. Elfers and Decker (553) recommend the use of 5-μcellulose acetate membrane filters impregnated with sodium formate for high-volume sample collections. These filters do not provide the high fluoride blank encountered with the use of glass fiber filters. According to Habel (554), who uses 1-μmembrane filter supported on a sintered quartz plate for separating particulate fluorides, significant losses of fluoride occur when the filter device is not heated. For separating gaseous and particulate fluorides, Mandl et al. (555) prefer the bicarbonate-coated tube method in conjunction with a membrane filter Acropore AN-800. Also, gases may be passed through a citric acid treated filter paper (Whatman No. 52) to collect particulate fluoride, then through sodium hydroxide treated filter paper (Whatman No. 4) for retaining gaseous fluorides. Atmospheric fluorides have been collected also on calcium formate filters and Millipore LSWP or AAWP filters, each followed by sodium carbonate impregnated filters (556).

The relatively large concentrations of the highly reactive HF in stack gases create some technical problems. Dorsey and Kemnitz (557) employ a heated glass probe to convert the HF in the stack gases to less reactive SiF_4 before the gases are filtered through paper impregnated with sodium carbonate.

An apparatus for the continuous monitoring of atmospheric gaseous fluorides, originally developed by Chaikin et al. (558) at the Stanford Research Institute, has been improved on by Ivie et al. (559) to permit 6 months of unattended, troublefree maintenance. A stream of aspirated gaseous sample is split into halves. One half is drawn over a film of sodium carbonate to absorb the fluoride, but not the other half. The two split streams impinge separately on a strip of filter paper sensitized with a fluorescing reagent (magnesium oxinate). The difference in the quenching of fluorescence by the split streams, which is monitored with a built-in photometer and continually recorded, is a measure of the HF concentration. Clayton (560) also uses a fluorimetric method, described by Feigl and Heisig (561), for measuring atmospheric fluorides in the 20–75 ppm range. Others to describe continuous monitoring techniques for atmospheric fluorides are MacLean et al. (562), Brown (563), Liberti and Mascini (564), and Thompson et al. (565).

The fluoride electrode is now being adapted for the determination of atmospheric fluorides after their absorption by appropriate procedures. The lowest measurable concentration is about 0.07 $\mu g\ F/m^3$, and F levels of 1 to $50 \mu g/m^3$ can be measured with an overall standard deviation of about 3.5%. Liberti and Mascini (566) absorb gaseous and total fluoride in ambient air and stack effluents in a cell containing citrate solution monitored with the fluoride electrode. The flow of gases (1 liter/min) and the absorbing citrate solution (0.3 ml/min) are adjusted to provide a fluoride concentration of 10^{-5} to $10^{-4} M$ in the cell. Fluoride concentration of 30 to 1000 $\mu g/m^3$ have been measured.

Sampling and analysis of gaseous and particulate emissions have been reviewed by Peters et al. (567).

4. Plant Materials

It is conventional even now to ash plant material with a fluoride fixative, distill the ash with acid, and determine fluoride in the distillate by a suitable procedure (87,568–572). Some investigators report calcium oxide to be a better fixative than magnesium oxide (572). This may be due to the greater solubility of calcium oxide in water, thus better availability of the calcium ions to fix fluorine. The problems of possible loss of fluorine during open ashing, despite the use of fixatives, and the poor recoveries of fluorine by distillation due to the formation of refractory compounds during ashing of plant materials rich in silica, have been discussed before. It would be ideal to resort to fusion of the ash and to collect an appropriately large volume of the distillate as suggested by Remmert and Parks (33) as well as by Rowley et al. (32). This practice has been preferred by several workers, including Hardin et al. (34), Gwirtsman et al. (574), Oelschläger (35), and Buck and Reusmann (575). The fusion step would be mandatory in the case of plant materials rich in silica. Yamamoto et al. (576) report better recoveries with sulfuric acid distillation than with perchloric acid distillation.

A blue-green product of fusion may indicate the presence of sodium permanganate, formed by the reaction of sodium hydroxide with manganese dioxide in the plant ash. During distillation with perchloric acid, the manganate ions undergo internal oxidation-reduction, forming MnO_2 and permanganate. The latter oxidizes chloride ions to free chlorine, which interferes with subsequent volumetric or spectrophotometric methods. Lockwood (577) suggests the addition of sodium nitrite to the distillate before fluoride analysis. However when the amount of permanganate formed is large (samples of some plant leaf ash have been found to contain as high as 0.3% MnO_2), Deutsch (578) considers the problem to be greater and recommends reduction of the permanganate in the distillation flask itself by the addition of hydrogen peroxide before the distillation step.

By the foregoing fusion procedures, although fluoride is released from refractory products of ashing, very gelatinous silica that retards the distillation of fluoride is also produced, thereby necessitating prolonged distillation. An analyst involved in the routine analysis of any single type of plant material should try distillation of fluoride with and without fusion. If the results are the same, the fusion step may be omitted. Perhaps the prolonged distillation could be avoided by prior removal of silica (and aluminum), as in the case of analysis of rock samples (98,492). Application of the reverse extraction technique to the isolation of fluorine from plant ash, with and without fusion, would seem to merit investigation.

The oxygen flask technique has been employed for combustion of plant materials preparatory to fluoride analysis. Thomas and Amtower (579) submit the evaporated washings and ash following combustion to diffusion and colorimetric analysis with SPADNS reagent. Levaggi et al. (50) and Peters and Baxter (51) employ the fluoride electrode following combustion of plant materials in the oxygen flask. Van Gogh (147) and Kakabadse et al. (30) determine fluorine in plant materials following destruction of the organic matter by pyrohydrolysis.

Weinstein et al. (580) have reported their trials on bypassing the time-consuming ashing and fusion steps by distilling fluoride directly from ground plant tissue by the semiautomated Technicon Autoanalyzer method of Weinstein et al. (132), in which the lanthanum-alizarin complexan reagent is employed in the final step. The authors find the results promising. Results employing 200 to 250 mg of tissue compare favorably with those obtained by ashing and fusion. The results are low when larger sample sizes of tomato, orchard grass, and gladiolus were employed. The present writer is a little wary about this application of spectrophotometric procedures to the determination of fluoride in diffusates or distillates of unashed samples. This situation is similar to the report by Singer and Armstrong (183) that their diffusion colorimetric method applied to *unashed* sera gave fluoride values similiar to those obtained with their microdistillation colorimetric method applied to *ashed* serum samples. It is now realized that such agreement was attributable entirely to an unfortunate coincidence of several experimental circumstances. Fallacies in this approach have been recently published by Venkateswarlu (171). Raman and Venkateswarlu (184) have encountered abnormally high fluorine values when unashed plant materials were analyzed with the Singer-Armstrong (183) diffusion colorimetric method. The diffusates on ashing and rediffusion have not revealed more than a small fraction of the fluorine obtained before with the same colorimetric reagent. Obviously some volatile component of the unashed plant material that diffuses over, other than the fluoride, behaves like fluoride toward the zirconium-eriochromecyanin R reagent. However the chances for such contamination of the distillate in the Weinstein procedure might be considerably smaller because of the very short distillation-diffusion time ($<$ 5 min, compared to the 22-hr diffusion at 60 °C in the former procedure). Furthermore, the lanthanum-alizarin complexan reagent employed in the Weinstein technique has a considerably greater tolerance toward interfering substances than the zirconium-eriochromecyanin R reagent, particularly from products of diffusion from unashed biological materials. To establish the soundness of the method proposed by Weinstein et al., however, one should be able to obtain similar results with the colorimetric reagent and with the fluoride electrode after separation by their technique of fluoride from unashed plant materials. The diffusion colorimetric methods may seem to be satisfactory for fluoride analysis of plant materials in certain situations [e.g., when the samples are naturally rich in fluorine, as in tea, or contaminated with atmospheric fluorides (581)]; the high levels of fluorine mask the interferences. To prevent incomplete disintegration of powdered plant material in perchloric acid due to aggregation of the powder, Hall (493) recommends suspending the finely ground sample in 0.1% agar solution before submitting it to diffusion.

Timber (telephone and power transmission poles) is treated with certain formulations containing fluorides. To determine the level of fluoride attained in timber, the fluoride is leached from sections of the wood with a sodium hydroxide solution and after neutralization is determined directly with the lanthanum-alizarin complexan reagent without any further separation of fluoride (582). The method succeeds in this case because of the overwhelmingly large amounts of fluoride involved ($> 0.005\%$, i.e., > 50 ppm).

Although the fluoride electrode has been used to determine fluoride in the diffusates or distillates of plant materials, it is now being advantageously employed for determining fluoride directly in solutions of plant ash and even in extracts of unashed plant materials. Reusmann and Westphalen (583) ash the plant material, fuse the ash with sodium hydroxide, dissolve the product in water, neutralize with acid, add citrate buffer, and measure the fluoride with the fluoride ion electrode. Based on this procedure, Buck and Reusmann (584) have described a semiautomatic method that permits 60 analyses per day by two analysts. Jacobson and Heller (585) carry out sequential extractions of dried, pulverized plant materials with $0.05N$ sulfuric acid, then with $0.05N$ ammonium hydroxide. After neutralization and addition of acetate buffer and sodium citrate solutions, the combined extracts are read with the fluoride electrode. In a recent method reported by Galloway et al. (586), fluoride in plant materials is extracted by digestion with $2N$ potassium hydroxide. Fluoride in the extract is determined with the fluoride electrode by the standard addition technique after acidification of the extract with phosphoric acid to prevent interferences from fluoride-complexing cations in the plant digests. Baker (587) uses fluoride electrode to determine fluoride in an acetate buffer solution of the product of fusion of plant materials; no citrate is added. Hukushima et al. (588) prescribe two successive fusions, first with sodium carbonate and later with sodium hydroxide, for fluorine analysis of plant materials. Jacobson and McCune (589) have described a semiautomatic technique involving alkaline fusion, acidification and distillation at 170 °C, and lanthanum-alizarin complexan spectrophotometric finish.

The results obtained by the extraction-fluoride electrode procedures not involving ashing have been found to be similar to those yielded by conventional procedures that entail ashing, fusion, and distillation or diffusion followed by analysis with spectrophotometric methods or with the fluoride electrode. The fluorine in the plant materials investigated in this research is essentially inorganic or may exist as organic compounds that are extremely labile under the conditions of extractions. The latter is, perhaps, unlikely. A higher fluorine value by ashing procedures compared to the values obtained by procedures not involving ashing might be indicative of the presence of organic fluorine in the plant material and would require further qualitative and quantitative studies. Hall (493) points out that in the differentiation of organic from inorganic fluorine, the effects of silica in plant tissue and soils may confuse the results. Silica in roots, stems, and leaves of many plants is reported to be highly variable. Blaeberry (*Vaccinium myrtillus*) contains 0.09% SiO_2 and white bent (*Nardus stricta*) contains 6.64% SiO_2 dry weight (590). Methods for identification and determination of fluoroorganic compounds in plants have been described by Wade et al. (591), Lovelace et al. (592), Cheng et al. (593), Peters and Shorthouse (594,595), and Weinstein et al. (596). Peters and Shorthouse (597,598) have also reported some observations on elusive organic fluorine compounds in plants.

5. Animal Tissues

A. HARD TISSUES

Hard tissues acquire considerable amounts of fluoride during the life of the animal and contain more fluorine than any other tissues or body fluids. This makes fluorine analysis of hard tissues so much the easier.

Fluorine in hard tissues has been determined by volumetric and spectrophotometric methods following isolation by distillation or diffusion of ashed samples. No addition of fixatives is required during ashing. Tooth enamel, which contains very little organic matter, can be analyzed by the isolation techniques already described, without the need for ashing.

The fluoride ion electrode technique has been employed in fluorine analysis of hard tissues following isolation of fluoride by diffusion or distillation (599). Singer and Armstrong (600) have described a highly simplified technique for determining fluoride in bone ash without having to isolate fluoride. First 5 mg of bone ash is dissolved in 1 ml of 0.25*M* hydrochloric acid; the solution is neutralized and buffered at pH 4.7, the volume is built to 5 ml, and the fluoride concentration is determined with the fluoride ion electrode. Results obtained by this technique were found to agree very well with those from a diffusion colorimetric method (183) over a wide range of fluoride concentrations in bone ash (0.036–0.635%). Concentrations of calcium and magnesium considerably in excess of those in bone tend to produce negative errors, presumably because of the complexation of fluoride.

In such situations, the authors suggest using smaller ash samples if the fluoride content is sufficiently high or, alternatively, preparing fluoride standard solutions containing calcium and magnesium at levels corresponding to those encountered in the test samples.

These problems have been more successfully handled by McCann (601) by using citrate buffer to complex calcium and magnesium ions, instead of the acetate buffer used by Singer and Armstrong. After 1 to 20 mg enamel is dissolved in 1 ml of 0.5*M* perchloric acid, 4 ml of 0.5*M* trisodium citrate is added, and the fluoride is measured with an ion-specific electrode. This system overcomes the interference of even aluminum. More than 50 determinations may be made using two or three electrodes and electrode switch. Close agreement has been found between the results of analysis of enamel, dentin, bone, and apatite by this new procedure and by a diffusion colorimetric method. The technique has been employed by Brudevold et al. (602) and also, with slight modifications, by Larsen et al. (603) to determine fluoride in enamel biopsies (0.2 mg). Levine (604) has calculated amounts as low as 0.2 ng F in enamel biopsies; the solution is placed in a specially designed microcell and the sensitivity of the fluoride electrode is sharpened by placing the electrode for a specified time in fluoride-free buffered solution, permitting the potential to reach 230 mV. Zober and Schellmann (605) have determined fluoride in bone ash using the total ionic strength adjustment buffer (TISAB).

Munksgaard and Brunn (606) have determined fluoride in enamel biopsies by gas chromatography: 4 to 6 μl of perchloric acid is placed on an enamel surface with a modified Hamilton syringe and aspirated back. Six biopsies can be etched from the same area with a depth of 5 μ.

A spectrophotometric procedure not involving diffusion or distillation for the determination of bone fluoride has also been reported by Ruzicka et al. (607). The bone ash is dissolved in perchloric acid; phosphate, which interferes with the colorimetric reagents, is removed by precipitation with silver perchlorate, and the filtrate is analyzed for fluoride with zirconium-xylenol orange reagent. The authors report a 2% loss of 25 to 50 μg F added to 0.02 g of bone ash and 0.1 g of ash of crowns of bovine incisors, 0.1 g of calcium phosphate, and 0.1 g of 80% phosphoric acid, because of adsorption of fluoride on silver phosphate precipitate. Perhaps the simplest of all the procedures published thus far for the determination of fluorine in bone ash, this approach should be very practical if its reliability is confirmed by others. The method may be applicable to most bone samples containing normal levels of fluorine, but we do not know how much the loss of fluoride due to adsorption on silver phosphate would influence the results of analysis of low-fluoride bones. A 2% loss of 50 μg amounts to a loss of an absolute amount of 1 μg F. This problem is not encountered in the reverse extraction procedure; less than 1 μg F in bone ash (F 0.003%) of rats raised on a special fluoride-low diet have been accurately determined with a spectrophotometric finish (15).

It is a common practice to ash bone samples before fluorine analysis, to express the fluorine concentrations as parts per million of bone ash (mineral matter). Though satisfactory for analysis of bones with normal and higher fluoride content, such a practice may introduce a significant positive bias in the analysis of low-fluorine bones of experimental animals raised on special low-fluorine diets. When 10-mg samples of low-fluoride calcium phosphate (608) are ashed in a muffle furnace, the samples have been found to acquire significant amounts of fluorine from the furnace and the environment: 0.01, 0.02, 0,04, 0.10, and 0.20 μg F with 2, 4, 6, 8, and 20 hr of ashing at 550 °C, respectively. Expressed in concentrations, the final values would be 1, 2, 4, 10, and 20 ppm F for the corresponding periods of ashing, compared to 0.02 ppm F in the unashed sample. Ashed left-halves of calvaria of mice raised on a low-fluoride diet have been found to contain 2.5 to 3.8 times as much fluoride as the unashed right-halves (1.95 to 2.5 ppm F) (609). It is likely that the calcium phosphate in the ash, like the fluoride-low calcium phosphate preparation, has acquired extraneous fluoride during ashing in the furnace. However further investigations are required to rule out totally the possibility that the higher results following ashing of bone samples are due only to contamination during ashing, not to the presence of organic fluorine compounds in the bone samples. Peters et al. (610) have reported a stray finding of organic fluorine in high-fluoride cattle bone; this observation remains to be confirmed.

For determining fluoride in unashed bone samples (e.g., whole tibia or humerus of rats), the reverse extraction technique (15) has been found most appropriate. It facilitates rapid solution of the mineral matter of the bone because of the strong acid medium, yet it prevents the loss of fluoride as volatile HF by prompt conversion of the HF to nonvolatile diphenyldifluorosilane. The preliminary extraction is carried out for 4 hr in 50% perchloric acid to ensure total disintegration of the bone tissue, which is necessary for quantitative results. The fluorine content may be expressed on the basis of bone ash (determined on the other right or left bone) or in terms of the calcium content, which can be found by analysis of the aqueous phase following extraction of fluoride. If it is proposed to analyze calcium by atomic absorption, it may be safer to use hydrochloric acid instead of perchloric acid during the preliminary extraction of bone fluoride. Fluoride in unashed bone has been determined following distillation (599) and diffusion (611) techniques. Fluoride in unashed enamel has been determined by an enzymatic method by McGaughey and Stowell (444), by a diffusion colorimetric method by Singer and Armstrong (182), with the fluoride electrode by McCann (601).

Fluorine in bone has been ascertained by neutron activation analysis. The limit of detection is 1 μg F by instrumental analysis and 0.1 μg F following radiochemical separation (612,613). Electron probe studies on fluorine in bone tissue (614) and in sound, carious, and developing teeth have been conducted (615).

B. SOFT TISSUES

Compared to the hard tissues, the soft tissues contain significantly less fluorine, calling for a more cautious approach in their analysis. Aorta and placenta have more fluoride than other soft tissues (616–618), presumably because they develop zones of calcification in which circulating fluoride can be trapped (619).

The common practice has been to ash the soft tissues in the presence of a fixative and separate fluoride by distillation or diffusion, completing the analysis by a suitable procedure involving titrimetry, spectrophotometry, or more recently the fluoride electrode.

Singer and Armstrong (183) have described a diffusion colorimetric method for the determination of fluorine in unashed soft tissues such as muscle and liver. When undried samples were employed, highly variable results were obtained and the diffusates were turbid. When dried powders were employed for analysis with the foregoing diffusion-colorimetric method, however, the authors obtained results giving excellent agreement with those obtained with a microdistillation procedure employing ashed samples: 0.45 and 0.47 ppm F for rat liver, 0.22 and 0.22 ppm F for beef liver, 0.14 and 0.14 ppm F for beef muscle, respectively. Similar agreement between the results was found in the case of serum samples as well. Yet, strangely enough, the method has been shown to yield some fallacious results and to display inadequacies with regard to fluorine analysis of serum (171). Therefore the results of analysis of unashed, dried soft tissues obtained with the same method must be accepted with caution. It would be safer to determine the fluoride content of the diffusates with the fluoride electrode than with the colorimetric reagents. Osis et al. (620) have employed the Singer-Armstrong diffusion colorimetric procedure for fluorine analysis of dietary and fecal samples; the results are similar with ashed and unashed samples.

The reverse extraction technique seems to be promising for isolating and concentrating fluoride from soft tissues. The recoveries of radioactive fluoride (^{18}F) from homogenates of rat kidney, muscle, and liver in one study have been nearly quantitative (95.5, 98.5, and 80%, respectively). The method has been employed for determining subnanomole amounts of fluoride in the rat tail tendon and in microtome sections of rat epiphyseal plates (15). Further work is required to develop the precise methodology applicable to all soft tissues. Some problems arising from the presence of extractable lipids need to be satisfactorily resolved.

It is to be expected that the effects of fluoride are a consequence of intracellular levels of fluoride rather than of the extracellular levels, therefore measurement of intracellular fluoride concentrations would be more meaningful than measurement of the whole tissue fluoride concentrations by the methods described thus far. Drescher and Suttie (621) and Quissell and Suttie (622) have been able to calculate the intracellular levels of fluoride levels in cultured mammalian cells exposed to different concentrations of fluoride labeled with ^{18}F. Following selected incubation periods, the cells are separated by centrifugation, resuspended in a small portion of the same supernatant to which ^{14}C-inulin is added (inulin does not enter the mammalian cells), and again separated by centrifugation. Based on measurements of ^{18}F and ^{14}C in the cell pellet and the supernatant, on the moisture content of the pellet, and on the investigator's knowledge of the specific activity of fluoride added to the medium, the amount of intracellular water and the fluoride concentration are calculated.

C. BODY FLUIDS

a. Urine. Kidneys play an important role in the homeostasis of fluoride by promoting urinary excretion of significant amounts of circulating fluoride, which may have been elevated following ingestion of fluoride. As a result, the fluoride level is higher in urine than in the other body fluids. Determination of fluoride in urine is more reliable and simpler than in the case of other body fluids.

The conventional practice is to ash the urine samples before the isolation of fluoride by distillation. This practice is followed by some investigators even now, but direct distillation of unashed urine has been shown to be adequate (623). Direct diffusion colorimetric methods employing unashed urine have also been described and found satisfactory by several workers (183,623–628). Fluoride in urine has been determined spectrophotometrically following separation on ion-exchange resins (629,630). More recently, the fluoride ion electrode is being increasingly employed for the direct determination of fluorine in urine, after adjusting the pH and the ionic strength between the urine samples and the standards. Results obtained by these procedures are in good agreement with those obtained by the conventional ashing, distillation/diffusion techniques.

Cernik et al. (631) have reported that not all fluorine in urine (collected from workers in a factory manufacturing hydrofluoric acid) responded readily to measurement with the fluoride ion electrode. More fluoride was found following diffusion of urine samples with acid and analysis of the diffusates spectrophotometrically or with the fluoride electrode. Similar results were obtained by direct measurement with the fluoride electrode, provided the urine samples were digested for a few minutes with 0.5M perchloric acid and the pH adjusted by the addition by 15% sodium acetate. The authors recommend such predigestion with perchloric acid in routine analysis of urine samples with the fluoride electrode. Tusl (632) was unable to confirm the need for such perchloric acid digestion of urine samples for direct fluoride determination with the electrode. In the course of analysis of urine samples from workers in the aluminum industry, Havaaldsen and Ostby (633) have obtained slightly lower results with the fluoride electrode than with the diffusion method. EDTA-treated urine samples gave better results with the fluoride electrode. Neefus et al. (634) have reported good agreement in results (0.3 to 13.7 ppm F) when urine samples from workers in aluminum, explosives and hydrofluroic acid factories were analyzed by a distillation colorimetric method and by the fluoride electrode using TISAB.

For determining total fluoride in urine following administration of flufenamic acid to man and animals, Dell and Fielder (635) carry out fusion of urine samples with sodium peroxide before distillation.

b. Blood Serum, Saliva, and Milk. Circulating fluoride in the body is partly excreted by way of urine by the kidneys and to a large extent is rapidly sequestered by the skeleton; thus the levels of fluoride in serum and other body fluids are very low. Determination of blood fluoride has been one of the most difficult, challenging, and controversial problems in the fluorine analysis of biological materials.

One of the earliest methods, which gave results somewhat comparable to those obtained by the more recent accepted methods, is due to Smith and Gardner (636). The authors encountered partial loss of fluorine during open ashing of whole blood, presumably as volatile iron fluoride. To avoid such a loss they carried out a preliminary distillation of blood with sulfuric acid, ashed the distillate, redistilled from perchloric acid, and determined the fluoride in the distillate by thorium nitrate titration. The method is rather cumbersome, involving distillation of as much as 50 ml of blood and requiring correction factors for the loss of fluorine during the double-distillation procedure.

The first successful method for the determination of fluorine in blood is the microdistillation procedure developed by Singer and Armstrong (28). Also more practical, it involves ashing 5 ml of serum or plasma (rather than whole blood) with 50 mg of fluoride-low magnesium oxide (637), followed by microdistillation and spectrophotometric determination. The use of plasma instead of whole blood for fluoride analysis is preferred because of the unequal distribution of fluoride between plasma and the erythrocytes on volume basis (638), as well as the possible variations in hematocrit values for different blood samples. Furthermore, plasma is a representative extracellular fluid with constant electrolyte composition. Variations in plasma fluoride concentration would reflect physiological and biochemical phenomena related to fluoride metabolism and transport. Singer and Armstrong report a normal plasma fluoride concentration of about 0.1 to 0.2 ppm. Hall (315), Takaesu (639), and Sita and Venkateswarlu (640) have employed diffusion instead of distillation for isolation of fluoride and have obtained more or less similar values for normal animal blood plasma.

Singer and Armstrong (183) have described a 22-hr, 60 °C, perchloric acid diffusion colorimetric method for the determination of fluorine in *unashed* plasma. The results by this new procedure were similar to those obtained with their previous microdistillation procedure employing *ashed* plasma (0.13 and 0.15 ppm F in bovine plasma by the new and the previous methods, respectively).

However Taves (641) has reported that the results of normal plasma fluoride concentrations obtained by the Singer-Armstrong diffusion procedure are about 10 times higher than the true concentrations ($0.7 \mu M$ F), which he believes are obtained with his fluorimetric method (190) following a 6-hr hydrochloric acid, room-temperature diffusion of fluoride also from *unashed* plasma. This report was followed by publication of several papers that nevertheless failed to resolve the uncertainties (7,29,642,643).

Taves (7) reported the existence of two fractions of fluorine in normal human serum, (*a*) the "exchangeable" fluorine, which readily exchanges with ^{18}F and is available for diffusion, and (*b*) the "nonexchangeable" fluorine, which is not available for diffusion until after ashing. The nonexchangeable fluorine concentration is obtained by subtracting the exchangeable fluoride value (diffusion of unashed serum) from that of the total fluorine (diffusion of ashed serum). Singer and Armstrong (29) believed that their diffusion procedure, which is more drastic than the Taves diffusion procedure, measures both the exchangeable and nonexchangeable fluorides together and therefore reflects total fluorine rather than just the exchangeable fluoride, which is why their method gives higher values than the Taves method. Singer and Armstrong (189) reaffirmed this belief more recently in connection with analysis of fluorine in serum ultrafiltrates.

Based on analyses of human and bovine sera, ultrafiltrates, and the serum protein concentrates by as many as 10 different approaches (Table II), however, it has been demonstrated only very recently by Venkateswarlu (171) that the higher values obtained by spectrophotometric determination of fluoride of unashed body fluids, believed by Singer and Armstrong to be total fluorine values, are in fact a consequence of lack of specificity of the colorimetric reagents for fluoride determination under the experimental conditions. The higher values so obtained reflect ionic fluoride plus interfering substances, the latter masquerading as nonionic fluorine. Furthermore, fluorine in representative fluoroorganic compounds and in nonionic fluorine concentrates prepared from human sera is not readily available for diffusion.

TABLE II

Fluoride Analysis of Bovine and Human Sera, Ultrafiltrates, and Protein concentrates by 10 Procedures[a]
(Mean fluoride Concentration, μM F)

	Total fluorine following combustion of the sample (number of analyses)		Analysis of pooled diffusates ($HClO_4$, 60°C, 22 hr) of unashed samples by five selected methods (number of analyses)					Ionic (plus ionizable) fluoride (number of analyses)		
Biological sample[b]	Oxygen bomb, reverse extraction, fluoride electrode (A)	Open ashing reverse extraction, fluoride electrode (B)	Zirconium eriochrome cyanin R (pH 1.4) (C)	Cerium alizarin complexan (pH 4.3) (D)	Fluoride electrode direct (pH 5.0) (E)	Reverse extraction ($HClO_4$ 20%), fluoride electrode (F)	Oxygen bomb, reverse extraction, fluoride electrode (G)	Calcium phosphate adsorption, fluoride electrode (H)	Reverse extraction, fluoride electrode (J)	Hanging drop fluoride electrode (K)
BS-1	4.0[c] (5)	6.1[d] (3)	7.9 (7)	5.8 (3)	0.8 (1)	0.8 (3)	1.0 (2)	0.7 (7)	0.6 (4)	
BS-1-UF			5.1 (3)							0.6[e] (1)
BS-2	2.3 (2)	1.6 (2)	18.9 (4)	10.5 (4)	1.9 (4)	2.7 (4)	1.9 (2)		1.1 (4)	1.2 (4)
BS-2-UF	1.5 (2)	—	7.4 (4)	2.6 (4)	1.7 (4)	1.9 (2)	1.9 (2)		1.4 (4)	1.7 (2)
BS-2-PC	3.6 (2)									
HS-1	4.1[c] (4)	3.9[d] (4)	13.2 (4)	2.3 (4)	1.7 (2)	2.5 (3)	—	1.3 (4)	1.3 (3)	
HS-1-UF	1.1[c] (2)		5.2 (7)							1.5 (2)
HS-2	1.8 (2)	1.2 (3)	9.0 (4)	10.0 (4)	1.6 (2)	1.6 (2)	0.9 (2)		0.8 (4)	0.7 (4)
HS-2-UF	0.6 (2)		3.7 (4)	7.4 (4)	1.8 (2)	1.8 (3)	1.3 (2)		0.7 (4)	0.8 (3)
HS-2-PC	2.8 (2)									

[a] Standard errors omitted to avoid overcrowding of data.

[b] BS–bovine serum; HS–human serum; UF–ultrafiltrate; PC–protein concentrate left in the filtration cone after ultrafiltration of serum.

[c] No reverse extraction, fluoride in the bomb washings read directly with fluoride electrode.

[d] No reverse extraction, fluoride determined spectrophotometrically after diffusion.

[e] Regular fluoride electrode, macro procedure (Singer and Armstrong, *Biochem. Med.*, **8,** 415, 1973).

The continued use of the diffusion colorimetric procedures for the determination of total fluorine and nonionic fluorine employing unashed body fluids would yield more misleading information, and this practice should be discontinued. In the light of these findings, the concept developed by Singer and Armstrong (189) that a large part of the fluorine in sera is ultrafilterable and bound, but not to protein of molecular weights of 25,000 or greater nor to metallic cations, does not appear to be valid.

Sereda (644) has reported using a highly basic resin (AB-17X6) for separating fluoride in urine, blood, and cerebrospinal fluid, preparatory to the determination of the fluoride by thorium nitrate titration. The blood of normal subjects has been reported to contain 0.035 to 0.14 mg%—that is, 0.35 to 1.4 ppm or 18.4 to 73.7μM F, a value that is enormously high compared to 0.5 to 3.0μM ionic fluoride values obtained by recent accepted methods. Undoubtedly several interfering substances are present in the eluates from the resin column, and they behave like fluoride during the thorium nitrate titration. The values obtained by Sereda are even higher than those obtained with the Singer-Armstrong diffusion colorimetric method (183). Cox and Dirks (645) resort to ion-exchange separation of fluoride from serum before diffusion and final photometric measurement with cerium-alizarin complexan. Without the ion-exchange step they obtain slightly higher results; they conclude that in this case, besides HF, other acids (e.g., acetic) diffuse over and vitiate the colorimetric analysis. Paez and Gil (646) determine serum and urine fluoride by a procedure combining Taves diffusion at room temperature (56) and the Singer-Armstrong zirconium-eriochromecyanin R colorimetric procedure for fluoride. Although the merit of this combined procedure may be the prevention of diffusion of interfering substances from unashed body fluids, the colorimetric procedure that is incorporated is not sensitive enough for reliable measurement of fluoride in the diffusates from *normal* serum samples.

While exploring the application of the calcium phosphate adsorption technique of Venkateswarlu and Sita (16) to the determination of fluorine in serum by a method not involving ashing or diffusion, Venkateswarlu et al. (8) encountered two fractions of fluorine in serum: (*a*) one fraction exchangeable with ^{18}F, adsorbable onto calcium phosphate (calcium hydroxyapatite) presumably by the replacement of the surface hydroxyl groups by the fluoride ions, therefore considered "ionic," and (*b*) the other nonexchangeable, not available for adsorption on calcium phosphate, therefore considered to be "nonionic." The nonionic fluorine fraction that remains in the supernatant after adsorption of serum with calcium phosphate, is revealed after ashing and is represented by organic compound(s) containing covalently bound fluorine.

It is now believed that the ionic and nonionic fluorine fractions in serum are similar to the exchangeable and nonexchangeable fluorine fractions described by Taves. However there is a subtle difference between "ionic" and "exchangeable" fluorine fractions. Whereas "ionic" fluoride is isolated by adsorption under neutral pH conditions, "exchangeable" fluorine is isolated by diffusion under highly acidic conditions. Therefore the exchangeable fraction would include ionic fluoride and any additional fluoride released from acid-labile, nonionic fluorine fractions, if present in the sample. Such compounds have not been found in normal sera according to Fry and Taves (63) and Venkateswarlu (15), but they have been reported for the first time by Taves et al. (647) to be present in the plasma of patients receiving methoxyflurane anesthesia. Theoretically the fluoride value obtained by the calcium phosphate adsorption technique should reflect ionic plus ionizable fluoride; because of the insignificantly small fraction of ionizable fluoride in serum, however, it reflects serum ionic fluoride for all practical purposes.

The reverse extraction technique (15), like the diffusion technique, measures both the ionic fluoride and the acid-labile fluorine in the sample. In the case of normal serum samples, both procedures measure ionic fluoride.

Since physiological processes are dependent on enzyme activity, which could be influenced by ionic fluoride rather than by nonionic fluorine, measurement of ionic fluoride in body fluids assumes considerable importance. Progress in this area has been considerably facilitated by the availability of the fluoride ion electrode.

Gron et al. (648) are among the first to explore the use of the fluoride ion electrode for direct measurement of ionic fluoride in body fluids like saliva. The pH of 5 ml of freshly collected saliva is adjusted to 4.7 to 4.8 by dropwise addition of $1M$ acetic acid. The fluoride concentration is measured directly with the fluoride electrode and calculated from an appropriate calibration curve obtained with saline standards of comparable ionic strength. According to Ericsson (649), there has been no difference in the results when fluoride in saliva is determined with the fluoride electrode and with a diffusion colorimetric method (cerium-alizarin complexan).

Barnes and Runcie (650) have used the fluoride electrode for measurement of serum fluoride, diluting 10 ml of plasma approximately twofold in the process of precipitation of plasma proteins and addition of buffer. As a result, the fluoride concentration of the final solution may fall below $10^{-6}M$, which is near the limit of sensitivity of the fluoride electrode. Fry and Taves (63) avoid such excessive dilution of fluoride. They measure fluoride in 1 ml of serum directly with the fluoride electrode after adding 2 drops of sodium acetate buffer (pH 4.8). They prepare the fluoride standards in a low-fluoride composite dog serum from animals that had been maintained on a mineral and vitamin-supplemented milk diet and distilled water. The results of fluoride analysis of human sera obtained with this procedure agree well with those obtained by the diffusion fluorimetric method (over the range 2–4μM F) and by Gran's plot calculation (over the range 4.5–13.3μM F).

Hall et al. (387) have described a method for the determination of fluorine in serum and urine with the fluoride ion electrode by the single-addition-of-known technique. This is similar to the technique based on Gran's plot calculation, and its success depends on a strictly Nernstian response of the electrode over the entire range of measurements. The authors recommend several precautions in measuring low levels of fluoride as in serum. They draw attention to the day-to-day variations in the background potential with different lots of buffer and with different electrodes, and to the need for weeding out electrodes that do not exhibit a strictly Nernstian response. Recoveries of added fluoride resulting in a concentration increment of 10μM F over the base level concentrations of 3 to 5μM F in samples of rabbit plasma are quantitative.

Ericsson (649) has used a diffusion colorimetric method to determine fluorine in milk samples following ashing. Dirks et al. (651) have reported two fractions of fluorine in milk: one responds to the fluoride electrode, and the other is revealed by gas liquid chromatography, which involves acidification of milk samples. Simpson and Tuba (652) have described a rather elaborate procedure for determining fluoride in milk. Starting with 10 ml of milk, the authors carry out three separate centrifugations to remove fat, casein, albumin, and globulin and also introduce a 1–10 dilution with water, resulting in a considerable decrease in the final fluoride concentration, which may be even below the sensitivity of the electrode itself. It is now possible to determine ionic fluoride directly in a few microliters of milk, as described in the following paragraph.

The method of Fry and Taves (63) has been adapted by Venkateswarlu (608) for simple direct determination of ionic fluoride in 5-μl samples of body fluids (serum, saliva, milk, etc.) with the hanging drop fluoride electrode. The low-fluoride serum required for making the fluoride standards is obtained by adsorbing the native ionic fluoride of bovine serum with calcium phosphate. Low-fluoride saliva and milk for making the saliva and milk standards are also prepared similarly. This approach is simpler and less expensive than raising experimental animals on special low-fluoride diets for harvesting low-fluoride serum. The recoveries of fluoride (0.5 to 4μM) added to various samples of serum, saliva, and milk are satisfactory (90 to 110%). In the case of samples containing very low levels of fluoride ($< 0.5\mu M$), which cannot be measured with an adequate degree of confidence, a known amount of fluoride (0.5 to 1.0μM) is added to the sample and the fluoride concentrations determined with the hanging drop fluoride electrode. The original concentration of the sample is calculated from this determination by subtracting the amount of fluoride added.

The results of fluoride analysis with the hanging drop fluoride electrode procedure requiring only 5 μl of sample per determination are in agreement with those obtained by other fluoride electrode macro procedures based on

Gran's plot calculations, reverse extraction, calcium phosphate adsorption, and ultrafiltration requiring 1- to 5-ml samples for each analysis. The results also indicate that the ultrafiltration step proposed by Singer and Armstrong (189) for measuring serum ionic fluoride with the electrode may be safely omitted. This hanging drop fluoride electrode technique should find useful application in clinical biochemistry laboratories and fluoride investigations with small experimental animals. Fuchs et al. (653) have proposed a "single-known-addition method combined with slope-by-dilution technique" for determining serum fluoride with the electrode. Potentials are taken with the fluoride electrode under three different circumstances: (*a*) 1000 μl of serum (5% TISAB), (*b*) 900 of μl of *a* plus 100 μl of standard fluoride solution with 5 to 10 times the fluoride concentration expected in the serum, and (*c*) 500 μl of *b* plus 500 μl of 5% TISAB. The method of calculation has been explained. Using an electrode switch and six electrodes, 18 samples can be analyzed per day.

6. Dental Plaque

There has been considerable interest in the determination of fluoride in bacteria, particularly in the area of dental caries. A large part of the work is concentrated on the dental plaque, a mucous film harboring bacteria on tooth surfaces. When the electrode is applied directly to the plaque material (254), to its fluid after separation following centrifugation (655–657), or to its suspension in saline (658), a small fraction of total plaque fluorine responds to the electrode. Higher fluoride values are obtained when the plaque is extracted in the cold with 0.5*M* perchloric acid (659) or with solutions containing metal chelating agents (660). Even higher values are obtained when the plaque is digested in hot perchloric acid (661,662). These differences are related to the different types of combination in which fluorine is present in the plaque material. These aspects are now being investigated.

7. Miscellaneous Samples

Determination of fluorine in diverse samples have been reported, including dentifrices (663,666) multivitamin preparations (667,668) and other pharmaceutical products (669,670), table salt (671,672), feed stuffs (673–677) and food products (678–680), beverages (681,682), hexafluorophosphates (683,684), tungsten (685), selenium (686), mineral fluorides (687), calcium and transitional metal orthophosphates (688), phosphoric acid (689), fuming nitric acid (690), chromium plating baths (691), pickling bath liquors (692), nuclear fuel reprocessing solutions (693), coal (694,695), and crude oil (696).

VI. SOURCES OF ERROR IN FLUORIDE ANALYSIS AND THEIR CONTROL

1. Determination and Control of the Fluoride Blank

The experimental fluoride blank has a bearing on the validity of the results of fluoride analysis, and its magnitude and range of variation must be given close attention. The overall blank is contributed by the apparatus, the reagents, and the environment.

A. THE APPARATUS BLANK

New distillation stills provide a high fluoride blank (1 to 2 μg F/50 ml of the distillate). Following repeated use of the stills, however, the blank values decrease to about 0.5 μg F/50 ml distillate. It is therefore desirable to carry out distillation in a new still for about 20 hr and check the distillate blanks. A low blank obtained during successive blank distillations ensures that the still is ready for routine use in fluoride analysis. It is desirable to have at least two sets of distillation stills, reserving one exclusively for distilling samples low in fluorine. This requires appropriate numbering of the components of the stills to prevent mixups during washing. Polypropylene diffusion cells, polyethylene diffusion bottles, and polypropylene tubes for reverse extraction, all provide far lower fluoride blanks than the distillation stills.

Evaporation of solutions should be carried out as far as possible in platinum dishes, which can be cleaned by boiling in dilute nitric acid solution, followed by several rinsings with double-distilled water, and flashing. Use of porcelain and silica dishes for evaporation of solutions more often results in poor recoveries of fluoride.

All reagents should be kept in well-stoppered polyethylene bottles. Sulfuric, perchloric, and phosphoric acids of the required concentrations may be kept in Pyrex bottles and stored in a refrigerator.

All glassware and polyethylene materials should be cleaned by scrubbing with a detergent solution, followed by thorough washing with tap water, 1 hr soaking in acid (25% nitric acid), thorough washing with distilled water, and finally washing with double-distilled water (three rinsings). Distillation flasks should be cleaned by boiling in them 10% sodium hydroxide solution to dissolve any gelatinous silicic acid that might be formed from time to time.

All apparatus must be dried and stored in dust-proof cabinets.

B. THE REAGENT BLANK

The major portion of the reagent blank is contributed by the acids employed in procedures for the separation of fluoride from the samples and also by the fixatives used for ashing the samples. Some of the reagents contain fluoride as a contaminant in amounts far exceeding the permissible limits for microanalysis of fluoride. Fortunately, adequate methods are now available for removing fluoride contamination in the reagents.

PREPARATION OF FLUORIDE-LOW REAGENTS

Fluoride-Low Acids. Sulfuric, phosphoric, and perchloric acids diluted with water (1–1) are further diluted (1–3) and boiled down with a few glass beads to the original volume (1–1). This operation may be repeated as often as necessary until the fluoride content of the acid is satisfactorily low. As an alternative, the diluted acid (1–1) may be submitted to Willard-Winter steam distillation, until three successive 50-ml portions of the distillate are found to contain very little fluoride.

The simplest and the most efficient procedure involves a few successive extractions of the acid (of the required concentration) with an equal or even smaller volume of an immiscible organic solvent (benzene, toluene) containing 2 to 5m*M* silanol (diphenylsilanediol, trimethylsilanol, triethylsilanol). Usually almost all the fluoride contamination is removed in two or three 1-hr extractions on a mechanical shaker. (See the following subsection and ref. 608.) The extremely low level of fluoride in the purified acid can be confirmed by employing the reverse extraction technique for the determination of fluoride (15).

Fluoride-Low Fixatives. According to a procedure described by Von Fallenberg (88), a substoichiometric amount of sodium carbonate solution is added to an excess of calcium salt solution; the fluoride ions present as an impurity are adsorbed onto the calcium carbonate precipitate, which is discarded. The calcium in the filtrate is precipitated as calcium carbonate by further addition of sodium carbonate solution. The calcium carbonate is washed and dissolved in acetic acid to obtain calcium acetate, or it may be decomposed overnight at 1100 to 1200 °C to obtain calcium oxide (697). Venkateswarlu and Narayanarao, who had success with the foregoing procedure only when the initial fluoride contamination in the calcium salt was low, not when excessive, have resorted to distillation for removal of fluoride (698). A similar approach has been described also by Weddle and Maurer (699.)

Venkateswarlu et al. (637) have prepared fluoride-low calcium oxide and magnesium oxide fixatives by decomposing the corresponding nitrates after removal of fluoride contamination by magnesium oxide adsorption techniques (179,180). With two successive adsorptions, it has been possible to reduce the ^{18}F added to solutions of calcium acetate, potassium chloride, sodium chloride, magnesium sulfate, and sucrose solutions to 0.00, 0.02, 0.06, 0.15, and 0.00%, respectively. When the magnesium oxide adsorption technique is inapplicable, salts can be purified by recrystallization as in the case of potassium dihydrogenphosphate (637) and ammonium chloride (700). Venkateswarlu and Sita (16) have prepared fluoride-low calcium phosphate by mixing stoichiometric amounts of fluoride-low calcium nitrate and potassium dihydrogenphosphate solutions and adjusting the pH. More recently, Venkateswarlu (608) has described a very simple procedure for preparing fluoride-low calcium phosphate, starting from reagent grade calcium phosphate from the shelf. The compound is dissolved in perchloric acid and the fluoride is removed by extracting three times with benzene containing diphenylsilanediol; the calcium phosphate freed from fluoride is reprecipitated by simple addition of ammonium hydroxide. Taves (701) has prepared low-fluoride magnesium chloride by acidifying and extracting the mixture six times with hexamethyldisiloxane; the solution is neutralized with sodium hydroxide solution.

C. EXTRANEOUS CONTAMINATION

A significant degree of fluoride contamination is possible if the air in the laboratory contains fluoride-bearing dust, as for example during "cooking" body fluids or tissue homogenates with fixatives on a flame or during prolonged evaporation in open dishes of large volumes of solutions, with only traces of fluoride. All possible precautions should be taken to keep the amount of dust in the air as low as possible. No hydrofluoric acid bottle should ever be opened in a laboratory engaged in fluoride analysis. It is also necessary to guard against contamination of air from inadvertent use of fluorine-containing sprays in and around the laboratory.

A very important source of fluoride contamination may be the lining material of the muffle furnace in which ashing is carried out. To find whether a particular furnace contributes to fluoride contamination of samples, fluoride analysis of appropriate amounts of low-fluoride fixatives (calcium oxide, calcium phosphate, magnesium oxide, etc.) is performed before and after ashing for a specified time (6 to 12 hr) in the furnace; the magnitude in the difference between the results in the two cases indicates the degree of contamination due to ashing in the furnace. As in the case of the new distillation stills, it is necessary to run a new furnace at a high temperature for several days and to check the consistency of a low-fluoride blank. Hall (493) advocates the use of a simple methanol spirit lamp instead of a muffle furnace for ashing and fusion of biological materials and soil samples alike.

2. Control of Fluoride Analytical Results

The following program is suggested to ensure checks and counterchecks on the results of fluoride analysis. The overall fluoride blanks obtained from day to day should be consistent and satisfactorily low. As far as practicable, a few fluoride recoveries should be run daily alongside the samples; the amount of fluoride added directly to the samples or at different stages of the procedure should be related to the anticipated levels of fluoride in the samples.

If the blanks or the recoveries are unsatisfactory, the results should be discarded. Results of replicate analyses should check within 10% of the mean; better precision is possible in most cases. It is highly desirable to analyze each type of material at least once, by two or more procedures that differ significantly in the methods of pretreatment of samples, isolation, and final measurement of fluorine. Any significant differences in the results obtained with the different methods should be carefully scrutinized for possible clues to the divergence in results. The results should be interpreted within the limits of accuracy, precision, and sensitivity of the method employed. These practices will greatly help in avoiding pitfalls in fluoride analysis and wasteful interpretations based on wrong results.

VII. PURIFICATION AND CONCENTRATION OF RADIOACTIVE FLUORINE, ^{18}F.

Radioactive fluorine, ^{18}F (half-life 109.7 min) can be prepared by a variety of nuclear reactions (702–707). In some instances ^{18}F requires separation from other radioactive isotopes formed concomitantly in the course of its production and from the parent target materials from which it is produced. Radiochemical separations of ^{18}F have been carried out by coprecipitation (708), distillation (709–711), solvent extraction (712), and ion exchange (713,714). Scholz and Sodd (715) have described a kit for the rapid isolation and preparation of ^{18}F in a medically usable form; the procedure is based on an exchange technique reported by Kozyreva-Aleksandrova et al. (716) in which hydrous zirconium oxide is employed as the ion exchanger. The yield of this 30-min procedure is about 80% of the ^{18}F in the material received from the reactor. The reverse extraction technique (15) should provide an excellent means for rapid and quantitative separation of ^{18}F. The technique is highly specific for fluoride, and ^{18}F could be concentrated into a volume no larger than a few microliters to obtain the "hottest" possible preparations, should they be needed.

Radioactive fluorine, ^{18}F, has been employed in the evaluation of fluoride analytical methods by Singer and Armstrong (29,183), Taves (56,190), Venkateswarlu et al. (8), and Venkateswarlu (15,41,171).

Acknowledgment

The author is indebted to Prof. D. Narayanarao and Dr. C. G. Pandit in India and to Prof. Wallace D. Armstrong in the United States for their personal interest in his investigations and their invaluable support and encouragement over the past 25 years. The author is deeply appreciative of several stimulating and creative discussions on the subject with Prof. Armstrong. The author's work has been supported in various ways by the Indian Council of Medical Research, New Delhi, by the Lady Tata Memorial Trust, Bombay, and by the National Institute of Dental Research, Bethesda, Maryland.

The author thanks M. K. Raman for his kind help in preparing the references.

References*

*C.A., Chemical Abstracts; A.A., Analytical Abstracts; F.A., Fluoride Abstracts, The Kettering Laboratory, University of Cincinnati, Cincinnati, Ohio.

1. G. J. Cox, "Fluorine and Dental Caries," in *A Survey of the Literature of Dental Caries*, National Academy of Sciences, Washington, D.C., 1952, p. 325.
2. *Fluorides and Human Health*, Y. Ericsson, Ed., World Health Organization, Geneva, 1970.
3. J. S. C. Marais, *Onderstepoort J. Vet. Sci. Ani. Ind.*, *20*, 67 (1944).
4. *Fluoride in Medicine*, T. L. Vischer, Ed. Huber, Bern, 1969.
5. *Symposium on Carbon-Fluorine Compounds: Chemistry, Biochemistry and Biological Activities*, R. A. Peters, Ed., Ciba Foundation Symposia, London, 1971.
6. D. W. Allmann and M. Benac, *J. Dent. Res.*, *55*, Special Issue B, Abstr. No. 523 (1976).
7. D. R. Taves, *Nature*, *217*, 1050 (1968).
8. P. Venkateswarlu, L. Singer, and W. D. Armstrong, *Anal. Biochem.*, *42*, 350 (1971).
9. H. H. Willard and O. B. Winter, *Ind. Eng. Chem.* (*Anal. Ed.*), *5*, 7 (1933).
10. I. Tananajev, *Zh. Prikl. Khim*, *5*, 834 (1932).
11. L. Singer and W. D. Armstrong, *Anal. Chem.* *26*, 904 (1954).
12. R. Belcher, M. A. Leonard, and T. S. West, *J. Chem. Soc.*, 1958, 2390.
13. R. Belcher, M. A. Leonard, and T. S. West, *Talanta*, *2*, 92 (1959).
14. M. S. Frant and J. W. Ross, Jr., *Science*, *154*, 1553 (1966).
15. P. Venkateswarlu, *Anal. Chem.*, *46*, 878 (1974).
16. P. Venkateswarlu and P. Sita, *Anal. Chem.*, *43*, 758 (1971).
17. R. Bock and H. J. Semmler, *Fresenius' Z. Anal. Chem.*, *230*, 161 (1967).
18. G. W. Busch, R. C. Carter, and F. E. McKenna, *The Analytical Chemistry of the Manhattan Projects*, *M. N. E. S. Div. VIII*, Vol. 1, McGraw-Hill, New York, 1950, p. 226.
19. F. E. McKenna, *Nucleonics*, *8*, 24–33, 40–49, 51–58 (1951).
20. P. J. Elving, C. A. Horton, and H. H. Willard, In *Fluorine Chemistry*, Vol. II, J. H. Simmons, Ed., Academic Press, New York, 1954, p. 51.
21. C. A. Horton, in *Advances in Analytical Chemistry and Instrumentation*, Wiley-Interscience, New York, 1962, p. 151.
22. A. M. G. Macdonald, in *Comprehensive Analytical Chemistry*, Vol. 1C, C. L. Wilson and D. W. Wilson, Eds., Elsevier, New York, 1962, p. 319.
23. A. M. G. Macdonald, in *Experimental Pharmacology*, Vol. 20, Part 2, O Eichler et. al. Eds., Springer-Verlag, Berlin, 1970, p. 1.
24. N. S. Nikolaev, S. N. Suvorova, E. I. Gurovich, I. Peka, and E. K. Korochemnaya; *Analytical Chemistry of Fluorine*, Wiley (Halsted), New York, 1973.
25. B. Ikert, *Chem. Ztg.*, *63*, 754 (1939). (C.A. 34:2151, 1940.)
26. Z. Gregorowicz, S. Kowalski, and I. Szalonek, *Chem. Anal.* (Warsaw), *10*, 889 (1965).
27. A. Brüning and H. Quast, *Z. Angew. Chem.*, *44*, 656 (1931).
28. L. Singer and W. D. Armstrong, *Anal. Chem.*, *31*, 105 (1959).

29. L. Singer and W. D. Armstrong, *Arch. Oral Biol.*, *14*, 1343 (1969).
30. G. J. Kakabadse, B. Manohin, J. M. Bather, E. C. Wellaer, and P. Woodbridge, *Nature*, *229*, 626 (1971).
31. F. J. McClure, *Ind. Eng. Chem.* (*Anal. Ed.*), *11*, 171 (1939).
32. R. J. Rowley, J. G. Grier, and R. L. Parsons, *Anal. Chem.*, *25*, 1061 (1953).
33. L. F. Remmert and T. D. Parks, *Anal. Chem.*, *25*, 450 (1953).
34. L. J. Hardin, W. H. MacIntire, and M. E. Tubb, *J. Assoc. Off. Agr. Chem.*, *37*, 552 (1954).
35. W. Oelschläger, *Landwirtsch. Forschu.*, *21*, 172 (1968). (F.A. 15:183, 1966–1968.)
36. P. Venkateswarlu and D. Narayanarao, *Anal. Chem.*, *26*, 766 (1954).
37. H. V. Churchill, R. W. Bridges, and R. J. Rowley, *Ind. Eng. Chem.* (*Anal. Ed.*), *9*, 222 (1937).
38. G. Brunisholz and J. Michod, *Helv. Chim. Acta*, *37*, 874 (1954).
39. F. A. Smith and D. E. Gardner, *Fluoride Studies*, U.S. Atomic Energy Commision, Unclassified Document, UR-36, 1948.
40. W. D. Armstrong, personal communication, 1951.
41. P. Venkateswarlu, *Anal. Biochem.*, *68*, 1975.
42. W. Schöniger, *Mikrochim. Acta*, 123, (1955).
43. A. M. G. Macdonald, in *Comprehensive Analytical Chemistry*, Vol. IB, C. L. Wilson and D. W. Wilson, Eds., Elsevier, Amsterdam, 1960, p. 547.
44. W. Schöniger, *Mikrochim. Acta*, 869 (1956).
45. C. A. Johnson and M. A. Leonard, *Analyst*, *86*, 101 (1961).
46. E. C. Olsen and S. R. Shaw, *Microchem. J.*, *5*, 101 (1961).
47. R. Levy and E. Debal, *Microchim. Acta*, 224 (1962).
48. M. Fernandopulle and A. M. G. Macdonald, *Microchem. J.*, *11*, 41 (1966).
49. G. Tolg, *Z. Anal. Chem.*, *194*, 20 (1963).
50. D. A. Levaggi, W. Oyung, and M. Feldstein, *J. Air Pollut. Control Assoc.*, *21*, 277 (1971).
51. J. A. Peters and K. J. Baxter, *Bull. Environ. Contam. Toxicol.*, *11*, 177 (1974).
52. H. Soep, *J. Chromatogr.*, *6*, 122 (1961).
53. H. Trutnovsky, *Mikrochim. Ichnoanal. Acta*, 499, (1963). (A.A., 11:3140, 1964.)
54. R. Mavrodineanu and J. Gwirtsman, *Contrib. Boyce Thompson Inst.*, *18*, 419 (1956).
55. A. L. Waldo and R. E. Zipf, *J. Lab. Clin. Med.*, *40*, 601 (1952).
56. D. R. Taves, *Talanta*, *15*, 969 (1968).
57. J. F. McClendon and W. C. Foster, *Ind. Eng. Chem.* (*Anal. Ed.*), *13*, 280 (1941).
58. R. Wickbold, *Angew. Chem.*, *66*, 173 (1954).
59. P. B. Sweetser, *Anal. Chem.*, *28*, 1766 (1956).
60. F. Martin and A. Floret, *Chim. Anal.* (Paris), *41*, 181 (1959).
61. F. Martin, A. Floret, and M. Diller, *Bull. Soc. Chim. Fr.*, 1961, 460.
62. F. Ehrenberger, *Mikrochim. Acta*, 192 (1959).
63. B. W. Fry and D. R. Taves, *J. Lab. Clin. Med.*, *75*, 1020 (1970).
64. W. I. Rogers and J. M. Wilson, *Anal. Biochem.*, *32*, 31 (1969).
65. H. Meyer, *Analyse und Konstitutionsermitlung organischer Verbindungen*, 5th ed., Springer, Berlin, 1931, p. 184.
66. P. J. Elving and W. B. Ligett, *Ind. Eng. Chem.* (*Anal. Ed.*), *14*, 449 (1942).
67. R. Belcher and A. M. G. Macdonald, *Microchim. Acta*, 510 (1957).
68. R. Belcher, A. M. G. Macdonald, and A. J. Nutten, *Mikrochim. Acta*, 104 (1954).
69. R. E. Banks, F. Cuthbertson, and W. K. R. Musgrave, *Anal. Chim. Acta*, *13*, 442 (1955).
70. E. Schröder and U. Waurick, *Plaste Kautsch.*, *1*, 9 (1960). (A.A., 8:206, 1961.)
71. A. Stepanow, *Ber. Bunsenges. Phys. Chem.*, *39*, 4056 (1906).
72. T. H. Vaughan and J. A. Nieuwland, *Ind. Eng. Chem.* (*Anal. Ed.*), *3*, 274 (1931).
73. R. D. Strahm, *Anal. Chem*, *31*, 615 (1959).
74. M. F. Little, F. S. Casciani, and J. Rowley, International Association for Dental Research, 46th General Meeting, 1968, Abstr. No. 467.
75. F. L. Benton and W. H. Hamill, *Anal. Chem.*, *20*, 269 (1948).
76. W. B. Ligett, *Anal. Chem.*, *26*, 748 (1954).

77. R. D. Chambers, T. F. Holmes, and W. K. R. Musgrave, *Analyst*, *89*, 369 (1964).
78. B. Pecherer, C. M. Gambrill, and G. W. Wilcox, *Anal. Chem.*, *22*, 311 (1950).
79. J. N. Wilson and C. Z. Marczewski, *Anal. Chem.*, *45*, 2409 (1973).
80. L. C. Clark Jr., F. Becattini, S. Kaplan, V. Obrock, D. Cohen, and C. Becker, *Science*, *181*, 680 (1973).
81. W. Wohler, *Pogg. Ann.*, *48*, 87 (1839).
82. C. R. Wagner and W. H. Ross, *Ind. Eng. Chem.*, *9*, 1116 (1917).
83. D. S. Reynolds, W. H. Ross, and K. D. Jacob, *J. Assoc. Off. Agr. Chem.*, *11*, 225 (1928).
84. W. D. Armstrong, *Ind. Eng. Chem.* (*Anal. Ed.*), *5*, 315 (1933).
85. J. K. Wilkeyson, *Anal. Chem.*, *47*, 2053 (1975).
86. W. K. Gilkey, H. L. Rohs, and H. V. Hansen, *Ind. Eng. Chem.* (*Anal. Ed.*), *8*, 150 (1936).
87. E. W. Scott and A. L. Henne, *Ind. Eng. Chem.* (*Anal. Ed.*), *7*, 299 (1935).
88. T. Von Fellenberg, *Mitt. Lebensm. Hyg.*, *39*, 124 (1948).
89. R. Mavrodineanu and J. Gwirtsman, *Contrib. Boyce Thompson Inst.*, *17*, 489 (1954).
90. H. H. Willard, T. Toribara, and L. N. Holland, *Anal. Chem.*, *19*, 343 (1947).
91. L. N. Devonshire and H. H. Rowley, *Proc. Okla. Acad. Sci.*, *34*, 155 (1953).
92. W. B. Estill and L. C. Mosier, *Anal. Chem.*, *27*, 1669 (1955).
93. A. C. Kuyper, *Anal. Chem.*, *28*, 922 (1956).
94. D. S. Reynolds, J. B. Kershaw, and K. D. Jacob, *J. Assoc. Off. Agr. Chem.*, *19*, 156 (1936).
95. H. V. Churchill, *Ind. Eng. Chem.* (*Anal. Ed.*), *17*, 720 (1945).
96. E. Eberius and W. Kowalski, *Z. Erzbergbau Metallhüttenw.*, *5*, 131 (1952). (C.A., 46:6546, 1953.)
97. G. Pietzka and P. Ehrlich, *Angew. Chem.*, *65*, 131 (1953).
98. H. R. Shell and R. L. Craig, *Anal. Chem.*, *26*, 996 (1954).
99. F. S. Grimaldi, B. Ingram, and F. Cuttitta, *Anal. Chem.*, *27*, 918 (1955).
100. J. Angot and N. Mevel, *Chim. Anal.* (Paris), *45*, (1963).
101. R. C. Calkins, *Anal. Chem.*, *34*, 837 (1962).
102. W. D. Armstrong, *Ind. Eng. Chem.* (*Anal. Ed.*), *8*, 384 (1936).
103. H. Ballczo and O. Kaufmann, *Mikrochim. Acta*, *39*, 13 (1952).
104. H. Kubota, *Microchem J.*, *12*, 525 (1967).
105. F. C. Seel, *Chem. Lab. Betr.*, *18*, 546 (1967).
106. D. S. Reynolds and W. L. Hill, *Ind. Eng. Chem.* (*Anal. Ed.*), *11*, 21 (1939).
107. L. H. Andersson and B. Gelin, Forsvarets Forskningsanstalt, Stockholm, Sweden, Report 1, 1967. (F.A. 15:81, 1966–1968.)
108. J. E. Deutschman, H. Boden, R. M. Duncan, and H. Simard, 61st Annual Meetings of Air Pollution Association, St. Paul, Minn., 1968, Abstr. No. 68–87.
109. H. Ballczo and O. Kaufmann, *Mikrochemie*, *28*, 237 (1951).
110. H. Ballczo, G. Doppler, and A. Lanik, *Mikrochim. Acta*, 809 (1957).
111. J. Samachson, N. Slovik, and A. E. Sobel, *Anal. Chem.*, *29*, 1888 (1957).
112. M. A. Wade and S. S. Yamamura, *Anal. Chem.*, *37*, 1276 (1965).
113. R. F. Bohlander, U.S. Atomic Energy Commission Document No. NLCO-1039, 1969.
114. B. J. MacNulty, G. F. Reynolds, and E. A. Terry, *Analyst*, *79*, 190 (1954).
115. P. Venkateswarlu, "Studies on Metabolism and Transport of Fluoride," Ph.D. thesis, University of Minnesota, 1962.
116. H. Stegemann and G. F. Jung, *Z. Physiol. Chem.*, *315*, 222 (1959).
117. R. J. Hall, *Analyst*, *85*, 560 (1960).
118. W. Büttner, S. Schülke, and S. Soyka, *Dent. Zahnärztl. Z.*, *18*, 24 (1963).
119. Von J. Bäumler and E. Glinz, *Mitt. Lebensm. Hyg.*, *55*, 250 (1964).
120. Von J. Bäumler, *Caries Res.*, *1*, 281 (1969).
121. F. J. Frere, *Anal. Chem.*, *33*, 644 (1961).
122. H. W. Wharton, *Anal. Chem.*, *34*, 1296 (1962).
123. L. Singer and W. D. Armstrong, *J. Dent. Res.*, *41*, 910 (1962).
124. R. J. Rowley and G. H. Farrah, *Am. Ind. Hyg. Assoc. J.*, *23*, 314 (1962).

125. M. M. Golutvina, N. N. Kononykina, and V. P. Ryabikin, *Gig. Sanit.*, *36*, 63 (1971). (F.A. 15:124, 1969–1971.)
126. C. W. Chang and C. R. Thompson, *Microchem. J.*, *8*, 407 (1964).
127. H. W. Wharton, *Anal. Chem.*, *35*, 406 (1963).
128. R. Tamburini Jr. and S. M. Zucas, *Revi. Fac. Farm. Bioquim. Univ. Sao Paulo*, *2*, 135 (1964). (F.A. 15:82, 1963–1965.)
129. F. J. Frere, *Microchem. J.*, *6*, 167 (1962).
130. L. Greenland, *Anal. Chem.*, *35*, 386 (1962).
131. J. L. Stuart, *Analyst*, *95*, 1032 (1970).
132. L. H. Weinstein, R. H. Mandl, D. C. McCune, J. S. Jacobson, and A. E. Hitchcock, *Contrib. Boyce Thompson Inst.*, *22*, 207 (1963).
133. R. H. Mandl, L. H. Weinstein, J. S. Jacobson, D. C. McCune, and A. E. Hitchcock, In Technicon Symposia, 1965, *Automation in Analytical Chemistry*, Mediad, New York, 1966, p. 270.
134. D. R. Taves, *Anal. Chem.*, *40*, 204 (1968).
135. E. B. R. Prideux and E. C. Roper, *J. Chem. Soc.*, 1926, 898.
136. J. C. Warf, W. D. Cline, and R. D. Tavebaugh, *Anal. Chem.*, *26*, 342 (1954).
137. C. V. Banks, K. E. Burke, and J. W. O'Laughlin, *Anal. Chim. Acta*, *19*, 239 (1958).
138. C. D. Susano, J. C. White, and J. E. Lee, Jr., *Anal. Chem.*, *27*, 453 (1955).
139. J. G. Surak, D. J. Fisher, C. L. Burros, and L. C. Bate, *Anal. Chem.*, *32*, 117 (1960).
140. A. R. Gahler and G. Porter, *Anal. Chem.*, *29*, 296 (1957).
141. R. H. Powell and O. Menis, *Anal. Chem.*, *30*, 1546 (1958).
142. S. S. Yamamura, M. E. Kussy, and J. E. Rein, *Anal. Chem.*, *33*, 1655 (1961).
143. G. S. Rao and M. S. Subramanian, *Talanta*, *9*, 947 (1962).
144. E. G. Berns and P. W. Van der Zwaan, *Anal. Chim. Acta*, *59*, 293 (1972).
145. D. C. Hoel, D. T. Fanale, and R. O. Clark, *Anal. Chim. Acta*, *23*, 83 (1960).
146. W. C. Schumb and K. J. Radimer, *Anal. Chem.*, *20*, 871 (1948).
147. H. Van Gogh, *Pharm. Weekbl.*, *101*, 881 (1966).
148. W. J. Kirsten and Z. H. Shah, *Anal. Chem.*, *47*, 184 (1975).
149. W. J. Kirsten, *Anal. Chem.*, *48*, 84 (1976)
150. R. P. Curry and M. G. Mellon, *Anal. Chem.*, *29*, 1632 (1957).
151. H. Gamsjäger and E. Schwarz-Bergkamf, *Microchem. J. Symp.*, Series 2, 359 (1962); *Microchemical Techniques*, N. D. Cheronis, Ed., Wiley, New York, 1962.
152. M. A. Volodina, T. A. Gorshkova, O. P. Morozova, and E. A. Tsukerman, *Zavod. Lab.*, *39*, 674 (1973). (A.A. 26:848, 1974.)
153. K. D. Moffett, J. R. Simmler, and H. A. Potratz, *Anal. Chem.*, *28*, 1356 (1956).
154. D. Aulagnier, H. Chermetta, C. Martelet, J. P. Schlienger, and J. Tourset, *Bull. Soc. Chim. Fr.*, 1971, 1135.
155. H. Chermette, C. Martelet, D. Sandino, and J. Tousset, *Anal. Chem.*, *44*, 857 (1972).
156. Y. Kakita and H. Goto, *Anal. Chem.*, *34*, 618 (1962).
157. W. E. Dahl, *Anal. Chem.*, *40*, 416 (1968).
158. F. L. Moore, *Anal. Chem.*, *35*, 1032 (1963).
159. Y. B. Kletenik and I. A. Bykhovskaya, *Zh. Anal. Khim.*, *25*, 351 (1970). (A.A. 21:1083, 1971.)
160. S. Bazille, *Contribution à l'Étude Toxicologique du Fluor*, Paris, Henry, 1935.
161. R. Fabre, *J. Pharm. Chim.*, *27*, 467 (1938).
162. W. Funasaka, M. Kawane, and T. Kojima, *Mem. Fac. Eng., Kyoto Univ.*, *18*, 44 (1956).
163. I. Zipkin, W. D. Armstrong, and L. Singer, *Anal. Chem.*, *29*, 310 (1957).
164. A. C. D. Newman, *Anal. Chim. Acta*, *19*, 471 (1958).
165. H. M. Nielsen, *Anal. Chem.*, *30*, 1009 (1958).
166. J. J. Nickl and C. von Braunmuehl, *Z. Anal. Chem.*, *246*, 313 (1969).
167. T. L. Zalevskaya and G. L. Starobinets, *Zh. Anal. Khim.*, *24*, 721 (1969). (A.A. 19:387, 1970.)

168. D. I. Eristavi, I. Sh. Shatirishvili, and L. F. Mekvabishvili, *Tr. Gruz. Politekh. Inst.*, *4*:132, 33 (1969). (A.A. 20:183, 1971.)
169. G. L. Starobinets and E. P. Zakhaetseva, *Ser. Khim. Nauk.*, *6*, 106 (1970). (A.A. 22:22264, 1972.)
170. J. Coursier and J. Saulnier, *Anal. Chim. Acta*, *14*, 52 (1956).
171. P. Venkateswarlu, *Biochem. Med.*, *14*, 368 (1975).
172. J. A. Broomhead and N. A. Gibson, *Anal. Chim. Acta*, *26*, 265 (1962).
173. J. Gera and J. Tomkowiak, *Chim. Anal.*, *14*, 135 (1969). (A.A. 18:3952, 1970.)
174. H. Seiler and T. Kaffenburger, *Helv. Chim. Acta*, *44*, 1282 (1961).
175. E. Gagliardi and G. Pokorny, *Mikrochim. Ichnoanal. Acta*, 699 (1965). (A.A. 13:4134, 1966.)
176. L. S. Bark, R. J. T. Graham, and D. M. McCormick, *Anal. Chim. Acta*, *35*, 268 (1966).
177. T. Deguchi, *J. Chromatog.*, *108*, 409 (1975).
178. M. Tokutomi and S. Kamiya, *Japan Anal.*, *21*, 81 (1972).
179. P. Venkateswarlu and D. Narayanarao, *Indian J. Med. Res.*, *41*, 473 (1953).
180. P. Venkateswarlu and D. Narayanarao, *Indian J. Med. Res.*, *42*, 135 (1954).
181. P. Venkateswarlu and D. Narayanarao, *Indian J. Med. Res.*, *45*, 273 (1957).
182. P. Venkateswarlu and D. Narayanarao, *Indian J. Med. Res.*, *45*, 369 (1957).
183. L. Singer and W. D. Armstrong, *Anal. Biochem.*, *10*, 495 (1965).
184. M. K. Raman and P. Venkateswarlu, unpublished results, 1966.
185. B. Vialatte, J. N. Barrandon, S. Alexandrov, I. N. Bourrelly, C. Cleyrergue, N. Deschamps, and H. Jaffrezic, *Radiochem. Radioana. Lett.*, *5*, 59 (1970).
186. M. Fedoroff and L. Debove, *CR Hebd. Séances Acad. Sci.*, Paris, Ser. C, *275*, 1189 (1972).
187. J. A. Fresen, F. H. Cox, and M. J. Witter, *Pharm. Weekbl.*, *103*, 909 (1968).
188. P. Venkateswarlu, unpublished results, 1974.
189. L. Singer and W. D. Armstrong, *Biochem. Med.*, *8*, 415 (1973).
190. D. R. Taves, *Talanta*, *15*, 1015 (1968).
191. P. Venkateswarlu, in *Trace Element Metabolism in Animals*, Vol. 2, W. G. Hoekstra, J. W. Suttie, H. E. Ganther, and W. Mertz, Eds., University Park Press, Baltimore, 1974, p. 442.
192. H. Chermette, C. Martelet, D. Sandino, M. Benmalek, and J. Tousset, *Anal. Chim. Acta*, *59*, 373 (1972).
193. H. Chermette, M. Perrousset, and J. Ratelade, *Anal. Chim. Acta*, *70* 217 (1974).
194. W. D. Armstrong, *J. Am. Chem. Soc.*, *55*, 1741 (1933).
195. R. J. Rowley and H. V. Churchill, *Ind. Eng. Chem.* (*Anal. Ed.*), *9*, 551 (1937).
196. W. M. Hoskins and C. A. Ferris, *Ind. Eng. Chem.* (*Anal. Ed.*), *8*, 6 (1936).
197. D. Dahle, R. U. Bonnar, and H. J. Wichmann, *J. Assoc. Off. Agr. Chem.*, *21*, 459 (1938).
198. H. A. Williams, *Analyst*, *71*, 175 (1946).
199. F. A. Smith and D. E. Gardner, *Arch. Biochem.*, *29*, 311 (1950).
200. P. Venkateswarlu and A. N. Ramanathan, *Indian J. Med. Res.*, *40*, 549 (1952).
201. H. H. Willard and C. A. Horton, *Anal. Chem.*, *22*, 1190, 1194 (1950).
202. J. Schmidt and R. Ortloff, *Z. Chem.*, *5*, 236 (1965).
203. G. Banerjee, *Anal. Chim. Acta*, *13*, 409 (1955).
204. W. Selig, *Analyst*, *93*, 118 (1968).
205. E. Hluchan, M. Sedlak, and M. Jenik, *Chem. Zvesti*, *23*, 219 (1969). (C.A. 71:27181, 1969.)
206. G. I. Il'inskaya and L. G. Golubeva, *Tr. Vses. Nauch.-Issled. Proekt. Inst. Alyum., Magn. Elektron. Prom.*, No. 74: 36 (1971). (C.A. 77:92399, 1972.)
207. R. F. Milton, H. F. Liddell, and J. E. Chivers, *Analyst*, *72*, 43 (1947).
208. R. F. Milton, *Analyst*, *74*, 54 (1949).
209. J. B. Rickson, *Analyst*, *75*, 84 (1950).
210. F. W. Cheng, *Mikrochim. Acta*, 1105 (1967).
211. S. M. Bein, *J. Dent. Res.*, *22*, 123 (1943).
212. H. Ballczo and O. Kaufmann, *Mikrochemie vs. Mikrochim. Acta*, *38*, 237 (1951).
213. J. H. Saylor and M. E. Larkin, *Anal. Chem.*, *20*, 194 (1948).
214. V. M. Brodskaya, A. F. Kuteinikov, and G. A. Lanskoi, *Byul. Nauch. Tekhnol. Inf. Miner Geol.*

Okhra. Nedr. SSSR., No. 4, 97 (1960). (C.A. 58:14, 1963.)
215. M. M. Vinnik and M. L. Chepelevetskii, *Soobshah. Nauch.-Issled. Rab. Nov. Tekh. Nauch. Inst. Udobr. Insektofungitsid.*, No. 10, 44 (1958). (C.A. 53:19665, 1959.)
216. M. G. Lubyanskaya, *Zavod. Lab.*, 22, 921 (1956). (A.A. 4:896, 1957.)
217. F. Cianflone and L. I. Braddock, *Anal. Chem.*, *26*, 425 (1954).
218. W. L. Everson and E. M. Ramirez, *Anal. Chem.*, *39*, 1771 (1967).
219. W. I. Awad, S. S. M. Hassan, and M. B. Elsayes, *Mikrochim. Acta*, *4*, 688 (1969).
220. R. P. Singh and T. S. B. Narasaraju, *Z. Anal. Chem.*, *247*, 52 (1969).
221. H. F. Combs and E. L. Grove, *Talanta*, *17*, 599 (1970).
222. A. R. Casal and A. F. D. de Namor, *An. Asoc. Quim. Argent.*, *58*, 161 (1970). (A.A. 20:3790, 1971.)
223. L. Moraru, *Stud. Cerc. Chim.*, *19*, 767 (1971).
224. L. Moraru, *Stud. Cerc. Chim.*, *19*, 779 (1971).
225. R. Valach, *Talanta*, *11*, 973 (1964).
226. J. H. De Boer, *Chem. Weekbl.*, *21*, 404 (1924).
227. E. Elvove, *Public Health Rep.* (*U.S.*), *48*, 1219 (1933).
228. J. M. Sanchis, *Ind. Eng. Chem. Anal. Med.* (*Anal. Ed.*), *6*, 134 (1934).
229. W. L. Lamar and C. G. Seegmiller, *Ind. Eng. Chem.* (*Anal. Ed.*), *13*, 901 (1941).
230. J. P. Boonstra, *Rec. Trav. Chim. Pays-Bas*, *70*, 325 (1951). (F.A. 15:492, 1958.)
230a.
H. E. Bumsted and J. C. Wells, *Anal. Chem.*, *24*, 1595 (1952).
231. S. Megregian and F. J. Maier, *J. Am. Water Works Assoc.*, *44*, 239 (1952).
232. S. Megregian, *Anal. Chem.*, *26*, 1161 (1954).
233. W. Oelschläger, *Z. Anal. Chem.*, *191*, 408 (1962).
234. E. Bellack and P. J. Schouboe, *Anal. Chem.*, *30*, 202 (1958).
235. M. Kamada and M. Ota, *Bull. Chem. Soc. Japan*, *28*, 148 (1955).
236. M. Kamada and T. Onishi, *J. Chem. Soc., Japan, Pure Chem. Sec.*, *80*, 275 (1959).
237. V. A. Nazarenko and E. A Biryk, "Soverm Metody. Anal. Metall., M., Metallurgizdat," 1955, 188. Zh. Khim. Abstr. no. 25937 (1956). (cited from A. M. G. Macdonald, Ref. 23).
238. L. Krahulec, *Ceskoslov. Hyg., Epidemiol., Mikrobiol., Immunol.*, *4*, 376 (1955). (C.A. 49:15621, 1955.)
240. H. Sano, *Japan Anal.*, *5*, 289 (1956).
241. R. Valach, *Talanta*, *8*, 443 (1961).
242. A. D. Wilson and J. R. Cooke, *Analyst*, *91*, 135 (1966).
243. I. G. Izhak, U.S.S.R. Patent 132,432, 1960.
244. I. G. Izhak, *Zavod. Lab.*, *28*, 907 (1962).
245. G. Gad and K. Naumann, *Gas. Wasserfach*, *81*, 183. (C.A. 32:4259, 1936.)
246. A. Jendrassik and S. Papp, *Magy. Chem. Foly.*, *49*, 137 (1943).
247. I. M. Kolthoff and M. E. Stansby, *Ind. Eng. Chem.* (*Anal. Ed.*), *6*, 118 (1934).
248. E. J. Dixon, *Analyst*, *95*, 272 (1970).
249. N. A. Talvitie, *Ind. Eng. Chem.* (*Anal. Ed.*), *15*, 620 (1943).
250. J. M. Icken and B. M. Blank, *Anal. Chem.*, *25*, 1741 (1953).
251. H. Nommik, *Fluorine in Swedish Agricultural Products, Soil and Drinking Water*, Esselte Aktiebolag, Stockholm, 1953.
252. J. J. Lothe, *Anal. Chem.*, *28*, 949 (1956).
253. D. Revinson and J. H. Harley, *Anal. Chem.*, *25*, 794 (1953).
254. E. Hluchan and J. Mayer, *Chem. Zvesti*, *17*, 569 (1963).
255. K. Toei, T. Hayami, M. Miyata, and Y. Shimoishi, *Bull. Chem. Soc. Japan*, *39*, 638 (1966).
256. M. A. Shemyakina, *Vop. Prikl. Geokhim.*, *1*, 128 (1966). (C.A. 66:111243, 1967.)
257. I. B. Moizhes and I. A. Stolyarova, *Metody Zimotec, Anal. Mineral'n., Syr'ya Sb.*, 1963. (C.A. 61:11321, 1964.)
258. J. L. Lambert, *Anal. Chem. 26*, 558 (1954).

259. S. K. Yasuda and J. L. Lambert, *Anal. Chem.*, *30*, 1485 (1958).
260. H. F. Liddell, *Analyst*, *78*, 494 (1953).
261. H. F. Liddell, *Analyst*, *79*, 752 (1954).
262. M. C. Mehra and J. L. Lambert, *Microchem. J.*, *18*, 226 (1973).
263. W. L. Hinze and R. E. Humphrey, *Microchem. J.*, *20*, 246 (1975).
264. H. Ballczo, *Mikrochim. Acta*, 1968, 205.
265. B. J. MacNulty, G. J. Hunter, and D. G. Barrett, *Anal. Chim. Acta*, *14*, 368 (1956).
266. W. E. Thrun, *Anal. Chem.*, *22*, 918 (1950).
267. P. Venkateswarlu, W. D. Armstrong, and L. Singer, *Indian J. Med. Res.*, *54*, 455 (1966).
268. M. J. Price and O. J. Walker, *Anal. Chem.*, *24*, 1593 (1952).
269. G. S. Beveridge, G. J. Hunter, and B. J. MacNulty, *Anal. Chim. Acta*, *9*, 330 (1953).
270. B. J. MacNulty and G. J. Hunter, *Anal. Chim. Acta*, *9*, 425 (1953).
271. H. Green, *Brit. Cast Iron Res. Assoc. J.*, *8*, 838 (1960).
272. A. F. Kuteinikov, V. M. Brodskaya, and G. A. Lanskoi, *Zh. Anal. Khim.*, *17*, 87 (1962).
273. M. Shimizu, *Hirosaki Med. J.*, *15*, 389 (1963).
274. B. J. MacNulty and L. D. Woollard, *Anal. Chim. Acta*, *14*, 452 (1956).
275. M. Deguchi, *Hiroshima Daigaku Kogakubu Kenkyu Hokoku*, *13*, 55 (1966). (C.A. 65:12859, 1966.)
276. W. D. Armstrong, *Ind. Eng. Chem.* (*Anal. Ed.*), *5*, 300 (1933).
277. J. J. Fahey, *Ind. Eng. Chem.* (*Anal. Ed.*), *11*, 362 (1939).
278. M. Kortum-Seiler, *Angew. Chem.*, *A59*, 159 (1947).
279. H.-D. Cremer and W. Voelker, *Biochem. Z.*, *324*, 89 (1953).
280. J. C. Bartlet, D. Kavagh, and R. A. Chapman, *Can. J. Technol.*, *33*, 348 (1955).
281. D. Monnier, Y. Rusconi, and P. Wenger, *Helv. Chim. Acta*, *29*, 521 (1946).
282. M. D. Foster, *Ind. Eng. Chem.* (*Anal. Ed.*), *5*, 225 (1933).
283. R. S. Ingols, E. H. Shaw, W. H. Eberhardt, and J. C. Hildebrand, *Anal. Chem.*, *22*, 799 (1950).
284. T. Ivanov and E. Koen, *Khig. Zdrav.*, *9*, 80 (1966). (C.A. 65:9609, 1966.)
285. M. L. Nichols and A. C. Condo, Jr., *Anal. Chem.*, *26*, 703 (1954).
286. S. Paslawska, *Gdansk. Tow. Nauk. Rozpr. Wydz.*, *3*, 125 (1970). (C.A. 75:83879, 1971.)
287. A. K. Majumdar and D. Chakraborti, *Anal. Chim. Acta*, *55*, 450 (1971).
288. G. Steiger, *J. Am. Chem. Soc.*, *30*, 219 (1908).
289. H. E. Merwin, *Am. J. Sci.*, *28*, 119 (1909).
290. D. Monnier, R. Vaucher, and P. Wenger, *Helv. Chim. Acta*, *33*, 1 (1950).
291. E. D. Schall and H. G. Williamson, *J. Assoc. Off. Agr. Chem.*, *38*, 454 (1955).
292. F. I. Brownley, Jr., and E. E. Sellers, *J. Am. Water Works Assoc.*, *49*, 1234 (1957).
293. A. K. Babko and P. V. Khodulina, *J. Anal. Chem. USSR*, *7*, 281 (1952) (English transl.).
294. J. S. Alonso, *Stud. Geol.*, *1*, 87 (1970).
295. R. J. Bertolacini and J. E. Barney, II, *Anal. Chem.*, *30*, 202 (1958).
296. L. Fine and E. A. Wynne, *Microchem. J.*, *3*, 515 (1959).
297. K. Hayashi, T. Danzuka, and K. Ueno, *Talanta*, *4*, 126 (1960).
298. A. L. Hensley and J. E. Barney, II, *Anal. Chem.*, *32*, 828 (1960).
299. M. J. Nardozzi and L. L. Lewis, *Anal. Chem.*, *33*, 1261 (1961).
300. C. D. West, R. L. Birke, and D. N. Hume, *Anal. Chem.*, *40*, 556 (1968).
301. R. P. Curry and M. G. Mellon, *Anal. Chem.*, *28*, 1567 (1956).
302. S. Skramovsky and L. Velikova, *Cas. Cesk. Lek.*, *63*, 299 (1950). (C.A. 46:6996, 1952.)
303. M. Jean, *Anal. Chim. Acta*, *57*, 438 (1971).
304. Z. Marczenko and L. Lenarczyk, *Chem. Anal.* (Warsaw), *15*, 607 (1970).
305. Z. Marczenko and L. Lenarczyk, *Chem. Anal.* (Warsaw), *15*, 753 (1970).
306. R. Belcher and T. S. West, *Talanta*, *8*, 853 (1961).
307. M. A. Leonard and T. S. West, *J. Chem. Soc.*, 1960, 4477.
308. R. Belcher and T. S. West, *Talanta*, *8*, 863 (1961).

309. R. Greenhalgh and J. P. Riley, *Anal. Chim. Acta*, *25*, 179 (1961).
310. S. S. Yamamura, M. A. Wade, and J. H. Sikes, *Anal. Chem.*, *34*, 1308 (1962).
311. G. F. Brunzie and R. T. Pflaum, *Proc. Iowa. Acad. Sci.*, *69*, 186 (1963).
312. M. Hanocq and L. Molle, *Anal. Chim. Acta*, *40*, 13 (1968).
313. M. Hanocq and L. Molle, *Anal. Chim. Acta*, *42*, 349 (1968).
314. C. A. Johnson and M. A. Leonard, *J. Pharm. Pharmacol.*, *13*, 164T (1961).
315. R. J. Hall, *Analyst*, *88*, 76 (1963).
316. A. G. Davies and J. K. Foreman, *Proceedings of the Society of Analytical Chemists Conference*, Nottingham, England, 1965, p. 167. (C. A. 65: 19310, 1966).
317. R. S. Manly, D. H. Foster, and D. P. Harrington, *Technicon Symposium*, *Automation in Analytical Chemistry*, Medied, New York, 1966, p. 250.
318. S. Hirano, H. Fujinuma, and T. Kasai, *Bunseki Kagaku*, *15*, 1339 (1966). (C.A. 67:36297, 1967.)
319. S. G. Iyer, P. H. Iyer, and C. Venkateswarlu, *Indian J. Chem.*, *11*, 1326 (1973).
320. J. P. S. Haarsma and J. Agterdenbos, *Talanta*, *18*, 747 (1971).
321. P. Venkateswarlu, unpublished results, 1969.
322. A. G. Macejunas, *J. Am. Water Works Assoc.*, *61*, 311 (1969).
323. L. Moraru, *Stud. Cerc. Chim.*, *19*, 757 (1971)
324. H. H. Willard and C. A. Horton, *Anal. Chem.*, *24*, 862 (1952).
325. J. Bouman, *Chem. Weekbl.*, *51*, 33 (1955).
326. W. A. Powell and J. H. Saylor, *Anal. Chem.*, *25*, 960 (1953).
327. J. C. Guyon, B. E. Jones, and D. A. Brotton, *Mikrochim. Acta*, *6*, 1180 (1968).
328. G. H. Schenk and K. P. Dilloway, *Anal. Lett.*, *2*, 379 (1969).
329. T. L. Har and T. S. West, *Anal. Chem.*, *43*, 136 (1971).
330. J. Harms and G. Jander, *Z. Elektrochem.*, *42*, 315 (1936).
331. Y. Israel, B. Barnas, and A. Yahalom, *Anal. Chim. Acta*, *36*, 526 (1966).
332. C. L. Grant and H. M. Haendler, *Anal. Chem.*, *28*, 415 (1956).
333. O. Klug, *Femip. Kut. Intez. Kozl.*, *6*, 367 (1962). (F.A. 15:15, 1959–1962.)
334. A. Bellomo, V. Marsala, and O. Klug, *Chem. Anal.* (Warsaw), *9*, 879 (1964). (F.A. 15:15, 1963–1965.)
335. J. P. Dowdall, D. V. Sinkinson, and H. Stretch, *Analyst*, *80*, 491 (1955).
336. R. Flatt, *Angew. Chem.*, *50*, 329 (1937).
337. S. T. Talipov, I. L. Tedorovich, and N. P. Shestakova,)(*Tr. Sredneaziat. Gosudarst. Univ.*)) (Tashkent), 33,) 83 (1952). (C.A. 50:1526, 1956.)
338. A. Langer, *Ind. Eng. Chem.* (*Anal. Ed.*), *12*, 511 (1940).
339. H. G. Petrow and L. K. Nash, *Anal. Chem.*, *22*, 1274 (1950).
340. R. Schliebs, *Z. Anal. Chem.*, *150*, 321 (1956).
341. I. M. Kolthoff and C. J. Sambucetti, *J. Am. Chem. Soc.*, *81*, 1516 (1959).
342. I. M. Kolthoff, E. J. Meehan, and C. J. Sambucetti, *Anal. Chim. Acta*, *21*, 233 (1959).
343. I. M. Kolthoff, E. J. Meehan, and C. J. Sambucetti, *Anal. Chim. Acta*, *22*, 351 (1960).
344. I. M. Kolthoff, E. J. Meehan, and C. J. Sambucetti, *Anal. Chim. Acta*, *22*, 253 (1960).
345. C. R. Castor and J. H. Saylor, *Anal. Chem.*, *24*, 1369 (1952).
346. J. K. Johannesson, *Chem. Ind.* (*London*), 1957, 480. (C.A. 51:13646, 1957.)
347. W. E. Harris, *Anal. Chem.*, *30*, 1000 (1958).
348. I. L. Liplavk, *Sb. Tr. Fiz.-Khim. Lab. Vses. Nauch.-Issled. Inst. Okhr. Tr.* (*Sverdlovks*), 5 (1962). (C.A. 60:4686, 1964.)
349. G. W. Low, Jr., and E. H. Pryde, *J. Am. Chem. Soc.*, *61*, 2237 (1939).
350. B. B. Baker and J. D. Morrison, *Anal. Chem.*, *27*, 1306 (1955).
351. B. B. Baker, *Anal. Chem.*, *30*, 1085 (1958).
352. O. H. Howard and C. W. Weber, *J. Am. Ind. Hyg. Assoc.*, *23*, 48 (1964).
353. S. Megregian, *Anal. Chem.*, *29*, 1063 (1957).
354. British Patent, 1,156,915, applied for July 10, 1965.

355. D. R. Turner, *Anal. Chem.*, *33*, 959 (1961).
356. D. L. Manning and J. C. White, *Anal. Chem.*, *25*, 1648 (1953).
357. R. Haul and W. Griess, *Z. Anorg. Chem.*, *259*, 42 (1949).
358. B. J. MacNulty, G. F. Reynolds, and E. A. Terry, *Nature*, *169*, 888 (1952).
359. C. E. Shoemaker, *Anal. Chem.*, *27*, 552 (1955).
360. A. Iwase, *Nippon Kagaku Zasshi*, *30*, 1138 (1959). (C.A. 55:4235, 1961.)
361. A. M. Bond and T. A. O'Donnell, *Anal. Chem.*, *40*, 1405 (1968).
362. H. E. Zittel and T. M. Florence, *Anal. Chim. Acta*, *40*, 27 (1968).
363. H. Berge and L. Bruegmann, *Anal. Chim. Acta*, *63*, 175 (1973).
364. K. Srinivasan and G. A. Rechnitz, *Anal. Chem.*, *40*, (1968).
365. R. E. Mesmer, *Anal. Chem.*, *40*, 443 (1968).
366. T. B. Warner, *Anal. Chem.*, *41*, 527 (1969).
367. M. S. Frant and J. W. Ross, Jr., *Anal. Chem.*, *40*, 1169 (1968).
368. M. S. Frant, *Galvotechnik*, *63*, 745 (1972).
369. H. Zentner, *Chem. Ind.*, *10*, 480 (1973).
370. A. F. Berndt and R. I. Stearns, *Anal. Chim. Acta*, *74*, 446 (1975).
371. J. J. Lingane, *Anal. Chem.*, *39*, 881 (1967).
372. J. J. Lingane, *Anal. Chem.*, *40*, 935 (1968).
373. C. Hrzdorf, *Z. Anal. Chem.*, *245*, 67 (1969).
374. W. E. Bazzelle, *Anal. Chim. Acta*, *54*, 29 (1971).
375. E. Heckel and P. F. Marsh, *Anal. Chem.*, *44*, 2347 (1972).
376. T. Anfalt, D. Dyrssen, and D. Jagner, *Anal. Chim. Acta*, *43*, 487 (1968).
377. T. Eriksson and G. Johansson, *Anal. Chim. Acta*, *52*, 465 (1970).
378. G. Muto and K. Nozaki, *Japan Anal.*, *18*, 247 (1969).
379. P. A. Evans, G. J. Moody, and J. D. R. Thomas, *Lab. Pract.*, *20*, 644 (1971).
380. E. W. Baumann, *Anal. Chim. Acta*, *42*, 127 (1968).
381. T. B. Warner, *Water Res.*, *5*, 459 (1971).
382. L. G. Bruton, *Anal. Chem.*, *43*, 579 (1971).
383. M. A. Peters and D. M. Ladd, *Talanta*, *18*, 655 (1971).
384. D. Jagner and V. Pavlova, *Anal. Chim. Acta*, *60*, 153 (1972).
385. J. R. Melton, W. L. Hoover, and J. L. Ayers, *J. Assoc. Off. Anal. Chem.*, *57*, 508 (1974).
386. T. Anfalt and D. Jagner, *Anal. Chim. Acta*, *53*, 13 (1971).
387. L. L. Hall, F. A. Smith, O. H. Lopez, and D. E. Gardner, *Clin. Chem.*, *18*, 1455 (1972).
388. A. L. Liberti and M. Mascini, *Anal. Chem.*, *41*, 676 (1969).
389. W. Selig, *Mikrochim. Acta*, 87 (1973).
390. A. L. Schick, *J. Assoc. Off. Anal. Chem.*, *56*, 1369 (1973).
391. N. Shiraishi, Y. Murata, G. Nagagawa, and K. Kodama, *Anal. Lett.*, *6*, 893 (1973).
392. R. A. Durst, *Mikrochim. Acta*, 1969, 611.
393. R. A. Durst and J. K. Taylor, *Anal. Chem.*, *12*, 1483 (1967).
394. R. A. Durst, *Anal. Chem.*, *40*, 931 (1968).
395. R. A. Durst, *Anal. Chem.*, *41*, 2089 (1969).
396. E. A. Materova, G. P. Grinberg, and M. M. Evstifeeva, *Zh. Anal. Khim.*, *24*, 821 (1969).
397. N. V. Bausova, V. G. Bamburov, L. I. Manakova, and A. P. Sivoplyas, *Zh. Anal. Khim.*, *28*, 2042 (1973).
398. H. Freiser, H. J. James, G. D. Carmack, B. M. Kneebone, and R. W. Cattrall, British Patent 1,375,785, applied for March 30, 1972.
399. H. Hirata and M. Ayuzawa, *Chem. Lett.*, 1974, 1451.
400. J. H. Fremlin, J. L. Hardwick, and J. L. Suthers, *Nature*, *180*, 1179 (1957).
401. W. Van der Mark and H. A. Das, *J. Radioanal. Chem.*, *23*, 7 (1974).
402. H. A. Das, W. H. Kohnemann, G. D. Wals, and J. Zonderhuis, *J. Radioanal. Chem.*, *25*, 261 (1975).
404. E. I. Onstott and W. P. Ellis, *Anal. Chem.*, *28*, 393 (1956).

406. H. Siegel, *Trans. Am. Nucl. Soc.*, *11*, 468 (1968).
407. J. M. Peixoto-Cabral and H. Gotte, *Z. Naturforsch.*, *106*, 440 (1955).
408. I. A. Carmichael and J. E. Whitley, *Analyst*, *94*, 737 (1969).
409. G. J. Atchinson and W. H. Beamer, *Anal. Chem.*, *28*, 237 (1956).
410. J. M. Bakes and P. G. Jeffery, *Analyst*, *91*, 216 (1966).
411. J.-P. Garrec and A. Fourcy, U.S. Atomic Energy Commission Document No. CEA-CONF-1747, 1970. (Nuclear Science Abstracts 25:29011, 1971.)
412. P. E. Wilkniss, *Anal. Chem.*, *41*, 421 (1969).
413. G. H. Andersen, F. M. Graber, V. P. Guinn, H. R. Lukens, and D. M. Settle, *Proceedings of a Symposium on Nuclear Activities, Techniques, and Life Sciences*, Amsterdam, 1967, p. 99.
414. D. H. Reitief, P. E. Cleaton-Jones, J. Turkstra, and W. J. De Wet, *Arch. Oral Biol.*, *16*, 1257 (1971).
415. R. Henkelmann, H. Starek, and H.-J. Born, *Radiochim. Acta*, *11*, 101 (1969).
416. W. F. Naughton and W. A. Jester, *Nucl. Appl. Technol.*, *9*, 851 (1970).
417. J. Wing and M. A. Wahlgren, *J. Radioanal. Chem.*, *3*, 37 (1969).
418. J. J. Lauff, E. R. Champlin, and E. P. Przybylowicz, *Anal. Chem.*, *45*, 52 (1973).
419. J. Perdijon, *Talanta*, *21*, 1047 (1974).
420. J. S. Hislop, *Rep. U.K. At. Energy Auth.*, AERER 7646, 1974.
421. D. Brune, S. Mattsson, and K. Liden, U.S. Atomic Energy Commission Document No. AE-333, 1968.
422. E. Engelman and A. C. Scherle, *Int. J. Appl. Radiat. Isot.*, *22*, 415 (1971).
423. P. E. Wilkniss, *Radiochim. Acta*, *11*, 138 (1969).
424. P. K. Wilkniss and V. J. Linnenbom, *Limnol. Oceanogr.*, *13*, 530 (1968).
425. S. Ohno, M. Suzuki, K. Sasajima, and S. Iwata, *Analyst*, *95*, 260 (1970).
426. L. Kosta and J. Slunecko, *Anal. Chem.*, *42*, 831 (1970).
427. M. M. Dorosh, N. P. Mazyukevich, A. M. Parlag, and V. A. Shkoda-Ulyanov, *At. Energy*, *35*, 339 (1973).
428. H. Rudolph, J. J. Krausharr, R. A. Ristinen, and R. R. Meglen, in *Trace Substances in Environmental Health*, Vol. 7, *Symposium*, D. D. Hemphill, Ed., University of Missouri, Columbia, 1974, p. 313.
429. G. E. Coote, N. E. Whitehead, and G. J. McCallum, *J. Radioanal. Chem.*, *12*, 491 (1972).
430. J. M. Bewers and F. C. Flack, *Analyst*, *94*, 1 (1969).
431. G. Demortier, *Radiochem. Radioanal. Lett.*, *16*, 329 (1974).
432. A. Anttila and J. Keinonen, *Int. J. Appl. Radiat. Isot.*, *24*, 293 (1973).
433. D. M. Lee, J. F. Lamb, and S. S. Markowitz, *Anal. Chem.*, *43*, 542 (1971).
434. J. W. Mitchell, E. Bloom, and C. E. Gooden, *J. Radioanal. Chem.*, *25*, 221 (1975).
435. J. K. Foreman, *Analyst*, *94*, 425 (1969).
436. J. L. Lambert, *Anal. Chem.*, *25*, 271 (1953).
437. W. D. Armstrong and L. Singer, *Anal. Chem.*, *26*, 1047 (1954).
438. M. L. Cabello-Tomas and T. S. West, *Talanta*, *16*, 781 (1969).
439. R. V. Hems, G. F. Kirkbright, and T. S. West, *Talanta*, *17*, 433 (1970).
440. I. Mori, *Japan An.*, *20*, 282 (1971).
441. D. Klockow, H. Ludwig, and M. A. Girando, *Anal. Chem.*, *42*, 1682 (1970).
442. H. Stetter, *Chem. Ber.*, *81*, 532 (1948).
443. H. W. Linde, *Anal. Chem.*, *31*, 2092 (1959).
444. C. McGaughey and E. C. Stowell, *Anal. Chem.*, *36*, 2344 (1964).
445. C. McGaughey and E. C. Stowell, *J. Dent. Res.*, *45*, 76 (1966).
446. M. A. McKee, E. C. Stowell, and C. McGaughey, International Association for Dental Research, 47th General Meeting, Houston, Texas, 1969, Abstr. No. 428.
447. F. Geike, *Z. Anal. Chem.*, *255*, 134 (1971).
448. R. Bock and S. Strecker, *Z. Anal. Chem.*, *266*, 110 (1973).

449. E. Cropper and N. A. Puttnam, *J. Soc. Cosmet. Chem.*, *21*, 533 (1970).
450. H. A. Russel, *Z. Anal. Chem.*, *252*, 143 (1970).
451. D. W. Groff, Abstracts of Papers, HIST 18, 156th American Chemical Society National Meeting, Atlantic City, New Jersey, 1968.
452. J. MacGee and K. G. Allen, *Anal. Chem.*, *42*, 1672 (1970).
453. B. Gutsche, R. Herrmann, and K. Ruediger, *Z. Anal. Chem.*, *258*, 273 (1972).
454. L. G. Spears and N. Hackerman, *J. Gas Chromatogr.*, *6*, 392 (1968).
455. O. Pitak, *Chromatographia*, *3*, 29 (1970).
456. W. S. Pappas and J. G. Million, *Anal. Chem.*, *40*, 2176 (1968).
457. A. Brachet-Liermain, L. Ferrus, and J. Caroff, *J. Chromatogr. Sci.*, *9*, 49 (1971).
458. J. E. Peterson, *Bull. Environ. Contam. Tocol.*, *13*, 751 (1975).
459. K. Yamanouchi, R. Murashima, and K. Yokoyoma, *Chem. Pharm. Bull. Tokyo*, *23*, 1363 (1975).
460. J. L. Hardwick and C. J. Martin, *Helv. Odontol. Acta*, *11*, 62 (1967).
461. P. Swift, *Brit. Dent. J.*, *123*, 326 (1967).
462. N. L. Gregory, *Anal. Chem*, *44*, 231 (1972).
463. J. R. McNally Jr., G. R. Harrison, and E. Rowe, *J. Opt. Soc. Am.*, *37*, 93 (1947).
464. A. K. Chakraborty, K. C. Singhal, and B. K. Banerjee, *Technology* (Sindri), *3*, 303 (1966).
465. I. Schoenfeld, *Appl. Spectrosc.*, *24*, 359 (1970).
466. J. Czakow, *Chim. Anal.* (Warsaw), *18*, 891 (1973).
467. R. B. Ioffe and L. S. Leonova, *Zh. Prikl. Spektrosk.*, *17*, 699 (1972); *Ref. Zh.*, *Khim.*, 19GD, 1973 (9), Abstr. No. 9G116.
468. I. A. Berezin, O. F. Degtyareva, and P. P. Shevchenko, *Zh. Prikl. Spektrosk.*, *4*, 170 (1966); *Ref. Zh.*, *Khim.*, 19GD, 1966 (19) Abstr. No. 19G67.
469. A. V. Karyakin, E. N. Savinova, and T. P. Andreeva, *Zh. Anal. Khim.*, *24*, 468 (1969).
470. A. M. Bond and T. A. O'Donnell, *Anal. Chem.*, *40*, 560 (1968).
471. C. L. Luke, *Anal. Chim. Acta*, *43*, 245 (1968).
472. G. Loch, *X-Ray Spectrom.*, *2*, 125 (1973).
473. U. Jecht, *Chem. Ztg.*, *96*, 104 (1972).
474. Davies Halide Meter, Bulletin 1148, Davies Instrument Division, Newark, New Jersey.
475. G. O. Nelson, *Am. Ind. Hyg. Assoc. J.*, *29*, 586 (1968).
476. A. M. Sazonov, *Geofizi. Appar.*, 1968, 109. (C.A. 69:73631, 1968.)
477. V. C. West, *J. Dent. Res.*, *47*, Suppl. Part 1, 1024 (1968).
478. J. W. Mandler, R. B. Moler, E. Raisen, and K. Rajan, *Thin Solid Films*, *19*, 1 (1973).
479. L. Porte, J. P. Sandino, J. Gera, J. P. Thomas, and J. Tousset, *J. Radioanal. Chem.*, *16*, 493 (1973).
480. W. Van der Lugt, *Fluoride*, *4*, 15 (1971).
481. J. O. Lay, *Metallurgia*, *51*, 109 (1955).
482. P. L. Hanst, *Appl. Spectrosc.*, *24*, 161 (1970).
483. P. J. Walitsky and R. H. Hopkins, *Appl. Spectrosc.*, *23*, 492 (1969).
484. M. Grasserbauer and K. Weinrotter, *Z. Anal. Chem.*, *263*, 184 (1973).
485. H. Malissa, M. Grasserbauer, and K. Weinrotter, *Z. Anal. Chem.*, *266*, 19 (1973).
486. E. I. Esmail, C. J. Nicholls, and D. S. Urch, *Analyst*, *98*, 725 (1973).
487. M. Fleischer, *Recent Estimates of the Abundance of the Elements in the Earth's Crust*, Washington D.C. (U.S. Geological Survey Circular No. 285, 1953).
488. W. H. MacIntire, L. S. Jones, and L. J. Hardin, *J. Assoc. Off. Agr. Chem.*, *33*, 653 (1950).
489. P. Venkateswarlu, "*Biochemical Investigations on Fluorosis*," D.Sc. Thesis, Andhra, 1955.
490. P. J. Ke and L. W. Regier, *J. Fish. Res. Bd. Ca.*, *28*, 1055 (1971).
491. B. S. Marshall and R. Wood, *Analyst*, *94*, 493 (1969).
492. C.-C. Chu and J. L. Schafer, *Anal. Chem*, *27*, 918 (1955).
493. R. J. Hall, *Analyst*, *93*, 461 (1968).

494. F. Seel, E. Steigner, and I. Burger, *Angew. Chem.* (*Int. Ed.*), *3*, 424 (1964).
495. W. Rechenberg, *Z. Anal. Chem.*, *263*, 333 (1973).
496. A. Hall and J. N. Walsh, *Anal. Chim. Acta*, *45*, 341 (1969).
497. W. H. Ficklin, U.S. Geological Survey, Professional Paper No. 700-C, 1970, p. 186.
498. C. R. Edmond, *Anal. Chem.*, *41*, 1327 (1969).
499. B. I. Polyakov and L. E. Kondrat'evskaya, *Uzbek. Khim. Zh.*, 16 (1969); *Ref. Zh.*, *Khim.*, 19GD, 1969 (13) Abstr. No. 13G188.
500. G. G. Taneeva and N. A. Freze, *Zavod. Lab.*, *36*, 665 (1970).
501. N. Shiraishi, T. Hisayuki, and K. Kodama, *Japan Anal.*, *23*, 453 (1974).
502. A. C. D. Newman, *Analyst*, *93*, 827 (1968).
503. R. L. Clements, G. A. Sergeant, and P. J. Webb, *Analyst*, *96*, 51 (1971).
504. M. Shiraishi, Y. Murata, and K. Kodama, *Japan Anal.*, *23*, 247 (1974).
505. J. E. Harwood and D. J. Huyser, *Water Res.*, *2*, 639 (1968).
506. O. M. Smith and H. A. Dutcher, *Ind. Eng. Chem.* (*Anal. Ed.*), *6*, 61 (1934).
507. A. M. Bond and M. M. Murray, *J. Physiol.*, *116*, 18 (1952).
508. R. Kh. Zamanov and R. Sh. Rafikova, *Sb. Nauch. Tr. Samarkand Med. Inst.*, *37*, 298 (1967). (C.A. 69:54212, 1968.)
509. O. B. Dirks and F. H. Cox, *Caries Res.*, *1*, 295 (1967).
510. R. E. Shoup, *Anal. Chem.*, *29*, 1216 (1957).
511. J. F. Grutsch, W. H. Nebergall, J. C. Muhler, R. B. Fischer, and H. G. Dey, *J. Dent. Res.*, *32*, 463 (1953).
512. M. J. Taras, H. D. Cisco, and M. Garnell, *J. Am. Water Works Assoc.*, *42*, 583 (1950).
513. T. Wierzbicki, W. Pawlita, and H. Pieprzyk, *Chem. Anal.*, *16*, 1079 (1971). (C.A. 76:89871, 1972.)
514. L.-H. Li and C.-C. Huang, *Hua Hsueh Tung Pao*, 1964, 433. (C.A. 62:2618, 1965.)
515. V. L. Zolotavin and W. M. Kazakova, *Zavod. Lab.*, *31*, 297 (1965). (C.A. 62:12900, 1965.)
516. F. S. Kelso, J. M. Matthews, and H. P. Kramer, *Anal. Chem.*, *36*, 577 (1964).
517. H. Fujinuma, S. Hirano, and K. Itsuki, *Japan Anal.*, *23*, 148 (1974).
518. T. S. Light, R. F. Mannion, and K. S. Fletcher, III, *Talanta*, *16*, 1441 (1969).
519. A. T. Palin, *Water Water Eng.*, *71*, 371 (1967).
520. M. Ya. Shapiro and V. G. Kolesnikova, *Zh. Anal. Khim.*, *18*, 507 (1963).
521. N. Shiraishi, S. Yoshikawa, and T. Hisayuki, *Bunseki Kagaku*, *17*, 1027 (1968). (C.A. 69:113228, 1968.)
522. A. H. Meyling and J. Meyling, *Analyst*, *88*, 84 (1963).
523. K. H. Quentin and A. Rosopulo, *Z. Anal. Chem.*, *241*, 241 (1968).
524. A. E. Hey and S. H. Jenkins, *Water Res.*, *3*, 901 (1969).
525. W. M. Shaw, *Anal. Chem.*, *26*, 1212 (1954).
526. J. J. Morrow and J. C. Henry, *J. Am. Water Works Assoc.*, *59*, 245 (1967).
527. T. G. Thompson and H. J. Taylor, *Ind. Eng. Chem.* (*Anal. Ed.*), *5*, 87 (1933).
528. C. D. Anselm and R. J. Robinson, *J. Mar Res.*, *10*, 203 (1951).
529. K. Grasshoff, *Second Technicon Symposium*, Medied, New York, 1965 p. 304.
530. K. M. Chan and J. P. Riley, *Anal. Chim. Acta*, *35*, 365 (1966).
531. R. A. Kletsch and F. A. Richards, *Anal. Chem.*, *42*, 1435 (1970).
532. R. H. Babcock and K. A. Johnson, *J. Am. Water Works Assoc.*, *60*, 953 (1968).
533. N. P. Kelada, N. I. McClelland, and K. H. Mancy, 160th National Meeting of the American Chemical Society, Chicago, 1970. Division of Water, Air and Waste Chemistry, Sec. B, Abstr. No. 76, 1970.
534. E. F. McFarren, B. J. Moorman, and J. H. Parker, U.S. Public Health Service Publication No. 1895, 1969.
535. P. J. Ke and L. W. Regier, *Anal. Chim. Acta*, *53*, 23 (1971).
536. E. J. Duff and J. L. Stuart, *Talanta*, *19*, 76 (1972).
537. T. A. Palmer, *Talanta*, *19*, 1141 (1972).

538. D. Daugherty and J. D. Johnson, *ESE Notes*, *6*, 3 (1969). (Water Pollution Abstr., 42:2303, 1969.)
539. N. Takemota, K. Mashiko, and M. Setoguchi, *Nippon Kaisui Gakkai-Shi*, *23*, 271 (1970). (C.A. 73:101879, 1970.)
540. E. Robinson, *Am. Ind. Hyg. Assoc. Q.*, *18*, 145 (1957).
541. R. Mavrodineanu and R. R. Coe, *Contrib. Boyce Thompson Inst.*, *18*, 173 (1955).
542. W. Leithe and G. Petschl, *Z. Anal. Chem.*, *226*, 352 (1966).
543. D. L. R. Bailey, C. M. Killick, and P. E. Trott, Report from the Warren Spring Laboratory, Stevenage, England, LR 129(AP), 1970.
544. K. P. Panin, *Gig. Sanit.*, *32*, 73 (1967); English transl. *Hyg. Sanit.*, *32*, 398 (1967).
545. B. S. Marshall and R. Wood, *Analyst*, *93*, 821 (1968).
546. P. W. West, G. R. Lyles and J. Miller, *Environ. Sci. Technol.*, *4*, 487 (1970).
547. S. W. Chaikin, A. C. Wilbur, and T. D. Parks, 126th Meeting of the American Chemical Society, Abstracts of Papers, 1954, p. 10B.
548. J. P. Nielsen and A. D. Dangerfield, *Arch. Ind. Health*, *11* 61 (1955).
549. D. M. Paez and J. A. Vasquez, *Rev. Asoc. Bioquim, Argent.*, *33*, 152 (1968). (A.A. 18:2823, 1970.)
550. S. N. Suvorova, A. M. Vorob'ev, and G. V. Rabovskii, *Gig. Sanit.*, *28*, 48 (1963). (F.A. 15:81, 1963–1965.)
551. M. R. Pack, A. C. Hill, and H. M. Benedict, *J. Air Pollut. Control Assoc.*, *13*, 374 (1963).
552. C. M. Killick, Report of the Warren Spring Laboratory, Stevenage, England, LR 130(AP), 1970.
553. L. A. Elfers and C. E. Decker, *Anal. Chem.*, *40*, 1658 (1968).
554. K. Habel, *Staub Reinhalt. Luft*, *28*, 283 (1968). English transl., *28*, 26, (1968).
555. R. H. Mandl, L. H. Weinstein, G. J. Weiskopf, and J. L. Major, Second International Clean Air Congress of the International Union of Air Pollution Prevention Associations, Washington D.C., 1970, *Proceedings Digest*, 1970, p. 107.
556. T. Okita, K. Kaneda, T. Yanaka, and R. Sugai, *Atmos. Environ.*, *8*, 923 (1974).
557. J. A. Dorsey and D. A. Kemnitz, *J. Air Pollut. Control Assoc.*, *18*, 12 (1968).
558. S. W. Chaikin, T. D. Parks, and C. I. Glassbrook, U.S. Patent 2,741,544, April 24, 1956, Stanford Research Institute. (C.A. 50:9071, 1956.)
559. J. O. Ivie, L. F. Zielenski, M. D. Thomas, and C. R. Thompson, *J. Air Pollut. Control Assoc.*, *15*, 195 (1965).
560. G. D. Clayton, American Gas Association, *Proceedings of the 35th Annual Convention*, 1953, p. 892.
561. F. Feigl and Heisig, *Anal. Chim. Acta*, *3*, 561 (1949).
562. D. C. MacLean, L. H. Weinstein, and R. H. Mandl, *Contrib. Boyce Thompson Inst.*, *24*, 9 (1967).
563. H. Brown, 62nd Annual Meeting of the Air Pollution Control Association, New York, 1969, Paper No. 69–119.
564. A. Liberti and M. Mascini, Second International Clean Air Congress of the International Union of Air Pollution Prevention Associations, Washington D.C., 1970, *Proceedings Digest*, 1970, p. 140.
565. C. R. Thompson, L. F. Zielenski, and J. O. Ivie, *Atmos. Environ.*, *1*, 253 (1967).
566. A. Liberti and M. Mascini, *Fluoride*, *4*, 49 (1971).
567. E. T. Peters, J. E. Oberholtzer, and J. R. Valentine, Report of the Environmental Protection Agency, EPA-R2-72-126, 1972.
568. O. B. Winter and L. Buttler, *J. Assoc. Off. Agr. Chem.*, *16*, 105 (1933).
569. W. E. Crutchfield, Jr., *Ind. Eng. Chem.* (*Anal. Ed.*), *14*, 57 (1942).
570. R. Truhaut, *Arch. Oral Biol.*, Suppl. 1963, 5.
571. R. Debiard, A. Fourcy, and J. P. Garrec, *CR Hebd. Séances* (*Paris*), *264*, Ser. D, 2668 (1967).
572. F. Yamazoe, *Nippon Dojo-Hiryogaku Zasshi*, *41*, 37 (1970). (C.A. 73:42297, 1970.)

573. T. Yamamoto, Y. Suketa, K. Takahashi, and E. Nagakane, *J. Hyg. Chem.* (Tokyo), *15*, 90 (1969). (C.A. 73:42297, 1970.)
574. J. Gwirtsman, R. Mavrodineanu, and R. R. Coe, *Anal. Chem.*, *29*, 887 (1957).
575. M. Buck and G. Reusmann, *Fluoride*, *4*, 5 (1971).
576. T. Yamamoto, Y. Suketa, E. Mikami, and M. Omura, *Eisei Kagaku* (*J. Hyg. Chem.*), *21*, 23 (1975).
577. H. C. Lockwood, *Analyst*, *62*, 775 (1937).
578. S. Deutsch, *Anal. Chem.*, *27*, 1154 (1955).
579. M. D. Thomas and R. E. Amtower, *J. Air Pollut. Control Assoc.*, *19*, 439 (1969).
580. L. H. Weinstein, R. H. Mandl, D. C. McCune, J. S. Jacobson, and H. E. Hitchcock, *J. Air Pollut. Control Assoc.*, *15*, 222 (1965).
581. R. Debiard and M. L. Dupraz, *Chim. Anal.*, *48*, 384 (1966).
582. A. I. Williams, *Analyst*, *94*, 300 (1969).
583. G. Reusmann and J. Westphalen, *Staub, Reinhalt. Luft*, *29*, 413 (1969).
584. M. Buck and G. Reusmann, *Fluoride*, *4*, 5 (1971).
585. J. S. Jacobson and L. I. Heller, Second International Clean Air Congress of the International Union of Air Pollution Prevention Associations, Washington, D.C., *Proceedings Digest*, 1970, p. 108.
586. H. L. Galloway, R. E. Shoaf, and C. H. Skaggs, *Am. Ind. Hyg. Assoc. J.*, *36*, 721 (1975).
587. R. L. Baker, *Anal. Chem.*, *44*, 1326 (1972).
588. M. Hukushima, H. Hukushima, and T. Kuroda, *Japan Anal.*, *21*, 522 (1972).
589. J. S. Jacobson and D. C. McCune, *J. Assoc. Off. Anal. Chem.*, *55*, 991 (1972).
590. B. Thomas and N. Trinder, *Emp. J. Exp. Agric.*, *15*, 237 (1947).
591. R. H. Wade, J. M. Ross, and H. M. Benedict, *J. Chromatogr.*, *14*, 37 (1964).
592. J. Lovelace, G. W. Miller, and G. W. Welkie, *Atmos. Environ.*, *2*, 187 (1968).
593. J. Cheng, Y.-O. Yu, G. W. Miller, and G. W. Wilkie, *Environ. Sci. Technol.*, *2*, 367 (1968).
594. R. A. Peters and M. Shorthouse, *Phytochemistry*, *11*, 1337 (1972).
595. R. A. Peters and M. Shorthouse, *Phytochemistry*, *11*, 1339 (1972).
596. L. H. Weinstein, D. C. McCune, J. F. Mancini, L. J. Colvito, D. B. Silberman, and P. Van Leuken, *Environ. Res.*, *5*, 393 (1972).
597. R. A. Peters and M. Shorthouse, *Nature*, *216*, 80 (1967).
598. R. A. Peters and M. Shorthouse, *Nature*, *231*, 123 (1971).
599. A. M. Pochomis and F. D. Griffith, *Am. Ind. Hyg. Assoc. J.*, *32*, 557 (1971).
600. L. Singer and W. D. Armstrong, *Anal. Chem.*, *40*, 613 (1968).
601. H. G. McCann, *Arch. Oral Biol.*, *13*, 475 (1968).
602. F. Brudevold, H. G. McCann, and P. Gron, *Arch. Oral Biol.*, *13*, 877 (1968).
603. M. J. Larsen, M. Kold, and F. R. Van der Fehr, *Caries Res.*, *6*, 193 (1972).
604. R. L. Levine, *J. Dent. Res.*, *52*, 1249 (1973).
605. A. Zober and B. Schellmann, *Z. Klin. Chem. Klin. Biochem.*, *13*, 197 (1975).
606. E. C. Munksgaard and C. Bruun, *Arch. Oral Biol.*, *18*, 735 (1973).
607. J. A. Ruzicka, H. Jakschova, and L. Mrklas, *Talanta*, *13*, 1341 (1966).
608. P. Venkateswarlu, *Clin, Chim. Acta*, *59*, 277 (1975).
609. P. Venkateswarlu, unpublished results, 1974.
610. R. A. Peters, M. Shorthouse, and P. F. V. Ward, *Biochem. J.*, 113, 9P (1969).
611. C. W. Weber and B. L. Reid, in *Trace Element Metabolism in Animals*, Vol. 2, W. G. Hoekstra, J. W. Suttie, H. E. Ganther, and W. Mertz, Eds. University Park Press, Baltimore, 1974, p. 707.
612. G. C. Goode, C. M. Howard, A. R. Wilson, and V. Parsons, *Anal. Chim. Acta*, *58*, 363 (1972).
613. W. Van der Mark and H. A. Das, *J. Radioanal. Chem.*, *13*, 107 (1973).
614. C. A. Baud and S. Bang, *Proceedings of the Sixth International Conference on X-Ray Optics and Microanalysis*, Osaka, Japan, 1971, University of Tokyo Press, Tokyo, 1972, p. 835.
615. S. Suga, in ref. 614, p. 847.

616. Y. Ericsson and S. Ullberg, *Acta Odontl. Scand.*, *16*, 363 (1958).
617. D. E. Gardner, F. A. Smith, H. C. Hodge, D. E. Overton, and R. Feltman, *Science*, *115*, 208 (1952).
618. F. A. Smith, D. E. Gardner, N. C. Leone, and H. C. Hodge, *Am. Med. Assoc. Arch. Ind. Health*, *21*, 330 (1960).
619. Y. Ericcson and L. Hammarstrom, *Gerontologia*, *9*, 150 (1964).
620. D. Osis, E. Wiatrowski, J. Samachson, and H. Spencer, *Clin. Chim. Acta*, *51*, 211 (1974).
621. M. Drescher and J. W. Suttie, *Proc. Soc. Exp. Biol. Med.*, *139*, 228 (1972).
622. D. O. Quissell and J. W. Suttie, *J. Cell Res.*, *82*, 59 (1973).
623. R. F. Miller and P. H. Phillips, *Anal. Chem.*, *25*, 172 (1953).
624. G. H. Farrah, *Am. Ind. Hyg. Assoc. J.*, *25*, 55 (1964).
625. H. A. Derner, *Am. Ind. Hyg. Assoc. J.*, *28*, 357 (1967).
626. J. Bernuchon and M. Derome-Tremblay, *Arch. Mal. Prof. Med. Tra. Securité Sociale*, *25*, 577 (1966). (F.A. 15:219, 1966–68.)
627. J. Tusl, *Chem. Zvest.*, *23*, 295 (1969). (C.A. 72:30335, 1970.)
628. American Conference of Governmental Industrial Hygienists, in *Manual of Analytical Methods Recommended for Sampling and Analysis of Atmospheric Contaminants*, 1967, p. 1. (F. A. *15*: 165, 1969)
629. N. A. Talvitie and L. W. Brewer, *Am. Ind. Hyg. Assoc. J.*, *21*, 287 (1960).
630. G. Rausa and R. Trivello, *Ig. Mod.*, *63*, 89 (1970). (A.A. 22: 253, 1972.)
631. A. A. Cernik, J. A. Cooke, and R. J. Hall, *Nature*, *227*, 1260 (1970).
632. J. Tusl, *Anal. Chem.*, *44*, 1693 (1972).
633. R. Haavaldsen and A. M. Ostby, *Nordisk Hyg. Tidskr.*, *49*, 46 (1968). (F.A. 15:225, 1966–1968.)
634. J. D. Neefus, J. Cholak, and B. E. Saltzman, *Am. Ind. Hyg. Assoc. J.*, *31*, 96 (1970).
635. H.-D. Dell and J. Fielder, *Z. Anal. Chem.*, *270*, 278 (1974).
636. F. A. Smith and D. E. Gardner, *J. Dent. Res.*, *30*, 182 (1951).
637. P. Venkateswarlu, W. D. Armstrong, and L. Singer., *Indian J. Med. Res.*, *54*, 1156 (1966).
638. C. H. Carlson, W. D. Armstrong, and L. Singer, *Am. J. Physiol.*, *199*, 187 (1960).
639. Y. Takaesu, *Jap. J. Dental Health*, *16*, 13 (1966).
640. P. Sita and P. Venkateswarlu, *Bull. Inst. Postgrad. Med. Ed. Res., Chandigarh, India*, *1*, 5 (1967).
641. D. R. Taves, *Nature*, *211*, 192 (1966).
642. L. Singer and W. D. Armstrong, *Nature*, *214*, 5093 (1967).
643. D. R. Taves, *Nature*, *215*, 1308 (1967).
644. G. A. Sereda, *Gig. Tr. Prof. Zabol.*, *12*, 54 (1968). (F.A. 15:172, 1966–1968.)
645. F. H. Cox and O. B. Dirks, *Caries Res.*, *2*, 69 (1968).
646. D. M. Paez and B. A. Gil, *Rev. Asoc. Bioquim Argent.*, *38*, 54 (1973). (A.A. 26:2179, 1974.)
647. D. R. Taves, B. W. Fry, R. B. Freeman, and A. J. Giles, *J. Am. Med. Assoc.*, *214*, 91 (1970).
648. P. Gron, H. G. McCann, and F. Brudevold, *Arch. Oral Biol.*, *13*, 203 (1968).
649. Y. Ericsson, *Caries Res.*, *3*, 159 (1969).
650. F. W. Barnes and J. Runcie, *J. Clin. Pathol.*, *21*, 668 (1968).
651. O. B. Dirks, J. M. P. A. Jongeling-Eijndhoven, T. D. Flissebaalje, and I. Gedalia, *Caries Res.*, *8*, 181 (1974).
652. W. J. Simpson and J. Tuba, *J. Oral Med.*, *23*, 104 (1968).
653. C. Fuchs, D. Dorn, C. A. Fuchs, H.-V. Henning, C. McIntosh, F. Scheler, and M. Stennert, *Clin. Chim. Acta*, *60*, 157 (1975).
654. G. N. Jenkins, W. M. Edgar, and D. B. Ferguson, *Arch. Oral Biol.*, *14*, 105 (1969).
655. W. M. Edgar, *Caries Res.*, *6*, 80 (1972).
656. W. M. Edgar and A. Tatevossian, in *Tooth Enamel*, Vol. 2, R. W. Fearnhead and M. V. Stack, Eds., Wright, Bristol, 1971.
657. A. Tatevossian and C. T. Gould, *Arch. Oral Biol.*, in press, 1976.

658. J. M. Birkeland and G. Charlton, *Caries Res.*, *10*, 72 (1976).
659. P. Gron, K. Yao, and M. Spinelli, *J. Dent. Res.*, *48*, 799 (1969).
660. W. M. Edgar and G. N. Jenkins, 50th General Meeting, International Association for Dental Research, Las Vegas, 1972, Abstr. No. 173.
661. W. M. Edgar, *Helv. Odontol. Acta*, *17*, 50 (1973).
662. J. M. Birkeland, *Caries Res.*, *4*, 243 (1970).
663. N. Shane and D. Miele, *J. Pharm. Sci.*, *57*, 1260 (1968).
664. G. Koch, *Svensk. Tandlak. T.*, *62*, 377 (1969). (Oral Res. Abstr. 5:1401, 1970.)
665. E. J. Bushee, D. K. Grissom, and D. R. Smith, *J. Dent. Child.*, *38*, 279 (1971).
666. R. W. Erischsen, *Meddrnorsk Farm. Selsk.*, *35*, 56 (1973). (A.A. 26:247, 1974.)
667. L. O. Guerello, *Proanalisis*, *2*, 68 (1969). (C.A. 74:79744, 1971.)
668. B. C. Jones, J. E. Heveran, and B. Z. Senkowski, *J. Pharm. Sci.*, *58*, 607 (1969).
669. G. Bussmann and W. Hanni, *Pharm. Acta Helv.*, *42*, 41 (1967).
670. M. Hanocq, *Mikrochim. Acta*, 707 (1972).
671. J. M. Rafols Rovira, *Quim. Ind. Bilbao*, *15*, 44 (1968). (A.A. 17:1741, 1969.)
672. K. Venkatarayachari, R. J. Rao, and M. G. C. Naidu, *Salt Res. India*, *6*, 79 (1969).
673. L. Torma and B. E. Ginther, *J. Assoc. Off. Agr. Chem.*, *51*, 1181 (1968).
674. C. W. Louw and J. F. Richards, *Analyst*, *97*, 334 (1972).
675. L. I. Pleskach, *Zavod. Lab.*, *40*, 382 (1974). (A.A. 27:3636, 1974.)
676. J. R. Melton, W. L. Hoover, J. L. Ayers, *J. Assoc. Off. Anal. Chem.*, *57*, 508 (1974).
677. K. Ranfft, *Z. Anal. Chem.*, *269*, 18 (1974).
678. R. B. Klemm and M. E. Ambrose, *J. Assoc. Off. Anal. Chem.*, *53*, 3 (1970).
679. Y. Ericsson, I. Hellstrom, and Y. Hofvander, *Acta Paediatr. Scand.*, *61*, 459 (1972).
680. A. L. Schick, *J. Assoc. Off. Anal. Chem.*, *54*, 798 (1973).
681. C. Martin and S. Brun, *Trav. Soc. Pharm, Montpellier*, *29*, 161 (1969). (C.A. 72:120040, 1970.)
682. W. P. Ferren and N. A. Shane, *J. Food Sci.*, *34*, 317 (1969).
683. V. S. Archer and F. G. Doolittle, *Anal. Chem.*, *39*, 371 (1967).
684. W. Selig, *Mikrochim. Acta*, 337 (1970).
685. B. A. Raby and W. E. Sunderland, *Anal. Chem.*, *39*, 1304 (1967); *40*, 939 (1968).
686. U. Westerlund-Helmerson, *Anal. Chem.*, *43*, 1120 (1971).
687. J. C. Van Loon, *Anal. Lett.*, *1*, 393 (1968).
688. E. J. Duff and J. L. Stuart, *Talanta*, *19*, 76 (1972).
689. W. C. Hanson and D. J. Lloyd, *Chem. Ind.* (*London*), 41 (1972).
690. E. F. Croomes and R. C. McNutt, *Analyst*, *93*, 729 (1968).
691. M. S. Frant, *Plating*, *54*, 702 (1967).
692. J. R. Entwistle, C. J. Weedon, and T. J. Hayes, *Chemistry*, *5*, 433 (1973).
693. H. H. Moeken, H. Eschrich, and G. Willeborts, *Anal. Chim. Acta*, *45*, 233 (1969).
694. H. E. Crossley, *J. Soc. Chem. Ind.*, *63*, 284 (1944).
695. J. Thomas and H. Gluskoter, *Anal. Chem.*, *46*, 1321 (1974).
696. E. Hohmann, K. Gottlieb, and U. Mueller, *Erdoelverarls, Kohleforsch. Erdgas Petrochem. Brennst.-Chem.*, *26*, 647 (1973). (A.A. 27:182, 1974.)
697. W. Oelschläger, *Z. Anal. Chem.*, *228*, 361, 1967.
698. P. Venkateswarlu and D. Narayanarao, *Indian J. Med. Res.*, *42*, 197 (1954).
699. D. A. Weddle and R. L. Maurer, *Anal. Chem.*, *26*, 584 (1954).
700. G. S. Beveridge, B. J. MacNulty, G. F. Reynolds, and E. A. Terry, *Analyst*, *79*, 267 (1954).
701. D. R. Taves, *Nature*, *220*, 582 (1968).
702. R. B. Bernstein and J. J. Katz, *Nucleonics*, *11*, 46 (1953).
703. H. O. Banks, Jr., *Nucleonics*, *13*, 62 (1955).
704. L. G. Stang Jr., W. D. Tucker, R. F. Doering, A. J. Weiss, M. W. Greene, and H. O. Banks, *Radioisotopes in Scientific Research*, Vol. I, *Proceedings of the Conference*, *Paris*, R. C. Extermann, Ed., Pergamon Press, New York, 1958, p. 50.

705. D. R. Stranks, *Inorganic Synthesis*, Vol. 7, J. Kleinberg, (Ed.,) McGraw-Hill, New York, 1963, p. 150.
706. V. I. Levin and L. S. Kozyreva, *Radiokhimiya* (*English transl.*), *5*, 34 (1963).
707. G. A. Nagy and K. Berei, *J. Inorg. Nucl. Chem.*, *26*, 659 (1964).
708. H. J. Dworkin and P. La Fleur, *Symposium on Radioactive Pharmaceuticals*, Rep. No. CONF 65111 U.S. Atomic Energy Commission, 1965, p. 636.
709. C. H. Carlson, L. Singer, D. H. Service, and W. D. Armstrong, *J. Appl. Radiat. Isot.*, *4*, 210 (1959).
710. T. Nozaki, Y. Tanaka, A. Shimamura, and T. Karasawa, *Int. J. Appl. Radiat. Isot.*, *19*, 27 (1968).
711. C. C. Thomas, J. A. Sondel, and R. C. Kerns, *Int. J. Appl. Radiat. Isot.*, *16*, 71 (1965).
712. L. H. Bowen and R. T. Rood, *J. Inorg. Nucl. Chem.*, *28*, 1985 (1966).
713. K. Beg and F. Brown, *Int. J. Appl. Radiat. Isot.*, *14*, 137 (1963).
714. Y. Maruyama, *J. Nucl. Sci. Technol.*, *4*, 185 (1967).
715. K. L. Scholz and V. J. Sodd, *Int. J. Appl. Radiat. Isot.*, *23*, 465 (1972).
716. L. S. Kozyreva-Aleksandrova, V. I. Levin, Y. F. Ivanov, and V. G. Zalesskii, *Sov. Radiochem.*, *8*, 521 (1966).

Determination of Heparin and Related Sulfated Mucopolysaccharides

L. B. JAQUES, *Hemostasis-Thrombosis Research Unit, Department of Physiology, University of Saskatchewan, Saskatoon, Saskatchewan, Canada*

Acknowledgments

In the preparation of this chapter I have been assisted by former and present associates responsible for development of many of the methods described, particularly Dr. C. P. Dietrich, Dr. L. M. Hiebert, Dr. L. W.

Kavanagh, Dr. N. M. McDuffie, Dr. H. B. Nader, Dr. A. S. Perlin, Dr. M. E. Silva, Dr. T. K. Sue, Mrs. Sandra Wice, and Dr. A. Wollin. Sections contributed by them are indicated at the beginning of the section by the contributor's name(s) in brackets without reference numbers: for example, (C. P. Dietrich) for Section II.1, (A. S. Perlin) for Section III.1.

Abbreviations

AMP	Acid mucopolysaccharide
SMP	Sulfated mucopolysaccharide
ChA	Chondroitin A = chondroitin sulfate A = chondroitin 4-sulfate
ChB	Chondroitin B = chondroitin sulfate B = dermatan sulfate
ChC	Chondroitin C = chondroitin sulfate C = chondroitin 6-sulfate
Ch	Chondroitin
HA	Hyaluronic acid
Hep	Heparin
Hpt A	Heparitin A = heparitin sulfate A } = heparitin sulfate = heparan sulfate
Hpt B	Heparitin B = heparitin sulfate B } = heparitin sulfate = heparan sulfate
Hpt C	Heparitin C = heparitin sulfate C } = heparitin sulfate = heparan sulfate
Hpt D	Heparitin D = heparitin sulfate D = low molecular weight heparin in commercial heparitin sulfate
KS	Keratosulfate = keratan sulfate
i - SO_4	Inorganic sulfate

Commercial heparitin sulfate (heparan sulfate) is a mixture of Hpt A, Hpt B, Hpt C, and Hpt D.

I. INTRODUCTION

The *determination of heparin* was reviewed by Jaques and Bell (79) in Volume VII of this series in 1959. The same volume contained a chapter on *lipoprotein lipase*, by E. D. Korn (99), and these materials provided workers with adequate descriptions of methods of the extraction and measurement of heparin in tissues and tissue fluids, as known at the time. A number of new developments have made it necessary to supplement these accounts.

The chapter by Jaques and Bell (79) discussed the principles and problems of anticoagulant assays for heparin. The need was pointed out for a collaborative study of the assay of heparin conducted on an unknown heparin sample with tests for nonparallelism or *nonidentity*. Such a study has been conducted on an international basis, and the results were reported by Bangham and Woodward (3). This study involved 13 laboratories receiving coded samples. It was learned that any qualified laboratory could obtain an assay value with anticoagulant tests for a heparin preparation against itself with 2% variation, but when different heparin preparations were compared, values might differ by as much as 50%.

During the 18 years that have intervened since the preparation of the previous chapter, there has been a growing awareness that the wide spread in values for heparin in physicochemical parameters in ultracentrifuge, light-scattering, and viscosity measurements indicates polydispersity of heparin rather than experimental error in the measurement. However no satisfactory resolution and demonstration of individual molecular weight components was reported until the recent serendipitious achievement of Nader et al. (126), using isoelectric focusing. In more than 30 commercial heparins, they demonstrated 21 fractions of increasing chain lengths with an interval of a sulfated hexasaccharide between each. Many commercial heparins contained these species in both monomer and dimer form (31). Furthermore, a considerable number of the fractions had no anticoagulant activity by the USP assay procedure, whereas others had twice the anticoagulant activity of the original heparin preparation. The lowest members of the series with no anticoagulant activity had the same metachromatic and antihemostatic actions as the original material. From rat and primate skin, Horner (65) has isolated heparin of molecular weight 1.1×10^6, macromolecular heparin. This heparin, while showing anticoagulant activity, did not behave like commercial heparin, activating lipoprotein lipase *in vitro* or releasing the enzyme *in vivo*. In fact it was an inhibitor *in vitro* and failed to release the enzyme *in vivo*. Hence methods of measurement for heparin that are not dependent on biological activities and do characterize the molecular species are important for studies of this substance and will provide information on unsuspected biological relations.

Although heparin is a trace substance in the body, it is the biological anion polyelectrolyte that possesses the unique properties of this class of compounds in the highest degree. An extensive literature now documents its ability to affect biological actions of enzymes, hormones, ions, proteins, viruses, unicellular organisms, and plant and animal cells, both isolated and *in situ* (73,140). The great success that resulted from the introduction of heparin as an anticoagulant for thrombosis and vascular surgery (71) has obscured and retarded recognition of the fundamental biological roles of heparin and its other favorable clinical actions. It must be emphasized that *anticoagulant activity cannot be equated with heparin, nor heparin defined by anticoagulant activity*.

Because of its clinical importance as a drug, its readily measured biological activities (best known are inhibition of blood coagulation and release of lipoprotein lipase), and its apparently relatively low molecular weight, heparin has received much more attention than the other sulfated mucopolysaccharides (SMPs). However the study of heparin is intimately joined to that of the other AMPs. For measurement in tissue extracts, the trace amounts of heparin must be separated from the other AMPs—hyaluronic acid, chondroitins, and heparitins, which in most tissues occur in much higher amounts. In some biological situations, moreover, the same function exercised by heparins in some cells or species seems to be performed by chondroitins in other cells or species [e.g., in basophilic leucocytes and mast cells ((76), Section VII)]. Chondroitins were first found in cartilage (hence the name), but like heparin, they probably have a much more active role in the regulation of body functions. The identification of the heparitins with cell membranes projects these substances into a regulatory role. Hence to collect useful biological information on the compounds requires simultaneous determinations of total SMPs, the individual SMPs, and appropriate histological parameters.

Fortunately our increased knowledge of biological anionic polyelectrolytes in the past 10 years (parelleling the development of the extensive industrial use of synthetic anionic polyelectrolytes) has resulted in methods that allow us to cope with these refractory substances. The methods involve chromatography on gels and columns, electrophoresis, and analyses by nuclear magnetic resonance (132) and appropriate bacterial enzymes (28). An important technical development for the successful use of the necessary chromatographic procedures has been the development very recently of densitometers with sufficient accuracy, reliability, and discrimination for densitometry to rival if not replace (because of its greater sensitivity) spectrophotometry. Finally, the extraction and separation of these compounds is no longer a strictly empirical procedure, owing to the understanding of the role of inorganic electrolytes in the reactions of these compounds provided by J. E. Scott (156) through his theory of "*critical electrolyte concentration.*"

II. NATURE OF ACID MUCOPOLYSACCHARIDES

Acid mucopolysaccharides are polymers of amino sugars bearing a uronic acid and possibly sulfate radicals. They are almost invariably prepared as sodium salts, although a wide range of salts of alkali metals and organic amines has been prepared. The chief nonsulfated mucopolysaccharide is hyaluronic acid, present in vitreous humor, umbilical cord and skin. Chondroitin A and chondroitin C are sulfated mucopolysaccharides of cartilage and provide an essential component of collagen. Chondroitin B is associated with skin. With micro methods now available, these compounds are being found in many tissues. Heparin (with other SMPs) is associated with mast cells.

Nomenclature for this group of substances has been variable and certainly inconsistent. The first names used (chondroitin, heparin) were derived from the tissue from which the material was prepared. However heparin is associated with the liver (Greek, *hepar*) only in the dog. As indicated previously, the chondroitins probably have significant functions in tissues other than cartilage. "Heparin" and "chondroitin" are such well-established terms that confusion would result from their replacement. As discussed under molecular structure (Section II.1), these provide the main structural types and the main stem words. Additional names based on the first tissue in which the substance was identified, such as dermatan, indicate a degree of difference not supported by the examination of these compounds with specific enzymes. Dermatan sulfate occurs as a component of membranes in most tissues (146). Examination with specific enzymes also does not support the distinctions conferred by the terms chondroitin 4- and 6-sulfate. It therefore appears that the older nomenclature of chondroitins A, B, and C is preferable. The name chondroitin "sulfate" is a survival of the erroneous idea that these compounds are ethereal sulfates of sugars. Increasing numbers of authors are dropping the "sulfate." The use of "sulfate" with a polymer is not in line with biochemical practice (e.g., nucleic acid is not called "nuclein phosphate"). Such nomenclature is incorrect for the *N*-sulfamate group in heparins and heparitins (144). Naming of enzymes is difficult, since enzymes attacking chondroitin sulfate are not necessarily sulfatases. Heparan sulfate refers to compounds with less sulfate than heparin. Workers now appear to be adopting chondroitin A, chondroitin B, and chondroitin C, reserving chondroitin without modifier to designate the nonsulfated compound (167). "Sulfate" must be retained in the name keratosulfate, since "keratin" refers to a protein, but this name may be temporary, pending resolution of the nature of the substances involved. For the whole group of compounds, terms such as polyaminosaccharides and glycosaminoglycans, fail to distinguish bacterial polysaccharides and antibiotics containing amino sugars. Acid mucopolysaccharides (AMPs) and sulfated mucopolysaccharides (SMPs) appear to provide an adequate description and have the further advantage of continuous use. The term "heparinoids" has been used chiefly for the products of sulfation or resulfation of polysaccharides (usually after some depolymerization). The term has had this association because the compounds were originally prepared for testing as substitute anticoagulants. If the anticoagulant idea is dropped, however, the term is very useful for compounds such as polysulfated chondroitin, and $N-{}^{35}S$-labeled heparin, whether semisynthetic or naturally occurring.

The heparitins bear a special relationship to heparin; they were first observed by Jorpes and Gardell (92) as material to be discarded in heparin manufacture. It is not surprising that this material includes considerable amounts of nonanticoagulant low molecular weight heparin, appearing as fraction D (heparitin D) in the heparin discards. Three other fractions in the commercial heparitin (sulfate) have been identified (33,34,36) bearing some similarities to heparitins excreted in the mucopolysaccharidoses: heparitin A, heparitin B, and heparitin C appear to be associated with cell membranes and are distinct chemically and biologically from both the heparins and chondroitins.

Commercial heparin is used in all hospitals and animal laboratories because it is effective in controlling blood coagulation and thrombosis and is relatively free from toxicity. After 40 years, the amount used appears to increase each year as new applications are found. Extensive analyses of commercial heparins produced by various manufacturers for clinical use during the past 20 years (84,95) have demonstrated the considerable variation between such preparations. A surprising range of values has become evident for all chemical and biological parameters studied. Various possibilities have been established in the literature as potential sources for differences in biological activities between commercial heparin preparations (79,83): (*a*) the preservative; (*b*) accompanying protein, polypeptide, or peptides; (*c*) accompanying polysaccharides; (*d*) accompanying low molecular weight substances; (*e*) acid groups of the heparin and accompanying polysaccharides—carboxyl, 0-sulfate, *N*-sulfate; (*f*) molecular weight; (*g*) net electrical charge; (*h*) structural differences in chain length.

Such differences must be expected by investigators. The art of heparin manufacture is to prevent the appearance of extraneous materials in the product, or to remove them and to obtain a reproducible product as judged by some particular biological assay. Heparin being such a selective and effective complexing agent, however, it seems that the best that can be done in this regard is for one accompanying substance to be replaced by another. Hence agents such as platelets, which are exquisite detectors of foreign substances (this is one of their important physiological roles), will detect such differences.

In view of the many possibilities for different heparin preparations showing considerable differences for different activities, it appears essential that those using commercial heparin either clinically or experimentally record in their report the exact heparin preparation used (75). This requires the manufacturer's name, the batch number on the bottle, and, if the information is available from the manufacturer, the tissue and species of origin, type of extraction used, and final formulation. Complete information is not always available, however, the date of purchase can be a useful substitute, since it gives those working in the field a clue to the nature of the preparation involved. Unless this recording practice is followed, the literature will become even more confused.

1. Molecular Structure and Size (C. P. Dietrich)

Heparin and other acidic mucopolysaccharides are polydisperse compounds composed of amino sugar and uronic acid residues. The amino sugar is glucosamine in heparins, heparitins, keratosulfate, and hyaluronic acid and galactosamine in chondroitins. The uronic acid is mainly iduronic acid in heparin, heparitin C, heparitin D, chondroitin B, and glucuronic acid in the other compounds, except keratosulfate where it is replaced by galactose. In

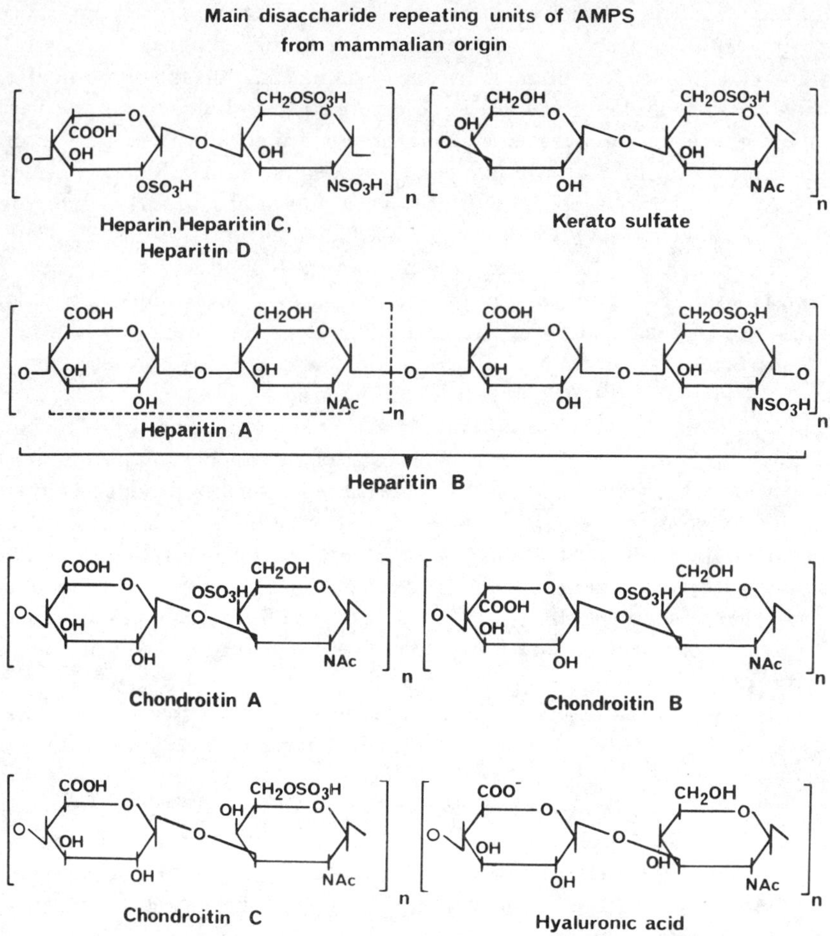

Figure 1

most AMPs the hexosamine moiety is sulfated, and the degree of sulfation varies from 1 to 3 sulfates per disaccharide repeating unit. Figure 1 represents the *main* disaccharide repeating units of the AMPs from mammalian origin, as determined by a variety of chemical, physical and biological methods. These are 2-sulfoiduronic acid, 2,6-disulfoglucosamine for heparins and heparitins C and D; galactose, 6-sulfoacetylglucosamine for keratosulfate; glucuronic acid, *N*-acetylglucosamine for heparitin A, and this disaccharide and also glucuronic acid-*N*-*O*-disulfoglucosamine for heparitin B; glucuronic acid, 6 sulfo-*N*-acetylgalactosamine for chondroitin C; glucuronic acid, 4 sulfo-*N*-acetylgalactosamine for chondroitin A; glucuronic acid, *N*-acetylglucosamine for hyaluronic acid; iduronic acid, 4-sulfo-*N*-acetyl galactosamine for chondroitin B. The reducing group of the amino sugar is linked to C_4 of the uronic acid in the α-configuration for heparin and the heparitins and in the β configuration for chondroitins, hyaluronic acid, keratosulfate. In the α-configuration in heparin and heparitins, and in the β-configuration in the other compounds, C_1 of the uronic acid is linked to C_4 of the amino sugar in heparin, heparitins, and keratosulfate, and to C_3 in chondroitins and hyaluronic acid. It is important to bear in mind that disaccharide units other than the ones shown in Figure 1 also occur within the same polysaccharide chain. For instance, heparin contains about 20% disulfated disaccharide units (37,161); heparitin A is composed of two distinct disaccharide repeating units, and heparitin B has at least four distinct disaccharide repeating units (160,67). Chondroitins A, B, and C also have variable amounts of other disaccharides (49,121,120,182). Chondroitin B contains in its structure a dodecasaccharide composed of chondroitin A units (119,122). The percentage amounts of these disaccharide units and the chain length of the AMP varies according to the tissue of origin. As an example, synovial cartilage contains AMPs composed of disaccharide units of chondroitin sulfates B and C (167), whereas articular and growth cartilages contain AMPs with disaccharide units of chondroitin sulfates A and C (124). The relative proportion of these disaccharides varies according to the layer of the cartilage examined.

As polymers, molecular weight is an important distinguishing characteristic, and since AMPs are polydisperse compounds, the meaningful values for molecular weights of a given preparation are those for the range and mode (largest component, therefore most representative). An average plus-or-minus

TABLE I
Molecular Weights[a] of SMP Samples from Electrophoresis in Polyacrylamide Gel [modified from (33)]

Compound[b]	Range	Mode value (largest component)
Chondroitin A	2,280–47,500	11,200
Chondroitin B	7000–40,000	18,400
Chondroitin C	10,300–131,000	33,500
Heparitin A	?	170,000
Heparitin B	10,800–51,000	33,000
Heparitin C	9300 ± 500	9,300
Heparitin D (low mol. wt. heparin)	1320–7800	3,370

[a] Values are for these particular samples. Values for both tissues and species of origin.
[b] Chondroitins A, B, and C from Miles Laboratories. Heparitins A, B, C, and D from commercial heparitin (Upjohn).

standard deviation is misleading because the components are usually not present in a normal frequency distribution. Values for molecular weights by electrophoresis in acrylamide gel (cf. Section VI.5) are reported in Table I as range and mode (greatest frequency) values for samples we have examined. As emphasized in the footnote to the table, the values appear to vary with both the tissue and species of origin. Exceptional is heparitin C, which is homogeneous (mol. wt. 9300). Heparin presents another special case, since there is no individual large component or single cluster of values, and different lots show differences in range and distribution. In solution, heparin appears to approximate a random Gaussian coil (43).

2. Standards

Each batch of commercial heparin is standardized in terms of inhibition of a blood coagulation system by comparison with a reference preparation based on the International Units so derived. *The heparin unit is 1/130 mg of the International Heparin Preparation.* The present (1975) Standard Preparation in use is the Third International Standard. Heparin preparations that have been compared with the International Standard or derived reference standards are therefore expressed in International Units. Since the ratio of activity of test heparin to activity of reference heparin can differ in different test systems, it is advisable to qualify International Units by naming the test system used. However Bangham and Woodward point out that for commercial heparins, the error involved in the use of coagulation systems (due to changes in substrate) is greater than that between systems, thus this refinement is not of great value. An alternative procedure is to express quantities of heparin in milligrams or micrograms. With commercial heparin received in solution, this can be calculated approximately on the basis of 1 unit = 0.01 mg. It would be preferable to express the concentration of active drug (i.e., quantity of anhydrous free acid, which is only 60 to 70% of the sodium salt), but this requires determination of volatile and base components of the heparin (94), which is not practical routinely. When heparin and other AMPs are available in dry powder form, it is important to realize these are markedly hygroscopic compounds and that weight changes can occur with storage.

A commercial heparin must be used as a reference compound for biological studies and results expressed in "equivalent" units or milligrams (micrograms). Until the heparin under study has been isolated as the pure mucopolysaccharide and compared directly with the reference commercial heparin in the test system, however, such "equivalent" units or milligrams must be considered provisional and subject to later correction.

Reference standards for all AMPs prepared by Matthews et al. (115) have for investigators a role similar to that of the International Standard Heparin

Preparation. They suffer from the same handicap, namely, that until the specific AMPs under study has been isolated and compared directly with the standard in the test used for measurement of the AMPs—chemical or biological (color reaction, NMR spectra, etc; anticoagulant activity), values must be considered provisional and subject to correction of as much as 50%. Thus for many purposes it is convenient to use a commercial heparin as reference for all SMPs. A sufficient quantity should be obtained for analyses to characterize this "reference heparin" and for providing adequate stock. Results are initially expressed in units (by the test used), then converted to micrograms of the SMP from the ratio value found for Standard SMP Preparation to Reference heparin. Perhaps the chief value of this procedure is psychological: the laboratory workers are aware that the results obtained are in activity units, converted to c.g.s. units by an arbitrary factor, thus subject to later correction.

Specific activity refers to the potency of heparin in a test system relative to that of the International Standard Heparin Preparation and is expressed in units per milligram. The International Standard Heparin Preparation, by definition, has a specific activity of 130 units/mg. Hence a comparison of a sample in any test system with International Standard Heparin will give the specific activity in International Units for that test system: (USP anticoagulant test British Pharmacopoeia (BP) anticoagulant test, lipoprotein lipase test, microelectrophoresis test, and so on). This is the most convenient way to compare activities in different biological and chemical systems for different heparin preparations and different SMPs. It is comparable to the convenience of expressing activities of acids in normalities.

3. Complexing Properties of AMPs

The remarkable ability of heparin to form complexes with amines and metals with chemical and biological properties different from those of the parent substances was recognized by A. Fischer (47,48), who described the complexing of proteins with heparin. Jaques (72), in extending Fischer's conclusions regarding the reaction with proteins, showed that the same principles applied to the reaction of heparin with dyes. In 1949 Jaques et al. (86) introduced octylamine, the first of an extensive series of long-chain amines used for precipitation of heparin and SMPs. J. E. Scott (153), in developing the use of these compounds, particularly cetylpyridinium chloride (CPC), demonstrated that the precipitations followed the same rules. An understanding of the properties of the complexes of heparin and related SMPs with proteins, dyes, organic cationic detergents, and other alkylammonium compounds is important in understanding both the practical techniques used and the biological importance of heparin and SMPs.

As discussed in the previous review (79), it is possible to demonstrate the combination of heparin with any protein, if a parameter suitable to the complex involved is used for measurement. The reaction follows a simple mass law relationship, and sufficient dissociation occurs to ensure that at equilibrium there is free heparin and free protein present in the solution. In considering the reaction of heparin with individual proteins, two values must be distinguished: (*a*) the relative proportion of acidic and basic groups in heparin and protein which determines the amount of heparin in the resulting compound, and (*b*) the dissociation constant for the complex, which determines the amount of protein required to remove all heparin from the system under study. The dissociation constant of the heparin compound is relatively low for protamine and relatively high for gelatin. Proteins of the blood coagulation system appear to hold the middle ground. Thus protamine competes successfully with the blood coagulation proteins for heparin, whereas when the heparin-gelatin compound is added to the coagulation system, the full activity of heparin present is observed. Of great practical importance are factors that affect the dissociation constant. A shift in pH to the alkaline side increases dissociation of the compound, but complex formation can be demonstrated up to pH 10. A strong anion such as trichlorocetate causes dissociation of the heparin-protein compound even at extremely acid pH values. Inorganic salts increase dissociation, and this effect increases with the valency of the ions (e.g., calcium chloride). Denaturation and lipide increase the affinity of protein for heparin. The reaction of heparin with protein is an expression of its high activity as an anion. Heparin reacts in an identical manner with alkaloids (quinine, brucine), basic dyes (thionine, toluidine blue, acriflavine), and complex organic bases (benzidine, histamine, and quaternary ammonium salts). Of the latter, the long-chain alkylammonium compounds, CPC, and cetyltrimethylammonium bromide (Cetavlon) have found many uses with heparin and SMPs.

The complexing properties of polyelectrolytes are responsible for their ability to hold counterions and coions. For polyanions, the counterions are cations such as sodium, alkylammonium ions, basic dyes, proteins. The coions are such anions as sulfate, chloride, acetate, organic acids, acid dyes, and proteins. The presence of co-ions and the difficulty of removing them was demonstrated by Schmitz and Fischer (150), who prepared the lead salt of heparin. When they attempted to remove the lead with hydrogen sulfide, they found that a black residue (lead sulfide) was not removable. Counterions and coions are distributed at different distances (binding forces) to the shell of charges on the polyanion molecule. Hence exchange is not immediate even for inorganic ions. Only an ion-exchange column can displace negative coions equivalent to about 15% of the negative charges on heparin. The polyelectrolytic nature of heparin interferes with the ordinary tests for accompanying inorganic ions, and ion-exchange or decomposition of heparin by combustion (in the oxygen flask) is required before these tests can be applied.

A. EFFECT OF ELECTROLYTE CONCENTRATION ON COMPLEXES

The importance of electrolyte concentration in dissociation of complexes formed by heparin was demonstrated by Fischer (47) and Jaques (72). This has been given a quantitative and theoretical interpretation by J. E. Scott (156) in the concept of the "critical electrolyte concentration." This allows analyses that are immediately applicable in practical techniques. The following account is a summary of Scott's presentation. The reader is referred to the original review and publications for the detailed documentation.

For the reaction of polymer P of z negative charges, with polymer cation R^+ (protein, dye, organic bases, detergents), in the presence of inorganic cation M^+, the ion-exchange equation is

$$P^{z-}zM^+ + zR^+ \rightleftharpoons P^{z-}zR^+ + zM^+ \qquad [1]$$

$$\frac{[P^{z-}zR^+]}{[P^{z-}zM^+]} = K\frac{[R^+]^z}{[M^+]^z} \qquad [2]$$

The left-hand expression is the ratio of precipitated to unprecipitated forms of the polymer P. The values for this ratio were plotted against the ratio of the two cations (organic and inorganic) for different values of z with the results presented in Figure 2. At high values of z there is a very sharp change in the value of the ratio precipitated to unprecipitated, at a critical value of $[R^+]/[M^+]$. Direct experiment has confirmed that there is a change from completely precipitated to completely unprecipitated at the critical value that is very easily observable and quantifiable. Extension of [2] shows that

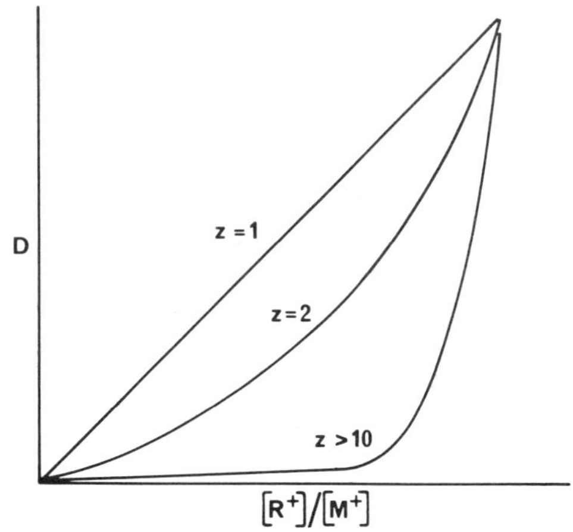

Figure 2. Ratio of concentrations of precipitated to unprecipitated polyanion versus ratio of concentrations of organic to inorganic cation for different values of z, the valency of the polyanion (156).

there is a maximum value of the critical electrolyte concentration for a given polyanion determined by its chemical structure, and the dependence on molecular weight is not very pronounced above quite moderate values of z. As pointed out by Scott, there is a very close analogy between precipitation of polyanions from solution by R^+ and absorption to an ion exchanger, because detergents in solution are in micelles. This solution maintains an equilibrium concentration of R^+ that is remarkably steady at ambient electrolyte concentrations greater than $0.1M$ ($0.001M$ for a C_{16} homolog).

Anion-cation interactions, which represent an electrostatic interaction, are evidently capable of introducing specificity into precipitation by cationic detergents and so on. Since grossly hydrophobic groups are absent in the polyanions (AMPs and nucleic acids), the hydrocarbon chains in micelle and precipitate are in almost identical milieus and interact only with each other. Hence the only interactions that change with different polyanions in this group are those between opposite charges. Three groups,–CO_2^-, $= PO_4^-$, and–OSO_3^-, make up a large proportion of biopolymer-bound anions. Comparing the properties of complexes involving these groups, it is evident that the critical electrolyte concentration values associated with the sulfate-ester groups are much higher than those of polycarboxylates and polyphosphates, in spite of considerable chemical variety elsewhere in the polymers. The difference is greater if Mg^{2+} is substituted for Na^+. The pattern is completely unaffected by major changes in the structure of the cationic "head" in the detergent molecule, no matter whether the cationic charge is diffused by resonance, present on more than one atom, or located on sulfur instead of on nitrogen. Precisely similar patterns are found if polycations are the precipitants.

It was this pattern that first drew attention to the practical value of critical electrolyte concentration phenomena in the fractionation of cetylpyridinium complexes of sulfated and nonsulfated polyanions from connective tissue. The easy separation of hyaluronate from chondroitins (e.g., from skin) was one of the earliest successes, but there have been many subsequent applications of the principle to animal, plant, and microbial materials. In addition, the extreme insolubility of the complexes and the simplicity with which the polyanions could be recovered from them were additional practical reasons for the widespread adoption of these techniques and their rapid incorporation into industrial methods—for example, for the manufacture of much of the world's heparin.

It seems that this pattern is determined not by the organic cation R^+, but by M^+. Bungenberg de Jong (13), on the basis of much work on the binding of cations to polyanions, postulated that polarizing cations such as Mg^{2+} were able to lose a part of their solvation shell to bind more strongly to "polarizable" anions such as–COO^-. The–OSO_3^- group, being less "polarizable" than water, could not displace the solvation shell, therefore remained at a distance, less strongly held. Extending this argument, since the bulky "-onium" ions, being neither polarizing nor solvated, cannot distinguish "polarizable" from "nonpolarizable" anions, they bind to both indiscriminately. Lacking greater affinity for–CO_2^- or PO_4^- than–OSO_3^-, they are more easily displaced from–CO_2^- and = PO_4^- by ions that have such an affinity.

Values reported by J. E. Scott for critical electrolyte concentrations of NaCl with AMP-polycation complexes are given in Table II. The critical electrolyte concentration provides a measure for comparison of affinities for the interaction of polyanion with organic cation. The affinity for a given complex depends not only on the strength of the electrostatic interaction but also on the closeness of fit of the two molecules. This property can allow strong binding by charge transfer and Van der Waals forces. The addition of a nonelectrostatic contribution to the binding energy makes it much more

TABLE II
Critical Electrolyte Concentrations for AMP-Polycation Complexes (156)

Detergents		NaCl (*M*)		
Cationic group	Hydrophobic group	Hyaluronate	Chrondroitin A	Heparin
$-N^+(CH_3)_3$	*n*-Dodecyl	—	—	0.90
	n-Hexadecyl	0.2	—	1.10
$-N^+$ 	*n*-Dodecyl			1.35
	n-Hexadecyl	0.23	1.0	1.55
$-N^+(CH_3)_2-[CH_2]_3-N^+(CH_3)_3$	*n*-Octadecyl	0.22	0.7	1.05
$-S-C^+(N(CH_3)_2)_2$	*n*-Dodecyl	0.13	0.55	0.8
$-S^+(CH_3)_2$	*n*-Dodecyl	0.15	0.65	1.05
"Neutral complexes"	pH			
Trypsin	4.9	0.025	0.1	0.25
Crude papain	4.7	0.1	0.3	0.75
Protamine sulfate	7.0	0.25	1.25	1.6
Polylysine HBr	7.0	0.54	1.15	1.76
ECTEOLA-cellulose	5.0	0.4	0.9	1.2

difficult to reverse the reaction (i.e., to split the complex by competition with Na^+, etc.). This then makes possible considerable specificity in these polyanion-organic cation interactions and the degree of specificity can be measured through the critical electrolyte concentration. Scott has shown that such fit may be due not only to conformations normally present but also to conformations of the polyanion molecule induced by the probe molecule under the condition of the experiment (methyl green with DNA and RNA). Steric hindrance can decrease affinity. Through gross steric hindrance with the sugar phosphate chains of nucleic acids, the tetramethylisothiouronium groups of alcian blue result in a lower affinity (critical electrolyte concentration) for nucleic acids than for SMPs, whereas the corresponding quinolinic phthalocyanine dye without the tetramethylisothiouronium groups (thus allowing intercalation in nucleic acid molecules) shows a much greater affinity for nucleic acid (critical electrolyte concentration with DNA double and with RNA fourfold that with heparin). Another possible factor is the presence of chelated metals in the organic cation: Al^{3+}, Fe^{3+}, Cr^{3+}, Zn^{2+}. The positive charge of hematoxylin is due to the chelated Al^{3+}. The critical electrolyte concentration for the complexes of this dye with polyanions is heparin $<$ chondroitins $<$ hyaluronic acid $<$ nucleic acids or the reverse of that with alkylammonium compounds (Table II).

B. REACTIONS WITH DYES AND METACHROMASIA

The reactions of heparin, SMPs, and nucleic acids with basic dyes are the typical complexing behaviors exhibited by these polyanions with complex cations. They are distinctive only in that the color reveals to the eye much of the nature of the process, and the formations of these complexes have practical uses in the demonstration and measurement of these compounds. The dye-polyanion complexes show the critical electrolyte concentration effect. It must be emphasized that dyes provide a great range of types of reaction. Thiazine, oxazine, azin, triphenylmethane, and phthallocyanin dyes provide different hydrophobic groups. The cationic group in any one series of dyes can range, for example, in the thiazine dyes (Figure 3) from the two primary amine groups of thionine to the corresponding tertiary amine groups of methylene blue, with azure A and toluidine blue intermediate, bearing one tertiary and one primary amine group. There are corresponding differences in affinity, leading us to expect a great range of patterns and specificities in these reactions. The general patterns are observed (e.g., precipitation of polyanion-dye complex can be perceived with these reactions), yet it would be wrong to consider the special features of the reaction with a given dye and polyanion to be universally applicable. Scott has termed the enormously extensive catalog of dye-polymer interactions in histology a goldmine of potentially specific interactions, awaiting systematic analysis in terms of biophysics and molecular biology.

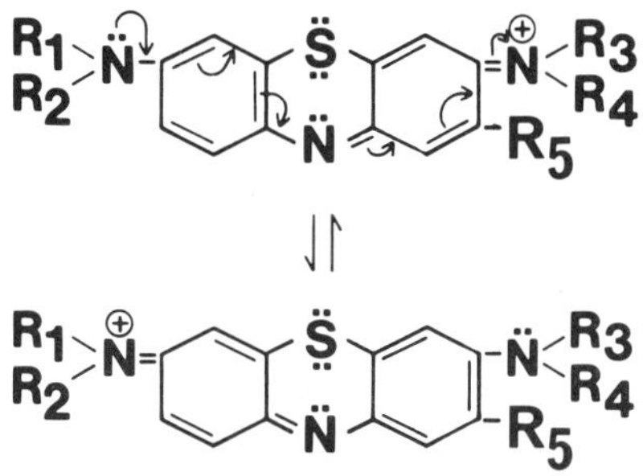

Figure 3. Structural formula for thiazine dyes = absorption maxima from ref. 108 given in parentheses. For toluidine blue, R_5 = CH_3; for other thiazine dyes, R_5 = H. R_1, R_2, R_3, R_4 = H: thionine (598–589 nm); R_1 = CH_3; R_2, R_3, R_4 = H: azure C (607–614 nm) = monomethylthionine; R_1, R_2 = CH_3; R_3, R_4 = H: azure A (625–632 nm), asymmetrical dimethyl thionine: toluidine blue (620–622 nm); R_1, R_2, R_3 = CH_3; R_4 = H: = azure B (645–650 nm) (trimethyl thionine); R_1, R_2, R_3, R_4 = CH_3: methylene blue (664–666 nm).

Ehrlich (44) in his seminal paper described structures stained by aniline dyes in a color different from that of the dye itself in dilute solution. This phenomenon is generally known as metachromasy or metachromasia. Lison (110) showed that the effect could be produced by mixing solutions of dye *in vitro* with known chemical compounds, particularly polysaccharide polysulfuric acids. The color change is presumably due to change in the light absorption spectrum of the dye. Bank and Bungenberg de Jong (4) showed with a number of polyanions a band shift in the dye solution, and this work was extended by Jaques et al. (80), who reported the production of a metachromatic color on addition of heparin to a large number of dyes (bismarck brown, azure A, brilliant cresyl blue, cresyl violet, brom cresyl blue, neutral red, basic fuchsin, pyronin, acriflavin) with absorption bands covering the visual and ultraviolet regions. In these cases the shift in the color of the dye indicated a shift of the absorption band to a shorter wavelength, a blue shift. [A red shift is observed with intercalating dyes (156).] Since as a rule the solid dye shows approximately its metachromatic color in contrast to the orthochromatic color observed in dilute solutions in water, alcohol, and acetone, it has generally been assumed that metachromasia indicates an association of dye molecules. This association may not necessarily be the same for all dye-polyanion complexes.

Two dyes are particularly important in relation to heparin, alcian blue and toluidine blue. As already indicated, given a suitable electrolyte concentration, alcian blue can be used for sulfated mucopolysaccharides without interference by nucleic acids. Much practical experience has revealed that toluidine blue possesses a special affinity for heparin. Both these dyes present special problems. The structure of alcian blue was a commercial secret and it required an investigation by Scott to establish that it consists of four isomers of the copper phthalocyanin with an *S*-methylenetetramethyl-isothiouronium group at the 3(4)-positions of each isoindole ring. The original manufacturer has now discontinued manufacture. Toluidine blue is one of a series of thiazine dyes (Figure 3). Preparation and separation of the individual members of this series is difficult. All the dyes complex with SMPs and nucleic acids through cationic groups, and the different degrees of substitution provide a wide range of affinities. The thiazine nucleus is formulated both as the isoquinoid and paraquinoid structures (108); thus resonance between these forms is possible. Toluidine blue, which blocks the anticoagulant activity of heparin, the same as protamine, has been used frequently for this purpose. The other members of the series exhibit this property proportional to the number of primary amino groups, with toluidine blue and azure A equivalent. This is true of most parameters used for examining the complexes. However the complexes of heparin with azure A and with toluidine blue have special features. They are not dissociated at electrolyte concentrations effective with the complexes of heparin with the other dyes, nor of other SMPs with toluidine blue and azure A. The complex with toluidine blue shows a more intense color than that with azure A under the same conditions. Not only do the complexes of heparin with toluidine blue and azure A possess higher affinities of the reactants, but the region of the spectrum involved (600 and 620 nm for the dyes, to 540–500 nm for the complexes) is that of maximum color sensitivity for the human eye. Hence the special role of toluidine blue for the histology of these compounds and the special role of polychrome methylene blue (a mixture of all the azures) for blood stains. But although complexing with heparin as judged by precipitation and prevention of anticoagulant activity is a constant property of azure A and toluidine blue, and a color change (blue to red) with heparin in aqueous solution is a constant property of azure A, this is not the case for toluidine blue. Authentic toluidine blue does not give metachromasia with heparin. This property appears during storage, with a slight change in absorption to resemble azure A more closely. Yet this is not simply a conversion of toluidine blue to azure A. Since a suitable sample of toluidine blue gives a higher E_0 for the metachromatic band than azure A.

Three absorption bands termed α-, β-, and μ-bands, can be seen for these dyes with polyanions in dilute aqueous solutions. The α-band (for values see the legend of Figure 3) is that for the dye monomer in solution. The β-band is given by the dye dimer, and the μ-band appears on adding the polyanion, hence is the metachromatic band. For toluidine blue, the α-band is at 620 nm, the β-band at 540 nm, and the μ-band with added heparin at 500 nm. The color change on adding heparin to dilute toluidine blue solution shows the following properties (181). The color change is definitely different from those given by the same dye with acid-base and oxidation-reduction shifts. Addition of heparin to the dye in solution decreases the α- and β-bands characteristic of the dye and produces the μ-band at a shorter wavelength. The position of the μ-band is characteristic for the substance producing metachromasy—about 120 nm less than the μ-band in the presence of heparin and heparinoids, 95 nm for dermatan. When heparin concentration is increased 10 times the concentration required for equilibrium, there is no shift in position of the μ-band. The color change can be seen with very low concentrations of heparin and heparinoids (1 μg/ml) in aqueous solution, on paper chromatograms, and in agarose gel, and a simple mass law relationship obtains. The μ-band is suppressed by excess heparin and a concentration excess of 3 orders of magnitude is required. The color change is suppressed by inorganic electrolytes and increasing temperature, but suppression from an increase in ionic strength and hydrogen ion in the medium is much greater with other mucopolysaccharides than with heparin and heparinoids. Dehydration suppresses metachromasy in aqueous solution with a shift of light absorption to the α-band. With the products obtained from heparin by enzymatic degradation, the monosaccharide and disaccharide fragments react slightly with the dye, but there is no metachromasy, whereas full metachromatic activity as compared to heparin is given by the *heparin hexasaccharide.* Dye-polyanion interaction that does not lead to metachromasia occurs in the precipitation of dye-polyanion complex from solution at higher concentrations. This interaction is not destroyed by ethanol, dehydration, or moderate concentrations of electrolytes. The limiting concentration of electrolyte depends on the nature of the polyanion. Finally, a μ band is produced by adding azure A to 3*M* NaCl

According to Wollin and Jaques (181), the metachromatic phenomenon observed with toluidine blue on adding heparin in a manner consistent with these observations, is due to the resonance in the dye molecules, which allows a partial charge to form on each end of the dimer of the dye to permit loose interaction with the negatively charged heparin molecule. This interaction is stabilized by the presence of water molecules, and the metachromatic band shift is the result of this interaction by way of the amino groups, which are part of the dye's chromophore. This interaction partially blocks the N-group in the thiazine dyes from being part of the conjugation system of the dye chromophore, thus reduces π-electron delocalization. The loss in electron mobility causes the dye molecule to absorb light energy at a shorter wavelength. The strength of the binding to the N-group is reflected in the band shift. The exceptional strength of binding of heparin and toluidine blue-azure A is due to the suitable arrangement of anionic sites for interaction with the dye-dimer. The dye-dimer can bridge between two sulfate groups across the coil of the polysaccharide chain. The asymmetrical distribution of methyl groups in azure A and toluidine blue means that a dimer can be formed with equal partial charges. The production of a metachromatic absorption band with the dye in 3*M* NaCl indicates that polyanions are not essential for metachromasy, as is usually thought. This band shift probably results from a great number of ions forming a cage around the partial charge of the dye-dimer, thus inducing an effect on the conjugation system similar to that produced by the polyanion.

A further factor in the development of the metachromatic color is the support medium. As indicated previously, reduction of water concentration by addition of alcohol to heparin-toluidine blue complexes in aqueous solution causes the color to revert to that of the dye alone. On the other hand, metachromatic staining is obtained with chromatograms (paper, agarose, etc.) and tissue sections with dye in acetone, alcohol, and so on. Probably the traces of water in the support medium or the fact that the dye-heparin complex is attached to the support suffice to stabilize the dye-dimer-heparin association and loss of electron mobility.

Toluidine blue then has a unique relationship to heparin, as can be seen in the many references to it in the practical sections of this chapter. In purchasing this dye, the investigator is cautioned not to limit himself to one source and sample but to conduct comparative tests with heparin in dilute aqueous solution and under the conditions of use. For many purposes complex formation is all that is necessary, but often a dye sample that gives a good metachromatic color can furnish valuable additional information (provided the observer has normal color vision). Chromatography on paper of the dye preparations may demonstrate more than one component, and these may be separated by column chromatography. This procedure may or may not improve the performance of the dye.

C. REACTIONS OF HEPARIN WITH PROTEINS

The metachromatic reaction of heparin with toluidine blue in which there is a visual change in the properties of the dye can be a prototype of the important property of this group of compounds—the ability to modify the properties of polycations and ampholytes. For proteins, the properties modified can be enzymic, hormonal, immunological, and so on (79,140). For the biochemist and medical scientist, the most valuable use of heparin and related substances may be as specific protein reagents. Because of this ubiquity of interactions with proteins, however, suitable control tests must be conducted when heparin is used in biological experiments.

One protein shows a special affinity for heparin. *Platelet factor 4* is a basic protein that combines so strongly with heparin that the complex migrates electrophorically as a unit at about half the rate of heparin (129). The presence of this protein in blood and variations in the quantity present in plasma (due to platelet breakdown) probably has been responsible for difficulties in determining heparin in plasma discussed in the review just cited.

4. Degradation Reactions of Heparin and Stability

Degradation reactions of AMPs are important from the standpoint of shelf storage and stability of these compounds. More specific information is available for heparin, since its biological properties provide sensitive indicators. Charles (16) found that heparin retained its potency when heated to 100

°C, either in a dry state or in an aqueous solution. Scott and Charles (152) showed that heparin lost its anticoagulant activity when heated (80 °C) in acid and alkaline solutions. This destruction proceeded much more rapidly in acid than in alkaline solutions, since a pH of 2.5 was as destructive as 0.25*N* alkali. If the pH of aqueous heparin solutions is adjusted to pH 7.0 to 7.2, they can be sterilized by autoclaving. Since the solutions do not have any significant buffering power, however, allowance must be made for pH changes on lyophilizing. The effect of pH on the anticoagulant activity of heparin is discussed further in Section IV.2. Koh and Barucha (97,98) have reported data for the stability of various heparin preparations. The neutral sodium (11.6% Na), potassium, lithium, and choline salts of heparin were completely stable for the 100 days of test. Heparinic acid (prepared by passage of sodium heparin through a cation-exchange resin) begins to decompose almost immediately, "making its isolation even by lyophilization very difficult." The stability of the sodium acid salts depends on the amount of Na removed:- $t_{1/2}$, - $\propto$, 25, 16 days for 8.9, 7.4, 6.2% Na, stored in the powder form in a vacuum dessicator over P_2O_5 at room temperature. Helbert and Marini (59) concluded from an analysis of the kinetics of self-hydrolysis that the reaction is complex.

The neutral sodium salt has been stored for up to 20 years in the cold without loss of anticoagulant activity. The only problem associated with the neutral compounds is hygroscopicity, which is responsible for absorption of water and the resulting dilution of specific activity. It is essential to remove neutral compounds from cold storage the day before opening for removal of aliquots and to allow them to reach temperature equilibrium with the room.

Although acid hydrolysis has frequently been used for the elucidation of structure, acid hydrolysis of neutral compounds is accompanied by extensive degradation and transformation of carbohydrate moieties. Perlin (186) avoided this with heparin by using methylation. Degradation with ascorbate was first observed by Robertson et al. (142) with mucin (crude hyaluronic acid). Its mechanism is still unknown. The technique has been used by Horner (65) to degrade macromolecular heparin to lower molecular weight heparin and by Dietrich et al. (31) to degrade the dimers of pork intestinal mucosal heparin to monomers.

III. PROCEDURES FOR ANALYSIS OF HEPARIN AND AMPS

As polyanion electrolytes, heparin and AMPs present special problems in chemical analysis, compared to neutral carbohydrates or proteins. The methods for fractionation and characterization described in Section VI have become available only recently; thus mixtures of the SMPs are frequently isolated. In fact, as described in Section II.1, hybrid molecules are found. As much as one-third of the sample can be volatile, another third appearing as ash. An essential requirement when chemical composition for AMPs is to be compared is to determine the percentage volatile, keeping the water content constant: "% volatile" should be determined by removal *in vacuo* at 110 °C, using a Fischer pistol. These substances as dry powders are highly hygroscopic and have a strong negative electrostatic charge. For frequent access, the most practical storage is in vials over $CaCl_2$ under slight vacuum at the temperature of the balance room.

It was indicated in Section II that tertiary structure is an important factor in the biological effects of heparin and AMPs. This and transformations occurring in the sugar components during degradation mean that physical methods that can be applied to the original polymer provide most important information. Measurements of viscosity and ultracentrifugal analysis have been widely used for determinations of molecular size of polymers. With polyelectrolytes, the electrical charge has offered difficulties in the application of these methods. Problems have been successfully overcome by carrying out determinations in which the charges are swamped by means of pH and/or ionic strength; the results have been reviewed by Ehrlich and Stivala (43). The difficulty posed by heterogeneity of heparin was met by Stivala by using a fraction of relatively uniform molecular weight. Values of 12,600, 12,500, and 12,900 were obtained from sedimentation equilibrium (103), intrinsic viscosity (107), and low-angle X-ray scattering (163). Similar studies are required for all the individual molecular weight components of heparin. Optical rotation with optical rotatory dispersion (ORD), and nuclear magnetic resonance provide procedures for analyzing structural components without degradation. Results obtained with ORD, with references to the procedures used, are reviewed by Ehrlich and Stivala (43). Results obtained on examination of heparin and SMPs by proton magnetic resonance have been described by Perlin (132–134). The information obtained from this source and from ^{13}C NMR was necessary background information for Section II.1. The methods used are reviewed by A. S. Perlin below.

With an anionic polyelectrolyte, the counterions and coions make a large contribution to the chemical analysis, although their determination is frequently neglected; as a result, other chemical data may be meaningless, therefore wrongly interpreted. Kavanagh and Jaques (94) reported that analysis of a number of commercial heparins revealed that sodium was almost the only base present. This is easily determined by flame photometry or atomic absorption spectroscopy. The investigators showed that this approach gives a more reliable and useful figure than the determination of ash, which has been frequently used. The same procedures can be applied to the potassium, barium, and calcium salts and should be used when these salts are prepared. Ammonium and alkylammonium bases are frequently used counterions with heparin. If not completely removed, they will increase significantly the nitrogen value. Ammonium ion can be determined by direct distillation in the micro Kjeldahl apparatus. We have found an ultramicro form of this apparatus useful. Chloride, sulfate, acetate, and carbonate salts, used in processing heparin, are not infrequently present in preparations. Phosphate will be present if the initial extraction of tissue is made with phosphate buffer. Kavanagh and Jaques (94) found as much as 5% acetyl in commercial heparins determined by the Kuhn-Roth procedure. Since this amount frequently was not found on PMR examination, the part removed by lyophilization with D_2O (see Section III.1) probably represents a short-chain aliphatic acid, present as a coion. Heparin interferes with direct tests for inorganic ions. For example, addition of silver nitrate to a heparin solution with added sodium chloride does not result in the precipitation of silver chloride. Hence combustion of the heparin is required before analysis for chloride, sulfate, and phosphate. The alternative is a titration procedure, which can detect all the acid groups, both attached to heparin and present as coions. The latter can be distinguished by absorption on an IRA-400 column. The titration can be followed potentiometrically or conductometrically. We have described each of these procedures. The specific test for the uronic acid component is the CO_2 liberated with mineral acid. As a gasometric method, "the measurement is based, as in gravimetric analysis, on direct observation of the amount of substance obtained, independent of comparison with standard solutions by titration, colorimetric or otherwise" (137). Determination of sulfate in these compounds was reviewed early in this series by Dodgson and Spencer (42). Therefore we include only two recent procedures used by us with these compounds and based on principles that appeared after 1957.

The color reactions for sugars, although older, rely on the same principles as mass spectrography of organic compounds. The sugars are destroyed by heating in concentrated sulfuric acid to produce furfurals. The products of destruction react with an added aromatic chromogen to give a color for measurement. Only a small percentage of the sugar ring appears as a furfural. Different destruction products will result from different uronic acids, amino sugars, hexoses, sulfated and unsulfated and so on, to react with different chromogens. A reference standard must be run at the same time. For the orcinol and naphthoresorcinol reactions, this is commonly glucuronic acid. It should be pointed out that the result is not "% glucuronic acid" but "milligrams of glucuronic acid for equivalent color." Values obtained with heparin can be as much as twice the value found for amino sugar for the same sample and the probable true uronic acid value (94), given the nature of the products from the sulfated iduronic acid present. These procedures have been reviewed in elsewhere in this series—the carbazole reaction for uronic acids by Dische (40) and the reactions for amino sugars by Gardell (52). Modifications of the carbazole test that improve the specificity of the carbazole reaction have been described by Bitter and Muir (9) and by Galambos (50).

Jaques et al. (84) and Kavanagh and Jaques (94) have reported the results of extensive chemical analyses of commercial heparins obtained over a 30-year period. Variability in values for all chemical parameters was observed. Although variations in analytical values for carbon, hydrogen, nitrogen, sulfur, and oxygen and for uronic acid equivalent compared to amino sugar, can be explained on the basis of the problems just outlined, what is surprising is the great variability in optical rotation evident in recent preparations when manufacturers believed they were obtaining greater purity. This indication of variation in the monosaccharide units between different preparations is substantiated by NMR examination (Section III.1).

1. Nuclear Magnetic Resonance Spectroscopy (A. S. Perlin)

PRINCIPLE

Heparin furnishes proton (^{1}H) and carbon-13 (^{13}C) NMR spectra, each of which readily distinguishes this polymer from other glycosaminoglycans and also differentiates between heparin specimens of varying composition. That is, variations in the nature of constituent residues, the position and/or configuration of glycosidic bonds, the number and location of substituent groups, all contribute to give distinctive chemical shift and spin-spin coupling patterns. Several classes of resonance signal are especially informative in this respect and, generally, are also easily identifiable in NMR spectra: those due to anomeric nuclei (H-1 or C-1), to the sulfamino or acetamido grouping (H-2, C-2, methyl, and carbonyl groups), to carbons involved in interresidue bonding, to a carbon (and its appended proton) that bears an *O*-sulfate group, and to carboxyl carbons. Quantitative aspects of the NMR method are highly satisfactory; the procedure is nondestructive and is rapid and convenient for experimental purposes.

Other applications of NMR spectroscopy in this area include the characterization of modified forms of heparin and AMPs, the description of fragments produced by enzymic or chemical degradation, and studies of interactions with cations and proteins.

METHOD

The methods employed are those in general use for high-resolution NMR spectroscopy (8,20,70,105) and have been described in particular for applications to macromolecules in refs. 10 and 147.

NMR spectrometers are to be found in virtually all chemical research laboratories and in many of those engaged in biochemical or medical research. Spectroscopic service is available also commercially. Proton spectra are more easily and commonly obtained than those of ^{13}C, owing to experimental limitations imposed by the low natural abundance and sensitivity of the ^{13}C nucleus. With the relatively recent introduction of pulse Fourier transform (FT) instruments, however, ^{13}C spectroscopy has been gaining rapidly in importance (105,147).

An operating frequency exceeding 200 MHz is especially advantageous for ^1H spectra of heparin and AMPs (134), because of the superior signal separation obtainable with such powerful spectrometers. Nevertheless, the more widely available 60- and 100-MHz instruments suffice for many purposes. A substantial enhancement in the usefulness of ^1H studies in general can be realized by adopting the FT mode of operation. Thus instead of the 50 to 100 mg of sample required for continuous wave spectra, measurements on one-fiftieth this scale are feasible. This modification has the added advantage of reducing viscosity, hence line broadening, although line widths of ^1H spectra may also be reduced effectively by use of elevated temperatures (60 to 90 °C).

FT spectrometers operating in the 15–25 MHz range (i.e., utilizing magnets of 60- to 100-MHz proton instruments) are becoming widely available for ^{13}C spectroscopy. Because ^{13}C resonance occurs over a much wider range (about 20 times that of ^1H), line broadening is normally a less vexing problem in ^{13}C spectra; thus these relatively low-frequency instruments have proved to be effective for most applications to date. ^{13}C spectra are usually recorded in a highly simplified format consisting of a group of narrow singlets by using proton noise-decoupling: that is, all coupling between ^{13}C and ^1H is eliminated by broad band irradiation at the proton frequency. Spectra in which coupling between the ^{13}C nuclei and protons is observed may be obtained most effectively by using a "gated" pulse operation (51). As with ^1H spectra, such coupling may provide useful structural information (54,133,51).

Quantitative measurements of peak intensities are made by electronic integration, which is a conventional facility of NMR spectrometers. Such measurements are less satisfactory for FT than for continuous wave spectra because the responses of nuclei under FT conditions vary more widely with differences in relaxation rates.

Deuterium oxide (99.7 to 99.95 atom %) is employed as the solvent for ^1H spectroscopy of heparin and AMPs. Solution preparation involves a prior proton-exchange treatment with D_2O (by successive lyophilizations in a moisture-free atmosphere), to minimize interference by the HOD signal. It may also be useful to alter the position of this signal by varying the temperature over the range 50–80 °C to avoid overlap with other signals. In the FT procedure, the HOD peak may be effectively eliminated by choosing a suitable radiofrequency pulse sequence (20), since the solvent protons have much longer relaxation times than do those of the polymer. D_2O is also used commonly as the solvent in ^{13}C studies because most instruments function with a deuterium "lock" system. A preliminary H-D exchange is not necessary in this instance and, indeed, H_2O itself can be the solvent if an external lock (e.g., C_6D_6) is employed.

The sample solutions must be free of sediment, gel-like fragments, and paramagnetic materials, which can adversely affect the resolution. A careful filtration or centrifugation is usually adequate.

Both ^{1}H and ^{13}C spectra are referenced with respect to tetramethylsilane (TMS). An external reference consisting of TMS in a coaxial capillary is used frequently for both nuclei. In proton spectra this entails a large bulk susceptibility correction of $-$ 0.4 to $-$ 0.5 ppm. A water-soluble internal standard, such as sodium 3-trimethylsilylpropionate-2,2,3,3-d_4 (TSP) (which resonates 4 to 5 Hz upfield of TMS), may be used instead, but it complicates recovery of the sample. In this laboratory (162), tetramethyl tin is used as a capillary reference for protons, it is more satisfactory than TMS at elevated temperatures (60 to 80 °C), and its chemical shift is 2 to 3 Hz downfield of TMS in chloroform.

COMMENTS

NMR spectroscopy has been used extensively to differentiate between heparin samples that contain appreciable acetamidodeoxyhexose (ADH) residues and those containing only a trace proportion of such residues (85,132,134,136). It appears that the latter have been obtained from beef lung, whereas hog mucosal heparins are characterized by higher content of ADH. Such differences are readily observed from variations in the intensity of the methyl proton signal (at 2.2 ppm) or that of the methyl carbon peak (at 25 ppm).

Although this distinction between heparins is easily made, the content of ADH is low [$<$ 10% (?)] as compared with other mucopolysaccharides. Heparan sulfates may contain at least 2 to 3 times as much ADH, whereas one-half the molecular structure of hyaluronic acid or the chondroitins is made up of this type of residue. Hence the latter polymers in turn may be differentiated from heparins and from each other based on the response of the acetamido signal—even with the relatively low resolution obtainable in 60-MHz proton spectra.

Other evidence regarding the number of different kinds of constituent residue and their relative proportions is obtained most directly from signals due to anomeric nuclei (H-1 in the 5-ppm region; C-1 in the 100-ppm region). Nevertheless, structural details sometimes are obscured by signal overlap. For heparins, this complication interferes with a straightforward NMR analysis of the minor uronide constituent (i.e., D-glucuronic acid). However ^{13}C resonances due to uronide carboxyl groups appear to furnish a useful measure of the relative contents of the L-*ido* and D-*gluco* isomers (55).

Deshielding induced by *O*-sulfate groups—of an αH by some 0.3 ppm., or an appended carbon by about 5 ppm—is helpful in locating the position of such substituents. The value of this step is exemplified in the ^{1}H spectra of enzymic degradation products from heparin (135), or in the differentiation by ^{13}C spectroscopy of sulfation at C-4 or C-6 in the chondroitin series (151).

NMR spectroscopy readily detects a variety of impurities in heparin and other AMP preparations, and this is an important property. Several instances of this have been reported, dealing with such species as strongly adsorbed small organic molecules (162), acetate ion (85,134,136), paramagnetic material (36), and contaminants from chromatographic supports (36).

The utilization of spin-spin coupling data has been restricted by the substantial broadening of proton signals on the one hand and, on the other, by the long instrument time required to obtain good proton-coupled ^{13}C spectra. Despite these limitations, however, a variety of applications has been made—for example, the relatively narrow line widths of the H-1 and H-5 signals of the main type of uronic acid residue in heparin and chondroitin B is indicative (134) of an *α-ido* configuration and a lC(L) conformation. A similar conclusion is afforded by the large coupling between $^{13}C - 1$ and ^{1}H of the L-iduronic acid residues in these polymers (54,133). This contrasts with the relatively small corresponding couplings (10 Hz less) for the D-glucuronic acid residues of chondroitins A and C, which accords with a *β*-configuration (and a 1C(D) conformation) for these constituents (54).

Because they are polyanionic, heparin and AMPs are notably affected by changes in pH and by the presence of cationic species. NMR spectra are sensitive to such effects and, more particularly, they permit direct observation of the extent to which individual nuclei are influenced. For example, as the pH is lowered, the most striking changes involve a downfield shift for H-5 of uronic acid residues (133) and an upfield shift for their carboxyl carbons (54,133). Complexing of calcium ion with heparin is evidenced by marked effects on the chemical shifts of H-1 and H-5 of the uronic acid residues (133). Similarly, the introduction of a lanthanide cation (Gd^{3+}) grossly broadens ^{13}C signals by virtue of the anomeric and carboxyl carbons of uronide residues. An important feature of the latter interaction is high stereoselectivity for anomers possessing an axial C-1, O-1 bond (14).

2. Sulfate and Carboxyl Groups

A. POTENTIOMETRIC TITRATION OF HEPARINIC ACID

PRINCIPLE

The procedure was described by Helbert and Marini (59), who used it to follow the changes with self-hydrolysis of heparinic acid. After thorough dialysis, Na-heparin is passed through a column of Dowex-50 in the hydrogen form. An aliquot of this is titrated for acid groups. Another aliquot is first passed through a column of IRA-400 in the hydroxyl form to remove inorganic anions, then titrated. The difference is due to the coion acid groups.

Procedure. Dowex-50 × 4 (200–400 mesh) is converted from the sodium to the hydrogen form, boiled in several changes of distilled water until color throwout is absent, and finally washed on a column with distilled water. IRA-400 is converted repeatedly from the chloride to the hydroxyl form and back again with KOH and KCl. It is converted to the hydroxyl form and washed on a column with distilled water until free of excess hydroxyl ion. Heparin is dialyzed at 4 °C for 48 hr against frequent changes of distilled water, followed by lyophilization. A column with a quantity of Dowex-50 about 5 times that estimated to be necessary on the basis of the exchange capacity of the resin and the quantity of exchangeable sodium is prepared in the hydrogen form. An accurately weighed quantity of the dialyzed and lyophilized heparin is dissolved in a few milliliters of distilled water. This solution is percolated slowly through the Dowex-50 column. The effluent is collected in a volumetric flask in ice water. A portion of the effluent is passed through a column of IRA-400 in the hydroxyl form.

Titrations are made with 1.000*N* KOH, into a water-jacketed titration cup designed to accommodate volumes of 5 to 10 ml. The contents are stirred with a motor-driven glass agitator, with the surface of the liquid continously flooded with nitrogen saturated with water vapor. Water at 20 ± 0.1 °C is circulated rapidly through the jacket of the titration cup; 4.00 ml of a specimen is transferred to the clean, dry titration cup, and 4 to 5 min with rapid stirring is allowed for final temperature equilibrium. Titrations are performed by adding successive increments of 1.000*N* KOH to effect pH changes of 0.2 unit up to pH 4.0 and changes of 0.5 unit from pH 4.0 to 11.5. Micrometer burettes with tips below the surface are used for the titrant; also required are a Radiometer pH meter, type PHM 4 C with glass electrode type G-202-B (low sodium error) and a type K-100 calomel reference electrode. From the resulting titration curves can be obtained equivalents of base at the inflection points for neutralization of sulfate and carboxyl groups. The difference between the titration values before and after passing through IRA-400 gives the quantity of anion as co-ion.

COMMENT

Kuettner and Lindenbaum (100) used a sensitive automatic titration apparatus. They lost 20% of the original heparin on dialysis in Visking cellophane casing; therefore, to obtain data on the total heparin, they omitted the dialysis step.

B. CONDUCTIMETRIC TITRATION

PRINCIPLE

The description follows that of Casu and Gennaro (15). For determination of the carboxyl content, the dialyzed sodium salt is titrated with 0.1*M* HCl with a conductimeter and estimated by the amount of standard acid for the first inflection point.

For sulfate plus carboxyl groups, the dialyzed sodium salt is converted to the free acid with Amberlite IR-120 (H^+) and titrated with 0.1*M* NaOH. The first inflection point is equivalent to the sulfate content, the second to the sulfate and carboxyl content.

Procedure. Before the conductimetric analyses, each sample is dialyzed for 4 hr against running, distilled water, in standard dialysis tubes (CRAMI, Milan), This duration is best for eliminating common inorganic salts and acetates without producing any significant self-hydrolysis of heparin. The dialyzed solutions are carefully brought to the original pH with NaOH, then recovered by evaporation of 40 °C. The conductimetric analyses are performed either on an automatic instrument (Metrohm model E 365 B) or on a manual instrument (Derritron model E 3924), using a standard cell (total volume 200 ml) or a semimicro cell (30 ml).

Direct Conductimetric Titration of the Carboxyl Groups. Samples (100 mg, or 5 mg using the semimicro cell) of heparin or chondroitin (Na salt, undried) are dissolved in 120 ml of distilled water (25 ml for semimicro), and the solution is introduced into the conductimetric cell, adjusted to 150 ml (50 ml), and directly titrated with 0.1*M* HCl. The carboxyl content in the original mucopolysaccharide is determined from the amount of standard acid corresponding to the first inflection point and expressed on a dry basis. The results are calculated as follows: $COO^-\% = A \times 4.4 \times 100/w$ (mg), where A = milliliters of 0.1*M* HCl corresponding to the first inflection point, 4.4 = equivalent weight (mg) corresponding to 0.1 meq of COO^-, and w (mg) = dry weight of the sample.

Conductimetric Titration of the Sulfate and Carboxyl Groups. Samples (100 mg) are dissolved in 30 ml of distilled water. The solution is passed through a column (15 × 150 mm) of Amberlite IR-120 (H^+) resin until the effluent becomes neutral and is collected as the acid form of the mucopolysaccharide, directly into the conductimetric cell, up to a volume of 150 ml. For small samples (5 mg), volumes are reduced six-fold and the measurements made in a semimicro cell. The sample is titrated with 0.1*M* NaOH. The sulfate content in the original sample is determined from the first equivalence point and evaluated by extrapolation of the first two branches of the conductance titration curve. The amount of standard base required to neutralize the carboxyl groups is determined by subtracting the amount required to neutralize the sulfate groups from the amount corresponding to the second inflection point (total acid equivalent). The sulfate-to-carboxyl molar ratio is determined from the ratio of the amount of base required to neutralize the sulfate to that required to neutralize the carboxyl. The results are calculated as follows:

$$SO_3{}^-\% = \frac{A \times 8.0}{w(\text{mg})} \times 100$$

$$COO^-\% = \frac{(B - A) \times 4.4}{w(\text{mg})} \times 100$$

and molar ratio $SO_3/COO^- = A(\text{B} - A)$, where A = milliliters of 0.1*M*

NaOH corresponding to the first inflection point, B = milliliters of 0.1M NaOH corresponding to the second inflection point, 8.0 = equivalent weight (mg) corresponding to 0.1 meq of SO_3^-, 4.4 = equivalent weight (mg) corresponding to 0.1 meq of COO^-, and w (mg) = dry weight of the sample.

COMMENTS

Excess inorganic salts in the sample interfere with the analysis—acetate encumbers the titration of carboxyl, and chlorid that of sulfate. Inorganic sulfate and sodium acetate contamination can be detected by infrared spectrophotometry in aqueous solution. Inorganic acetates can also be detected by PMR spectroscopy. It was observed that although contamination with inorganic acetate caused the expected shift of the equivalence point for the carboxyls, it did not affect the inflection point for the sulfate groups. The shift of carboxyl equivalence point is also accompanied by an increase in the slope of the carboxyl branch of the conductance curve, but in the presence of inorganic acetate it seems no longer possible to detect the original equivalence point for the uronic acid groups of the mucopolysaccharide. Hence before conducting the titration, it is necessary to desalt the sample by dialysis or gel-filtration. As indicated in Section III.2.A, however, it is necessary to check that heparin is not lost during dialysis. After dialysis, coions will still be titrated (chloride with sulfate; acetate, phosphate, and bicarbonate with carboxyl). They are removed only by ion exchange.

C. DETERMINATION OF SULFATE BY ATOMIC ABSORPTION SPECTROPHOTOMETRY (A. WOLLIN)

PRINCIPLE

The sample is decomposed with bromine-nitric-perchloric acid solution, followed by hydrochloric acid. The sulfate is precipitated with barium chloride in the presence of lanthanum chloride. The barium sulfate is then dissolved in EDTA-Na solution and the quantity of barium measured by atomic absorption spectrophotometry (179). Scale: $> 5\ \mu g$.

APPARATUS AND REAGENTS

An atomic absorption spectrophotometer is set for the measurement of barium. We have used a Perkin-Elmer model 303 with instrument settings: barium hollow-cathode lamp, wavelength setting 5536 R, slit 3, air 30 psi, 6.5 liters/min; acetylene 8 psi, 8.5 liters/min; solution flow rate 3.2 ml/min (constant during measurement), expansion scale 2—10 × (meter response 1).

Double-distilled deionized water, and the following analytical grade chemicals are used: sodium chloride (40 mg/ml), bromine, nitric acid (concentrated), perchloric acid (70 to 72%), and hydrochloric acid (concentrated). For the standard solution K_2SO_4 was dried over P_2O_5 in a Fischer pistol at 69 °C, 1000 μ pressure, and a 2 mg/ml solution was prepared. Next 15% $BaCl_2$, 5% lanthanum chloride solution, and 11.72 g of LaOH were dissolved in 100 ml of double-distilled water, 50 ml of HCl (37%) was slowly added. After cooling, the volume was brought to 200 ml with double-distilled water. Finally 1 g of EDTA tetrasodium salt with 1 g of NaOH were dissolved in double-distilled water and the volume adjusted to 200 ml (EDTA-Na).

Procedure. The digestion is conducted simultaneously in a fume hood on six samples or more. The sample in powder or solution is placed in a 30-ml Kjeldahl flask; 0.5 ml of NaCl (40 mg/ml) is added, and the flask neck is washed down with approximately 0.5 ml of double-distilled water. Then 0.05 ml of bromine, 3 ml of concentrated nitric acid, and 2 ml of perchloric acid (70%) are added to the flask and mixed. The Pyrex glass reflux column (12.5 × 1.35 cm cylinder filled with glass beads), described by Diuguid and Johnson (41) is inserted into the neck of the Kjeldahl flask and the flask is placed in a Kjeldahl digestion rack. After 15 min heat is applied slowly to drive off the excess bromine, and the temperature is increased to a moderate boiling. After approximately 30 min, when the boiling has slowed down, the heat is increased to maximum. At this point the flasks are closely watched and the Kjeldahl flask removed from the hot plate as soon as all the liquid has evaporated from the bottom. After the Kjeldahl flask has cooled, 1 ml of concentrated HCl is added and the flask is returned to the heating rack at a low temperature for about 15 min. The temperature is gradually increased until all the liquid is evaporated from the bottom of the flask, which is removed immediately from the hot plate.

After the Kjeldahl flasks have cooled, the contents of each are dissolved in double-distilled water and quantitatively transferred into 12-ml graduated centrifuge tubes. The transferred solutions should not exceed 9 ml. Then 0.5 ml of 5% lanthanum chloride is added, followed by 1 ml of 15% $BaCl_2$, the volumes are brought up to 12 ml, and the contents mixed. The solutions are left standing at room temperature for at least 4 hr and centrifuged for 10 min. The white precipitates of $BaSO_4$ and the centrifuge tube walls are washed twice with double-distilled water to remove excess barium. The $BaSO_4$ precipitates are dissolved in 10 ml of EDTA-Na (this volume can be reduced if desired). The samples must not be left dissolved in EDTA-Na if it becomes necessary to interrupt the procedure. It was found that a loss occurred on storage of such solutions.

The clear solutions are measured for barium in the atomic absorption spectrophotometer. Standard samples are evenly spaced among the unknown samples, and a new standard curve is prepared for each day.

COMMENTS

The method is accurate for amounts of sulfur above 5 μg; it is simple and can be used for routine determinations. The relationship for percentage absorption with concentration of standard solution is linear. The reproducibility of the values for individual samples is relatively good—maximum scatter for heparin less than 1%. Results with sulfanilamide, thiourea, ammonium sulfate, and azure A were in good agreement with the theoretical values.

D. DETERMINATION OF SULFATE BY CHROMATOGRAPHY (H. B. NADER AND C. P. DIETRICH)

PRINCIPLE

The method is based on the separation of the $i - SO_4$ (inorganic sulfate) (formed from the SMPs by acid hydrolysis) from other contaminating ions by chromatography, followed by complexation with toluidine blue for estimation. Scale: 2–50 μg.

Procedure. First 10 to 50 μg of SMP is dissolved in 20 μl of HCl (final concentrations, 0.04*N* for labile sulfate and 8*N* for total sulfate) and transferred to capillary tubes, which are sealed and heated at 100 °C (boiling water bath) for 2.5 hr (labile sulfate) and 5 hr (total sulfate). Appropriate standards of 2 to 20 μg of potassium sulfate and a reagent blank are subjected to the same procedure. After the hydrolysis the solutions are transferred to small test tubes (culture tubes; 0.3-mm diameter). The capillary tubes are washed twice with 20 μl of water to ensure total transfer of the solution. The combined solution (60 μl) is evaporated to dryness under vacuum and NaOH. The residue is resuspended in 20 μl of water and again dried. This operation is repeated once, and the residue is resuspended in 20 μl of 0.5*N* Na_2CO_3, spotted on Whatman No. 1 paper and subjected to descending chromatography in isobutyric acid-1*M* NH_4OH 5–3, v/v for 6 hr. The chromatogram is dried and dipped in a solution of 0.1% toluidine blue in absolute ethanol for 15 min, and the paper is destained with absolute alcohol (3 washes, 15 min each, with agitation) and allowed to dry. The quantity of $i - SO_4$ is measured by densitometry at 600 nm. An alternate procedure is to cut the chromatogram containing the stained spot of $i - SO_4$, immerse it in a solution of 2*M* Na_2SO_4 for 2 hr with agitation, and measure the resulting solution by spectrophotometry at 650 nm. Appropriate blanks of same dimensions, cut from the stained chromatogram, are run throughout the procedure. The error of both methods is in the order of ±3.5%.

COMMENTS

The method can be used for the determination of sulfate in the presence of interfering ions such as phosphate and borate. Most of the ions have a different chromatographic migration in this solvent system. An added advantage is the accurate measurement of labile sulfate. In the methods described in the literature, the partially degraded SMPs interfere with this determination. In the present method the by-products of the hydrolysis remain at the origin of the chromatogram.

E. DETERMINATION OF URONIC ACID FROM CARBON DIOXIDE LIBERATION (A. WOLLIN)

PRINCIPLE

When heated in 12% hydrochloric acid, uronic acids and their polymers lose 1 mole of carbon dioxide for each carboxyl group (173). The reaction is conducted in sealed tubes submerged in a bath of boiling toluene for 5 hr. The CO_2 is transferred to 0.5*N* NaOH in a Van Slyke manometer (137). After degassing, the CO_2 is liberated with 2*N* lactic acid and the pressure measured for a volume of 2.0 ml. The CO_2 is reabsorbed with 5*N* NaOH and the pressure again measured. The difference is the P_{CO_2} from which milligrams of carbon and uronic acid are calculated.

EQUIPMENT

Decomposition bulbs are prepared from soda-glass tubing, 7 mm OD, 5 mm ID, in lengths of 8 to 10 cm. After immersion in acid cleaning mixture for 3 hr, the tubes are rinsed with distilled water, dried, and a bulb blown on the end about 15 mm in diameter (~ 1 ml volume). Tubes are numbered with glass-marking ink, dried in the oven, and stored in a dessicator.

Spatula with handle of glass tubing that is a sliding fit in the sample tube.

Capillary pipettes. Disposable *micropipettes* with an 0.5- or l-ml syringe for acid.

Wire tray for decomposition bulbs.

In the fume hood, a *toluene boiling bath* with reflux condenser and two cold fingers in the cover.

Van Slyke-Neill manometric apparatus, with a length of pressure tubing (~ 12 cm) firmly attached (wired) to the gas inlet of the reaction chamber. 12% (w/w) HCl. 0.5*N* NaOH. 2*N* Lactic acid.

Procedure. The bulb is weighed, and the sample is placed in it without contaminating the tube walls. This is done by holding the tube horizontally, inserting the special spatula with the sample, rotating the spatula horizontally to deposit the sample, returning the instrument to the original position, and withdrawing. Solutions are introduced by capillary pipette with the same care and dried off without splashing. The tube is reweighed, and 0.2 ml of HCl (12% w/w) is placed in the bulb. The tube is sealed off in the flame with a narrow prolongation of at least 2 cm. The sealed tubes are secured vertically in a rack, which is immersed in boiling toluene (110 °C) for 5 hr. The rack and tubes are cooled in water.

For estimation of the CO_2 liberated, the neck of the tube is covered with adhesive tape and inserted as far as the bulb in the pressure tubing on the Van Slyke apparatus. With the stopcock open to the bulb, the system is exhausted three times. To check for leaks, the mercury reservoir is closed off and the mercury level in the manometer observed; 2 ml of 0.5*N* NaOH is admitted to the reaction chamber. The decomposition bulb is connected to the chamber by way of the stopcock, and the tube of the decomposition bulb in the rubber tubing is crushed with pliers. The system is again checked for leaks. The mercury reservoir is lowered and raised 10 times to transfer all CO_2 to the NaOH solution. The transfer is assisted by tapping the bulb during exhaustion and warming it in the hand during the mercury upstroke. The undissolved gases are ejected, the NaOH level is brought to the midmark of the reaction chamber, and 1 ml of 2*N* lactic acid is introduced. The solution is stirred for 1.5 min, brought to the 2.0-ml mark, and the pressure is read (P_1). Next 0.5 ml of 5*N* NaOH is introduced and the CO_2 is reabsorbed by moving the mixed solutions up and down 3 times. The level is brought up to the 2.0-ml mark and the pressure (P_2) is recorded. The temperature of the water jacket is taken, the solution is ejected from the reaction chamber and sucked out of the cup by vacuum, and 2 cc of lactic acid is admitted to the chamber and ejected, followed by acidified distilled water. After ejection of the water, the mercury is lowered to the 50-cc mark, then slowly brought up again, leaving the apparatus ready for the next determination. The procedure is repeated with the reagents to obtain the blank value, Pc. $P_{CO_2} = P_1 - P_2 - Pc$. Volume of CO_2 at 0 °C and 760 mm Hg is calculated from the equation

$$V_0^0, 760 = P_{CO_2} \times \frac{ia}{760(1 + 0.00384t)} \times \left(1 + \frac{S\alpha^1}{A - S}\right),$$

where i = empirical reabsorption factor (1.017), a = fixed gas volume (2.0), t = temperature of analysis, S = volume of aqueous solution in apparatus, and α^1 = ratio in which CO_2 distributes between equal volumes of aqueous and gas phases. The second expression corrects to 760 mm Hg and 0 °C, and the third expression corrects for carbon dioxide dissolved in the aqueous phase in the chamber. The V_{CO_2} is converted to millimoles by dividing by

22.26 (volume in milliliters occupied by 1 mmole of CO_2) and to milligrams of uronic acid by multiplying by 234 (mol. wt.).

$$\text{mg. uronic acid} = \frac{234}{22.26} \times V_{CO_2}(\text{corr.})$$

COMMENTS

Deposition of the sample in the bulb is critical. Any contamination of the walls will result in charring during sealing and formation of extraneous CO_2. Tubes and samples must be dry or solutions reduced to dryness before adding HCl, since this concentration is also critical.

The method can also be used for determination of carbonates. The sidewall above the bulb is softened in the flame, a pocket of about 0.5 ml is blown, and the HCl is kept in the pocket until sealing is completed. The tube is placed in an incubator at 40 °C for 5 hr and the results compared with a tube at 110 °C. Since carbonate can be present as a coion in SMPs, it can be useful to determine the amounts of carbonate, but a larger sample (200 mg) is required.

The method is satisfactory for measurement of uronic acid in 5 to 20 mg of heparin. For smaller samples, we have conducted the decomposition in smaller tubes and determined the CO_2 in a Natelson gasometer. Glass tubes 4mm × 50mm were sealed at one end, acid cleaned, dried, and kept in a dry container with soda-lime guard tubes. Heparin (0.04 to 0.08 mg) was introduced carefully as previously in the weighed tube, which was reweighed. Then 0.1 ml of 12% HCl was added, and the tube was sealed. Tubes were kept for 5 hr in a toluene bath (110 °C) consisting of a three-neck flask with a wire down the central reflex condenser. The sample basket was suspended from the wire in the toluene. After cooling, the combustion tube was taped, inserted in a piece of Tygon tubing attached to the Natelson gasometer, and the tubing closed with a clamp. The determination proceeded as in the micro procedure.

IV. QUANTITATION OF HEPARIN AND SMPS

The determination or assay of heparin and SMPs was difficult in the past because investigators' knowledge of the chemical nature of these compounds was very limited. Furthermore, no single property or test is able to measure all the compounds equally. Thus there is no test corresponding to the ultraviolet absorption test, which has proved so useful with proteins and amino acids by demonstrating contamination with these substances. Likewise, *no* test (including anticoagulant activity) is specific for a single compound or subgroup for heparin and SMPs. Hence differentiation is essential. With this and appropriate reference standards described in Section II.2, it is possible, on determining the *ratio* of unknown to reference in the test, to report equivalent quantities of individual SMPs by the test. Tests using the complexing properties of these polyelectrolytes provide a number of ways of measurement, based on combination with dyes and with enzymes such as those of the clotting system.

1. Reactions with Dyes

The reaction of heparin and SMPs with dyes (Section II.3.B) can be used for the determination of heparin, heparinoids, and related SMPs in various ways, with varying degrees of specificity (79). For volumetric determinations of heparin and SMPs, simple forms of the procedure are the determination of the amount of dye precipitated by the polyelectrolyte or remaining in the solution. From the discussion in Section II.3.A, it is evident that control of electrolyte concentration is a critical factor in precipitation. The determination of the amount of dye is relatively simple, involving the light absorption of the dye or of its base form. Determination of these compounds by ascertaining the dye-anion polyelectrolyte complex can be confused by the accompanying phenomenon of metachromasia—the decrease in light absorption of the dye at its characteristic wavelength and the appearance of light absorption at a shorter wavelength. The increase in light absorption at the metachromatic band is not as great as the decrease, and the position of the band shift is not the same for different SMPs and for the physical state of the complex. Thus Jaques and Wollin (91) showed that for the complex of heparin-toluidine blue in agarose gel, the metachromatic band is at 500 nm in the wet slide and at 540 nm in the dry slide. Clearly, in using the reactions with dyes, investigators and technicians must be trained first to observe and record what they see in the tube, cuvette, chromatogram, and so on.

The detection of the dye complex in the presence of uncombined dye depends on the spectral resolving power of the instrument, and this results in the paradox that a color shift that is clearly evident to the human eye may give only marginal changes of reading in a spectrophotometer. Jaques and Bell (79) reported that considerable difficulties were encountered in measuring this phenomenon with spectrophotometers because of (*a*) variations in light path between the period of exposure of the sample and the reference solutions, (*b*) variations in the temperature of the solution at the time of measurement and variations in the dye solution produced by aging, (*c*) failure of the instrument to discriminate the increased light absorption at 505 nm in the presence of the greater decrease in light absorption at 625 nm, (*d*) variations in the position of the absorption maximum for the heparin-dye complex, and (*e*) the tendency of the heparin-dye compound to flocculate and precipitate. Jaques and Wollin (90) resolved some of these problems with the Beckman DK-2 spectrophotometer, using the dye solution as the water blank.

Since this instrument is a virtual double-beam instrument, problems *a* and *b* are balanced for dye and dye + heparin, and by careful temperature control; *c* is improved by measuring the increase in light absorption at 505 nm over that of the dye alone, *d* is assessed by recording the complete spectrum at every determination, and *e* is reduced by carrying out the reaction at a low concentration of electrolyte. The same procedure can be used with spectrophotometers with the same spectral resolving power insofar as they provide the features to control variables *a* to *d*. Given the great ability of the human eye to detect color changes in the 500–600 nm spectral region and the ability of the attached computer to allow for such physical factors as slight precipitation, a quantitative procedure based on this instrument has a number of benefits. Jaques et al. (80) exploited these advantages for quantitative assay for heparin by measuring the reaction of heparin with toluidine blue by the Lovibond Tintometer. In this instrument, the color is matched on the Color System of the International Commission on Illumination (CIE).

In the case of chromatograms, the dye-polyelectrolyte complex provides a marker of its position. Differentiation of the complexes by different SMPs is usually obtained by the chromatography procedure separating them. Hence quantitation can be made by the size of the area marked (Section IV.1.C). After removal of background dye, the measurement of optical density provides satisfactory quantitation. In this situation, in view of the change of the absorption band of the complex with physical conditions such as drying, and with different SMPs, a high degree of spectral differentiation appears to be a disadvantage. Here the ability of the instrument to measure and record separately each of the components on the chromatogram is of chief importance. Fractionation and characterization of AMPs are discussed in Section VI, with special attention to cellulose acetate electrophoresis in agarose gel in Section VI.4. Here we describe its use for routine quantitative microdeterminations in the following procedures: spectrophotometric determination of heparin (90), titration of heparin with Toluidine Blue, area measurement on paper chromatograms, optical densitometry of electrophoretic slides, and optical densitometry for films after electrophoresis and electrofocusing.

As pointed out by Jaques et al. (77), International Standard Heparin No. 2 appeared to give a relatively higher metachromatic reading than other heparins, compared to the results on anticoagulant tests. Hence other heparin preparations give values for the sodium salt of about 100 IU/mg on dye tests and not 130 to 160 IU/ml found on the USP assay.

A. SPECTROPHOTOMETRIC DETERMINATION OF HEPARIN (A. WOLLIN)

PRINCIPLE

Heparin is mixed with azure A (1 mg %) in 0.07 μ barbital buffer. The increase in optical density at the metachromatic absorption band produced

(505 nm) is measured in a Beckman DK-2 double-beam spectrophotometer, using the dye diluted without heparin as the water blank.

REAGENTS AND EQUIPMENT

Barbital buffer, pH 8.6, 0.05 μ, is prepared by dissolving barbitone (22.1 g) in 200 ml of boiling N/2 NaOH and diluting to 2 liters. The buffer solution must be prepared with the diethyl barbituric acid and allowed to stand 1 day before use.

Stock solution of 100 mg of azure A per 100 ml (biological stain certified by the Commission on Standardization of Biological Stains) is prepared. It is diluted 9–100 immediately before use.

Standard heparin (International Standard preferably) and samples of heparins and heparinoids are dissolved in water to give approximately 5 to 30 U/ml.

The visible region of the spectrum is recorded with a Beckman DK-2 ratio recording spectrophotometer (115 V, 3 A, 60 Hz), with the tungsten lamp and lead sulfide cell (slit width 0.02 to 0.22 mm) and a constant flow portable cooling unit (Blue M Electric Co.), with the recorder maximum set for 200% transmission.

Water at 15 °C is circulated through the temperature-regulated cell holder by a circulating pump, from a reservoir regulated to within ±0.5 °*C*. Clouding of the outside wall of the cuvette is prevented by passing a stream of air through the chamber. Absorption of dye on the cuvettes is reduced by using new or siliconized cuvettes. Accumulated dye is removed by rinsing with ethanol at the end of the day, and at longer intervals with ethanol-HCl (10–1). Any cuvette showing scratches or other damage must be discarded.

Procedure. Distilled water (5 ml) is placed in a 25-ml beaker; 1 ml of barbital buffer and 1 ml of azure A are added with stirring (magnetic stirring). The solution is poured into a dry Pyrex cuvette (10-mm light path). This vessel is placed in the Beckman DK-2 recording spectrophotometer and left for 1 min to cool to the required temperature (15 °C) before a reading is taken. This constitutes the reference solution. For the sample solution, a suitable quantity of heparin (2 to 12 units) is diluted with water to lower the salt concentration, 5 ml is accurately measured into a beaker, and barbital buffer and azure A are added as for the reference solution. After adequate mixing with the magnetic stirrer, some of the mixture is used to rinse out a Pyrex cuvette, and a sufficient quantity is taken and placed in the spectrophotometer. After 1 min in the sample holder, the two solutions are compared by scanning the wavelength region from 900 to 400 nm. The sample mixture is examined for precipitation, then discarded. The cuvette is rinsed and left filled with distilled water. The same reference solution may be used for a series of samples but not for longer than 2.5 hr. The percentage transmission at 505 nm is converted to units of heparin from a reference line obtained with 0.5 to 6.0 units of Standard Heparin. To assay an unknown solution, the sample is diluted until the transmission obtained when the solution is added to the dye is between 35 and 80%. Then two readings are made with different dilutions, giving values in this range. On the next day or later, two determinations are made with dilutions different from those of the first day, but also giving values in the 35–80% range. Values obtained when there is a visible precipitate in the cell after removal from the spectrophotometer are discarded. Values are averaged for a minimum of three determinations.

COMMENTS

The method has satisfactory sensitivity and reproducibility. It is useful for heparin preparations assaying 30 units/mg or higher but cannot be used for crude preparations of the drug. The amount of inorganic salts present is critical and may not exceed 0.14 g% (0.02*M*) of sodium chloride or the equivalent normality of other salts. Chondroitins and other mucopolysaccharides may not be present in the final solution to a degree greater than 0.0014 mg/ml (or 30% of the dry weight of the sample). Chondroitins A, B, C and heparitins A, B, C give an absorption maximum at about 530 nm and give satisfactory agreement with the Lambert-Beer law only at the lowest concentrations. A single measurement at 505 nm is sufficient, but valuable additional information in the shape and position of the absorption band is given by recording the complete spectrum. If the absorption band is broad or not at 505 nm, or if an additional band is formed, it is likely that interfering impurities are present and the result is not valid. For example, chondroitin sulfate in a heparin solution will widen the peak, and with higher concentration (approximately 0.02 mg/ml) will shift it to 510 nm.

B. TITRATION OF HEPARIN WITH TOLUIDINE BLUE

PRINCIPLE

Heparin is added to a toluidine blue solution until the blue color is changed to purple. The color obtained is matched with that obtained with reference heparin. Effects of pH and electrolyte from the sample are controlled by using 0.15*M* buffer at pH 8.5 Color matching to a reference heparin can be made on a tile plate, in a test tube comparator block, or in a Lovibond Tintometer (see ref. 79), depending on whether convenience or accuracy is more important.

REAGENTS AND EQUIPMENT

Dilute a stock solution of 100 mg % toluidine blue (certified biological stain) to 15 mg %.

Borate buffer, 0.10*M* H_3BO_3-0.15*M* NaCl-NaOH, pH 8.5, 0.15*M* NaCl.

Reference heparin, 20 units/ml.

Coors plate with depressions and/or test tubes with comparator block and/or Lovibond colorimeter (BDH) with light apertures partly masked to fit a glass cell, 9 × 27 × 38.5 mm ID.

Glass stirring rods.

Procedure. Into the depressions of a Coors plate or a similar glass plate on a white background measure 0.2 ml of borate buffer and 0.1 ml of dilute toluidine blue. From a graduated pipette, add the sample solution dropwise with stirring until the dye becomes purple. Add reference solution to adjacent depression until the same color is obtained: units per milliliter of heparin in the sample = milliliters of reference per milliliter of sample × units per milliliter of reference solution. For the comparator block, 2.0 ml of borate buffer and 1.0 ml of dye solution are taken in a test tube and the solution is added. Test tubes with graded amounts of reference heparin are prepared for comparison. If the salt content of the sample heparin is known, the volume of the references tubes is adjusted to volume of the sample with 0.15*M* NaCl to give the same electrolyte concentration. For examination with the Lovibond colorimeter, the heparin solution is measured into the colorimeter cell; 0.15*M* NaCl is added to a volume of 2.0 ml, and 2.0 ml borate buffer and 1.0 ml dilute toluidine blue 0. The solutions are mixed and the color match made in the Lovibond box. The procedure is repeated with decreasing quantities of solution. Reference curves are constructed for the red and blue colors with reference heparin (79, p. 260). Dilutions are found that give readings on the steep part of the curve. The assay is repeated in triplicate for two or more of these dilutions.

COMMENTS

This procedure in its various forms has many advantages as a fast, semiquantitative identification test for heparin; it is applicable to both crude and purified preparations and has been used for a very long time. The electrolyte and pH serve to suppress interference by proteins and change of dye color with other SMPS. Precipitation is not as great a problem for the eye as for spectrophotometers. Hence the procedure often proves useful with "crude" heparin preparations. Individuals who have color vision anomalies may experience difficulty. It is not unusual to obtain some metachromatic color with a small portion of the sample that is lost when more is added. This occurs when purified heparin is used because of depression of metachromasia either by excess heparin (79) or by high electrolyte concentration (72) in the sample; with crude preparations, the inhibiting or competitive effects of proteins and inorganic electrolytes on the reaction are responsible (79).

C. AREA MEASUREMENT ON PAPER CHROMATOGRAMS (79)

PRINCIPLE

Dr. H. J. Bell found that after ascending chromatography of heparin on Whatman No. 2 paper, the area of the spot obtained is proportional to the amount of heparin applied up to 50 units.

Procedure. To Whatman No. 2 paper, apply 10 μl of each heparin solution to be tested in 2-μl portions, allowing the spot to dry after each portion is applied. Subject the test specimens to ascending chromatography with water-EtOH-ammonia. Stain with azure A, 50 mg in 95% EtOH. After drying, measure the area of the red spot observed.

COMMENT

Further details are given in ref. 79.

D. OPTICAL DENSITOMETRY ON ELECTROPHORETIC AGAROSE SLIDES (WITH T. K. SUE, N. M. MCDUFFIE AND S. WICE)

PRINCIPLE

Apply 2-μl samples containing 0.1 to 0.5 μg of heparin or 0.2 to 1.0 μg of SMP to agarose gel on a microscope slide in parallel with 0.8 μg of reference heparin. This is subjected to electrophoresis at 6.3 V/cm for 10 to 25 min at 2 °C. The mucopolysaccharide in the gel is fixed by immersing the slide in 0.1% Cetavlon (cetyltrimethylammonium bromide) or 0.1% cetylpyridinium chloride (CPC) solution. The slide is stained with toluidine blue. The background is cleared by washing with 1% acetic acid, and the slide is air dried. The distance migrated is measured and reported as the ratio to the same distance for the reference heparin.

The microelectrophoresis procedure itself separates heparin from many other materials present in tissue extracts. Usually some chemical treatment of tissues and body fluids is required (Sections V to VII). Differentiation of heparin and related polyanions is accomplished at the stages of (1) electrophoretic migration, (2) fixation, (3) staining, (4) destaining, and (5) evaluation of the metachromatic spot. Optical density of the spot(s) is compared with that of slides prepared with a range of concentrations of standard heparin, either with a suitable densitometer or visually, to give the value in equivalent heparin units. From conversion factors obtained with reference standards, units are converted to milligrams of respective SMPs.

EQUIPMENT

The electrophoresis is conducted in a Lucite chamber with three compartments and a metal tray as lid and cold finger, described by Jaques et al.

(77). Overall dimensions are 20 × 8.6 × 4.4 cm for 5- and 7.5-cm slides and 31 × 11 × 9.5 cm for 15-cm slides; the central compartments are 12 and 18 cm long, respectively. The end chambers contain buffer and electrodes joined to the central chamber by bridges of buffer in agarose. The agarose slide completes the circuit. Crude pentane (petroleum ether) is placed above the slide and ice is placed in the metal lid. A temperature at 2 °C is maintained by boiling of the pentane and its condensation. Current is provided from a suitable power-pack, with voltage and amperage selected to give sufficient migration of the heparin without raising the temperature from 2 °C. Microscope slides are carefully cleaned, precoated, and coated with agarose, Glass slides 2.5 × 5.0 cm are generally used for quantitative work. Slides 5.0 × 5.0 cm, which allow five 5-mm slits 5 mm apart, and longer slides 5 × 7.5 and 5 × 15 cm, which give greater separation, are also used, particularly in comparison studies. It is most important in preparation of the agarose slides to select glass of good quality for microscopy and to ensure that the gel has uniform thickness and a concentration of 0.9%. If the agarose gel is too concentrated or the gel film too thick, there will be interference with removal of unwanted substances at washing stages. The thickness must be 1.0 to 1.1 mm for heparin. The agarose gel film furnished by Analytical Chemists Inc. (A.C.I.) Palo Alto, California, is satisfactory for this procedure because the films of this manufacturer have a gel thickness of 1 mm. For determination of chondroitins, and so on, hence of total sulfated mucopolysaccharides, it is advisable to reduce wash-out by using a slide 2 mm thick. Gels are prepared with 0.06*M* barbital buffer, pH 8.6. Other buffers used are: Tris buffer, pH 7.5; lithium-acetic acid buffer, pH 3.0 (64); 0.1*M* ethylene diamine acetate, pH 8.0; and 0.025*M* propylene diamine acetate, pH 8.9 and 10.0. To change buffer in the gel, a slide or film is left in the new buffer for 20 min. Since the agarose in the Lucite chamber must have the corresponding buffer, it is necessary to prepare a sufficient number of chambers with required buffers. With the A.C.I. film, the A.C.I. electrophoretic chamber is used. Special slide cutters made from razor blades are used to give 5-mm slits, 5 mm apart. Also required are 1- and 2-μl micropipettes, calipers, a plastic ruler, and a magnifier (2.5 ×) such as Ednalite Macroscope model 150A, for measuring the migration distance on the slides.

REAGENTS

Agarose [L'Industrie Biologique Française S.A., 35 à 49, Quai de Moulin de Cage, Gennevilliers (Seine), France]. 0.1 g is dissolved in 5 ml of 1% glycerol + 25 ml of barbital buffer + 70 ml of water. This is used to precoat slides. 0.9 G is dissolved at 65 ° C in 0.06 M barbital buffer. This is used for coating the slides. Each batch of agarose should be tested immediately on receipt by preparing slides to ensure that background toluidine blue can be washed out (absence of agaropectin).

Cetavlon (hexadecyl trimethylammonium bromide), Eastman Organic Chemicals T 5650, is made up as a 1% aqueous solution and diluted 1–10 before use.

CPC (cetylpyridinium chloride), Eastman.

Toluidine blue C.I. No. 52040 (formerly 925) certified by the Commission on Standardization of Biological Stains: 40 mg is dissolved in 20 ml of water and 80 ml of dry acetone added.

A *reference* sample is placed on all slides. For a suitable reference heparin, a commercial beef heparin must be used, which gives a single band on microelectrophoresis. We have used Heparin Leo, Lots 4275 and 4439, labeled 5000 IU/ml, obtained from Leo Pharmaceutische Production N. V. Emmen, the Netherlands. This is diluted in water to 50 Leo units/ml before use. For quantitative comparisons, it is necessary to have a set of slides prepared with graded amounts of *standard* Heparin (preferably International Standard Heparin), which by definition contains 130 IU/mg. Similar sets of slides for one or more reference chondroitins (see Section II.2) are required to check the conversion factor. For comparison and checking enzymes, good quality commercial chondroitins A, B, and C, heparitin, dextran sulfate, and other reagents, are required.

Manipulative Procedures. Each slide is cut from the gel immediately before application of the test solutions. With the cutter, cuts are made in the agarose on the slide, 2 cm from the end of the slide for the negative electrode, and strips of filter paper are inserted in the cuts. The solution is placed in the slot with a 2-μl micropipette as the filter paper is removed and the lips of the cut closed. Solutions for analysis should contain about 0.5 μg heparin or 1 μg SMPs; 0.05 to 0.10 ml of a solution in buffer or water is prepared, and precautions are taken to avoid change of concentration by evaporation. For satisfactory pipetting, there should not be more than 10 to 20 mg of solids in 0.1 ml (ratio of heparin to other solids $>$ 1–400). The reference solution is placed in the slots on the right side. Two 2.5-cm slides or one double slide are placed face down in the electrophoresis chamber, previously cooled to 2 °C. Current is passed through the slides for the predetermined time required to give a satisfactory migration of the material (about 3 cm traversed by the reference heparin solution). With barbital buffer this is usually 25 min at 90 to 130 V, 80 mA; with Tris buffer, 15 to 20 min at 160 V, 40 to 60 mA. If the A.C.I. equipment is used, electrophoresis time is 15 min at 90 V. Running time depends on the agarose, voltage supply, and so on.

Slides are placed immediately in 0.1% Cetavlon for fixation. The spot containing heparin or related mucopolysaccharides appears as a white haze. The slides are placed under 250-W infrared lamps for 2 hr with an air current from a fan. They are then stained by immersing for 15 min in toluidine blue solution, rinsed in 1% acetic acid, and left in 1% acetic acid to remove the background color. The heparin-dye compound remains insoluble under these conditions. Other mucopolysaccharides may show fading and can be removed, unless it is desired to examine them. Heparin gives a dense reddish-purple spot against the almost colorless background of the slide. Time for destaining and stability of the complex with other SMPs appears to depend on the batch of dye. We have found the dye at 92% dye content from Harleco, Philadelphia, (Item 364) to be satisfactory. For radioautographs, the microelectrophoresis slide is processed, given final drying, and exposed to Royal Blue Medical X-Ray film (Kodak, Toronto, Canada) for 3 to 5 days, depending on the type and amount of isotope used.

Observations. A great deal of useful information is obtained by simple inspection of the microelectrophoresis slides during processing; it is always recorded. Such information on the initial examination can help in deciding on changes in amount of sample taken and what further identification tests are required. Data are recorded for position and appearance of spots at fixation, at staining, after destaining, and after drying. At the fixation stage, the spots appear as a white cloud; large spots suggest that slides should be prepared with less sample. After staining with toluidine blue and before decolorizing with 1% acetic acid, hyaluronic acid is seen as purple spots with relative migration of 0.7 in agarose-barbital. The spots disappear on decolorizing. At this stage also, protein frequently shows up as clear areas in the blue-stained gel close to the point of application. This technique was found very useful in the examination of platelet factor 4 and heparin interaction (129). When protein is seen, duplicate slides can be fixed in alcohol and stained with amido black for quantitation of protein. Metachromatic colors are in general more brilliant in wet slides than in dry (91,180,181); hence it is important to record the appearance of the slides after staining and destaining. The relative colors are listed in Table III. The color shown by the dye on the slide is quite distinctive for these substances. Heparin, particularly when the slide is first prepared, gives a characteristic bright reddish-purple spot with a sharp outline. The spots with chondroitins A, B, and C are more bluish-purple. Keratosulfate appears on the slides as bright blue spots with an R.M. value of 0.67. Nucleic acids also give bright blue spots with an R.M. value of 0.74 to 0.83, which changes to 1.0 in ethylene-diamine or propylene-diamine buffer. This property can serve as a useful internal reference when differentiating chondroitins and heparitins but can mask part of the heparin spots, and many require removal by treating the sample with trichloracetic acid or with nucleases (see Section V.1.B.b.). Nucleic acids with any residual proteins can be largely removed by adding an equal volume of 20% trichloracetic acid at 4 °C to the solution of crude heparin (about 10 mg/ml

TABLE III
Appearance of Various SMP and Other Compounds in Agarose Electrophoresis Slides During Processing[a]

Compound	Fixation with Cetavlon or CPC	Stained with toluidine blue: Before decol-orizing	Stained with toluidine blue: Destained (1% acetic)	Dry
Hyaluronic acid	White precipitate	Dark blue	No spot	No spot
Commercial heparin	White	Red	Red	Purple
Heparitin A	White	Purple	Blue purple	Blue purple
Heparitin B	White	Purple	Blue purple	Blue purple
Heparitin C	White	Purple	Purple	Purple
Heparitin D	White	Purple	Purple	Purple
Chondroitins A, C	White	Purple	Blue purple	Blue purple
Chondroitin B	White	Purple	Purple	Blue purple
Keratosulfates	White	Blue	Blue	No spot
Nucleic acids (RNA, DNA)	White	Bright blue	Bright blue	Bright blue

[a] Results shown for barbital buffer, pH 8.6. Results are similar for other buffers, although colors are more blue for alkylamine buffers and for CPC fixation.

in an ice bath). After standing 15 min, the mixture is centrifuged at 4 °C. To the supernatant is added 1% NaCl and 5 volumes of acetone. After centrifuging, the precipitate is lyophilized.

Heparin preparations show considerable variability in electrophoretic mobility, with values from 0.81 to 1.00 in agarose-barbital as indicated in Table IV and reported in refs. 81,94, and 97. Hence these color differences are most helpful in identifying a small proportion of heparin in a mixture of SMPs. Identification of the SMPs in a mixture requires the use of different buffers in electrophoresis, as well as enzymes and determination of critical electrolyte concentration; it is obvious from the foregoing discussion, however, that the most useful procedure for establishing the tests that are required is the careful inspection of the electrophoresis slide at various stages of the process. This and the tests described below make it possible to identify SMPs when measured by microelectrophoresis on agarose.

Quantitation Procedures. An essentially linear relation is found between absorbance and SMP concentration up to 0.3 unit, and between log area of the spot and log of quantity of heparin applied up to 0.4 unit. About 0.1 unit of heparin applied is optimal for measurement and should be used for quantitative work. The optical density of the stained spot is compared visually with that shown in a series of slides prepared with graded amounts of the International Standard Heparin. Visual comparison should always be done to ensure each slide is of good quality. Tests slides are compared blind with the set of standard slides, and the matching value is recorded.

The optical density may also be read in a densitometer. Jaques et al. (77) reported that the only densitometer sufficiently sensitive to give adequate readings on these slides is the Chromoscan densitometer, and variations in integral reading occurred with the same electrophoresis slide and same spot when the position of the spot was changed in the slide holder. By the use of a modified slide holder and correction of the integral value by subtracting the base line, quite satisfactory reproducibility and accuracy are obtained. Extensive testing has shown that best results are produced when the following precautions are observed. Instrument settings are: cam D, slit 8 mm × 0.5 mm, specimen drive gear ratio of 1–9, filter No. 550 (540 nm), optical arrangement in the reflectance mode. The specimen holder is loaded so that the microelectrophoresis band of interest is in the top position and the agarose surface faces the light beam. The slide carriage is adjusted so that the microelectrophoresis band on the right-hand side is in the center of the incident light beam. The slide carriage is then drawn to the left until the light beam is 5 mm below the band on the slide. The base line is adjusted with "base line control" to a distance of 3.5 cm from the rim of the drum. Great care should be taken in setting the base line because considerable errors can be introduced, especially if the base line is set too low. Tolerance should never be more than ±1 mm between settings. A trace of 18 cm is made. If the pen is found to "plateau," indicating that the density is beyond the recorder, it is necessary to change the cam. After a satisfactory trace is obtained, this test is repeated and the integrator count recorded for the first 3 cm, the next 12 cm, and the last 3 cm. The sum of the count for the first and last portions is doubled and subtracted from the count for the midportion to give the correct integrator count. This is done in duplicate. Standard curves are constructed from the set of standard slides, prepared with graded amounts of International Standard Heparin or of a secondary standard similarly calibrated by comparison with the International Standard Heparin.

Results are reported in equivalent heparin units. These values can be converted to micrograms by multiplying by 8.0 for heparin and heparitins and by 17.2 for chondroitins, as determined by ourselves. If possible, the investigator should determine these factors using suitable standards under laboratory conditions. It has been found that the optical densities of the spots obtained with heparin, chondroitins, and heparitins when expressed in equivalent units are additive. Hence following identification by changes in migration in buffers (Table IV), the composition of the sample in different SMPs can be expressed as a percentage of the total. This has also been found to be true for the action of enzymes and electrolytes. There should be

TABLE IV

Effects of Buffers on Relative Electrophoretic Migration of SMP in Agarose (values relative to migration of reference[a] beef lung heparin)

			Heparins[a]				Heparitins				Chondroitins			
Buffer	M	pH	Beef lung	Pork muc.	Matt.	Macro mol.	A	B	C	D	A	B	C	KS
Barbital	0.12	8.6	0.99	0.90 0.97	0.88 0.99	0.56 0.67	(0.81)	0.83	0.86/ 0.93	1.04	0.88	0.88	0.88	0.67
Tris	0.12	7.5	0.98				(0.76)	0.90/ 0.95	0.91/ 0.98		0.93	0.93	0.93	0.69
Phthalate	0.0	3.0	1.00	0.93 1.04	0.87 0.90	0.15/ 0.54					0.90	0.88	0.75	0.81– 0.85
KCl-HCl	0.02	2.0	1.00	0.87– 1.03 (2)	0.87 0.92	0.29 0.38					0.66 0.90	0.71– 0.92	0.61– 0.79	0.70 0.82
HAc-LiOH	0.35	3.0	1.00	0.93 1.00		0.67	0.72	0.76	0.88	1.04	0.84	0.82	0.84	
Ethylene-diamine	0.1	8.0	1.00	1.00		0.39– 0.58	0.92	0.92	0.89	1.04	1.16	1.05	1.16	
Propylene-diamine	0.025	8.5– 10.0	1.00	0.98	0.95	0.75	1.11	1.12	0.92	1.08	1.35	1.25	1.35	1.20
			Percentage remaining after fixation with CPC (M NaCl)											
			100	100		76	39	70	0	0	0	0	0	?

[a] Reference: Leo Heparin No. 4439 and Upjohn Heparin Lot No. ZX320. Muc. = intestinal mucosa; Matt. = from reference collection of Matthews et al. (115); Macro mol. = macromolecular heparin of Horner (65); KS = keratosulfate/indicates alternative values depending on samples or runs; two values unconnected indicates two separate bands. Pork intestinal mucosa heparin usually gives two bands in barbital and acid buffers in varying proportions.

quantitative agreement in the results for the different tests. If there is no agreement, the SMP under examination is known to differ in these properties from the reference compounds.

Quantitation of SMPs Identified by Migration in Different Buffers with Agarose. The relative electrophoretic migration in agarose of different SMPs is changed by pH. We have found that an additional useful variable is complexing with the buffer cation. Relative electrophoretic migration values are listed in Table IV. In general, the heparins—including the low molecular weight heparin fraction of commercial heparin (heparitin D)—migrate relatively closely together. Macromolecular heparin is an exception. The heparitins (A, B, C) form another group, and the chondroitins (A, B, C) a third group. The degree of separation and direction is affected by the buffer. There is little separation with Tris buffer, more with barbital, and still more with the acid buffers. In these buffers, the order of migration is heparitins $<$ chondroitins $<$ heparins. With the diamine buffers (which form complexes with SMPs as well as providing an alkaline pH), the order is heparitins $<$ heparins $<$ chondroitins. It is therefore possible to distinguish these SMPs by comparing electrophoretic migration in different buffers in agarose.

In quantitation, 1 μl (for 5-cm slides) or 5 μl (for 15-cm slides) of the solution to be tested is placed in slits of two agarose slides, one with barbital buffer and one with propanediamine; 5 μl of a mixture of SMPs is placed in another slit of each slide. For the barbital this is composed of chondroitin A, heparitin sulfate, keratosulfate, and hyaluronic acid (1 mg/ml). The propanediamine slide takes heparitin sulfate, chondroitin A, and B (1 mg/ml). These two buffer systems give a preliminary indication of the composition of the unknown mixture. Slight variations in migration are observed among chondroitin B and heparitins obtained from different tissues. Nevertheless these variations are smaller than the relative distance migrated by the different SMPs. The relative amounts of the SMPs are quantitated by densitometry of the slide.

Quantitation of SMPs Identified with Enzymes. Identification on the basis of enzyme specificity is more reliable for identification of SMPs because it is based on identifying specific chemical structures. Bacterial enzymes are now available, and their use in combination with the microelectrophoresis procedure as described allows enzymatic identification on the microscale. Enzymes available are chondroitinase ABC, which breaks down chondroitin A, B, and C (145,182) and chondroitinase AC, breaking down chondroitins A and C; heparinase, breaking down heparin; and heparitinase I, which breaks down heparin and heparitins (heparitins A, B, C, D). Suitable controls with reference compounds are essential. Since the products resulting from the action of these enzymes are removed in the processing of the microelectrophoresis slides, a separate fractionation step is not required, as is the case when color reactions (carbazole, etc.) are used for measurement. The amount of a sulfated mucopolysaccharide present can be quantified from the amount removed by the specific enzyme as measured on the agarose electrophoresis slide. Preparation of enzymes is described in Section VI.6.

The SMP solution is incubated with 0.05 unit of chondroitinase AC, chondroitinase ABC, and crude extracts of heparin-induced *Flavobacterium heparinum* cells (100 μg). The incubation mixture contains 25 μl of mucopolysaccharide substrate, 10 μl of Tris-HCl buffer (pH 8.0 for chondroitinase ABC and pH 7.3 for chondroitinase AC), 10 μl of sodium acetate solution, and 10 μl of albumin solution. Enzyme is added in excess and incubated at 37 °C for 16 hr to ensure complete degradation of the substrate. After the incubation, 5 μl of each incubated mixture is subjected to agarose gel electrophoresis in the two buffer systems. The type of SMP present in the mixture is indicated by the disappearance of the spots by the action of the specific enzymes. Chondroitin A and chondroitin C have the same migration in both agarose systems and are susceptible to the action of the same mucopolysaccharidases. The distinction between these two SMPs is made by chromatography of the remaining 40 μl of incubates as described in Section VI.6. Chondroitin A produces Δ-di-4S by the action of chondroitinase AC, whereas chondroitin C produces Δ-di-6S. The chromatography of the other incubates also confirms the presence of the other SMPs by the type of disaccharides formed. The crude *F. heparinum* extract produces two compounds from heparitin sulfate—namely, glucosamine 2,6-disulfate and *N*-acetylglucosamine, which are used for identification.

Quantitation with Critical Electrolyte Concentration. The principle of critical electrolyte concentration was presented in Section II.3.A. The critical electrolyte concentration for insolubility of complexes of SMPs with organic cations depends on the charge and molecular weight of the SMPs. Fixation of agarose slides with CPC or Cetavlon in 1*M* NaCl is useful for distinguishing commercial heparins (164). Since chondroitin polysulfates are precipitated in this test, it must be supported by other tests when applied to tissue extracts for differentiation. The application of the principle was investigated in connection with microelectrophoresis as part of the microelectrophoresis procedure at the stages of fixation or staining (180). We find it convenient as a test to apply it to the fixation step by placing slides in a plastic jar containing CPC solution in NaCl at 37 °C. Molarity of NaCl is 0 to 1.5. After shaking for 5 hr in a Dubinoff shaker at 37 °C, the slides are removed and the CPC solution replaced by CPC in water. This is shaken at 37 °C for 1 hr.

The CPC solution is replaced with fresh CPC, the temperature lowered to room temperature, and the shaking continued overnight.

COMMENT

Heretofore, determination of sulfated mucopolysaccharides has been done by color reactions for constituent sugars (chiefly the carbazole reaction). Determination of SMPs by the microelectrophoresis method has advantages compared to use of the carbazole method as being more specific for sulfated mucopolysaccharides. Hyaluronic acid, an unsulfated mucopolysaccharide, produces an equal color value in the carbazole reaction but gives no reading on electrophoretic slides stained by toluidine blue. The microelectrophoresis procedure is more sensitive than the color tests. The microelectrophoresis test is 15 times as sensitive as the carbazole (30 times with heparin) and 50 times as sensitive as the aqueous metachromatic test for the SMP. Furthermore, the microelectrophoresis test is relatively insensitive to contamination by the protein, carbohydrate, and inorganic salt that are present in crude tissue samples, interfering with color development and changing the color seen with carbazole or with a metachromatic dye in aqueous solution. In the agarose film, the latter substances are separated from the mucopolysaccharide during migration, therefore do not affect the staining and color of the spot.

The microelectrophoresis method has marked advantages over carbazole and other carbohydrate color reactions for determination of specifity because of its sensitivity to SMPs and insensitivity to the influence of other substances. It has the added advantage of being a suitable basis for simple procedures for the identification and characterization of the individual SMPs. To apply the carbazole reaction in this way requires the use of column chromatography, elution at high salt concentrations, and desalination. With microelectrophoresis, information for identification is obtained simultaneously by examining the color of the spot and the degree of electrophoretic migration judged by the position of the spot, and comparing these observations with those seen with control slides prepared with reference compounds. This information is readily amplified by using different buffers in the gel and applying other tests (e.g., the critical electrolyte concentration for solubility of complexes). Specific enzymes (heparinase, chondroitin sulfatases ABC, AC, and B) can readily be used with microelectrophoresis to obtain directly further definite quantitative and qualitative information for each mucopolysaccharide. Again, this result can be achieved with the carbazole reaction only after using column chromatography, since the enzymatic degradation products also give a positive carbazole reaction.

Continuous experience with the method has demonstrated that it is an accurate, economical, and rapid procedure for examination and quantitation of heparin, because of the control of temperature and time and gel properties in the microelectrophoresis. The method has proved versatile and adaptable to many studies on the chemistry of heparin and on the distribution of heparin in tissues. It makes possible conducting on microgram quantities such tests as solubility of material in solvents, precipitation by chemical agents, determining the actions of enzymes, and examining extraction procedures on small tissue samples. It is equally accurate, economical, and rapid for examination and quantitation of chondroitins and heparitins for the same reasons. Since optical densities obtained with mixtures of these compounds prove to be additive, it is a quantitative procedure for analyzing mixtures of SMPs. Differentiation is readily accomplished by noting the difference in mobility of SMPs in different buffers and the action of enzymes, which can be quantitated in the microelectrophoresis slide.

Effective separation of a mixture of SMPs depends on the relative proportions of SMPs. With a proportion of 95 to 5%, to obtain sufficient optical density to measure the minor component, the amount of the major component will spread over this portion of the slide. Greater separation can be obtained by the use of a longer slide, as well as the use of enzymes and/or critical electrolyte concentration at the fixation stage, to remove the major component of the mixture and reveal the minor components. There are certain technical problems. Acid pH, particularly when acid is combined in the hot agarose solution for preparing the gel, affects gel strength. It is therefore preferable to make up the gel in barbital buffer and equilibrate individual slides 1 mm thick with other buffers as required. There is also some danger of wash-out at the destaining step for chondroitins and heparitins. The use of Tris buffer in the gel results in close migration distances for the SMPs. Tris buffer in 1-mm gel thickness is preferable for measurement of total SMPs (omitting Keratosulfate). Direct comparison of slides with heparin 59 fixed with Cetavlon and CPC showed no significant difference in optical density.

The optical density of the spot is not compared directly with that of the reference heparin spot on the slide because the variation in optical density between the two spots on the same slide is as great as for the spots for the same concentration in a series of slides. The optical density of the reference spot is compared to the range of values obtained with reference spots on standard slides, to ensure that there has been no serious change in conditions of slide preparation.

Visual comparison of microelectrophoresis slides with a series of reference slides with 0.02 to 0.6 unit of heparin gave a coefficient of variation of 15.5%. Difficulty was found in providing a densitometer with sufficient sensitivity and reproducibility. By specifying carriage position for the slide, slit size in the optical system, base line adjustment, background correction, and light filter, reproducibility of measurement could be achieved with the Chromoscan densitometer with 10% maximum error with low counts. The coefficient of variation was 6.2% when all precautions were taken.

E. OPTICAL DENSITOMETRY FOR HEPARIN FILMS AFTER ELECTROPHORESIS AND ELECTROFOCUSING (N. M. MCDUFFIE)

As described in Sections IV.1.D, V, VI, and VII, mixtures of SMPs may be separated in various gels into their component parts. Commercial heparins may be separated into 21 fractions. Quantitation calls for densitometers that can measure a variety of sizes of films and gels on slides and plates, distinguish and record the values for the individual components, and total and print out the percentages for the components of the mixture. As indicated earlier, isolation of a small section of spectrum is probably not an advantage. Instruments are now available with these features, although it is too soon to give an evaluation based on substantial experience. The following list is representative.

Biomed Soft Laser Scanning Densitometer
Biomed Instruments
5742 W. Dakin St., Chicago, Ill. 60634.

CAMAG electrophoresis scanner with computing integrator and analog recorder
CAMAG,
Homburgerstr. 24
CH-4132 Muttenz, Switzerland

Chromoscan 200/Scan 201
Joyce Loebl & Co.
Northwest Industrial Park
Burlington, Mass. 01803.

Corning Scanning Densitometer model 730
Corning Instruments
Medfield
Industrial Park
Medfield, Mass. 02052.

IL377 Densitometer with normalized integral format
Instrumentation Laboratory Inc.
Lexington, Mass. 02173.

Kontes Calculating Integrator model K21
Kontes
Vineland, N.J. 08360.

Microzone Computing Densitometer CDS-100 and
Microzone Computing Densitometer MCD
Beckman Instruments, Inc.
Clinical Instruments Division
Fullerton, Calif. 92634

Transidyne model RFT and PDQ Computing Integrator
Transidyne General Corporation
462 South Wagner Road
Ann Arbor, Mich. 48106

2. Anticoagulant and Allied Tests

Two applications of anticoagulant and related tests are monitoring drug dosage in heparin therapy and standardization of the drug for anticoagulant therapy. The first is properly the responsibility of the clinical hematologist. The second is the responsibility of the manufacturer as controlled by pharmacopoeias. However other investigators have occasion to use these techniques in their examination of heparin and SMPs, and it is evident from the literature that there is confusion with respect to what the procedures accomplish. In the earlier review in this series by Jaques and Bell (79), methods for the first use were simply indicated by a reference to a coagulation manual, whereas those for the second were described in detail. Both types of method have been called "assays" by their originators, although they represent assays based on quite different principles and serving different purposes. This chapter summarizes principles and reviews new developments for these "assays" under the headings *Hematology Procedures* and *Pharmacology Procedures*. This arbitrary division has the advantage of permitting the user to make rapid identification for the purpose required. A list of these procedures with references is given in Table V.

In all coagulation tests involving heparin, extra attention must be given to the special significance of pH and surfaces. Values for clotting times with most coagulation systems are essentially the same between pH 5.5 and 7.6, but the anticoagulant effect of heparin is markedly pH dependent and is minimal at pH 6.4 (73). A quantity of heparin sufficient to give an infinite clotting time in glass with blood at pH 7.5, produces scarcely a significant prolongation of clotting time with the blood at pH 6.4. There is a proportional decrease of anticoagulant activity as the pH is lowered from 7.5, indicating that a decrease in the coagulation system of only 0.1 pH unit can significantly decrease the anticoagulant action of heparin. Therefore it is essential to maintain careful control of the pH of the test system and of the heparin solution. Hardaway (56) states that "heparin is inactivated in the presence of acidosis." Hence a good blood flow is necessary when blood is taken for examination with heparin. Lyophilization generally results in a pH shift, since bicarbonate, citrate, and ammonium salts decompose. On reconstituting lyophilized plasma, all the material must be redissolved, and the pH measured and adjusted.

The "inactivation" of heparin anticoagulant activity by a slight acid pH shift in the test system points to the necessity of ensuring that the heparin

TABLE V
Anticoagulant and Related Test Procedures for Heparin

Identification	Substrate, etc.	Author	Reference
	A. Hematological		
Direct tests on blood or plasma			
Whole blood clotting time	Fresh blood	Lee and White (1913)	(11,170)
Silicone clotting time	Fresh blood taken in silicone	Jaques et al. (1946)	(170)
Partial thromboplasitn time (PTT)	Plasma	Langdell, Wagner and Brinkhous (1953)	(11,170)
Activated PTT	Plasma	Langdell et al; Proctor and Rapaport (1961)	(1,11)
Thrombin time	Plasma	Jim (1957); Lewis and Didesheim (1957)	(11)
Titration with thrombin	Plasma	Quick (1936)	(139)
Protamine titer	Blood or plasma	Waters, Markowitz, and Jaques (1938)	(11,79)
Toluidine blue titer	Plasma	Meneghini and Cervini (1949)	(79,93)
Factor Xa Inhibitor	Plasma	Yin, Wessler and Butler (1973)	(1,183)
Lipoprotein lipase release	Determination of LPL on Ediol	Hahn (1943)	(99)
Diamine oxidase release	Determination of DAO on ^{14}C putrescine	Tryding (1965)	(96)
Extraction of blood or plasma, with measurement by coagulation, metachromatic reaction, etc.			
Octylamine	Plasma	Jaques, Monkhouse, and Stewart (1949)	(79)
Phenol	Plasma	Monkhouse and Jaques (1949)	(79)
Calcium phosphate	Plasma	Nilsson and Wenckert	(79)
Trypsin	Plasma	Engelberg, Dudley, and Freeman (1955)	(79)
AMPs of white cells	Blood	Bassiouni (1955)	(79)
Tests on urine			
Spot test	5 mg % toluidine blue	Jaques, Napke, and Levy (1953)	(87)

TABLE V (*Continued*)

Identification	Substrate, etc.	Author	Reference
Metachromatic	5 mg % toluidine	Jaques, Napke, and Levy (1953)	(87)
Turbidimetric	2-Dimethylamine methyl-dibenzo-furan HCl	Marbert and Winterstein (1951)	(79) (79)
Benzidine	Precipitation with benzidine	Astrup (1944)	(79)
B. Pharmacological			
Coagulation tests in vitro			
Modified Howell	Fresh whole blood (dog or cat)	Jaques, Charles, and Kavanagh (1975)	(79); Section IV.2.B.a
Fischer	Whole chicken plasma + thromboplastin	Fischer and Schmitz (1932)	(79)
Jorpes et al.	Beef blood	Jalling, Jorpes, and Linden (1946)	(79)
USP	Recalcified citrated sheep plasma	Reinert and Winterstein (1939)	(79)
BP	Sulfated whole blood + thromboplastin	Adams (1950)	(79)
Antithrombin	Oxalated beef plasma + thrombin	Stueder and Winterstein (1951)	(79)
Antithrombin	Citrated beef blood + thrombin	Jaques and Charles (1941)	(79)
Tests in vivo			
Jorpes et al.	Sheep with indwelling catheters for Cl.T.	Jorpes, Blomback, and Blomback (1954)	(79)
Kuo et al.	8 blood samples from conscious, trained dogs for Cl.T., APTT, LPL	Kuo, Jaques, and Millar (1972)	(102)
Other tests	Conscious, trained dogs	Kuo, Jaques, and Millar (1972)	(102)
Other identification tests			
Metachromatic—Lovibond	Toluidine blue	Jaques et al. (1947)	(79,80)
Metachromatic—DK2	Azure A	Jaques and Wollin (1967)	Section IV.1.A

TABLE V (*Continued*)

Identification	Substrate, etc.	Author	Reference
Paper chromatography	Toluidine blue, $AgNO_3$, etc.	Jaques and Bell (1959)	Section IV.1.C
Electrophoresis	Agarose, acrylamide, cellulose acetate + toluidine blue	Various	Sections IV.1.D,VI
Isolation as crystalline barium salt	Barium salt	Charles and Scott (79,95)	

solution itself does not produce this effect. T. Astrup and Jaques have agreed that most of the reports on acid inactivation of heparin are of limited value because they fail to specify that exact neutralization was accomplished before testing. Furthermore, variations in this regard may be responsibile for reports of instability of heparin in intravenous fluids. O'Kuno and Nelson (130) examined the stability of anticoagulant activity of heparin in $2\frac{1}{2}$% dextrose, 5% dextrose, 10% dextrose, normal saline, and Ringer's lactate solutions. They observed a marked reduction in anticoagulant activity by both antithrombin and factor Xa inhibition procedures. The lost anticoagulant activity was fully recovered in 24 hr when measured by the factor Xa inhibition but not by the antithrombin procedure. These results might be attributable to differences in the pH effect in different test systems or to a lactone form of heparin proposed by A. F. Charles (17). The recent demonstration that iduronic acid is the main uronic acid of heparin supports the second possibility. Conversion of lactone to acid to salt and reverse will depend on time, temperature, and concentration of hydroxyl ion.

Although the effects on anticoagulant activity are not well documented for exposure of heparin to acid, there is universal agreement that there is a deleterious effect that is increased by heating and must be avoided. All solutions should be adjusted to pH 7.2; and if they are to be lyophilized, allowance must be made for the resulting pH change. At pH 7.0 to 7.2, heparin solutions can be sterilized by autoclaving. For storage as solids, salts should be precipitated or lyophilized from a neutral solution (pH 7.0 to 7.2).

Blood coagulation is activated by surface contact, and probably heparin *in vivo* is effective because it prevents this process, as indicated by its efficacy when combined in the surface layers of prostheses and by the fact that it is taken up specifically by endothelium (61,62). It should not be surprising, therefore, that the surfaces with which blood comes in contact themselves affect profoundly the inhibitory effectiveness of heparin on blood coagulation. Normal endothelium appears to be neutral or inert to blood coagulation. The same condition appears to hold for glassware completely covered with the silicone coating Dri-Film (SC87, General Electric Co., Silicone Products Dept., Waterford, N.Y.). Forty years ago when heparin first became available, all laboratory glassware vigorously promoted clotting of blood, and laboratory workers guarded carefully the few pieces of equipment in which clotting was slower. Today with the ubiquity of plastics and surface-coating agents in the laboratory, the worker has considerable difficulty preserving glassware with the original properties on which coagulation procedures were standardized. It is therefore necessary to describe the selection and preservation of test tubes used in the standard tests.

PREPARATION OF GLASSWARE FOR COAGULATION TESTS WITH HEPARIN

(WHOLE BLOOD CLOTTING TIME, HOWELL WHOLE BLOOD ASSAY FOR HEPARIN, ETC.)

Two times the required number of Pyrex glass tubes, 8×60 mm or 10×75 mm, are filled with 1 ml of distilled water. The tubes in which the water-air interface is at a height above or below that observed in most of the others are rejected. For Howell assays, the 1-ml level is marked permanently around the whole circumference with a glass file. The interface is inspected. A straight line demonstrates the presence of a water-repellent surface, which must be removed. The tubes are left overnight in acetone-NaOH (2 liters of acetone, 2 liters of water, 200 g of solid NaOH). They are rinsed with deionized water. The tubes are boiled with 10% cleaning solution (sulfuric acid-dichromate), rinsed, and boiled twice in deionized water. The air-water interface is inspected. This must be a miniscus. The tubes are then dried in a hot-air oven. This treatment leaves the glass surface with a desired negative charge, since to remove all sulfate from the glass requires repeating this procedure 20 times. All rinsing must be done with deionized or distilled water; detergents or tap water must not be used at any stage. As soon as the clotting test is completed, the tubes are emptied. The bare end of a cotton applicator is used to fish out clots, and the cotton end is used to remove any clot or protein adhering to the glass. The tubes are washed well with soft (deionized) water. Given the universal use of plastics and coatings in laboratory ware today, to maintain test tubes with the proper surface for determination of clotting times, in addition to observing the caution by Bowie et al. (11) never to wash these tubes with siliconed glassware, we find it necessary to repeat the treatments with acetone-NaOH and 10% sulfuric acid-chromate as in the initial preparation of the tubes.

A. HEMATOLOGICAL PROCEDURES

The hematological procedures are chiefly coagulation tests carried out in the clinical laboratory for the investigation of coagulation disorders, which are described in handbooks of methods for coagulation laboratories

(11,22,66,93,170) and in booklets by manufacturers of coagulation reagents—Warner Chilcott (General Diagnostics Division, Warner Chilcott Laboratories, Morris Plains, N.J.). Metrix (Clinical and Diagnostics Division, Armour Pharmaceutical Co., Chicago), Baltimore Biological Laboratories (BBL: Division of Becton Dickinson & Co.), Ortho (Ortho Diagnostics Raritan, N.J.), Hyland (Division of Travenol Laboratories, Inc., Costa Mesa, Calif.), Dade (Division of American Hospital Supply Corp., Miami).

Those used especially in the detection of the effect of heparin in the blood after its administration are the Lee and White clotting time (Cl.T), the partial thromboplastin time (PTT), the activated partial thromboplastin time (APTT), the thrombin time, the thrombin titration, the protamine/toluidine blue titer, and the factor Xa inhibition test. These tests are used in the first place to measure and follow the degree of hypocoagulability produced. Since there is a wide range in the degree of hypocoagulation that occurs in different patients subjected to the same concentration of heparin, since the relationship between coagulation time and concentration of heparin in the blood is not linear, and since it is believed that the therapeutic effectiveness and the danger of hemorrhage are equally related to the degree of hypocoagulability, it is evident that hypocoagulability is the significant parameter, rather than the actual concentration of heparin. With intermittent intravenous injections of heparin, the blood is incoagulable for most of the time, and the objective is to determine when coagulation is returning to a normal range and more heparin must be administered. With other methods of administration—subcutaneous, hourly intravenous (via a catheter), and continuous intravenous (catheter or mini-pump) injections—an approximately uniform degree of hypocoagulation is maintained. This effect is usually monitored by a coagulation test. Mini-dose heparin is a *subcutaneous* dose of heparin (about 5000 units) that produces a degree of hypocoagulation hardly detectable by most routine coagulation tests.

The clotting time shows a hyperbolic relationship to concentration of added heparin in the blood. This means that a minimal amount of added heparin is required for a significant change in the clotting time, whereas at a concentration about 3 times this minimum effective concentration, the clotting time becomes infinite (blood incoagulable). Within the range of measurable clotting times, there is a linear relation between log clotting time and heparin concentration. Moreover, these two limiting values for heparin concentration can be adjusted by technique. (This, of course, means that the technique must be carefully standardized, as described in coagulation manuals.) Methods for determination of the clotting time of whole blood (in silicone, glass, etc.) and of plasma (plasma clotting time, PTT, APTT, etc.) provide procedures for estimates of hypocoagulability for concentrations of heparin in blood and plasma of the order of 0.01, 0.1, and 1.0 unit/ml. With the thrombin time, adjustment of thrombin concentration likewise allows estimates for different orders of heparin concentration. Procedures that depend on neutralization of anticoagulant activity of heparin (protamine titer, toluidine blue titer, thrombin titer) in general are limited to heparin concentrations that render the blood incoagulable, although it is evident from the foregoing that the minimum heparin concentration required will be a function of the technical details of the coagulation procedure. Since coagulation test values change with heparin concentration, is it not possible to determine added heparin concentration in the blood from these values? This means constructing a reference curve from values determined by adding heparin to blood *in vitro*. Jaques and Riker (88) showed that for whole blood clotting times (in glass) to agree for the same concentration of heparin when added to blood inside and outside the body, it was necessary to mix the heparin with blood removed by the silicone technique. Therefore special techniques permit the construction of a suitable reference curve, although this is seldom done. In the case of the titration procedures, it is necessary to rely on the titration with heparin mixed with the blood outside the body to determine equivalence, although here also, the mixing should approximate that in the body as closely as possible, since changes in platelets and other components during withdrawal can make a difference in the equivalence value. Such effects are attributable to the presence of platelet factor 4, which when released from platelets binds heparin strongly and blocks its anticoagulant activity.

The log relation between clotting time and concentration of heparin in the blood means that pharmacodynamic parameters (e.g., half-life) calculated from coagulation test values will not agree with those calculated from concentrations of the drug in the blood. Hence they must be indicated as coagulation half-lives or drug response half-lives. The half-life for the drug itself is longer. There is probably a range of values for the half-lives of the different molecular weight species described in Section I.1, and different values for different drug responses (e.g., effect on platelets, endothelium, etc.).

The wide range in degree of hypocoagulation produced by a given dose of heparin (or given concentration of heparin in the blood) is not due to technical error when the tests are performed by skilled technical personnel in constant practice. Rather, this wide range reflects the nature of the anticoagulant action of heparin. As discussed in Section IV.2.B., heparin is an inhibitor of a number of steps in the coagulation mechanism. There are differences in rates for these steps between bloods of different individuals, and this is made more evident when heparin is added to the blood.

It is common but not necessarily valid to assume that an observed anticoagulant effect is due to heparin. Other SMPs can show marked anticoagulant activity, however, depending on the test system. Marbet and Winterstein (113) ascribed heparin properties to chondroitin B (β-heparin) because this substance markedly inhibited the coagulation system they used:

beef plasma + thrombin. Moritani and Ohta (123) showed that heparitins and chondroitin B have a synergistic action on antithrombin activity of heparin cofactor, antithrombin III. The increased coagulation times observed when heparin is administered orally with chelating agents, are due to the chondroitin A/C liberated into the blood by the chelating agent (166).

The titration procedures consist of determining the quantity of reagent that must be added to give a measurable clotting time (usually but not necessarily that for the subject without heparin). For such purposes, the original form of the thrombin time devised by Quick (139) can be more useful than the thrombin time. Protamine is itself an anticoagulant. Hence measurable clotting times require approximately equivalent concentrations of protamine and heparin. The equivalence is the least amount of protamine to give the normal clotting time. The actual value depends on the source and purity of the protamine, on the heparin preparation, and on the conditions of the coagulation test; thus protamine must be determined with a reference heparin under identical conditions. The principles and pitfalls in the use of protamine with heparin are more extensively reviewed in ref. 74. Toluidine blue can be used similarly to protamine. The procedure of Meneghini and Cervini (116) has been used by European workers. In this approach a constant quantity of commercial heparin is added to the blood sample, ensuring that the results of this test represent the combined effects of the reaction of the toluidine blue with the added heparin and SMPs and platelets in the blood sample, and referred to as measuring changes in heparin tolerance (23).

Other responses to heparin administration that are measured on plasma are the enzymes lipoprotein lipase and diamineoxidase (histaminase). The assay of lipoprotein lipase was reviewed by Korn (99). Diamine oxidase can be measured by the Conversion of ^{14}C-putrescine to ^{14}C - Δ^1-pyrroline, which is released in human and rat plasma but not in dog plasma. A $10^{-2}M$ solution of chloral hydrate should be added to inhibit destruction of the product. Factor Xa inhibition and some other procedures attempt to measure the anticoagulant effectiveness of heparin present in the plasma by its inhibitory action in a prepared coagulation system. In this respect, this and similar methods are intermediate between the coagulation tests described previously and procedures for the chemical estimation of heparin based on chemical extraction of the heparin and its measurement by its specific polyelectrolyte properties. It does not necessarily measure all aspects of the degree of hypocoagulability produced. On the other hand, heparin can be measured by chemical extraction followed by chemical measurement, independent of anticoagulant activity, and from Section I.1. it is evident that this route has interesting possibilities. The procedures listed for chemical extraction of heparin from blood and plasma are described elsewhere (79).

Of the various tests for heparin in urine, the simplest, and a very useful procedure, is to mix a drop of urine with a very dilute dye solution (1 mg % toluidine blue in distilled water) on a white plate, preparatory to conducting a titration as a described in Section IV.1.B. With the urinary electrolytes present, the red color of the dye is fairly specific for heparin. Whereas dog urine "enhances" the metachromatic action of heparin (87), however, rat urine suppresses it. Human urine appears to have no effect. Marbet and Winterstein devised a nephelometric procedure for urine (79). Urinary SMP can be fractionated by the addition of 9 volumes of acetic acid (90%). Chondroitins are insoluble in 20% and soluble in 90% acetic acid. Heparin and uroheparin are precipitated only by 90% acetic acid.

Three procedures are described. The Lee and White clotting time has provided most of the information on the anticoagulant effects of heparin and is still the procedure that yields, in competent hands, the most direct assessment of the effect on the total intrinsic coagulation process. The most popular test in clinical laboratories at the present time (1975) for monitoring heparin in blood appears to be the APTT test. The procedure of factor Xa inhibition is under extensive testing, and it or a similar test will become essential for many investigators.

a. Whole Blood Clotting Time

PRINCIPLE

A blood sample is taken carefully by venipuncture, transferred to a glass test tube, and examined at intervals until the blood has gelled. The time required to reach this stage is the whole blood clotting time. The range of concentrations of heparin added to blood that give measurable increased clotting time values, are those up to 3 times the concentration required to increase the clotting time value obtained without addition of heparin. The absolute heparin concentration for this range is determined by the technique used in taking the blood sample and by the conditions of the measurement. By selecting these with careful standardization of the variables the "measurable" concentration range can be preselected over a hundredfold range. The test measures the degree of hypocoagulability produced—that is, the degree of inhibition of intrinsic coagulation mechanisms.

Standardization of Conditions. The first essential is the preselection of conditions suitable for the purpose in hand. The requirements for a good venipuncture are a good blood flow, occlusion of the vein, preparation of the skin (alcohol, etc.), direct entry of the needle into the vessel, and removal of the blood as rapidly as the blood flow permits, to prevent the occurrence of any clotting in needle and syringe. Time must be spent in removing vasoconstriction and establishing these conditions, since they are the factors that determine whether the value obtained can be accepted or must be rejected.

The requirements for a good blood sample are as follows: (1) no eddies, negative pressure or mixing may occur during withdrawal of the sample, and (2) there must be sufficient volume to ensure that the initial blood entering the syringe and the last portion are not used for the clotting time determination. Jaques et al. (82) showed that because of the presence of thromboplastin picked up by the needle passing through the vein wall, the initial portion gives a clotting time value one-half that given by the midportion. To meet these requirements, a 5- or 10-ml blood sample is taken from the antecubital vein in man, 5 ml from the saphenous or brachial vein in the dog, 2 ml from the ear vein in rabbits, and 2 ml from the exposed heart in rats and other small animals. When it becomes necessary to use other vessels and samples, there will be admixture of tissue thromboplastin, and therefore extrinsic coagulation processes will participate.

Conditions significant in the measurement are the surface of the syringe and measuring tube, the volume of the tube, the temperature, and movement of the tube. A glass syringe is used, and such equipment must be kept apart from all siliconized glassware. Residual blood is to be washed out of syringes as soon as the test blood has been removed. Then the syringes are boiled thoroughly in distilled water, not only for sterilization but to destroy thrombin absorbed on the surface, which survives ordinary washing and can cause a 50% reduction in clotting time. Glass test tubes as purchased in 1975 are coated with silicone or other nonwettable material that must be removed (with acetone-NaOH) to avoid unnecessarily long coagulation times (as described in Section III.5). On the other hand, a completely silicone-coated system (syringe, test-tubes, etc.) may be used to detect very low degrees of hypocoagulation due to very small quantities of heparin (82). The coagulation time is increased with large tube diameter. Tubes for determination of clotting time are kept in a constant temperature water bath. Since the coagulation time of blood without heparin is shorter at 37 °C than at room temperature, whereas that with heparin is longer at 37 °C, it is usually preferable to use 37 °C.

Movement of the measurement tube is carefully standardized. As soon as the blood is placed in the tube, it is covered with a paraffin film (Parafilm) and inverted three times in a way that serves to cover the entire glass surface, at which time the stopwatch is started (preferably an electric watch with foot pedal). This treatment initiates or accelerates the reaction through glass contact activation. If the three-tube technique is used (1 ml of blood from the midportion in each of three tubes), the first tube is examined every 10 sec by gently tilting it without removal from the water bath. When the blood no longer runs, the examination is continued with the second tube and, when this clots, with the third tube. When the tube can be completely inverted without breaking up the blood, the stopwatch is stopped and the clotting time read and recorded. If only one tube is used, it is examined at intervals of 30 sec (or about one-twentieth of the expected clotting time) for the same end point. The clotting time end point can be registered by mechanical and electrical means using changes in viscosity, optical density, and other variables as parameters; a number of instruments based on such principles are on the market. The long coagulation times obtained with heparin mean that in most investigations, a number of samples must be followed simultaneously. An automated procedure is urgently needed, but since none of the coagulometers we have seen handle more than two or three multiple samples simultaneously, the human observer is much better from the standpoints of economy and reliability.

As indicated previously, much of the testing can be done with heparinized blood, prepared by adding the heparin to blood collected and handled by the silicone technique (82).

Procedure. Three 8 × 60 mm test tubes (carefully preselected and cleaned) are placed in a 37 °C water bath. After 10 ml of blood is taken by *clean* venipuncture, the needle is removed, 1 ml of blood is discarded, and 1 ml of blood is delivered gently into each tube. The tubes are capped with Parafilm and inverted three times, and a stopwatch is started. (The blood discarded and left in the syringe can be used for other coagulation tests, e.g., APTT and blood chemistry.) The first tube is examined at 10-sec intervals in the water bath by gentle tipping. When the blood no longer flows, the second tube is examined in the same way, followed by the third tube. When the third tube can be inverted with the blood remaining in a solid gel, the stopwatch is stopped to give the whole blood clotting time. This procedure is modified for specific investigations on the basis of the principles described above for standardization of conditions.

b. Activated Partial Thromboplastin Time

PRINCIPLE

Commercial partial thromboplastin reagent and plasma are incubated. After the required activation time, calcium chloride solution is added and the clotting time of the mixture determined.

REAGENTS

Kits for the determination are supplied by a number of manufacturers.

The activator is kaolin (Hyland), Celite (Warner-Chilcott), elladic acid (Metrix, Ortho, Dade).

Calcium chloride of specified molarity is provided by the manufacturer.

Initially there appeared to be some difficulty with reagents, but Alexander (1) compared the preparations of seven manufacturers and found them satisfactory. To ensure the best performance of each reagent, each manufacturer's instructions (i.e., activation time, molarity of calcium chloride, storage of reagents) must be followed exactly, since the optimal conditions are not the same.

Procedure

1. Reconstitute the partial thromboplastin reagent according to the manufacturer's directions.

2. Place calcium chloride of recommended molarity in a 37 °C water bath.

3. Place tubes containing 0.1 ml of APTT reagent in the water bath for the suggested time to reach 37 °C (usually 2 min, except for BBL's Fibrolet, which has already been incubated at 37 °C for 10 min and remains at 37 °C throughout)—up to 4 hr of testing.

4. Add 0.1 ml of the patient's plasma, which has previously been placed on ice. At the same time as the plasma is being added, a stopwatch is started. Mix briskly and replace the tube in the water bath.

5. After the required activation time, 0.1 ml of the $CaCl_2$ is blown into the mixture and a second stopwatch is simultaneously started. Mix well.

6. After approximately 25 sec of further incubation, the tube is removed from the water bath and is tilted gently, about once per second, under a good overhead light source until the clot end point appears—the first sign of fibrin strands. Stop timer and record time.

7. All tests are performed in duplicate. The average of the two times is taken to one decimal place, usually to the nearest 0.5 sec. If the two times do not agree within 10%, a third APTT must be performed and the average of the two times showing the best agreement taken. The procedure may be automated using a Fibrometer (Fibrometer Precision coagulation timer, BBL, Baltimore). Heating times must be adjusted. Add 1 min to the incubation and activation times.

COMMENTS

The APTT normal values change slightly depending on the reagent selected. Following are the normal values in seconds suggested by the manufacturers: Warner Chilcott, 25 to 45; Metrix, 30 to 40; BBL, 25 to 40; Ortho, 30 to 45; Hyland, 30 to 45; Dade, manual—less than 40 sec; Fibrometer—less than 35 sec. Use scrupulously clean or disposable tubes and pipettes that contain no trace of protein. For best reproducibility, the time should be taken at the same stage of clot formation in each instance—usually the first sign of fibrin strands.

c. Factor Xa Inhibition Test

PRINCIPLE

In the test devised by Yin et al. (183), activated factor X (Xa) is rapidly neutralized by its plasma inhibitor in the presence of trace amounts of heparin. At a specific time in the initial phase of the reaction between the plasma inhibitor and Xa in the presence of heparin, the amount of Xa neutralized is proportional to the concentration of heparin present in the test.

The sample in which the heparin is to be estimated is mixed with factor Xa, buffer, and an excess of plasma inhibitor (Standard Human Plasma) and incubated for 2 min. Then $CaCl_2$ and Cephalin, an anticoagulant-free bovine plasma, are added, and the clotting time of this mixture is recorded.

REAGENTS

The Sigma Heparin Determination Kit (Sigma Chemical Corp., Division of Sigma International Limited, St. Louis, Mo.) contains the following:

Trizmal Buffer, Sterile 0.02*M* Trizma Maleate, pH 7.5, at 25 °C. This material is stable at room temperature while sterile. Store at 0 to 5 °C after opening. Discard if turbidity develops.

Calcium chloride solution, sterile 0.025*M* calcium chloride. Stable at room temperature while sterile.

Plasma–CEF reagent–Cephalin in anticoagulant-free bovine plasma (AFBP). Store at 0 to 5 °C. Stable at least one year. Reconstitute by injecting 2 or 4 ml of 37 °C distilled water as indicated, through a rubber stopper. Gently shake to suspend the contents and keep the suspension at room temperature (20–25 °C). Suspension is stable for 8 hr at 20 to 25 °C or for 3 hr at 37 °C. Excess suspension may be frozen and thawed once for reuse.

Activated factor X (Xa), activated factor X stabilized in bovine serum albumin. Store at 0 to 5 °C. Stable at least one year. Reconstitute by injecting 1 or 2 ml of cold distilled water through a rubber stopper. Shake gently to suspend contents. Keep on ice for use in assay. (The Xa solution is best stored frozen at −60 °C.) Before use it should be thawed rapidly at 37 °C, then stored at 1 to 4 °C for at least 15 min but not longer than 5 hr.

Additional requirements are Standard Human Plasma (SHP) and heparin. For Standard Human Plasma, blood is collected at 3.8% (w/v) sodium citrate, such that the final ratio of blood to anticoagulant is 9–1. Whole blood is then centrifuged at 2500_g for 20 min at 4 °C. The plasma is separated, aliquoted, labeled as SHP, and stored at −20 °C. Heparin will be the laboratory standard, heparin under investigation, and so on.

EQUIPMENT

37 ± 0.5 °C water bath.
Two stopwatches.
10 × 75 mm Pyrex test tubes, Corning no. 9820.
0.2- and 0.5-ml Mohr long-tip pipettes calibrated to "deliver."
0.1 ml of Micro-Folin "to contain" pipettes.

Procedure. Blood is drawn from a patient by clean venipuncture into a 5.0-ml capacity B-D vacutainer containing 0.5 ml of 3.8% (w/v) sodium citrate such that the ratio of blood to anticoagulant is 9–1. The blood is mixed and centrifuged at 2500_g for 10 min at room temperature. The plasma is immediately removed into a clean, labeled tube and placed at 4 °C on ice until the assay is performed. All determinations are done within 4 hr of the taking of the specimen.

The assay consists of a primary reaction mixture (PRM) containing factor Xa, test plasma, buffer, and an excess of plasma inhibitor provided by SHP. These components are allowed to react at 37 °C for exactly 2 min, at which time the remaining Xa in the mixture is measured.

1. Select desired sensitivity range in accordance with method A or method B (Table VI).
2. To a 10 × 75 mm 12 × 75 mm glass tube, add the components of PRM in the order and amounts indicated in Table VI.
3. Place a second 10 × 75 or 12 × 75 mm test tube (clotting mixture tube) in the water bath to warm.
4. Exactly 90 sec after addition of factor Xa, accurately transfer 0.1 ml of the PRM to the bottom of the clotting mixture already in the water bath.
5. Exactly 110 sec after addition of factor Xa in step 2 (i.e., 20 sec after step 4), add 0.1 ml of $CaCl_2$ solution to the clotting mixture tube.
6. Exactly at 120 sec, forcefully blow into the clotting mixture tube 0.2 ml

TABLE VI
Quantities of Components for Factor XA Inhibition Procedures

	Method A[a]			Method B[b]	
PRM components	Standard range (0.25–3.0)	Deci-range (0.05–0.5)	Centi-range (0.01–0.2)	Standard range (0.5–10.0)	Deci-range (0.1–1.0)
Test Plasma see notes	0.1 ml 1/5 in SHP	0.1 ml undiluted	0.5 ml undiluted	0.1 ml 1/5 in SHP	0.1 ml undiluted
Standard human plasma (shp)	0.5 ml	0.5 ml	0.1 ml	0.1 ml	0.1 ml
Trizmal buffer	0.3 ml	0.3 ml	0.3 ml	0.7 ml	0.7 ml
Mix well and prewarm for one minute in 37° C water bath, then rapidly add factor Xa as follows and start stopwatch simultaneously. See Notes.					
Factor Xa	0.1 ml	0.1 ml	0.1 ml	0.1 ml	0.1 ml
Mix well and leave reaction mixture at 37° C. Leave pipette in tube. See notes.					

[a] Ranges in units per milliliter of plasma.

of plasma—CEF reagent and start a second stopwatch simultaneously; mix by agitation, and record the time when a solid clot is formed.

7. Using the clotting time obtained, read off the heparin concentration of the test plasma from the appropriate heparin calibration curve.

Plasma samples containing known concentrations of heparin employed in establishing heparin activity curves are prepared by mixing 0.1 ml of heparin solution of known concentration with 0.9 ml of normal plasma for the species under study; 0.1 ml of saline replaces the heparin in the zero heparin control samples. The mixtures are allowed to equilibrate at room temperature for 15 to 30 min before testing. To construct the calibration curves, refer to Table VI. Replace the test sample in the PRM with the plasma solutions containing known heparin concentrations. Prepare plasma samples by making serial dilutions in SHP starting with the maximum concentration of each range.

NOTES

Alexander (1) reports that when the following notes suggested by Yin et al. (183) were not strictly adhered to, poor reproducibility in results was obtained.

a. To achieve reproducible results, it is necessary to exercise extreme caution when adding the Xa to the reaction mixture in the tube. An accurately calibrated 0.1-ml "to contain" pipette should be used. Draw the Xa solution just above the 0.1-ml mark, carefully wipe the entire tip of the pipette that has been in contact with the Xa solution, and slowly drain out the excess Xa in the pipette. Lower the tip of the pipette into the reaction tube to just above the meniscus of the mixture. Taking care not to drip any Xa down the inner side of the test tube, rapidly blow out the Xa and rinse the pipette by sucking the reaction mixture up and down 10 times. This action will ensure proper mixing of the incubation mixture, but foaming must be avoided. Leave the same pipette in the reaction tube.

b. Just before transferring 0.1 ml of this mixture to the second tube, draw the incubation mixture up and down in the pipette twice.

c. It is suggested that the SHP be prepared in one's own laboratory because it has been demonstrated that some of the currently available commercial preparations of "normal" human plasma employed in other coagulation assays are adulterated with an unknown substance and are unsuitable for use in this heparin assay.

d. Always use a solid clot as the end point in this assay.

e. When testing a sample other than plasma for heparin activity, the substance must always be diluted first with SHP as in the preparation of a plasma sample of known heparin concentration. This keeps the plasma-activated factor X inhibitor level in excess.

f. The centiunit procedure is affected by a high concentration of fibrin split products in plasma and gives falsely elevated results, but the Standard and deciunit procedures are not affected.

COMMENT

A major problem at present is the price of the reagents, particularly to provide activated factor X for only one to three tests. The factor Xa solution must be stored frozen at −60 °C, and facilities for this are not available in many laboratories. Teien et al. (168) have replaced the final clotting time determination by a colorimetric procedure based on the amidolysis of Bz-Ile-Glu-Gly-Arg-*p*NA. Chen et al. (185) have modified the procedure to a one-stage procedure.

B. PHARMACOLOGICAL PROCEDURES

Tests comparing the activity of two or more heparin preparations in a test system are designated "pharmacological procedures." Usually the one heparin is the International Standard Heparin or a reference preparation (a laboratory standard that has been compared previously with the International Heparin Standard). Since these tests were developed by people involved in bioassays of insulin and other drugs, use was made of the statistical procedures developed in the field of biopharmaceutics. The basic requirement for these procedures is that "like" is being compared with "like." Jaques and Charles (81) compared four heparin preparations from different sources in tests with different coagulation systems. All the preparations were isolated as crystalline barium salts, and all appeared to have identical chemical composition (optical rotation, etc.). The four preparations did not have identical anticoagulant activity, however, but showed a series of values having the same order, but the relative anticoagulant potency was greatly different with different test systems. In fresh whole blood, the most active heparin (from dog liver) was 10 times as active as the least active, whereas in preserved plasma + thromboplastin (USP assay), there was only a threefold difference. Similar differences were observed by Marbet and Winterstein (113) for SMPs and by Nagasawa (128) comparing whale heparin with beef heparin. Walton et al. (176) reported that when preparations of heparin from the lungs and mucosa of ox, pig, and sheep were tested for potencies by the BP and USP assay procedures, a discrepancy of up to 40% was observed between estimates obtained by the two procedures on comparison of samples of heparin derived from lung and intestinal mucosa.

On the basis of the findings of Nader et al. (126) described in Section I.1, it is now possible to understand the basis for these apparent discrepancies, which for many years were either ignored or said not to exist. It is known that heparin preparations inhibit a number of steps in the coagulation sequence. Some known heparin preparations (e.g., dog and whale: refs. 89, 128) are much more active than commercial heparin when tested for inhibition of clotting of fresh whole blood. Commercial heparins are a mixture of anticoagulant and nonanticoagulant heparins, with differing relative proportions of the 21 different chain length heparins. Inhibitory specificity is presumably exact for one coagulation step, but it appears most unlikely that the same specific heparin chain length provides the most effective inhibitor for each step in the coagulation sequence. Rather, the most effective heparin chain length changes with the step in the coagulation sequence. This explains why different heparin preparations exhibit relative activities for inhibition of clotting of whole blood, of preserved plasma, and of plasma with thrombin, that differ greatly, reflecting, respectively, inhibition of the initial steps of the intrinsic coagulation system, of succeeding steps, and of the final common path for coagulation. Jaques et al. (83) summarized the marked variability found by our Saskatoon laboratory for all analytical values in commercial heparins obtained over the past 20 years. With variations in composition, it is not surprising to find considerable differences in anticoagulant activity on testing in different coagulation systems, or in the same coagulation system with variation in the reactants due to preparation and changes during storage and even during use.

The most important study of these problems has been the *Collaborative Study of Heparin from Different Sources* (3). Under the auspices of the World Health Organization Expert Committee on Biological Standardization, the 13 collaborating laboratories were supplied with coded heparin preparations made from the intestinal mucosa of hog, sheep and ox, from beef lung, and the USP reference heparin (prepared from intestinal mucosa and diluted with excipient). The beef lung heparin preparation was the International Standard Heparin Preparation No. 2. The collaborating laboratories compared the activity of the heparin preparations in pairs, using the assay procedures employed in their laboratories. Eleven procedures were used—mostly the methods of the U.S. Pharmacopoeia (1965) or the British Pharmacopoeia (1963) and their modifications. Nine of the procedures were coagulation tests and two of these were *in vivo* tests on the whole animal. The main result of the study (unexpected at the time) was that there was no difference between the methods in terms of reproducibility: all encountered the problem just described—namely, *any two heparin preparations were not alike for the test system.*

The collaborative study was planned to investigate the problems of comparing different preparations of heparin by different bioassay methods. Examination of the results, however, immediately drew attention to a statistical problem, the selection of a suitable procedure to test the results for statistical validity. It became apparent in the study that in a large number of the assays analyzed, there was little or no variation in the responses (usually clotting times) to a given dose level for any one preparation. No statistical inference should be attached to an analysis of variance carried out on data for which the responses are not normally distributed within each set. In view of this, and since the slopes of the log dose-response lines for some preparations were not consistently greater (or less) than those for other preparations, the distribution was examined of each of the sets of log potencies from assays in which the same assay method used for each laboratory—first including, then excluding, the estimates that came from assays exhibiting significant deviations ($p < 0.01$) from parallelism. This showed that the inclusions of the estimates from "nonparallel" assays made no significant difference to the mean and variance of most of the distributions. Results were therefore calculated as unweighted geometric mean potencies, and confidence limits were based on the direct estimate of variance of the log potencies.

The assay of the porcine mucosal heparin versus itself provided a measure of the accuracy of the methods used. The potency ratios for different methods were: BP, 1.016; USP, 1.006; clot protein, 0.982; thrombin titration, 0.991; Howell, 1.086; colorimetric-DK2, 1.007; colorimetric-Lovibond, 0.994; *in vivo* (clotting time), 1.011. The difference between confidence limits ($p = 0.95$) for the *in vitro* coagulation tests were from 0.075 to 0.109; thus it is evident that these tests are consistent within themselves. In fact, "the estimates are so precise that an analysis of variance is meaningless."

However when different heparin preparations were compared—porcine mucosal heparin versus sheep mucosal heparin, International Standard Heparin No. 2 (beef lung heparin), beef mucosal heparin, USP reference heparin, beef mucosal heparin, and USP reference heparin versus International Standard—estimates for the same method by different laboratories, or for different substrate batches, varied by as much as 40 to 50%. With this great variation for a given method, it was not possible to demonstrate marked differences between methods or between heparins of different origin (except for sheep mucosa heparin and the USP reference heparin). There was a tendency for the mucosal heparins assayed against International Standard Heparin No. 2 (beef lung heparin) to be consistently higher by the BP method than by the USP method. From the study, the following major recommendations were made.

1. The pig mucosal preparation included in this study should in due course replace the current International Standard when stocks are exhausted (because it is more representative of modern preparations than the current standard), with a potency assigned of 1370 IU/ampoule.

2. Estimates of the biological activity of heparin preparations should be based on several independent assays using different batches of substrate.

3. The biological assay methods currently used were devised some 15 to 20 years ago and are regarded in the light of modern knowledge of mechanisms of blood clotting as unsatisfactory in a number of respects. It is hoped that

research into the mechanism of action of heparin will lead to improved methods of assay.

The first recommendation was put into effect in 1974. The pig mucosal preparation described by Bangham and Woodward (3) is now International Standard Heparin No. 3. The second recommendation is presumably being acted on. Investigators producing new heparin preparations probably should go beyond this, however. In addition to doing independent assays on different batches of substrate, it would be well to compare the heparin(s) being used with a reference heparin by two or more different procedures. With the demonstration of the polymorphicity of heparin, it appears that identification of the fractions of clinical value would allow physicochemical standardization. In the mean time, current unsatisfactory methods can be improved. The first improvement is stated in recommendation 2. This is all that can be done with the USP and BP methods.

Methods that contain biological variability in the procedure are amenable to more improvement. Bangham and Woodward pointed out that this possibility is present in the Howell assay, since different substrate (freshly drawn blood) is taken each day. The report recommends that in any assay method for heparin, the test be performed on a number of substrate samples and results be averaged to give a "representative value." In the case of the Howell assay, all the assays have followed this last criterion in that repeated assays involve using blood from different cats or different dogs (species specified by the individual laboratory). In the assay as used, however, preliminary assays are run to determine the approximate dilution required to give a match with the 2.5 units/ml of International Standard Preparation serving as reference. Further tests on more animals are performed until a narrow dilution range of test heparin matches the action of the heparin standard in three assays. The results of any tests in which the match is not in this fairly narrow range are not used. On the basis of the conclusions of Bangham and Woodward, in discarding these assay values the original "approximate value" is being selected as being very close to the true value, and the assay procedure is continued until enough assays have been performed to give replicates of this value. Yet it is evident from the results reported by Bangham and Woodward that the initial "approximate value" selected can be as much as 50% too high or 50% too low compared to the representative value (or true value). It is therefore necessary to redesign the Howell assay. Having obtained a roughly approximate value, a sufficient number of tubes should be set up to give a wide range of dilutions of test heparin. Tubes containing standard heparin are present for comparison. This step should then be repeated for a sufficient number of animals to give a series of assay values with bloods of different dogs, which may vary by a factor of

50%. By taking an average of these values a true "representative value" will be obtained. Probably this can be accomplished with about the same number of assay tubes, and the same number of animals, as required in the past for Howell assays, but then all the experimental data are used and much truer values obtained. For this reason, Dr. L. W. Kavanagh has made changes in the Howell assay procedure to cover a wider range of concentrations for matching, as described next.

In vivo assays have a greater possibility of including in the design both biological variability and its measurement. Two laboratories measured the response of dogs to heparin *in vivo* in the collaborative study. Thus these assay procedures had the advantage of being compared in the development stage with the *in vitro* assay procedures in a major study and of being subjected to critical analysis. As indicated by Bangham and Woodward, the *in vivo* assay procedure of Rezansoff and Jaques was reexamined and data submitted for further criticism. Bangham and Woodward found that although the *in vivo* assay of heparin in the dog was cumbersome, it could be designed to ensure that analysis for parallel line assays could be carried out and statistically valid assays made. The procedure was therefore redesigned by Kuo et al. (102) and subjected to further testing. The *in vivo* procedure of Kuo et al. is a statistically valid assay procedure. It uses three coagulation parameters and one noncoagulation parameter (lipoprotein lipase release). It can be extended to other parameters.

a. A Modification of the Howell Assay for Heparin (L. W. Kavanagh)

PRINCIPLE

In the test the degree of clotting of three concentrations of heparin in fresh dog blood is matched for test and standard heparin solutions. The potency is assigned by determining the dilution of heparin that most closely corresponds to the reference heparin.

In test tubes calibrated for 1.0 ml, the analyst places 0.1, 0.2, and 0.3 ml of the Standard Heparin Solution (2.5 units/ml, International Standard Heparin) and the same amounts of 24 dilutions of the heparin under test. These 24 dilutions cover the complete range of anticoagulant activity for weaker to stronger than that of the test heparin. Fresh arterial dog blood is added to the 1-ml mark. After 6 hr the degree of clotting is recorded and the dilution matching that of the reference heparin is determined. This procedure is carried out on five dogs, and the five critical dilutions are averaged to give a mean activity. Activity is then expressed in units per milliliter.

Procedure. Preselected Pyrex culture tubes, 8 mm ID, with a mark cut at the 1-ml level, are prepared for use (see Section III.5). Three racks of 3×10 tubes each are set up for each assay. The standard is International Reference Heparin, 2.5 units/ml in 0.9% sterile saline intravenous solution with 0.3% tricresol added. A dog is anesthetized with sodium barbital, and one femoral artery is exposed and cleared carefully without trauma. The distal portion is ligatured. A bulldog clamp with rubber-sheathed jaws is placed on the proximal portion. A cut is made in the artery with iridectomy scissors. All blood in the artery is rinsed out with physiological saline as soon as the artery is opened. A clean polyethylene cannula or a siliconized glass cannula is inserted and tied in place. Blood is then added to each test tube to the 1-ml mark in the following order: row 1, row 2, row 3 of each rack. If clotting occurs in the cannula, the clot is removed between racks. Tubes are inverted to mix the contents after filling. Tubes are placed in a room temperature water bath, examined after 2 hr, and read after 6 hr. The degree of clotting is determined by tipping each tube and observing the clot formation. A code is used to describe the extent of clotting, ranging from complete to no clotting. After use, the blood is removed from the tubes and washed as described in Section III.5.

PREPARATION OF DILUTIONS

A preliminary assay is done to determine the approximate range of activity for the test heparin solution. Dilutions of 1–10, 1–25, 1–75, 1–100, 1–150, and 1–200 are tested. Then 0.1, 0.2, and 0.3 ml of each dilution or standard is pipetted into each group of three tubes and the volume is brought up to 0.3 ml with saline. A stock heparin solution is then prepared, about 62 units/ml by the preliminary assay, from which to make final dilutions. The midpoint of dilutions prepared for the assay is a 1–25 dilution, and the matching dilution of test heparin that matches the standard (2.5 units/ml) should lie in this range. If the activity of the test heparin is higher, a further dilution must be made. Dilutions are made of 2.000, 1.875, 1.750, 1.625, 1.500, 1.375, 1.250, and 1.125 ml with physiological saline to 20 ml, to provide sufficient of each dilution for five dogs. Dilutions were selected that could be prepared using automatic pipettes. The final dilutions vary by about 10%. Lamda pipettes (with no rinsing in between) are used to add the small volumes to the assay tubes. For the first rack, 0.1, 0.2, and 0.3 ml of each dilution is pipetted into three tubes and saline added to bring the volume to 0.3 ml. For the second rack, 0.05-, 0.1-, and 0.15-ml volumes of dilutions are used, and the third rack employs 0.025-, 0.05-, and 0.075-ml volumes. Two sets of standard heparin solution are included in each rack, using 0.1, 0.2, and 0.3 ml each time. Blood is added to the tubes in each rack in the following order: row 1, row 2, row 3.

READING OF TUBES

The degree of clotting in each tube is recorded in the following form: +++, ++, +, ±, −, −−, −−−. The dilution in which the degree of clotting in all three tubes most clearly matches that in the reference tubes is recorded for each trial. In reading the tubes one must note whether there is a gradation in response in the three different doses of reference. Since the clotting pattern for reference heparin varies, the pattern of the standard tubes closest in the rack to the dilution of the unknown selected should be used. The results of five assays on separate dogs are averaged to give a representative anticoagulant activity.

COMMENTS

In the previous modification, the tubes were left 2 hr before reading. Although the longer time generally improves the end point, this is not always the case: with some animals, the degree of clotting fails to show a gradation in the standard tubes, and it is necessary to read the tubes at 2 hr. There is a real variation in degree of clotting of the standard tubes for any one blood. Checking the standards in one rack is no indication of the degree of clotting in the other racks. The range of potencies determined by this method were not scattered as found previously and given in the report by Bangham and Woodward (3). Using sterile technique for the cannulation, and withdrawal of the blood, followed by administration of an antibiotic, it is possible to recover the animal. Using the femoral and carotid arteries, and so on, this allows drawing blood 4 to 6 times, thus standardizing reading time. Preliminary tests suggest that the assay might be adapted to human blood, using blood freshly collected from a donor into a blood pack.

b. An Improved *in Vivo* Assay for Heparin

PRINCIPLE

A test designed by Kuo et al. (102) provides statistically valid estimates of relative potency *in vivo*. Heparin is injected intravenously into conscious, trained dogs, and eight blood samples are taken for three coagulation tests and lipoprotein lipase estimation. The response is calculated as the area under the response curve, using log values for coagulation parameters. The test is repeated at 2-week intervals in five dogs. Test heparins and standard heparin are given at three dose levels on a random basis to all five animals to provide a multiple crossover assay. The standard method for parallel line assays is used, regression lines being calculated for $\log_2$ dose–$\log_{10}$ response. Analyses of variance are conducted for an estimate of potency. Fiducial limits are ±10% for anticoagulant parameters, 87.5% for antilipemic parameters.

Procedure. Heparin preparations are dissolved in 0.9% sterilized saline containing 0.3% cresol. The solutions are divided into suitable aliquots in sterilized vials, which are sealed and stored frozen. Usually four preparations plus reference are included in one test. Three doses, preferably values in geometrical progression in milligrams per kilogram are assigned for each heparin tested. These are administered for each heparin under test, together with the reference standard, to each of five dogs in random order. The order is selected by drawing cards, by computer, or from a table of random numbers.

Male dogs are screened by age (2 to 3 years old), weight (10 to 15 kg), normal Lee and White clotting time (4 to 6 min), and reaction to taking of blood sample. Hematological examinations (platelets, leucocytes, and hematocrit) are made bimonthly. Before each injection, the animal is fasted overnight, and it receives 24 mg of acepromazine maleate 1 hr before. Blood samples (5 ml) are removed before injection, at 5, 10, 15, 30, 45, 60, 90, and 120 min after injection, 4 ml is added to 3.8% sodium citrate, and 1 ml from the middle portion of the sample is used for determining the Lee and White clotting time at 37 °C. The citrated blood is centrifuged at 2100 rpm at 4 °C for 15 min for plasma for PTT, APTT, LPL, or other parameters for measurement (see Section IV.2,3). Values for coagulation parameters are recorded on a log scale. The areas under the response curves of log Cl.T, log PTT, and LPL are calculated according to the formula $5b + 5c + 10d + 15e + 15f + 22\frac{1}{2}g + 30h + 15i - 117\frac{1}{2}a$, where a, b, c, d, e, f, g, h, and i are values at preinjection, 5, 10, 15, 30, 45, 60, 90, and 120 min after injection. A programmed calculator is useful for this calculation. For ease of calculation of relative potencies, values for dose are converted to $\log_2$ and area values to $\log_{10}$. The mean regression lines, $\log_{10}$ response versus $\log_2$ dose, are calculated for each heparin preparation. Bartlett's test is used to confirm the homogeneity of variances. The results for all four heparins of a series are analyzed simultaneously according to a multiple-assay method (46). In a good assay, most of the variation "between doses" should be accounted for by the regression relationship between dose and response, with relative small variation due to differences in response "between preparations," "deviation from regression," and "deviation from parallelism." The p values for the analyses of variance are calculated. Because of their small magnitude in the data, the values for the mean square for "error" tend to magnify the values of p and cause the analysis of variance to detect extremely small deviations from parallelism. For these reasons, it is reasonable to accept $p = 0.05$ as the approximate criterion of significance for nonparallelism for these assays.

Relative potency is estimated by calculation of M and R, where M is the horizontal distance between two log dose–log response lines and R is relative potency. If two linear regression lines are parallel

$$M = \overline{X}_S - \overline{X}_T - \frac{\overline{Y}_S - \overline{Y}_T}{b}$$

where $\overline{X}$ and $\overline{Y}$ represent the mean dose and mean response, respectively, b is the common slope, and subscripts S and T refer to the standard and test preparations. In a parallel line assay, $R = X_S/X_T$. When the dose scale is logarithmic, log R is provided by M, which gives $\log X_S - \log X_T$. Calculation in this way gives a relative potency that is fractional when the potency of the test preparation is less than that of the standard. Since the dose scale is logarithmic to base 2, $M = \log_2 R$. Fiducial limits of R are estimated by calculating M_L and M_U, the lower and upper fiducial limits to M, using Fieller's theorem (45).

M_L or M_U

$$= \frac{M - (\bar{X}_S - \bar{X}_T) \pm \frac{t}{b}\left[(1-g)\left(\frac{1}{N_S} + \frac{1}{N_T}\right) + \frac{(M - X_S + Y_T)^2}{\sum\sum (X - \bar{X})^2}\right]^{1/2}}{1 - g}$$

where $g = (t^2 s^2)/[b^2 \sum\sum (X - \bar{X})^2]$, s^2 = mean S.S. for error in analysis of variance with f degrees of freedom, and t = value of student's t for $p = 0.05$ and f degrees of freedom. When g exceeds 1, the data are not suitable to calculate relative potencies or fiducial limits. Kuo et al. found that all g values were less than 0.02. They included them in the calculation of limits, although this is not necessary with such low values.

As described previously, M_L and M_U are transformed to R_L and R_U, the limits of relative potency. By definition, the potency of International Standard Heparin No. 2 is 130 units/mg; International Standard Heparin No. 3 is 1370 IU/ampoule. The potency in International Units of a reference heparin is determined by direct comparison in this assay. The potencies in International Units of heparins under test are obtained by multiplying their relative potencies by the appropriate factor. Results are reported as means with fiducial limits. Kuo et al. found that for coagulation parameters, potencies were estimated within ± 10% limits. The lipoprotein lipase measurements showed wider limits, averaging about ± 15%. Kuo et al. tested the reproducibility of this assay method by comparing the relative potency and its fiducial limits of the same heparin preparation within a series (within the same group of dogs) and between two series (two groups of dogs). For all parameters tested, the variations of "between preparations," "deviation from parallelism," "deviation from regression," and "between dogs" were nonsignificant; "regression" and "between doses" were highly significant. Within a series, the relative potency values measured by anticoagulant parameters were less than 1.03; in other words, the reproducibility of this assay system in terms of anticoagulant activity is greater than 97% (Cl.T, 98.3%; PTT, 97.3%; APTT, 97.7%). The relative potency measured by LPL was 1.1251, and the reproducibility, 87.5%. Between two series, the mean relative potency for all tests was about 0.94, which is considered satisfactory, but it will be observed that individual values ranged from 0.84 to 1.07. The mean reproducibility is 93.8%—92% for Cl.T, 98% for PTT, 92% for APTT, and 89.7% for LPL.

COMMENTS

For routine application, the total procedure as described is laborious and time-consuming. Kuo et al. examined the effect of reducing the number of doses, number of dogs, and number of blood samples. With four dogs, two doses (deleting the medium dose), and five blood samples (area under response at 0, 10, 30, 60, 90 min after injection), p values of analyses of variance for the four parameters were satisfactory with fiducial limits ± 16% by coagulation tests, and ± 20% by measurement of LPL activity. It does not appear to be advisable to reduce further the number of subjects. To use fewer than five blood samples would not represent a significant technical saving. Heparinoids, heparins of other species (whale, dog, etc.), and modified heparins, will require the extensive study described here and even modification of it. Heparinoids can show a prolonged anticoagulant effect for the same maximum increase in coagulation values; hence different sampling times may be required to cover the response. The coagulation parameters selected provide options. Kuo et al. tested whole blood clotting time, PTT, and APTT. Very similar results were obtained for the potency of the heparin preparations by PTT and APTT. The PTT test could be replaced by the Yin-Wessler procedure. If the plasma tests become the generally accepted tests for clinical control of heparin, the Lee and White clotting time determinations might also be dropped. However there is no great technical saving in this, the immediate information obtained is often useful, and Lee and White clotting times will doubtless continue to be used by many clinicians; thus this parameter may be retained.

In selecting doses of heparin to be used in the assay (as with other drugs) for the statistical tests, the lowest and highest doses used must give measurable values, that is, responses with the lowest doses must be significant and the responses with the highest dose must be measurable (not infinity). Furthermore, for greatest accuracy there should be a close match in the responses for the reference and test preparations. Since the doses required can be determined only approximately in preliminary tests, doses are assigned on the basis of equal weights. Kuo et al. found that this criterion does not always give a satisfactory range of doses, but they consider it to demonstrate the effectiveness of this assay procedure rather than being a serious defect. Full details of the experimental protocols, results, and statistical treatments in this study have been reported by Kuo (101).

Walton et al. (176) suggested that "when the assay system is as complex as the coagulation process, it is desirable that the assay should measure the effect on the overall process and essential that a positive correlation with the *in vivo* activity be established." There has been no precise indication of the relation of *in vitro* potency to the *in vivo* activity because there has not been any method of assessing accurately the *in vivo* potency. The Kuo assay procedure reported here allows for an accurate assessment, since it meets statistical requirements for reproducibility and accuracy described by Bangham and Woodward (3). Comparison with *in vitro* results has been made by Jaques et al. (83).

3. Procedures Based on Individual Components

Components used in the past for measurement of the compounds under study have been ester sulfate, uronic acid, and amino sugar. Problems associated with the use of these substances are discussed in Section III, where references are given. In applying these procedures to routine quantitation of heparin, and so on, separation of the individual SMPs is required as described in Sections V and VII.

V. EXTRACTION OF HEPARIN AND SMPS

1. Extraction from Tissues

Methods of extraction of heparin and SMPs can be grouped on the basis of whether the procedure separates the tissue polysaccharide fraction, to be followed by fractionation into individual polysaccharides, or whether the fractionation is to a considerable extent carried out in the extraction process. For the preparation of heparin, A. F. Charles and D. A. Scott (18) established the principle of making an initial aqueous extract from the tissue, to contain as much heparin as possible, while discarding the "meat." This extract can then be subjected to purification and fractionation. Charles (16) established that a suitable extract could be effectively obtained by allowing the tissue to autolyze, then extracting at 65 °C with $N/_2$NaOH containing 0.8*M* ammonium sulfate. The alkaline pH and high concentration of electrolyte prevented association (caused dissociation?) of protein and heparin, whereas the high concentration of electrolyte resulted in aggregation and precipitation of most of the denatured protein. Autolysis increased the yield of heparin, evidently making more heparin available. It has been speculated that this might be heparin bonded covalently. The main point is that more than 90% of the tissue can be discarded. Most industrial processes follow the principle of Charles and Scott. These investigators processed the alkaline aqueous extract through acid precipitation of a heparin-protein complex, removal of fats with solvents and of protein with trypsin, and further purification steps involving Lloyd's reagent, charcoal, benzidine, and crystallization as a barium salt. There are a great variety of combinations of these steps using different enzymes, complexing agents, and so on. The procedures are described by Jaques and Bell (79). The most popular step, complexing with CPC and related compounds, was reviewed by J. E. Scott (154).

From his studies on the effect of electrolyte concentration on the complexing properties of AMPs and other anionic polyelectrolytes (described in Section II.3.A) and analyzing the principles of the Charles-Scott procedure for the extraction of heparin in these terms, J. E. Scott (155) reduced the extraction of heparin to a simple two-stage procedure. For the initial extraction at 65 °C of the autolyzed tissue, potassium acetate is added to the tissue to a concentration of 2*M* at pH 5.7. The 2*M* potassium acetate extract obtained is placed in the cold to give a precipitate that is almost pure heparin. One cannot conceive of a simpler procedure to obtain a trace substance from tissue than this elegant tour de force. Furthermore, as Scott emphasizes, the second step can be used as a rapid fractionation step for heparin obtained by another extraction procedure.

The separation of the tissue polysaccharide fraction to be followed by its fractionation can be the basis for the extraction of any AMPs. The initial steps are the separation of a fat-free tissue powder, which is subjected to proteolytic digestion with dialysis followed by alcohol precipitation. As such it was described by Meyer (117) and Dorfman (148). It was adapted to heparin by Jorpes (79), Marx (114), J. E. Scott (157). The crude product, consisting of polysaccharides together with inorganic electrolyte and buffer salts, can be analyzed as described in Sections III, IV, and VII and subjected to fractionation for individual AMPs as described in Section VI for separation of individual mucopolysaccharides. The mixed polysaccharide fraction is readily prepared and fractionated in the laboratory. On analysis, this gives the total mucopolysaccharides irrespective of the mode of binding in the tissue, hence provides maximum information. For preparative purposes, it is necessary to preselect a tissue that is rich in the desired AMPs. Thus commercially available chondroitin B and hyaluronic acid have been prepared from hog skin, chondrotin A from whale cartilage, and chondroitin C from shark cartilage. The AMP varies with the tissues of origin.

The classical procedures for the extraction and preparation of heparin and SMPs were described by Jaques and Bell (79) and in the references just cited. This section is therefore limited to presentation of new, recent procedures. The methods of J. E. Scott with 2*M* potassium acetate are described in two forms as the newest procedure for heparin. The preparation of a crude AMP fraction from tissues with proteolytic enzymes is used universally in this field on the basis of contributions by many workers. We have described this with details as used in our own laboratories. The procedure of Dietrich, Nader, Michelacci, and Cassaro (Section V.1.C) is described as giving a preparative method for crude AMPs and their fractionation.

A. USE OF 2*M* POTASSIUM ACETATE WITH HEPARIN

The 2*M* potassium acetate procedure was devised by J. E. Scott and patented both as a means of extracting heparin from tissues and as a way of fractionating commercial heparin. Examples are given of the procedure for extraction from whole tissue and for separation of heparin from the SMPs obtained by proteolytic digestion of dehydrated, defatted tissue. The latter is described for use with small tissue samples.

The procedure is based on the finding that carboxylate salts of potassium having an activity coefficient at least as high as potassium chloride are exceptional in that their aqueous solutions function as heparin extractants at higher temperatures and as heparin precipitants at lower temperatures. The extracting function is accomplished by precipitating proteins almost completely at acid pH at the high temperature. Potassium acetate has value as a precipitant for heparin because the potassium salts of the other SMPs show much greater solubility than potassium heparin in 2*M* potassium acetate at low temperatures—approximately 3 mg/ml for chondroitin B and commercial heparitin and 0.5 mg/ml for chondroitin A/C, compared to 0.06 mg/ml for sheep lung heparin. This solubility difference can be used to separate the heparin from other SMPs. Where the concentration difference between heparin and the other SMPs is very great (e.g., heparin is 0.1% or less of the total SMPs), it is probably necessary to conduct the fractionation in stages. The limit on recovery of heparin is the absolute solubility of a particular heparin in 2*M* potassium acetate.

a. Extraction of Heparin from Tissues

PRINCIPLE

Minced tissue is extracted with 2*M* potassium acetate at pH 5.7 and 56 °C. The extract after clarification is left at 5 °C overnight. The sediment that settles out is collected as heparin of about 100 USP units per milligram.

Procedure. Excess fat is removed from the tissue and 2 kg is minced or blended. For autolysis, toluene is added and the tissue kept at 37 °C for 24 hr. Glacial acetic acid is added to 4*M* potassium acetate to pH 5.5. An equal volume (2-1) is mixed with the minced tissue, the mixture is raised to 56 °C, and this temperature is maintained for 2 hr with frequent stirring or shaking. The mixture is then rapidly filtered by centrifuging either through a bag filter or through No. 1 Whatman paper in a basket. The filtrate is collected in glass and if it contains a suspension of fine particles, these are redissolved by reheating to 56 °C. An upper layer of fat forming is removed by suction. The filtrate is passed by suction through a Hyflo filter pad in a Buchner funnel to give a clear, bright filtrate. This material is left at 5 °C overnight to develop a sediment, which is removed by centrifuging for 30 min at 1500 rpm at 5 °C. The walls of the centrifuge tubes and the precipitate are washed with 2*M* potassium acetate, pH 5.7, at 5 °C, then with alcohol and ether, and dried. Alternatively, after the wash with potassium acetate, the sediment is dissolved in 20 to 50 ml of water, a few milliliters of concentrated potassium acetate is added, then 2 volumes of ethanol. The precipitate is collected, washed with alcohol and ether, and dried. For preparative purposes, the potassium acetate may be recycled by taking the mother liquor from the first precipitation, adjusting salt concentration and pH, and using this for extraction of fresh raw material.

COMMENTS

The method was developed for beef lung and mucosa. As with other extraction procedures for heparin, minor changes are required for other species and tissues because of differences in the fat and proteins present and the amount of water contained in the tissue. Such differences may affect the expression of the aqueous phase from the meat or make difficult the removal of fat. Since other substances in the extract (lipids, bile salts?) may delay the low temperature precipitation, this stage may require 4 to 5 days at − 5 °C for satisfactory sedimentation.

b. Separation of Heparin from SMPs in Small Tissue Samples (O. M. S. Toledo and C. P. Dietrich)

PRINCIPLE

This is a description of the Scott fractionation procedure as adapted for small tissue samples. Scale: 1 g of tissue.

Procedure. One gram of dry tissue (previously defatted with acetone and dried, see Section V.1.B) is suspended in 10 ml of 0.05*M* Tris-HCl buffer, pH 8.0. The suspension is incubated at 37 °C for 24 hr with 10 mg of trypsin and a few drops of toluene. After the incubation, 1.98 g of anhydrous potassium acetate is added to the incubate and the mixture is shaken to ensure complete solution of the salt. The pH is adjusted to 5.7 and the suspension maintained in an oven for 2 hr at 60 °C. About 100 mg of Celite is added to the suspension, which is filtered in an oven at 60 °C. The filtrate is collected in a centrifuge tube, the volume (usually 6 to 7 ml) is recorded, and the material is kept at 3 °C overnight. The precipitate formed is collected by centrifugation at 3 °C. The supernatant is discarded and the precipitate dissolved in 0.5 ml of water. To this solution is added 2 volumes of alcohol. After standing for at least 4 hr in the cold, the precipitate formed is collected by centrifugation and dried. It is dissolved in 10 to 100 μl of water to be analyzed by agarose gel electrophoresis and other methods (Sections IV.1.D and VIII).

COMMENTS

As discussed earlier, the procedure utilizes the considerable difference in solubilities between heparin and other SMPs under these conditions. Since heparin is found in trace amounts in tissues, this procedure is very useful in removing the other SMPs (when present in much higher amounts), allowing the analysis of heparin present. With a very low percentage of heparin, it appears that 2.5*M* potassium acetate and rigid control of temperature at − 5 °C or lower increase this difference. The procedure has not yet been examined for recoveries of heparins of very low and very high molecular weight.

B. PREPARATION OF A CRUDE AMP FRACTION FROM TISSUES WITH PROTEOLYTIC ENZYMES

PRINCIPLE

Fresh tissue is defatted and dehydrated with organic solvents. Proteins are removed with proteolytic enzymes. If necessary, nucleic acids are removed. The SMPs are precipitated with EtOH/acetone-1% NaCl and dried.

a. Procedure of T. K. Sue and L. B. Jaques. Cut 0.3 to 3 g of tissue (wet weight) into small pieces and homogenize in an electric homogenizer. Transfer to a 50-ml centrifuge tube. Add 30 ml of acetone. Leave overnight with occasional stirring. Centrifuge and discard the supernatant. The residue is resuspended in 30 ml of a 1–1 mixture of isopropanol and petroleum ether, allowed to stand for a few hours, then centrifuged and resuspended in another 30 ml of isopropanol-petroleum ether mixture. After centrifuging, the residue is heated in a water bath to drive off the organic solvents. A small amount of 0.1*M* Tris-HCl buffer (pH 8.0, containing 0.01*M* $CaCl_2$) is added to the dehydrated, defatted tissue, and it is transferred to a bag of dialysis tubing. Pronase Grade B (Calbiochem, La Jolla, Calif.) is added (about 15 mg/g of tissue). The dialysis tube is placed in a jar containing about 600 ml of 0.1*M* Tris-HCl buffer (pH 8.0), which is put on a rotating platform inside an incubator. The mixture is left overnight, then examined for complete digestion. If not well digested, it is left longer and more pronase may be added. After digestion, the sample is transferred to a 50-ml centrifuge tube, heated in a water bath for 10 to 15 min, and centrifuged. The residue is discarded. Sodium chloride (about 1%) is added to the supernatant. After this is dissolved, 5 volumes of acetone is added. After standing at least an hour, the precipitate formed is collected by centrifuging. If desired, reprecipitation can be conducted with alcohol-NaCl. To remove nucleic acids and residual protein, the tissue extract is dissolved in a suitable amount of water (10 mg/ml). The solution of tissue extract and 20% trichloroacetic acid (TCA) is chilled in an ice bath. Add an equal volume of cold 20% TCA to the tissue extract solution and shake gently. Centrifuge at 10,000*g* for 15 min at 4 °C to obtain a clear supernatant. Discard the precipitate. Add 0.1 to 0.2 ml (1–35 of volume) of saturated NaCl solution to the supernatant. Add 5 volumes of acetone. Stand for 30 min (in cold), centrifuge at 2500*g* for 15 min. Discard the supernatant. If desired, wash the precipitate with alcohol, Lyophilize the precipitate. Analyze as discussed in Section VII.

b. Procedure of C. P. Dietrich, L. O. Sampaio and O. M. S. Toledo. Suspend 100 μg to 1 g of dry tissue (previously defatted and dried with acetone) in 20 ml of 0.05*M* Tris-HCl buffer, pH 8.0. Add 10 mg of trypsin to this mixture, which is incubated at 37 °C for 24 hr, with a few drops of toluene forming a layer at the surface. After incubation, the mixture is brought to pH 11.0 with concentrated NaOH and maintained for 6 hr at room temperature. The pH is then brought to 6.0 with HCl and the mixture is centrifuged. To the supernatant, 0.1 ml of 2*M* NaCl and 2 volumes of alcohol are added and kept overnight at 5 °C. The precipitate formed is collected by centrifugation and dried. The resulting powder is resuspended in 1 ml of 0.05*M* sodium acetate, pH 6.5, and DNAse and RNAse (1 mg of each) are added to the solution. After incubation for 24 hr at 37 °C with a layer of toluene, 0.1 ml of 2*M* NaCl and 2 volumes of ethanol are added to the solution, which is kept overnight at 5 °C. The precipitate formed is collected by centrifugation and dried. The resulting powder is dissolved in 0.5 ml of water, heated at 100 °C for 2 min, any residue present is removed by centrifugation and the supernatant is analyzed as discussed in Section VII.

C. EXTRACTION WITH SEPARATION OF AMPS BY AGAROSE GEL ELECTROPHORESIS (C. P. DIETRICH, H. B. NADER, Y. M. MICHELACCI, AND C. M. CASSARO)

PRINCIPLE

Defatted and dried tissue powder is digested with trypsin and pronase successively. The AMPs are precipitated with acid alcohol and separated by electrophoresis in agarose containing propanediamine acetate buffer. Strips of gel (0.50 cm) are frozen. The eluate obtained on thawing is concentrated and precipitated with alcohol, and the AMPs identified for pooling.

Procedure. Defatted and dried tissue powder (see Section V.1.B) is suspended in 20 parts (w/v) of 0.05*M* Tris-HCl buffer, pH 8.0. Trypsin (1 part in 200 parts of dried tissue, w/v) is added to this suspension and the mixture incubated for 24 hr at 37 °C with a layer of toluene. After incubation, pronase (1 part in 200 parts of dried, w/w) is added to the mixture; which is kept at 37 °C for another 24 hr. The pH of the solution is brought to 11 by the addition of NaOH, and this condition is maintained for 8 hr at room temperature. The pH is then brought to 6.0 and the mixture centrifuged. To the supernatant solution, 1.5. volumes of alcohol is added and the mixture kept for 24 hr at 5 °C. The precipitate formed is collected by centrifugation, resuspended in water, and dialyzed against 0.1*M* NaCl. The dialyzate is centrifuged and the supernatant precipitated with 2 vol of alcohol, washed twice with alcohol at 80%, and dried.

This dried precipitate is resuspended in the smallest possible volume of water (up to 100 mg of AMPs) and applied to Whatman 3MM (strips, 0.8 cm × 15 cm). (A trace of cresol red is added to the suspension before applying.) About five strips are wetted with the solution (usually 250 μl per strip), held together, and placed in a slot made in a 1-cm-thick agarose gel block (0.9.% agarose, 17 × 30 cm, with 0.05*M* propanediamine acetate buffer, pH 9.0). The gel is subjected to electrophoresis (at 5 °C in any appropriate chamber) at 100 mA until the dye indicator migrates about 20 cm from the origin. Each hour the buffers from the outer compartments are switched, to maintain their buffering capacity. After the run, the agarose gel block is cut in 0.5-cm strips, which are frozen. Upon thawing, the slices that have lost their gel structure are squeezed through cheesecloth, and the mucopolysaccharides in the eluate are concentrated and precipitated with 5 volumes of alcohol at 5 °C overnight. The resulting precipitate is collected by centrifugation, washed with 80% ethanol, dried, and dissolved in 1 ml of water. The overall recovery of the AMPs by this procedure varies from 65 to 87% in different fractionations. Control slices of gel treated by the same procedure but without AMPs yield about 1 mg of an insoluble material for 10 ml of eluate. This impurity (probably agarose) does not contain any detectable amount of sulfate, acetyl groups, uronic acid, or hexosamine.

The fractions obtained are analyzed by agarose gel electrophoresis in propanediamine and barbital buffers as described in Section VI.4. The fractions with the same electrophoretic migrations are combined. The pooled fractions that still contain more than one AMP in microelectrophoresis barbital buffer can be resubmitted to fractionation in large scale in this buffer.

COMMENT

Most of the AMPs can be fractionated and purified from extraneous materials by this method.

Hermelin et al. (60) have described a similar preparative procedure in which the electrophoresis medium is Pevikon (a copolymer of polyvinyl chloride and polyvinyl acetate). This is washed thoroughly with distilled water with decantation to obtain homogeneous granulation. The material is then dried and suspended in buffer. After decantation, the Pevikon is placed on the refrigerated plate of a high-voltage electrophoresis apparatus of the Pherograph type, measuring 350 × 150 × 15 mm. A 10-mm gutter is cut 50 mm long, 100 mm from the cathode end of the plate. The AMP solution to be separated is placed in this gutter, and dry Pevikon is added. Two gutters are cut, one on each side of the center gutter to take marked AMPs. The AMPs are colored by addition of procion red (I.C.I., Macclesfield, U.K.) and submitted to dialysis to remove excess dye and salts. Electrophoresis is conducted for 5 hr at 0 °C and a potential difference of 600 V, with the following buffers: glycocoll HCl, 0.05*M*, pH 2.0, barium acetate, 0.015*M*, pH 7.0; pyridine formate, 0.05*M*, pH 2.3. After migration, the Pevikon plate is cut in 1-cm strips. The AMPs are eluted with distilled water and identified with enzymes.

2. Extraction from Blood and Endothelium

The demonstration and measurement of heparin in blood and in endothelium deserves special consideration because it can be extremely important for many investigations. Such measurements have often been frustrating, for reasons that are now identifiable. Hematological procedures of measuring heparin in blood are listed in Table V. When the coagulation system (coagulation factors) in the blood is normal in nature and amount and the only coagulation change has been the addition of heparin (by administration to the subject), direct coagulation tests are the only means required and used for monitoring and even measuring heparin in blood. When coagulation factors change because of pathology or surgery, however, these tests are no longer reliable indices of quantity of heparin in plasma, and it is necessary to extract the heparin before estimation in a suitable test system. Furthermore none of the tests described in Table V is specific for heparin. Hence when anticoagulant that appears to be SMPs is found in the circulation, extraction is necessary to make possible identification by electrophoresis, enzymes, or other means (Sections IV.1.D, VI.1 to VI.6, and VII).

The determination of endogenous heparin in blood was reviewed by Jaques and Bell (79), who pointed out that extraction of normal blood for heparin gave only a slight anticoagulant activity because of the presence of sodium chloride, and so on, whereas the same procedure not only extracts added heparin from blood but quantitatively extracts heparin from the blood of dogs in anaphylactic shock, hence endogenous heparin released to the circulation. Their review of the literature demonstrated problems of quantity, differentiation, and distribution associated with endogenous heparin normally present in plasma. The quantity is probably less than 0.1 unit of anticoagulant activity per milliliter, which is the practical limit for detection by the various procedures used. Since quantities of chondroitins in plasma are greater than this, differentiation of the SMPs is essential. Heparin appears to be associated with the white cell layer, and heparin in plasma may be in a bound form. Horner (12) demonstrated the association of a large part of endogenous heparin with the platelets. The ubiquity of platelet factor 4 and its high affinity for heparin makes these problems more difficult, since increased breakdown of platelets *in vivo* or *in vitro* has a marked effect on results.

For extraction of heparin from blood, the distinction must be made between whole blood, white cells, platelets, and plasma (platelet-poor). Jaques and Waters (89) extracted whole blood by the Charles-Scott procedure. It appears to be advisable that for all studies, platelets be preserved carefully and removed separately, to ensure that platelet factor 4 does not offer problems with the other blood fractions. The red cell, white cell and platelet portions can be extracted separately by high electrolyte concentration (83; Section V.1.A) or by proteolytic digestion (Sections V.1.B, V.1.C). These procedures have not been found to be very successful with plasma. Platelet-poor plasma can be deproteinized with phenol (79). This is not satisfactory for heparin with rat or guinea pig plasma. Heparin can be satisfactorily precipitated from plasma with octylamine (79). After treatment to give a crude polysaccharide powder, the material can be quantitated, fractionated, and analyzed as described in Sections III, IV, and VI. In many situations it is more important to know whether any alteration of coagulation factors has occurred subsequent to heparin administration. Hence the most important new procedure is that of Thompson and Counts (169) for the removal of heparin and protamine from blood plasma.

Heparin and SMPs can be extracted from endothelium by the procedures used for other tissues. Since no procedures have been described for obtaining endothelium for biochemical studies, we have included a procedure developed in connection with determinations of heparin in endothelium.

A. REMOVAL OF HEPARIN AND PROTAMINE FROM BLOOD PLASMA

PRINCIPLE

In a procedure developed by Thompson and Counts (169), citrated plasma containing heparin is passed through a ECTEOLA-cellulose column; citrated plasma containing protamine is passed through a carboxymethyl-cellulose column, and through both columns when both heparin and protamine had been added. The plasma can then be used for determination of coagulation parameters—thrombin time, prothrombin time, partial thromboplastin time, and concentration of fibrinogen, prothrombin, and factors V, VII, VIII, IX, and XI.

PREPARATION OF COLUMNS

Powdered ECTEOLA-cellulose (100 g) is slurried in water and poured into a 10-cm Buchner funnel with filter paper. It is washed alternately with 200 ml of 0.5*N* NaOH, 0.5*N* HCl, and 0.5*N* NaOH. Between these cycles, 1-liter washes with distilled water are applied. Following the last water wash, the substituted cellulose is equilibrated with 0.05*M* Tris-HCl (pH 7.5) containing 0.1*M* NaCl (Tris-saline). The prepared cellulose can be used even after storage for several months at 4 °C. Columns are 5-in. Pasteur pipettes, siliconized with a 1% solution of Dry-Film SC-87 (Pierce Chemical Co., Rockford, Ill.) in petroleum ether. A small amount of siliconized glass wool is tapped into the narrow portion of the pipette with a thin glass rod. The pipette is then placed vertically into a 10 × 75 mm glass tube and filled with Tris-saline to avoid air bubbles. Approximately 1 ml of slurried ECTEOLA-cellulose is added and allowed to settle to a final packed height between 0.5 and 2 cm. Columns are transferred to empty test tubes and washed with 2 ml of Tris-saline before use. Columns with flow rates between 1 and 3 min/ml are selected and can be stored in buffer for several months at 4 °C.

Carboxymethyl-cellulose (100 g) is suspended in 500 ml of water and poured into a 20-cm Buchner funnel with filter paper. The cellulose is washed as described for ECTEOLA-cellulose and equilibrated with Tris-saline buffer. Columns are prepared in siliconized pipettes as described for ECTEOLA-cellulose except that the height is 1 to 2 cm. Alternatively, columns are prepared by layering a 1-cm bed height of carboxymethyl-cellulose on top of a 0.5-cm bed height of ECTEOLA-cellulose.

Procedure. To remove heparin, plasma samples containing heparin (0.5 to 1 ml, transferred in 9-in. siliconized Pasteur pipettes) are applied to the top of the cellulose after all the buffer wash has been included into the bed. When undiluted plasma appears in the effluent, a second aliquot (1 ml) of the sample is applied to the substituted cellulose and the effluent is collected in a 10 × 75 mm siliconized test tube, representing the adsorbed plasma sample. The entire procedure takes less than 5 min; multiple samples are prepared and run simultaneously.

To remove protamine, samples are applied to the carboxymethyl-cellulose columns and collected as described for heparin removal. For the removal of both protamine and heparin from the same plasma sample, the effluent plasma from the carboxymethyl-cellulose column is applied to an ECTEOLA-cellulose column. The columns with carboxymethyl-cellulose layered on top of ECTEOLA-cellulose give the same results as those obtained by sequential columns for removal of protamine and heparin.

COMMENTS

Thompson and Counts (169) report that values for prothrombin, partial thromboplastin, and thrombin times for plasma samples passed through the columns are in close agreement with values for the plasma without treatment, and the average activity for factors V, VII, VIII, IX, and X after treatment was at least 90% of that found in the starting plasma. When 1 ml of sample containing up to 300 units of heparin was absorbed by 1-cm ECTEOLA-cellulose columns, no heparin could be detected in the effluent by the factor Xa inhibition assay. ECTEOLA-cellulose used for plasma samples has the ability to adsorb certain clotting factors, but this potential problem is avoided by equilibrating the substituted cellulose with physiological ionic strength and pH, and using the plasma samples directly. After passage of plasma samples through this resin, a 10% decrease of factor IX activity occurs. Because a similar decrease is seen for all factors, including a fibrinogen assay not based on the kinetics of clotting, the decrease is best explained by dilution of plasma with small amounts of buffer. Some clotting factors (e.g., factor

XIa) bind to carboxymethyl-cellulose. This resin was found to have no affect on screening tests or assays of factors VIII and IX.

B. SEPARATION OF ENDOTHELIUM (L. M. HIEBERT)

PRINCIPLE

Large blood vessels are attached to dental wax, slit open, and washed. Cellulose-polyacetate film is applied to the endothelial surface. The cellulose acetate with attached endothelial cells is removed. The cellulose acetate is dissolved in acetone at − 20 °C, leaving a dry tissue powder containing polysaccharides and proteins (enzymes, etc.). This is used for determination of heparin and other substances.

Procedure. Pieces of vessel of 2.5 to 3 cm^2 are pinned to dental wax, slit open, and washed with oxygenated Locke's solution. A cellulose-polyacetate electrophoresis strip (Sepraphore III, Canlab No. 51003) is applied to the endothelial surface of the vessel, covered with a glass slide, and held for a few seconds. The cellulose acetate paper is then removed, taking the endothelial cells with it. An additional application of cellulose acetate paper follows, to ensure the removal of all endothelial cells. The imprint made by the endothelium on the paper (observed at the first application) is measured to estimate the area of endothelium obtained. Excess cellulose acetate paper is cut off and discarded. The cellulose acetate paper with endothelium is cut into small pieces and placed in a vial with 10 ml of acetone at −20 °C, agitated, and allowed to stand for 1 hr at −20 °C. This mixture is centrifuged at 2000*g* for 15 min and the supernatant is discarded. Next 10 ml of acetone at −20 °C is added to the residue, and the process is repeated. Then 1 to 2 ml of distilled water is added to the residue and the whole is freeze-dried. An additional 10 ml of acetone at −20 °C is added to the dried powder and agitated to remove any cellulose acetate paper remaining. After centrifuging, the residue is transferred with water to a weighed vial and freeze-dried to a dry powder. This powder weighs about 0.5 mg/cm^2 of endothelium.

Heparin and SMPs may be determined directly by dissolving the powder in 5 to 10 μl of water and examining by microelectrophoresis (on agarose slides) as described in Section IV.1.D. The dry endothelial powder may also be treated with organic solvents and digested with pronase as described in Section VII. This is necessary if the amount of heparin is less than 0.3 u/g of powder or less. Differentiation is carried out as described in Section VII.

COMMENT

The procedure has been used successfully with fresh aorta and vena cava of rats, rabbits, and dogs to demonstrate uptake of heparin from Locke's solution and blood *in vitro* and *in vivo* (61,62). It provides a dry endothelial tissue powder that can be used for determination of many polysaccharide and protein components of endothelium, including enzymes. Cellulose acetate may contain a trace of cellulose sulfate. This apparent SMP is easily identified; it migrates faster than heparin in barbital-agarose, presenting a difficulty only when trace quantities of heparin are being estimated.

3. Preparation of Radioisotope-Labeled Compounds

A. ^{35}S-CHONDROITIN B, ^{35}S-HEPARITIN, ^{35}S-CHONDROITIN-A (C. P. DIETRICH AND Y. M. MICHELACCI)

PRINCIPLE

Infant rats are injected with $Na_2{}^{35}SO_4$, and ^{35}S-chondroitin B and ^{35}S-heparitin are extracted from the skin, and ^{35}S-chondroitin A from the carcass.

Procedure. First 2 mCi of $Na_2{}^{35}SO_4$ (carrier free in 2.5 ml of isotonic saline) is injected subcutaneously into 14 rats, 24 hr old. After 24 hr the rats are killed by decapitation. The radioactive chondroitin B and heparitins are extracted from the skin and fractionated as described in Section V.1.A. The skinned rats are blended in a Waring blendor with 10 volumes of acetone. The resulting powder is washed once with acetone and dried. The chondroitin A that comprises 90% of the total AMP is extracted and purified as described in Section V.1.A.

COMMENT

Between 20 to 30 μc, with a specific activity of 1 mCi/g each SMP is usually obtained by this procedure.

B. N ^{35}S-HEPARIN (C. P. DIETRICH)

PRINCIPLE

Na-Heparin is converted to heparinic acid on a Dowex column and allowed to autohydrolyze. The partially N-desulfated heparin is combined with $^{35}SO_3$-trimethylamine. The trimethylamine is removed by hydrolysis and the uncombined SO_4 by dialysis.

Procedure. First 140 mg of heparin (Na salt) is passed through a column (1 × 20 cm) of Dowex-X8 H^+ resin. The heparinic acid formed is maintained at 56 °C for 4 hr. The volume is reduced to 5 ml and 2 volumes of alcohol is added in the presence of 2 drops of 2*N* NaCl. The precipitate formed is washed with alcohol and dried. This partially N-desulfated heparin is added to $^{35}SO_3$-trimethylamine complex at 0 °C. This complex is prepared by dropwise addition of 3 ml of anhydrous trimethylamine (Eastman Kodak) at − 40 °C, to $^{35}SO_3$ (10 mCi, 20 mCi/m*M*, New England Nuclear) also at −40 °*C*. After the addition of trimethylamine, the temperature is allowed to rise to 3 °C to evaporate the excess of this compound; 100 mg of partially N-desulfated heparin and 140 mg of Na_2CO_3 in 2.5 ml of water are added to the $^{35}SO_3$-trimethylamine complex. The mixture is incubated at 55 °C for 24 hr with agitation. After incubation the mixture is diluted to 10 ml and dialyzed under agitation against several changes of water. When almost no radioactivity is detected in the dialyzate (usually four changes), the N ^{35}S-heparin is removed from the dialysis bag and precipitated with 2 volumes of alcohol in the presence of NaCl.

COMMENT

Degradation of heparin may occur during the preparation of N ^{35}S-heparin, and this has been observed with the commercial product and reported to us. We obtain by the foregoing procedure 400 μCi of N ^{35}S-heparin with a specific activity of 5.3 mCi/g. This material has an anticoagulant activity of 150 IU/mg and is completely degraded by the *Flavobacterium heparinum* heparinases, forming the same products as obtained from the original nonradioactive heparin. It has also the same electrophoretic mobility as the nonradioactive heparin in agarose gel electrophoresis. For other details in the preparation of these compounds see ref. 118.

C. TRITIATED HEPARIN

PRINCIPLE

The ^{3}H is introduced into heparin by exchange in the presence of palladium catalyst.

Procedure. The procedure was initially described by Barlow and Cardinal (6). Their improved procedure is now carried out by New England Nuclear Corp. (see p. 14, of the firm's 1976 catalog) by gas exposure labeling by the Witzbach procedure. Heparin (1000 mg) is exposed to 15 Ci of tritium at 24 °C for 2 weeks. Labile tritium is removed by back exchange with water and 50% ETOH (repeated distilling of water, then 50% EtOH from the product). The sample is purified by usual methods for heparin (precipitating with CPC or Cetavlon, precipitating with alcohol, etc.). Purity of the labeling is checked by radiochromatogram scanning (5). The final product is about 77 μCi/mg. On electrophoresis and ultracentrifuge, radioactivity and metachromasia coincide in a single peak.

D. ^{51}CR-HEPARIN

PRINCIPLE

The procedure was described by Varga et al. (174). Heparin is reacted with $^{51}CrCl_3$ in 87% ethanol at 80 °C for 60 min. The product in $10^{-2}M$ EDTA is absorbed on a Sephadex G-50 column and eluted with water. A stable product, 120 μCi/mg, with unchanged anticoagulant and metachromatic activity is obtained.

Procedure. A half-milliliter of heparin solution (20 mg/ml) is placed in a 20-ml volume glass reflux vessel to which is added 0.1 ml of radioactive chromium (^{51}Cr) solution (120 μCi of $^{51}CrCl_3$ containing 0.5 μg Cr) and 6 ml of 96% ethanol. The reactants are mixed until the solution becomes opalescent, then heated to 80 °C and maintained at that temperature for 60 min. The ethanol and water are then removed in 10 to 15 min by air and low heat evaporation. Next 5 ml of $10^{-2} M$ EDTA solution is added and chromatography carried out on a Sephadex G-50 column. The column is eluted with distilled water and the first radioactive peak (^{51}Cr-heparin complex) collected. The second peak (^{51}Cr-EDTA complex) is discarded. Addition of further EDTA and rechromatography on Sephadex G-50 should give a single radioactive peak.

COMMENT

Varga (174) reports that the product is stable and can be stored in solution. It has the same metachromatic and anticoagulant activity as the original heparin. An Amberlite IRC column removes 15% of the radioactivity; hence this amount is present as coion. The remaining 85% is bound, probably to OH groups.

VI. FRACTIONATION AND CHARACTERIZATION OF SMPS (C. P. DIETRICH)

The description of the complexity of the SMPs indicates that no single method can be used to fractionate and characterize any one of these compounds. As implied in this statement, fractionation and characterization are almost synonymous for the SMPs. To simplify the organization of the material, the principles employed are covered briefly, followed by descriptions of representative procedures as tested in our hands.

Since the SMPs are polyanionic compounds, the general principles involved in their fractionation and characterization take advantage of the relative amount of negative charges in the polyanionic chain. Fractionation by complexing with quaternary amines, ion-exchange chromatography, and electrophoresis are the usual methods selected for the fractionation of these compounds. Complexing with quaternary amines was previously reviewed in this series by J. E. Scott (154). The usual method is based on the differential solubility of cetylpyridinium-AMP complexes in the presence of increasing salt concentrations (e.g., Lemperg et al., 104). The method has been widely used for the industrial preparation of acidic mucopolysaccharides and has been adapted in the past few years to microscale determination of AMPs (2).

Ion-exchange chromatography is one of the widely used methods for identification of AMPs. Ringertz and Reichard (141) fractionated hyaluronic acid, chondroitin A, and heparin by elution with NaCl-HCl mixtures from ECTEOLA-cellulose columns. The technique has been used in the study of AMPs from mast cells, cornea, and cartilage. Similar methods using DEAE-cellulose (111,159), DEAE-Sephadex (149), and Dowex 1 × 2 (148) have also been published. Tortolani and Romagnoli (172) have recently described the determination of molecular weights of AMPs by thin-layer chromatography on Sephadex G-100-cellulose (4–1). Distance of migration of chondroitins A, B, C, heparin and duodenal mucopolysaccharide varied linearly with the molecular weight.

Electrophoresis has many advantages in separating such high negatively charged compounds as AMPs and has been used with the Tiselius apparatus, chromatography paper, falling curtain and CPE (continuous particle electrophoresis) system, starch gel, cellulose acetate, agarose, and acrylamide gels and films. Cellulose acetate electrophoresis is undoubtedly the technique most widely used for the identification of AMPs. The method takes advantage of the different net charges of the AMPs and is usually combined with divalent cations that change selectively the migration of some AMPs. Within limits, the separation is independent of the polydispersity of the AMPs. One-dimensional electrophoresis for the separation of AMPs has been used with the following compounds: barium acetate at pH 7.0 and 8.0 (177), zinc acetate (138), copper acetate (57), calcium acetate (158), barbital, phosphate buffer, and 0.1*N* HCl (178). Two-dimensional electrophoresis in cellulose

TABLE VII
Relative Mobilities[a] of AMPs Electrophoresis in Cellulose Acetate with Various Electrolyte Solutions (57)

Electrolyte solution		Acidic glycosaminoglycans							
	pH	Hep	ChA	ChC	ChB	Hpt	KS	HA	Ch
0.1*M* pyridine 0.47*M* formic acid	3	1.1	1.0	1.00	0.91	0.87	0.85	0.63	0.64
0.1*M* phosphate	6.7	1.1	1.0	1.00	1.0	0.86	0.77[b]	0.68	0.72
0.1*M* barium acetate	8	0.61	0.93	1.00	0.78	0.61	0.83	0.74	0.74
0.1*M* cupric acetate	5.3	0.31[b]	1.1	1.00	0.94	0.70[b]	1.2[b]	0.43	0.76
0.1*M* zinc acetate	6.3	0.88[b]	0.97	1.00	0.84	0.76	0.93	0.68	0.74
0.1*M* lanthanum nitrate	2.6	0.87	1.0	1.00	—	0.74[b]	1.1	—	—

[a] Expressed as ratios of mobilities of AMPs to that of ChC in each electrolyte solution.
[b] Broad spot.

acetate strips with these compounds has been described (57). The relative mobilities of the AMPs in different electrolyte solutions are shown in Table VII.

In fractionating and characterizing SMPs by electrophoresis in gels, the results obtained depend not only on the amount of negative charges in the polyanion but on the relation of this and other properties to the medium itself. In column chromatography, and to some extent in complexation with quaternary amines, absolute charge of the SMPs seems to be the determinant of the fractionation. Compounds with similar absolute charge; such as the chondroitins, the heparitins, and keratosulfate, are not fractionated in these systems. In gel electrophoresis, molecular size is the determinant on acrylamide gel, net charge in cellulose acetate and agarose gels. Complexing activity is another factor when diamines or divalent cations are introduced into the system. The change in electrophoresis properties with streptomycin was suggested as distinguishing anticoagulant and nonanticoagulant heparin by the USP assay (7).

A recent development in the fractionation and characterization of SMPs was the introduction of agarose as a support for electrophoresis. The method was introduced in 1963 by Jaques and co-workers (78) and was applied for both quantitative (77) and preparative purposes (34). It has been further improved, and its versatility has been increased, with the use of different buffers (30) and by combining with enzymatic degradation of the SMPs (28) and using critical electrolyte concentrations (28,164). The method has still not been used by many workers in the field, but it has been shown to be extremely versatile in a variety of problems (26,27,125,127,165,166). Table VIII gives the electrophoretic migration of the AMPs in agarose with barbital and different diamines. The AMPs have different migration rates according to the buffer used. These changes in migration can be used as a criterion for the characterization of most AMPs.

The separation of AMPs by the electrophoretic methods described thus far are based on the differences of their net charge and ability to complex with different cations and organic amines. Another method that serves for identification and preparation of AMPs is polyacrylamide gel electrophoresis, which separates the AMPs as a function of their molecular sizes. The method has proved to be useful for the determination of the average molecular sizes of the different SMPs (33,63,68) and for the identification of heparitins (33). Table IX compares the relative electrophoretic migration of the heparitins in polyacrylamide with migration in agarose and cellulose acetate. Enzymatic methods allow the more specific identification of SMPs in very small amounts of tissues and fluids. A very recent development has been the use of specific mucopolysaccharidases for the identification of acidic mucopolysaccharides (24,25,109,145). Chondroitinase AC and chondroitinase ABC are available

TABLE VIII
Relative Electrophoretic Mobilities of AMPs in Agarose with Barbital and Diamine Buffers[a] (30,32)

AMP	Buffers[b]								
	B	C_2	C_3	C_4	C_5	C_6	C_7	C_8	C_{10}
Heparin (beef lung)	1.15	0.83	0.64	0.66	0.67	0.70	0.72	0.09	—[c]
Heparitin A	0.91	0.78	0.64	0.72	0.72	—[d]	—[d]	—[d]	0.78
Heparitin B	0.91	0.78	0.64	0.72	0.72	0.72	0.70	0.67	0.81
Heparitin C	0.94	0.73	0.55	0.62	0.62	0.62	0.70	0.70	0.67
Heparitin D	1.15	0.83	0.64	0.66	0.67	0.70	0.72	—[d]	0.72
Chondroitin A	1.0	1.0	1.0	1.0	1.0	1.0	1.0	1.0	1.0
Chondroitin B	1.0	0.89	0.83	0.82	0.82	0.82	0.82	0.82	0.91
Chondroitin C	1.0	0.94	0.96	0.95	0.95	0.91	0.91	0.91	0.96
Keratosulfate	0.8	—[d]	—[d]	—[d]	—[d]	—[d]	—[d]	—[d]	—[d]
Hyaluronic acid	0.61	0.65	0.72	0.75	0.75	0.68	0.75	0.68	—[d]

[a] Values relative to chondroitin A.
[b] B = barbital; C_2, C_3, . . . , C_{10} = ethylenediamine, propylenediamine, . . . , decanediamine acetate buffers; 0.05*M* pH 8.5, for C_2 to C_4; pH 10.0 for C_5 to C_8; and pH 10.3 for C_{10}.
[c] Did not migrate from point of application.
[d] Not determined.

commercially. A heparinase (37,161), two heparitinases (160), a chondroitinase B (119,120), and a chondroitinase C (121) have been recently isolated from *Flavobacterium heparinum* and purified from each other.

Another new development in the fractionation of AMPs was the finding that SMPs form complexes with ampholyte mixtures and precipitate at specific pH regions when submitted to electrofocusing. In the case of heparin, after 18 hr of focusing, 21 bands are visible at pH 4.2 to 3.0. Molecular

TABLE IX
Comparison of Relative Electrophoretic Mobilities of Heparitins and Heparin in Agarose, Cellulose Acetate, and Polyacrylamide Gels[a]

AMP	Agarose	Cellulose acetate	Polyacrylamide
Heparitin A	0.70	0.91	—[b]
Heparitin B	0.76	0.91	0.51
Heparitin C	0.78	0.91	0.83
Heparitin D[c]	1.0	1.00	1.0
Heparin[d]	1.0	1.0	0.75

[a] Data from M. E. Silva, C. P. Dietrich, and H. B. Nader, *Biochem. Biophys. Acta,* in press.
[b] Did not migrate from point of application.
[c] Heparin, mol. wt. 1320–7800.
[d] Beef lung heparin from Upjohn, mol. wt. 2200–33,200.

weight determinations have shown that these fractions differ from one another by stepwise increase of about 1800 in molecular size. The other SMPs exhibit different patterns of fractionation in regions other than the ones shown by heparin.

1. Ion-Exchange Chromatography (C. P. Dietrich)

PRINCIPLE

Mixtures of AMPs are applied to a DEAE cellulose column and eluted by a lithium chloride gradient. Scale: 0.5–5 mg.

Procedure. A mixture of 2 mg each of hyaluronic acid, chondroitin A, and heparin is applied in a 1 × 5 cm DEAE cellulose column previously equilibrated with 0.15*M* sodium chloride. The materials are eluted from the column by a LiCl gradient at a rate of 5 ml/hr. The mixing vessel and the reservoir each contain 100 ml of 0.2*M* and 3*M* lithium chloride, respectively. Fractions of 3 ml are collected and 0.5-ml aliquots are analyzed for uronic acid by the carbazole reaction. Hyaluronic acid, chondroitin A, and heparin are eluted after about 20, 50, and 75 ml, respectively. Several modifications of the steps outlined have been made (53), to improve the sharpness of the peaks and the reproducibility of the method.

2. Complexing with Quaternary Amines (C. P. Dietrich)

PRINCIPLE

AMPs are absorbed on a cellulose column impregnated with cetylpyridinium chloride. The AMPs are eluted with increasing concentrations of NaCl. scale: 100 μg–1 mg.

Procedure. One milliliter of cellulose suspension (10%) in water is introduced into a column of 3 mm diameter. After the cellulose is settled, the column is washed with 5 ml of 1% aqueous solution of CPC. The AMP mixture is added to the column and eluted successively with 1 ml of 1% CPC solution and 1 ml of solutions containing 0.05% CPC with increasing concentrations of NaCl (0.2*M* to 0.5*M*) or $MgCl_2$ (0.4*M* to 1.2*M*). The AMP is determined by the carbazole reaction (40). Hyaluronic acid is eluted with about 0.3*M* NaCl, chondroitin sulfate with 0.4 to 0.6*M* of $MgCl_2$, and heparin with 0.9 to 1.2*M* of $MgCl_2$.

COMMENTS

The method is useful for the determination of mixtures of some AMPs with clear differences of charge in microquantities. Nevertheless the variables must be carefully controlled, since small changes in the concentrations of the salts modify considerably the pattern of elution of the compounds.

3. Cellulose Acetate Electrophoresis (C. P. Dietrich)

PRINCIPLE

AMPs are applied to cellulose acetate membranes and subjected to electrophoresis. The dried membrane is stained with alcian blue. Individual AMPs are identified by migration relative to heparin reference (Table VII). Scale: 0.5–5 μg.

Procedure. A cellulose acetate membrane (Separax, Cellogel, Sepraphore III, etc., 100 × 100 mm or 100 × 25 mm) is soaked in 0.1*M* barium acetate buffer, pH 7.0, for 30 min, and excess of buffer is removed with the aid of a filter paper. The electrophoresis strip is then placed in position in an electrophoresis chamber already filled with the same buffer. After the application of the samples, a potential of 10 V/cm is then applied for 2.5 hr. After the run, the membrane is blotted dry and stained for 10 min in 0.1% alcian blue in 0.1% acetic acid and washed with 300 ml of 0.1% acetic acid for 20 min. SMPs are identified by relative migration distance.

COMMENTS

The method is most useful for identification of small quantities of AMPs in a relatively pure state. Crude mixtures should be avoided because of the small load capacity of the membranes; in addition, salts are usually a drawback in the characterization of the compounds. Sometimes it becomes very difficult to resolve a mixture of AMPs in which there is wide variation in the concentration of the individual AMPs in the mixture. Tortolani et al. (171) extract the toluidine blue of the spot from the cellulose acetate with 1 ml of 0.1*M* $CaCl_2$ at 70 °C for 5 min, add 2 ml of alcohol, and measure the absorption at 630 nm in a spectrophotometer. Hata and Nagai (58) stain the strip with 0.1% alcian blue 8GX in 0.1% acetic acid-0.02% sodium azide. After rinsing in 0.1% acetic acid twice, the dye-AMP spot and an adjacent blank of similar size are cut out and dropped into testtubes; 1 ml of 5% of CPC is added, and the tubes are placed in boiling water for 15 min. The optical density is read at 615 nm.

4. Agarose Gel Electrophoresis (C. P. Dietrich, S. M. C. Dietrich, L. O. Sampaio, and C. M. F. Cassero)

PRINCIPLE

SMPs are applied to agarose-coated slides containing barbital buffer (pH 8.5) and propanediamine buffer (pH 8.0 or 10.0) and subjected to electrophoresis. Slides are fixed with Cetavlon, dried, and stained with toluidine blue. SMPs are identified by migration relative to a reference heparin in the two buffers (Table VIII). Similar, more extensive procedures are described elsewhere—in Section IV.4.D as a quantitative micro method, and in Section V.1.C as a preparative procedure applied to tissue extracts. Scale: 0.1–10 μg.

Procedure. Agarose slides and chamber are prepared as described in Section III 4.D, except that different buffers are used with an agarose layer 2 mm thick. Two sets of chambers and slides are prepared with 0.06*M* barbital buffer, pH 8.5, and 0.05*M* propanediamine-acetate buffer pH 9.0 or 10.0, respectively. A minute amount of cresol red is added to the SMP solution. Up to 5 μl of concentrated solution is applied in a 4-mm slot. The dye indicator is used to control the migration of SMPs. For a slide 7.5 cm long, the dye should be allowed to run only to 2 cm distance from the origin with barbital and 4 cm from the origin with diaminopropane.

For staining the agarose gel prepared with propanediamine and the other diamines, the following procedure is used. After the fixation of the slide with 0.1% Cetavlon and drying, the slide is immersed in a staining solution containing 0.1% toluidine blue dissolved in acetic acid-ethanol-water (1–50–50, v/v/v/). The destaining procedure must be carefully controlled, since fading may occur with some SMPs if the slide is maintained for long periods in the solution. Usually three washes, 2 to 3 min each with agitation, is enough for complete destaining of the slide. For staining slides prepared in barbital buffer, a solution of 0.1% toluidine blue in 80% ethanol is used. After staining for 15 min, the slides are destained with 1% acetic acid in 50% ethanol. Usually three washes of 2 to 3 minutes each sufficient for the removal of the background.

COMMENTS

The method is most useful for the characterization of SMPs. It has a number of advantages over cellulose acetate electrophoresis. Thus it can be used for preparative purposes by scaling up the load of AMP and the thickness of the agarose blocks (Section V.1.C). It accepts heavily contaminated mixtures without change of the pattern of fractionation. It can be used for purification purposes (33; Section V.1.C). As an added advantage, because of the load capacity of agarose gel, small amounts of individual SMPs can be identified when in a mixture. For instance, urinary SMPs show the presence of four components in agarose gel, one of them as 2% of the total (35). The same SMPs on cellulose acetate show only the presence of the two main components (106). Compared to cellulose acetate electrophoresis, the method has a disadvantage of a longer development time. A complete operation takes at least 8 hr, whereas with the latter method, results can be obtained in 3 hr of operation. However the time factor is compensated for by leaving the slides overnight for fixation. The agarose method does not distinguish completely heparitin A from heparitin B.

5. Polyacrylamide Gel Electrophoresis (C. P. Dietrich and H. B. Nader)

PRINCIPLE

SMP is applied to polyacrylamide gel and subjected to electrophoresis. The gel is stained with toluidine blue and the migration measured relative to a reference heparin. Using reference SMP of known mean molecular weights, the relative migration value can be used to determine the molecular weight of the sample. Scale: 1–30 μg (can be used for preparative purposes).

Procedure. The polyacrylamide gel electrophoresis is performed in gel 0.2 cm thick on 5.0 × 7.5 or 10 cm slides. The gel is prepared as follows: 1.2 g of acrylamide (BDH or Eastman) containing 30 mg of bisacrylamide and 14 mg of ammonium persulfate are added to 20 ml of 0.06*M* barbiturate buffer, pH 8.6, held at 5 °C. The solution is subjected to vacuum for 5 min at 5 °C, and 25 μl of TEMED (N,N,N^2,N^2-tetramethylenediamine) is added. The resulting solution is quickly poured into glass molds (inside dimensions 40 × 70 × 1 mm) according to the procedure described by Vesterberg (175); 16 hr at room temperature is allowed for polymerization before use. Polyacrylamide slabs are stored immersed in buffer at 5 °C. A trace of cresol red is added to the sample, and 1 μl (5 to 10 μg of SMP) is applied on the polyacrylamide gel as a spot or band 3 mm wide. Electrophoresis is conducted at 2 V/cm for 50 min with electrodes in barbital buffer using the chamber described in Section IV.1.D. Migration distance is judged by the position of cresol red. The gel is stained with 0.1% toluidine blue in 1% acetic acid for 30 min and destained with 1% acetic acid with three washes of 30 min each.

For determination of molecular weights by the procedure of Hilborn and Anastassiadis (63). the above-mentioned gel and procedure are satisfactory. Hilborn and Anastassiadis used 6% Cyanogum gels, 18 × 10 × 0.5 cm., with 0.1*M* phosphate buffer, pH 11.5, containing 0.125*M* sodium formate. Gels were water -cooled, and electrophoresis was conducted in the Raymond-Wang apparatus. Distance migrated is determined relative to a reference heparin. Reference SMPs of Matthews et al. (Section II.2) are run. A plot of molecular weight × 10^3 on a log scale versus relative distance from the origin gives a straight line. The relative distance from the origin for the unknown sample is then transformed to the corresponding molecular weight from the regression line.

COMMENTS

The method is very useful for the determination of the molecular sizes and degree of polydispersity of pure SMPs and can be readily adapted for the preparation of the heparitins (33). The polydispersity of SMPs makes the method poor for identification in mixtures of different SMPs. It should only be used in the final steps of purification of the SMPs under special conditions and for the determination of molecular weights.

6. Enzymatic Degradation (C. P. Dietrich, M. E. Silva, and Y. M. Michelacci)

PRINCIPLE

AMPs are incubated with the chondroitinases ABC, AC, B, C, heparinase and heparitinases I and II. The reaction mixtures are analyzed by quantitative agarose gel electrophoresis for the SMPs remaining and subjected to paper chromatography for the characterization of the products formed by one of a number of procedures. Scale: 200 μg–1 mg.

PREPARATION OF THE MUCOPOLYSACCHARIDASES

The chondroitinase AC and chondroitinase ABC are commercially available and can be obtained from Miles Laboratories (Elkhart, Ind.). For the preparation of the other mucopolysaccharidases the following method is used. *Flavobacterium heparinum* (ATCC No. 13,125) is grown in trypticase soy broth without dextrose (27 g/liter, containing 150 mg of heparin (for the induction of the heparinase and the heparitinases) or 150 mg of chondroitin A, B, or C (for the induction of the chondroitinases)). The cells are harvested in the logarithmic phase of growth, washed with 0.1*M* ethylenediamine acetate buffer (pH 7.0), resuspended in 10 ml of 0.1*M* ethylenediamine acetate buffer (pH 7.0), and treated ultrasonically for 5 min in a Bronwill Sonicator at 3 °C. The suspension is then centrifuged at 10,000*g* for 30 min. The precipitate formed is washed once with 10 ml of buffer. The supernatants are combined, and 9 volumes of acetone is added at − 10 °C. The precipitate formed is collected by centrifugation at − 10 °C and freeze dried. This precipitate (100 mg) is resuspended in 1 ml of the same buffer, and the insoluble matter is removed by centrifugation. The resulting protein solution is applied as a band in an agarose gel block (1 cm thick, 20 × 30 cm) prepared in 0.1*M* ethylenediamine acetate, pH 7.0, in 0.9% agarose (for the fractionation of the heparinases) and pH 8.0 (for the fractionation of the chondroitinases). The agarose block is then submitted to electrophoresis (1.5 V/cm for 60 hr). After the run, the agarose block is cut into strips 0.5 cm wide and frozen. Upon thawing, the slices, which have lost their gel structure, are squeezed through a cheesecloth prewashed with distilled water. Egg albumin is added to the eluates (150 μg/ml), and these are assayed with chondroitin C and chondroitin B for the location of the chondroitinases AC, C, and B, and heparin and heparitin B for the heparinase and heparitinases. The fractions containing each one of the enzymatic activities are pooled and maintained at 5 °C. The enzymes are stable for more than 2 months at this temperature. (For further details see refs. 37,120,121, and 161.)

Procedure. To a mixture of AMPs (10 to 100 μg of each) 0.05 units (see Comment, 1) of one of the following enzymes is added in 0.1*M* ethylenediamine-acetate buffer (pH 7.0) in a final volume of 20 μl. The temperature, requirements, and time of incubation are given in parentheses: chondroitinase AC (37 °C, 4 hr), chondroitinase ABC (37 °C, 4 hr), chondroitinase B (30 °C, 8 hr), chondroitinase C (25 °C, 8 hr), heparinase (30 °C, 8 hr, 0.02 *M* $MnCl_2$) heparitinase I (37 °C, 8 hr), heparitinase II (37 °C, 8 hr). Crude *F. heparinum* extract (100 μg) induced with heparin may be used for heparinase, heparitinases, and chondroitinases. After incubation, enzyme action is demonstrated by disappearance of SMPs and by identification of products.

After incubation with enzymes, 5-μl aliquots are submitted to quantitative agarose gel electrophoresis (see Section IV.1.D) in barbital and diamine buffers. The disappearance of any of the SMPs by the action of the specific mucopolysaccharidases will indicate the types of SMPs present in the mixture. The lack of degradation by any of the enzymes indicates the presence of keratosulfate. Quantitative agarose gel electrophoresis before and after the specific enzyme makes it possible to report the amount of the specific SMPs present.

For identification of the products, the reaction mixtures are applied to Whatman No.1 paper and submitted to descending chromatography for 48 hr in 5–3 (v/v) mixture of isobutyric acid-1*M* ammonia. The chromatogram is then dried and the products characterized and quantitated by one of the methods given below. The main disaccharide product formed from each SMP by the action of the mucopolysaccharidases is listed in Table X. Table XI gives some properties of the disaccharides that can be used for identification.

1. The disaccharide products formed by the enzymes from the AMPs are visualized with the aid of a short-wave ultraviolet lamp and eluted from the paper; the hexosamine content is determined by a modified Morgan-Elson reaction (143) or uronic acid content by the carbazole reaction (40).
2. The amount of the disaccharides eluted from the paper can also be determined by absorbance at 230 nm at pH 2.0.
3. The paper is stained with silver nitrate reagent or toluidine blue, and the stained disaccharides are quantitated by densitometry.
4. When crude extracts are used, glucosamine 2,6-disulfate is formed from heparin and heparitins (*N*-acetylglucosamine from heparitin A). This compound is located in the chromatograms with the silver nitrate reagent.
5. Radioactive products obtained from enzymatic or chemical degradation of radioactive-labeled SMPs are chromatogrammed in isobutyric acid-1*M* NH_4OH (5–3, v/v) in Whatman No. 1 paper, or subjected to electrophoresis

TABLE X

Enzyme Disaccharide Products[a] from the Action of Mucopolysaccharidases

AMP	Products[b] Ch-ase AC	Ch-ase ABC	Ch-ase B	Ch-ase C	Heparinase	Heparitinase I	Heparitinase II
Chondroitin							
A	Δ Di-4S	Δ Di-4S	—	—	—	—	—
B	—	Δ Di-4S	Δ Di-4S	—	—	—	—
C	Δ Di-6S	Δ Di-6S	—	Δ Di-6S	—	—	—
Heparitin							
A	—	—	—	—	—	Δ Di-GNA (α)[c]	—
B	—	—	—	—	—	Δ Di-GNA (α)	Δ Di-DiS
C	—	—	—	—	Δ Di-TriS	—	—
D	—	—	—	—	Δ Di-TriS	—	—
Heparin	—	—	—	—	Δ Di-TriS	—	—
Keratosulfate	—	—	—	—	—	—	—
Hyaluronic acid	Δ Di-GNAc (β)[c]	Δ Di-GNAc (β)	—	Δ Di-GNAc (β)	—	—	—

[a] Only the main disaccharide product is given.
[b] Δ = unsaturated (contains double bond); Di = disaccharide; 4S, 6S = sulfate on C-4, C-6 of the hexosamine; GNAc = acetylglucosamine; DiS = disulfated; TriS = trisulfated.
[c] The disaccharides formed from hyaluronic acid have β^{1-3} linkages, whereas the disaccharides formed from heparitins have α^{1-4} linkages.

TABLE XI

Properties of Disaccharides Produced by Mucopolysaccharidases

Disaccharide[a]	Chromatographic[b] mobility (Di-4S as reference)	ε, 230 nm	Molar proportions: Hexosamine	Sulfate	Uronic acid
Δ Di-4S	1	5.1×10^3	1	1	1
Δ Di-6S	0.63	5.5×10^3	1	1	1
Δ Di-GNAc (α)	1.30	—	1	0	1
Δ Di-GNAc (β)	1.85	—	1	0	1
Δ Di-DiS	0.25	3.32×10^3	1	2	1
Δ Di-TriS	0.18	3.45×10^3	1	3	1

[a] Δ = unsaturated; Di = disaccharide; 4S, 6S = sulfate on C-4, C-6 of the hexosamine; GNAc = acetylglucosamine; DiS = disulfate; TriS = trisulfated.
[b] Isobutyric acid, 1*M* NH_3, 5:3, v/v.

in 0.3*M* pyridine acetate buffer, pH 4.2, in Whatman 3 MM paper. The radioactive materials are localized by radioautography with Kodak Royal Blue X-ray film (Eastman Kodak Co.) The areas of the chromatograms containing the radioactive materials are then cut out and counted in 10 ml of a solution of 5 g of diphenylozazole in 1 liter of toluene in a liquid scintillation spectrometer.

COMMENTS

1. A unit of enzyme is defined here as the amount of enzyme able to degrade 100 μg SMPs/min.

2. The method has the unique advantage of being able to characterize completely most of the SMPs. For a single SMP it will give the relative proportion of each one of the disaccharides and oligosaccharides present in the polymer chain. For a mixture of SMPs, it allows quantitative determination of each SMP or type of SMP present using the corresponding enzyme. It is not generally influenced by differences of molecular size or the net charge of the compounds. This is important for structural studies of the chondroitins and heparitins of different origins. Nevertheless it is the most time-consuming of the methods described, and a further disadvantage at present is that most of the enzymes required are not commercially available.

7. Isoelectrofocusing (H. B. Nader, C. P. Dietrich, and N. M. McDuffie)

PRINCIPLE

SMP is subjected to electrophoresis in acrylamide gel containing LKB ampholytes, pH 3 to 5. Heparin forms 21 bands corresponding to the chain lengths of the individual heparins in the sample. Scale: 20–100 μg.

Procedure. Polyacrylamide gel slabs (0.2 cm thick, 5 × 15 cm) are prepared essentially as described in Section VI.5 except that the buffer is replaced by 2% ampholytes, pH 3 to 5, from LKB, Sweden. Next 50 to 100 μg (2 μl) of sample is applied to a 2-mm strip of 3 MM filter paper, of the length required for the volume of sample. This paper is placed on the gel at 2 cm from the negative end. The electrodes are placed directly at the two ends of the gel, previously wrapped in Whatman 3 MM paper wetted with 1*M* phosphoric acid (positive electrode) and 1*M* NaOH (negative electrode). A potential of 1.5 V/cm is applied for 12 hr. The gels are stained as described in Section VI.5.

COMMENTS

Heparin is the only substance that gives a large series of bands (112). Heparitin D, which by the enzyme tests has the heparin structure but is of low molecular weight, gives bands in the low molecular weight region.

Heparitins A and B give a faint smear between pH 3.5 and 4.2. Chondroitins A, B, and C give multiple bands at very acid regions. Chondroitin polysulfates show some banding. Purification of heparin is required before subjecting it to electrofocusing. LKB ampholytes are necessary. The Biorad ampholytes are polysulfated, not polycarboxylated, and fail to give patterns with heparin.

The method is very useful to determine the number of molecular species of heparin from different sources and is potentially useful for preparation of these fractions for biological studies. Of the SMPs tested, only heparin gives a characteristic pattern not superimposed by any other SMPs.

VII. GENERAL MICRO SYSTEM FOR DETERMINATION OF HEPARIN AND SMPS IN TISSUE SAMPLES

This section summarizes procedures for study of heparin and SMPs that are related in having a similar action or function—namely, to provide in soft tissues a polyanion that acts like a drug by influencing enzymes such as those of the blood coagulation system and lipoprotein lipase and by conferring a negative charge on the endothelium and other tissues. This polyanion, occurring naturally in tissues, regulates biologically active trace substances through ion exchange. This function is closely allied with mast cells, although not limited to these cells. As pointed out by Padawer (131) and Jaques (76), this function is complementary to that of histamine/5-HT as biological cation. Large amounts of polyanion with a gross ion-exchange function are provided by hyaluronic acid in soft tissues and by chondroitins in bone and cartilage. In the procedures described based on toluidine blue staining, hyaluronic acid essentially is not measured, although the viscosity it imparts to tissue extract can be troublesome. The limitation to soft tissues excludes the chondroitins of bone and cartilage. Hence although investigators using the carbazole and hexosamine color reactions deal with AMPs (or glycosaminoglycans), the procedures reviewed provide methods for the SMPs as related to heparin function.

Chondroitins, heparitins, and heparins are present in soft tissues in microgram amounts. Heparitins are thought to be associated with cell membrane function. Chondroitins, which show some sharing of functions with both heparitins and heparins, are found in cell membranes. Dog mastocytoma cells contain both heparins and chondroitins. Variations in distribution between these components of mast cells may be biologically significant. T. K. Sue has pointed out that in the chicken and the rabbit, the two species that develop atherosclerosis readily with diet, there are few mast cells, and he has demonstrated that the mast cells of chickens and the basophilic leucocytes of rabbits (the significant basophilic cells of this species) contain chondroitins, not heparins. Absolute differences in mast cells of

TABLE XII

Procedure for Study of Heparin, Related SMPs, and Mast Cells in Small Tissue Samples

Tissue (3–5 mm dimensions) for histology Alcohol or Susa fixation, paraffin, sections Stains: H.&E., Polychrome, Toluidine Blue Microscopic examination: H&E, polychrome, for pathology Toluidine blue: count 20 fields for mast cells	Section of Chapter
0.3–4.5 g for chemical determination. Weigh, homogenize, defat, and dry with acetone, isopropanol - petroleum ether; weigh.	V.1C
Digest by pronase with dialysis or by trypsin.	V.1C
(remove nucleic acid with cold 15% trichloracetic acid or DNAse + RNAse)	V.1.C
Precipitate with acetone-EtOH-1% NaCl; dry; weigh.	V.1.C
Microelectrophoresis with heparin reference in agarose-barbital = total SMP in metachromatic units.	IV.1.D
Microelectrophoresis in agarose with barbital, propylene-diamine, acid buffers to identify SMP = units for each individual SMP. Convert to percentage metachromasia for individual SMP in mixture.	IV.1.D
(For confirmation, use Chondroitinases AC and ABC, heparinase, heparitinases I and II; for additional characterization, test CPC fixation with ↑ molarity of NaCl to determine critical electrolyte concentration.)	IV.1.D
Convert units to micrograms for each SMP (= μg of each SMP/g tissue.)	
Fractionate and separate individual SMP	IV.1.A.,V.I.1–5
Electrofocusing (number of heparin components; % distribution)	VI.7
Enzyme degradation (carbohydrate units)	VI.6
Acrylamide gel (representative value, molecular weights, range)	VI.5

different tissues of rats have also been demonstrated by Csaba and Baráth (21) for synthesis, uptake, and release of histamine/5-HT. Furthermore the chondroitin polysulfates (found, e.g., in aorta) exhibit activities similar to those of heparin with respect to enzymes. The anticoagulant activity and the synergistic action of chondroitin B and heparitins with heparin antithrombin was presented in Section IV.2.A.

By suitably combining the principles and procedures described in Sections III to VI, it is possible to identify and measure any combination of SMPs. An effective demonstration of this has been given by Izuka and Murata (69) in their identification of a trace amount of heparin in the chondroitin polysulfate fraction of AMP of human aorta. A scheme for the application of the principles and techniques previously described to the determination of heparin and SMPS in tissue samples appears in Table XII. First 1 to 5 g of tissue is taken, 0.1 to 0.4 g is removed for histological examination, and the remainder is homogenized, defatted, dried, deproteinized, and precipitated with acetone or alcohol. The resulting dry powder (about 1 mg per gram of tissue) is examined by microelectrophoresis for total SMPs and for differentiation and characterization of the individual SMPs present. Values for fractions are determined in units, then converted to micrograms with appropriate conversion factors. This procedure is satisfactory for the SMPs that constitute more than 10% of the total SMPs and for tissues that contain heparins or chondroitins or heparitins in moderate amounts. With tissues rich in heparin, minute samples can be lyophilized, homogenized with water, and applied directly for microelectrophoresis. In experiments on release of heparin by tissues in organ baths, the procedure is much simplified by replacing the Ringer's solution by an isotonic solution of ammonium carbonate at pH 7.8 (79). Lyophilizing the bath fluid serves to decompose the ammonium carbonate, making the appearance or disappearance of a few micrograms of heparin in the bath fluid easily detectable. Radioisotope labeling offers a further alternative for specific identification of administered drug (Sections V.3, VI.6). When only commercial heparin is to be examined, a more specific extraction procedure (e.g., with 2*M* potassium acetate: Section V.1.A) may be used, further modified, if desired, by extracting a portion of the fresh tissue (without autolysis). The resulting product will presumably be the "free heparin" portion of commercial heparin in tissue.

For a more complete characterization, the extracted SMPs must be separated from one another for further analyses. Usually the crude SMPs from a series of samples are combined for this. Alternatively, the material is prepared on a larger scale. The agarose method with two buffer systems can be used for semipreparative purposes (up to 200 mg of SMPs) as discussed in Section V.1.C. Once the SMPs are separated from one another they can be analyzed by the polyacrylamide method (Section VI.5) for molecular weight, digested with the more specific mucopolysaccharidases to ascertain the type and amount of disaccharides present, and subjected to chemical analyses. It is important to emphasize that these methods will give reasonable information about the nature of the mammalian SMPs. They can also be used as a starting point for the analyses of SMPs excreted in a variety of mucopolysaccharidoses and also of the SMPs present in lower forms. Nevertheless the latter compounds do not always follow the same rules. Chemical and physical differences are being found continuously (125).

The parameter used in the microelectrophoresis procedure is based on separation from other materials by electrophoresis, fixation with an alkylamine, and staining with toluidine blue. Apart from the first step, the parameter is very similar to that used in histology for mucopolysaccharides. The differences in metachromasia for different mucopolysaccharides in microelectrophoresis are similar to the differences in metachromasia seen in histology. The biochemical examination on the submicro level provides molecular biology data to explain these differences and is already supplying new evidence of the pharmacological action and biological significance of heparin and related SMPs.

References

1. H. Alexander, *Can. J. Med. Technol.*, *36*, 301 (1974).
2. C. A. Antonopoulos, S. Gardell, J. A. Szirnai, and E. R. De Tyssonsk, *Biochim. Biophys. Acta*, *83*, 1 (1964).
3. D. R. Bangham and P. M. Woodward, *Bull. WHO*, *42*, 129 (1970).
4. O. Bank and H. G. Bungenberg de Jong, *Protoplasma*, *32*, 489 (1939).
5. G. H. Barlow, *Biochim. Biophys. Acta*, *83*, 120 (1964).
6. G. H. Barlow and E. V. Cardinal, *Proc. Soc. Exp. Biol. Med.*, *123*, 831 (1966).
7. G. H. Barlow, N. D. Sanderson, and P. D. McNeill, *Arch. Biochem. Biophys.*, *94*, 518 (1961).
8. E. D. Becker, *High Resolution NMR*, Academic Press, New York, 1969.
9. T. Bitter and H. M. Muir, *Anal. Biochem.*, *4*, 330 (1962).
10. F. A. Bovey, *High Resolution NMR of Macromolecules*, Academic Press, New York, 1972.
11. E. J. W. Bowie, J. H. Thompson, P. Didisheim, and C. A. Owen, Jr., *Mayo Clinic Laboratory Manual of Hemostasis*, Saunders, Philadelphia, London, Toronto, 1971.
12. R. A. Bradshaw and S. Wessler, *Heparin: Structure, Function and Clinical Implications*, Plenum Press, New York, 1975.
13. H. G. Bungenberg de Jong in *Colloid Science*, Vol. 2, H. R. Kruyt, Ed., Elsevier, Amsterdam, 1949, p. 276.
14. B. Casu, G. Gatti, N. Cyr, and A. S. Perlin, *Carbohydr. Res.*, *41*, C6 (1975).
15. B. Casu and U. Gennaro, *Carbohydr. Res.*, *39*, 168 (1975).
16. A. F. Charles, Ph.D. thesis, University of Toronto, 1932.
17. A. F. Charles, private communication, 1935.
18. A. F. Charles and D. A. Scott, *J. Biol. Chem.*, *102*, 425 (1933).
19. A. F. Charles and D. A. Scott, *J. Biol. Chem.*, *102*, 431 (1933).
20. B. Coxon, *Adv. Carbohydr. Chem.*, *27*, 7 (1972).

21. G. Csaba and P. Baráth, *Acta Biol. Acad. Sci. Hung.*, *25*, 323 (1974).
22. P. De Nicola, *The Laboratory Diagnosis of Coagulation Defects*, Thomas, Springfield, Ill., 1956.
23. E. Deutsch, *Blutgerinnungsfaktoren*, Franz Dueticke, Vienna, 1955.
24. C. P. Dietrich, *Biochemistry*, *8*, 2089 (1969).
25. C. P. Dietrich, *Biochemistry*, *8*, 3342 (1969).
26. C. P. Dietrich, *Anal. Biochem.*, *51*, 345 (1973).
27. O. W. Cruz and C. P. Dietrich, *Proc. Soc. Exp. Biol. Med.*, *126*, 420 (1967).
28. C. P. Dietrich and S. M. C. Dietrich, *Anal. Biochem.*, *46*, 209 (1972).
29. C. P. Dietrich and S. M. C. Dietrich, *Cienc. Cult.*, *26*, 55 (1974).
30. C. P. Dietrich and S. M. C. Dietrich, *Anal. Biochem.*, *70*, 645 (1976).
31. C. P. Dietrich, H. B. Nader, and N. M. McDuffie, *An. Acad. brasil Cienc.*, *47*, *30* (1975).
32. C. P. Dietrich, N. M. McDuffie, and L. O. Sampaio, *J. Chromatogr.*, *30*, 299 (1977).
33. C. P. Dietrich and H. B. Nader, *Biochim. Biophys. Acta*, *343*, 34 (1974).
34. C. P. Dietrich, H. B. Nader, L. R. G. Britto, and M. E. Silva, *Biochim. Biophys. Acta*, *237*, 430 (1971).
35. C. P. Dietrich, H. B. Nader, and P. A. S. Mourão, *Biochem. Med.*, *8*, 371 (1973).
36. C. P. Dietrich, H. B. Nader, and A. S. Perlin, *Carbohydr. Res.*, *41*, 334 (1975).
37. C. P. Dietrich, M. E. Silva, and Y. M. Michelacci, *J. Biol. Chem.*, *248*, 6408 (1973).
38. C. P. Dietrich, M. E. Silva, and Y. M. Michelacci, (1975).
39. C. P. Dietrich, M. E. Silva, and Y. M. Michelacci, (1976).
40. Z. Dische, in *Methods of Biochemical Analysis*, Vol. 2, D. Glick, Ed., Wiley-Interscience, New York, 1955, p. 313.
41. L. I. Diuguid and N. C. Johnson, *Microchem. J.*, *12*, 371 (1967).
42. K. S. Dodgson and B. Spencer, in *Methods of Biochem. Analysis*, Vol. 4, D. Glick, Ed., Wiley-Interscience, New York, 1957, p. 211.
43. J. Ehrlich and S. S. Stivala, *J. Pharm. Sci.*, *62*, 517 (1973).
44. P. Ehrlich, *Arch. Mikrosk. Anat.*, *13*, 263 (1877).
45. E. C. Fieller, *J. Roy. Stat. Soc. Suppl.* *7*, 1 (1940).
46. D. J. Finney, *Statistical Method in Biological Assay*, 2nd ed., Charles Griffin, London, 1964.
47. A. Fischer, *Biochem. Z.*, *278*, 133 (1935).
48. A. Fischer and A. Schmitz, *Z. physiol. Chem.*, *210*, 129 (1932).
49. L. A. Franson, A. Anseth, C. A. Antonopoulos, and S. Gardell, *Carbohydr. Res.*, *15*, 73 (1970).
50. J. T. Galambos, *Anal. Biochem.*, *19*, 119 (1967).
51. O. A. Gansow and W. Shittenhelm, *J. A. Chem. Soc.*, *93*, 4294 (1971).
52. S. Gardell, in *Methods of Biochemical Analysis*, Vol. 6, D. Glick, Ed., Wiley-Interscience, New York, 1958, p. 289.
53. A. Hallem, *J. Chromatogr.*, *71*, 83 (1972).
54. G. K. Hamer and A. S. Perlin, *Carbohydr. Res.*, in press.
55. G. K. Hamer and A. S. Perlin, unpublished.
56. R. M. Hardaway, *Syndromes of Disseminated Intravascular Coagulation*, Thomas, Springfield, Ill., 1966.
57. R. Hata and Y. Nagai, *Anal. Biochem.*, *45*, 462 (1972).
58. R. Hata and Y. Nagai, *Anal. Biochem.*, *52*, 652 (1973).
59. J. R. Helbert and M. A. Marini, *Biochemistry*, *2*, 1101 (1963).
60. B. Hermelin, E. Deudon, A. Bruel-Groleas and J. Picard, *J. Chromatogr.*, *110*, 43 (1975).
61. L. M. Hiebert and L. B. Jaques, *Artery*, *2*, 26 (1976).
62. L. M. Hiebert and L. B. Jaques, *Thromb. Res.* *8*, 195 (1976).
63. J. C. Hilborn and P. A. Anastassiadis, *Anal. Biochem.*, *39*, 88 (1971).
64. A. A. Horner, *Can. J. Biochem.*, *45*, 1009 (1967).
65. A. A. Horner, *J. Biol. Chem.*, *246*, 231 (1971).

66. C. Hougie, *Fundamentals of Blood Coagulation in Clinical Medicine*, McGraw-Hill, New York. 1963.
67. P. Hovingh and A. Linker, *Carbohydr. Res.*, *37*, 181 (1974).
68. D. Hsu, P. Hoffman, and T. A. Mashburn, Jr., *Anal. Biochem.*, *52*, 382 (1973).
69. K. Izuka and K. Murata, *Experientia*, *29*, 655 (1973).
70. L. M. Jackman and S. Sternhell, *Applications of Nuclear Magnetic Resonance Spectroscopy in Organic Chemistry*, 2nd ed., Pergamon Press, Oxford, 1969.
71. L. B. Jaques, M.Sc. thesis, University of Toronto, Toronto, Canada, 1935.
72. L. B. Jaques, *Biochem. J.*, *37*, 189 (1943).
73. L. B. Jaques, *Progr. Med. Chem.*, *5*, 139 (1967).
74. L. B. Jaques, *Can. Med. Assoc. J.*, *108*, 1291 (1973).
75. L. B. Jaques, *Lancet*, *1*, 287 (1975).
76. L. B. Jaques, *Gen. Pharmacol.*, *6*, 235 (1975).
77. L. B. Jaques, R. E. Ballieux, C. P. Dietrich, and L. W. Kavanagh, *Can. J. Physiol. Pharmacol.*, *46*, 351 (1968).
78. L. B. Jaques, R. E. Ballieux, and C. van Arkel, *Acta Med. Scand.*, *174*, 717 (1963).
79. L. B. Jaques and H. J. Bell, in *Methods of Biochemical Analysis*, Vol. 7, D. Glick, Ed., Wiley-Interscience, New York, 1959, p. 253.
80. L. B. Jaques, M. Bruce-Mitford, and A. G. Ricker, *Rev. Can. Biol.*, *6*, 740 (1947).
81. L. B. Jaques and A. F. Charles, *J. Pharm. Pharmacol.*, *14*, 1 (1941).
82. L. B. Jaques, E. Fidlar, E. T. Felsted, and A. G. Macdonald, *Can. Med. Assoc. J.*, *55*, 26 (1946).
83. L. B. Jaques, L. W. Kavanagh, and S. H. Kuo, *Thromb. Res.*, *3*, 295 (1973).
84. L. B. Jaques, L. W. Kavanagh, and A. Lavallée, *Arzneim. forsch.*, *17*, 774 (1967).
85. L. B. Jaques, L. W. Kavanagh, M. Mazurek, and A. S. Perlin, *Biochem. Biophys. Res. Commun.*, *24*, 447 (1966).
86. L. B. Jaques, F. C. Monkhouse, and M. Stewart, *J. Physiol.* (London), *109*, 41 (1949).
87. L. B. Jaques, E. Napke, and S. W. Levy, *Circ. Res.*, *1*, 321 (1953).
88. L. B. Jaques and A. G. Ricker, *Blood*, *3*, 1197 (1948).
89. L. B. Jaques and E. T. Waters, *J. Physiol.*, *99*, 454 (1941).
90. L. B. Jaques and A. Wollin, *Can. J. Physiol. Pharmacol.*, *45*, 787 (1967).
91. L. B. Jaques and A. Wollin, *Anal. Biochem.*, *52*, 219 (1973).
92. J. E. Jorpes and S. Gardell, *J. Biol. Chem.*, *176*, 267 (1948).
93. J. Jurgens and F. K. Beller, *Klinischen Methoden der Blutgerinnungsanalyse*, Thieme, Stuttgart, 1959.
94. L. W. Kavanagh and L. B. Jaques, *Arzneim.-forsch.*, *23*, 605 (1973).
95. L. W. Kavanagh and L. B. Jaques, *Arzneim.-forsch.*, *24*, 1943 (1974).
96. Y. Kobayashi, J. Kupelian, and D. V. Mandsley, *Biochem. Pharmacol.*, *18*, 1585 (1969).
97. T. Y. Koh and K. R. Barucha, U.S. Patent 3,577,534, Continuation-in-part of U.S. Patent 3,506,642, 1971.
98. T. Y. Koh and K. R. Barucha, *J. Lab. Clin. Med.*, *80*, 47 (1972).
99. E. D. Korn, in *Methods of Biochemical Analysis*, Vol. 7, D. Glick, Ed., Wiley-Interscience, New York, 1959, p. 145.
100. K. E. Kuettner and A. Lindenbaum, *Science*, *144*, 1228 (1964).
101. Shu-hui Kuo, Ph.D. thesis, University of Saskatchewan, Saskatoon, Canada, 1971.
102. Shu-hui Kuo, L. B. Jaques, and G. J. Millar, *J. Pharm. Pharmacol.*, *24*, 858 (1972).
103. S. E. Lasker and S. S. Stivala, *Arch. Biochem. Biophys.*, *115*, 360 (1966).
104. R. K. Lemperg, S. E. Larsson, and S. O. Hjertquist, *Calcif. Tissue Res.*, *15*, 237 (1974).
105. G. C. Levy and G. S. Nelson, *Carbon-13 Nuclear Magnetic Resonance for Organic Chemists*, Wiley-Interscience, New York, 1972.
106. P. W. Lewis, J. K. Kennedy, and D. N. Raine, *Biochem. Soc. Trans.*, *1*, 844 (1973).

107. P. A. Liberti and S. S. Stivala, *Arch. Biochem. Biophys.*, *119*, 510 (1967).
108. R. D. Lillie, *H.J. Conn's Biological Stains*, 8th ed., Williams & Wilkins, Baltimore, 1969.
109. A. Linker and P. Hovingh, *J. Biol. Chem.*, *240*, 3724 (1965).
110. L. Lison, *Arch. Biol.*, *46*, 599 (1936).
111. D. F. Lloyd, G. P. Roberts, and K. O. Lloyd, *Biochem. J.*, *75*, 14 (1960).
112. N. M. McDuffie, C. P. Dietrich, and H. B. Nader, *Biopolymers*, *14*, 1473 (1975).
113. R. Marbet and A. Winterstein, *Helv. Chim. Acta*, *7*, 2311 (1951).
114. W. Marx, L. Marx, P. Rucker, L. Ruggen, and L. Freeman, *Proc. Soc. Exp. Biol. Med.*, *94*, 217 (1957).
115. M. B. Matthews, J. A. Cifonelli, and L. Rodén, *Acid Mucopolysaccharide Reference Standards*, Department of Pediatrics, the University of Chicago, Chicago (n.d.).
116. P. Meneghini and C. Cervini, *Boll. Soc. Ital. Biol. Sper.*, *25*, 1 (1949).
117. K. Meyer and E. Chaffe, *J. Biol. Chem.*, *138*, 491 (1941).
118. Y. M. Michelacci and C. P. Dietrich, *Biochimie*, *55*, 893 (1973).
119. Y. M. Michelacci and C. P. Dietrich, *Biochem. Biophys. Res. Commun.*, 973 (1974).
120. Y. M. Michelacci and C. P. Dietrich, *Biochem. J.*, *151*, 121 (1975).
121. Y. M. Michelacci and C. P. Dietrich, *J. Biol. Chem.*, *251*, 1154 (1976).
122. Y. M. Michelacci and C. P. Dietrich, *Biochem. J.*, *151*, 121 (1976).
123. S. Moritani and G. Ohta, *Japan. Circ. J.*, *37*, 619 (1973).
124. P. A. S. Mourão, S. Rozenfeld, J. Laredo, and C. P. Dietrich, *Biochim. Biophys. Acta*, *428*, 19 (1976).
125. P. A. S. Mourão, S. P. A. Toledo, H. B. Nader, and C. P. Dietrich, *Biochem. Med.*, *7*, 415 (1973).
126. H. B. Nader, N. M. McDuffie, and C. P. Dietrich, *Biochem. Biophys. Res. Commun.*, *57*, 488 (1974).
127. H. B. Nader, P. A. S. Mourão, S. P. A. Toledo, and C. P. Dietrich, *Clin. Chim. Acta*, *50*, 245 (1974).
128. K. J. Nagasawa, *J. Assoc. Of. Anal. Chem.*, *51*, 330 (1968).
129. N. Nath, S. Niewarowski and J. H. Joist, *J. Lab. Clin. Med.*, *82*, 754 (1973).
130. I. Okuno and C. A. Nelson, *J. Clin. Pathol*, *28*, 494 (1975).
131. J. Padawer, *Am. J. Anat.*, *1411*, 299 (1974).
132. A. S. Perlin, *Proceedings of an International Symposium on Macromolecules, Rio de Janeiro, 1974*, Elsevier, Amsterdam, 1975, p. 337.
133. A. S. Perlin, *Int. Rev. Sci.*, *Org. Chem.* (Ser. 2), *7*, 1 (1976).
134. A. S. Perlin, B. Casu, G. R. Sanderson, and L. F. Johnson, *Can. J. Chem.*, *48*, 2260 (1970).
135. A. S. Perlin, D. M. Mackie, and C. P. Dietrich, *Carbohydr. Res.*, *18*, 185 (1971).
136. A. S. Perlin, M. Mazurek, L. B. Jaques, and L. W. Kavanagh, *Carbohydr. Res.*, *7*, 369 (1968).
137. J. P. Peters and D. D. Van Slyke, *Quantitative Clinical Chemistry*, Vol. 2, *Methods*, Williams & Wilkins, Baltimore, 1932; Reprinted, 1963.
138. R. S. Prout, *Biochim. Biophys. Acta*, *177*, 157 (1969).
139. A. J. Quick, *Am. J. Physiol.*, *115*, 317 (1936).
140. W. Regelson, *Adv. Chemother.*, *3*, 303 (1968).
141. N. R. Ringertz and P. Reichard, *Acta Chem. Scand.*, *14*, 303 (1960).
142. W. V. B. Robertson, M. W. Ropes, and W. J. Bauer, *J. Biol. Chem.*, *133*, 261 (1940).
143. C. J. M. Rondle and W. T. J. Morgan, *Biochem. J.*, *61*, 586 (1955).
144. A. B. Roy and P. A. Trudinger, *The Biochemistry of Inorganic Compounds of Sulphur*, Cambridge University Press, London, 1970.
145. H. Saito, T. Yamagata, and S. Suzuki, *J. Biol. Chem.*, *243*, 1536 (1968).
146. L. O. Sampaio, S. P. A. Toledo, and C. P. Dietrich, *Cienc. Cult.*, *28*, 489 (1976).
147. J. Schaefer, in *Topics in Carbon-13 NMR Spectroscopy*, Vol. 1, G. C. Levy, Ed., Wiley-Interscience, New York, 1974, p. 150.

148. S. Schiller, G. A. Slover, and A. Dorfman, *J. Biol. Chem.*, *236*, 983 (1961).
149. M. Schmidt, *Biochim. Biophys. Acta*, *63*, 120 (1962).
150. A. Schmitz and A. Fischer, *Z. physiol. Chem.*, *216*, 261 (1933).
151. J. A. Schwarz and A. S. Perlin, *Can. J. Chem.*, *50*, 3667 (1972).
152. D. A. Scott and A. F. Charles, *J. Biol. Chem.*, *102*, 437 (1933).
153. J. E. Scott, *Biochim. Biophys. Acta*, *18*, 428 (1955).
154. J. E. Scott, in *Methods of Biochemical Analysis*, Vol. 8, D. Glick, Ed., Wiley-Interscience, New York, 1960, p. 145.
155. J. E. Scott, U.K. Patent 122784, U.K. Patent Office, London, 1971.
156. J. E. Scott, *Biochem. Soc. Trans.*, *1*, 787 (1973).
157. J. E. Scott, S. Gardell, and I. M. Nilsson, *Biochem. Soc. Proc.*, *67*, 7 (1957).
158. N. Seno, K. Anno, K. Kondo, S. Nagase, and S. Saito, *Anal. Biochem.*, *37*, 197 (1970).
159. J. E. Siebert, *J. Biol. Chem.*, *239*, 1310 (1964).
160. M. E. Silva and C. P. Dietrich, *Biochem. Biophys. Res. Commun.*, *56*, 965 (1974).
161. M. E. Silva and C. P. Dietrich, *J. Biol. Chem.*, *250*, 6841 (1975).
162. R. Simoneau and A. S. Perlin, unpublished.
163. S. S. Stivala, M. Herbst, O. Kratky, and I. Pilz, *Arch. Biochem. Biophys.*, *127*, 795 (1968).
164. T. K. Sue and L. B. Jaques, *Can. J. Physiol. Pharmacol.*, *51*, 994 (1973).
165. T. K. Sue and L. B. Jaques, *Proc. Soc. Exp. Biol. Med.*, *146*, 1006 (1974).
166. T. K. Sue and L. B. Jaques, *Toxicol. Appl. Pharmacol.*, *34*, 521 (1975).
167. S. Suzuki, H. Saito, T. Yamagata, K. Anno, N. Seno, Y. Kawai, and T. Furuhashi, *J. Biol. Chem.*, *243*, 1543 (1968).
168. A. N. Teien, M. Lie, and A. Ulrich, *Thromb. Res.*, *8*, 413 (1976).
169. A. R. Thompson and R. B. Counts, *J. Lab. Clin. Med.*, *88*, 922 (1976).
170. L. M. Tocantins and L. A. Kazal, *Blood Coagulation, Hemorrhage and Thrombosis, Methods of Study*, Grune & Stratton, New York, London, 1955.
171. G. Tortolani, S. Nobilio, P. G. Ferri, and E. Romagnoli, *Farmaco* (*Ed. Pr.*), *30*, *641*, (1975).
172. G. Tortolani and E. Romagnoli, *Anal. Biochem.*, *66*, 29 (1975).
173. M. V. Tracey, *Biochem. J.*, *43*, 185 (1948).
174. L. Varga, F. Antoni, M. Skalka, and I. Zedginidze, *Int. J. Appl. Radiat, Isot. 22*, 165 (1971)
175. O. Vesterberg, *Biochim. Biophys. Acta*, *257*, 11 (1972).
176. P. L. Walton, C. R. Ricketts, and D. R. Bangham, *Brit. J. Haematol.*, *12*, 310 (1966).
177. E. Wessler, *Anal. Biochem.*, *26*, 439 (1968).
178. E. Wessler, *Biochem. J.*, *122*, 373 (1971).
179. A. Wollin, *At. Absorpt. Newsl.*, *9*, 43 (1970).
180. A. Wollin and L. B. Jaques, *Can. J. Physiol, Pharmacol.*, *50*, 65 (1972).
181. A. Wollin and L. B. Jaques, *Thromb. Res.*, *2*, 377 (1973).
182. T. Yamagata, H. Saito, O. Habuchi, and S. Suzuki, *J. Biol. Chem.*, *243*, 1523 (1968).
183. E. T. Yin, S. Wessler, and J. V. Butler, *J. Lab. Clin. Med.*, *81*, 298 (1973).
184. Z. Yosizawa, T. Kotoku, P. Yamauchi, and M. Matsuno, *Biochim. Biophys. Acta*, *141*, 358 (1967).
185. A. L. Chen, E. J. Hershgold and D. E. Wilson: *J. Lab. & Clin. Med. 85*, 843–854 (1975).
186. A. S. Perlin and G. R. Sanderson. *Carbohyd. Res.*, *12*, 183 (1970).

Field Desorption Mass Spectrometry and Its Application in Biochemical Analysis

HANS-ROLF SCHULTEN, *Institut für Physikalische Chemie, Universität Bonn, Wegelerstr. 12, 5300 Bonn, West-Germany*

Abbreviations

FD Field desorption
FI Field ionization

EI	Electron impact ionization
CI	Chemical ionization
MS	Mass spectrometry, mass spectrum, mass spectra
LR	Low resolution
HR	High resolution
ehc	Emitter heating current
BAT (T^*)	Best anode temperature
Py	Pyrolysis
CA	Collisional activation

During the past 10 years mass spectrometry has been used increasingly in solving analytic problems in biochemistry. When employed in conjunction with other biochemical, chemical, and physical methods, such as nuclear magnetic resonance and ultraviolet and infrared spectroscopy, mass spectrometry has become a powerful, standard method for structural elucidation of compounds of biological origin. The characteristic properties of mass spectrometry are (1) high sensitivity, only microgram or nanogram quantities (or less) are required for analysis, and (2) the large amount of information obtained from a mass spectrum regarding the molecular weight, and the structure and the quantity of biochemicals, drugs, and environmental substances.

One of the most widely employed techniques for MS analysis is electron-impact mass spectrometry (EI-MS). This method uses the collision of a beam of electrons whose average energy exceeds the ionization potential of the molecule to generate ions from the gaseous sample molecules. The extensive experimental data and theoretical considerations of this classical MS Method have been compiled in a number of texts and reference books (e.g., 1–3), periodical reviews (e.g., 4), collections of mass spectral data (5), and computer-stored files (6). It is beyond the scope of this chapter to describe in detail the very well established EI technique which, roughly estimated, is employed in more than 90% of all organic mass spectrometric analyses. Two facts, however, should be stated very clearly. First, the EI method requires evaporation when solid samples are under investigation (thermal excitation). As a result, the EI analysis can be less effective for substances of low volatility that cannot be transferred into the gas phase without thermal decomposition. The EI mass spectra of these nonvolatile substances often contain ions of pyrolytic products and no molecular ion or large structurally diagnostic fragments. This holds in particular for biochemicals, and in general for compounds of biological origin. Second, the energy transferred by the colliding electron to the molecular ion $[M]^{+}$ in the ionization process is of the order of a few electron volts. This electronic excitation may cause extensive fragmentation and very small (or undetectable) molecular ion intensities. In these cases difficulties may arise in determining the molecular weight of a compound. This effect plays a major role in the investigation of the multicomponent mixtures frequently encountered in biochemical analysis.

To overcome these difficulties, modern ionization techniques for thermally and electronically soft ionization have been developed. A survey and comparison of 13 of these methods has recently been given by Milne and Lacey (7). For biochemical analysis, however, three methods stand out in utility and importance. Chemical ionization (CI), especially in combination with gas chromatography (GC), is a powerful tool in mass spectrometric analysis of natural products. Excellent reviews on CI-MS (8) and its structural applications have been published (7).

Of other methods, field ionization (FI) is mentioned only briefly, since the theory, physicochemical problems, and analytical applications of FI-MS have been treated in a fundamental monograph by Beckey (9). Both CI and FI-MS have in common considerably less electronic excitation than with conventional EI ionization. The transferred electronic energy is only of the order of a few tenths electron volts. If evaporation of the sample is possible, or at least a fraction of the sample molecules reaches the gas phase undecomposed, the CI and FI mass spectra are characterized by reduced fragmentation and high molecular ion intensities. However the thermal stress that is required for evaporation cannot be avoided with these methods and might cause complete decomposition of the sample before ionization.

Field desorption (FD) mass spectrometry (10), although only a recent offspring of FI, has been regarded as the method of choice for the analysis of nonvolatile and thermally labile substances. Thus we ask: What are the distinct features of FD-MS, what are the specific problems of the technique, and what is its capacity for the analysis of biochemicals, drugs and drug metabolites, biomacromolecules, and inorganic compounds?

I. PRINCIPLE OF HIGH-FIELD IONIZATION

Atoms or molecules are ionized in high electrostatic fields of the order of 10^7 to 10^8 V/cm, which are produced on tips, thin wires, and sharp edges. FI is an electronic process. The potential well of the most loosely bound electron in a molecule is distorted by the high external field in such way that a certain probability exists for the quantum mechanical tunneling of the electron into

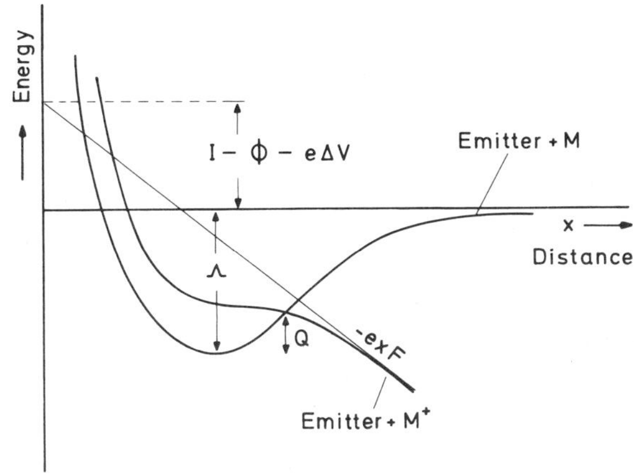

Figure 1. Interaction curves for a neutral molecule M and an ion $[M]^+$ with the surface of an emitter in the presence of the external electric field F (10^7 to 5×10^8 V/cm): Φ is the work function of the organic emitter, I is the ionization potential for the molecule M, ΔV is the potential drop between the electron acceptor level and the surface of the emitter, Λ is the desorption energy of the neutral molecule, and Q is the activation energy for desorption in the ionized state.

the emitter (for details see ref. 9). As a result of the FI process, the ionized molecules (or atoms) acquire little electronic excitation energy (about 0.1eV). The interaction for a neutral molecule M and an ion $[M]^+$ with the surface of an emitter is schematized in Figure 1. The energy Λ required for the evaporation of a molecule is considerably larger than the energy Q for its ionic desorption. Clearly the desorption of $[M]^+$ in the high field is performed under strongly reduced thermal stress.

II. DEFINITION OF FIELD IONIZATION AND FIELD DESORPTION MASS SPECTROMETRY

A convenient rule is that FI and FD can be distinguished by the way the sample is supplied to the emitter. If the sample molecules approach the emitter from the gas phase, (Figure 2*a*), one speaks of field ionization. If the sample is adsorbed to the emitter in a solid layer (Figure 2*b*) and is ionized in the adsorbed state, the process is called field desorption.

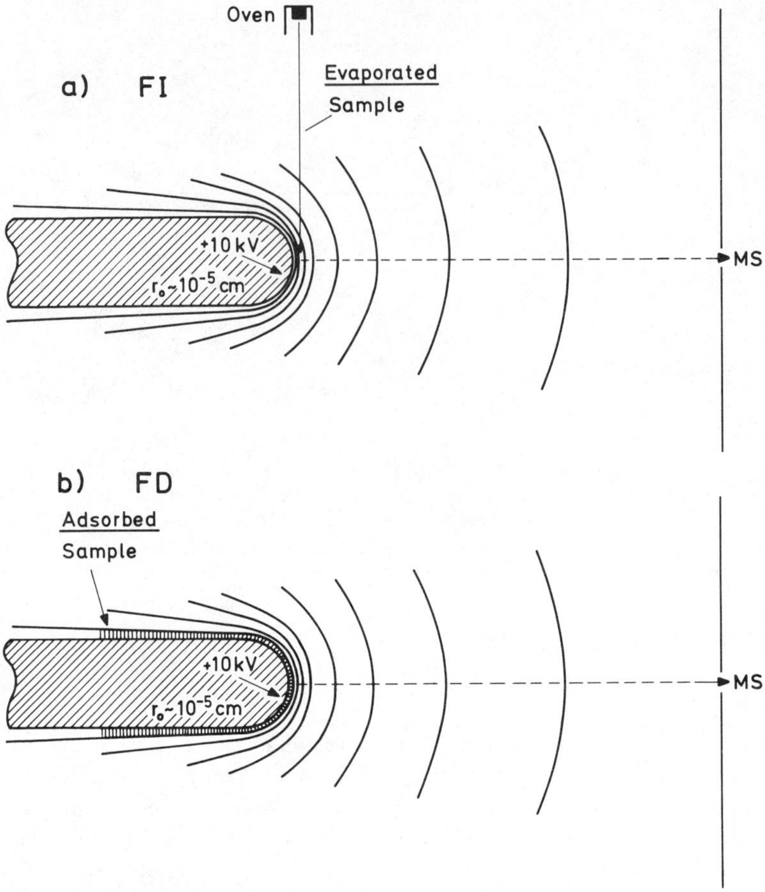

Figure 2. Supply of solid organic or inorganic samples to the place of ionization in the ion source of the mass spectrometer. (*a*) In the FI mode by evaporation from a microoven onto the emitter (i.e., supply by way of the gas phase). (*b*) In the FD mode by depositing the sample on the emitter surface from a solution; that is, the sample molecules are placed at the site of ionization (or might migrate to it by surface diffusion).

III. FIELD DESORPTION TECHNIQUE

1. Emitter

One of the most crucial points in FD-MS is the preparation of the FD emitter. So far the best results have been obtained with 10-μm (diameter) tungsten wires activated at 1200 °C with benzonitrile (11–13), Figure 3 shows the electrolytically cleaned emitter wire before and after activation. Microneedles of pyrolytic carbon (14) have been grown to a length of about 30μm. The dispersed distribution of the needles makes space for the adsorption of the sample solution, and the needle tips form a small-toothed comb. As Section V.6.D relates, the irregular morphology and complex material of the microneedles are two primary causes for difficulties in obtaining high reproducibility in FD-MS. In addition to a large number of emitter types (15), recently metal dendrites have been grown electrolytically on the tungsten wires and used as FD emitters (16,17). Whereas the carbon emitters take about a few hours for preparation, the metal emitters require minutes. On the other hand, little is known about the behavior of large organic molecules on metal surfaces under the conditions of FD-MS, and catalytic processes in the high field are likely to occur.

2. Sample Handling

Three different techniques for the coating of the FD sample onto the emitter have been described.

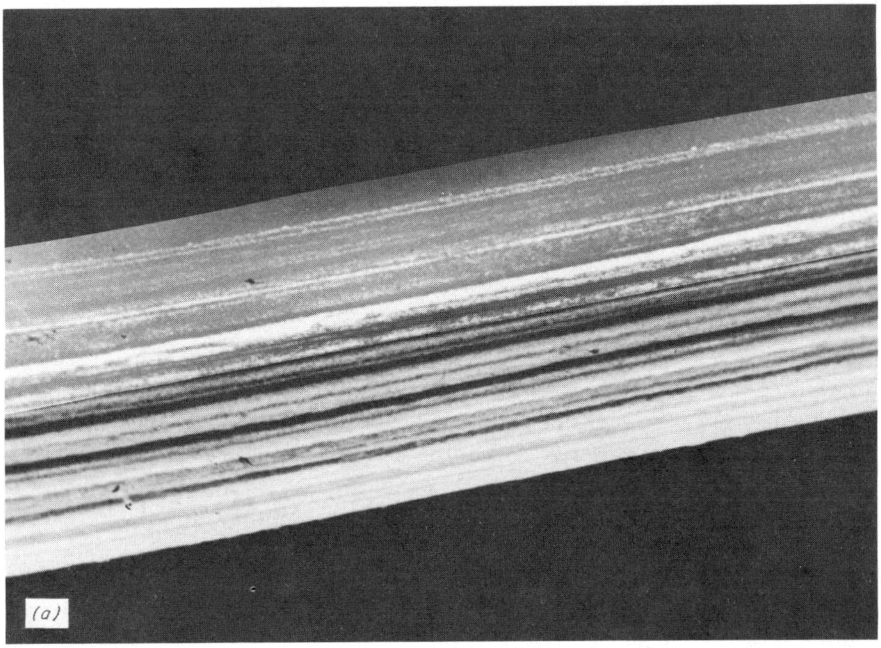

Figure 3. Scanning electron microscope photographs. (*a*) A lo-μ tungsten wire before activation. (*b*) The wire after activation with benzonitrile vapor at 1200°C. Length of the microneedles is about 30 μ. Photomicrograph courtesy of JEOL Ltd., Tokyo, Japan.

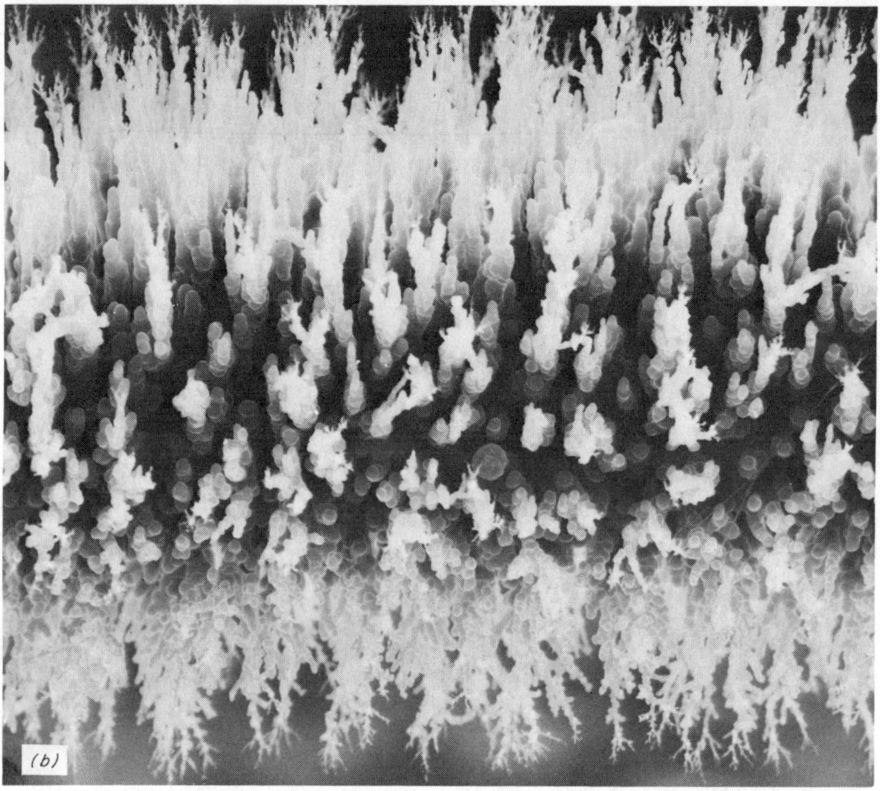

Figure 3. Continued.

First, the emitter can be dipped into the solution of the sample (Figure 4*a*). (The adsorbed layer of the sample from a 0.01*M* solution of D-glucose appears in Figures 5*a* and 5*b*. When a concentrated solution of the sample is available, the surface of the FD emitter can be covered by a thick layer of substance—a few micrograms of material—as illustrated for a nucleoside in Figure 5*c*.) If *qualitative* FD-MS is intended and enough sample material is available to prepare a sufficient amount (2 to 3 μl) of a solution of relatively high concentration, the dipping technique (10) is simple and fast.

Second, for *quantitative* investigations with FD-MS and sensitivity tests, the syringe technique (18) is preferred. This method allows a known amount of sample (i.e., known volume of solution of known concentration) to be deposited on the emitter surface (Figure 4*b*) Special variants of this technique

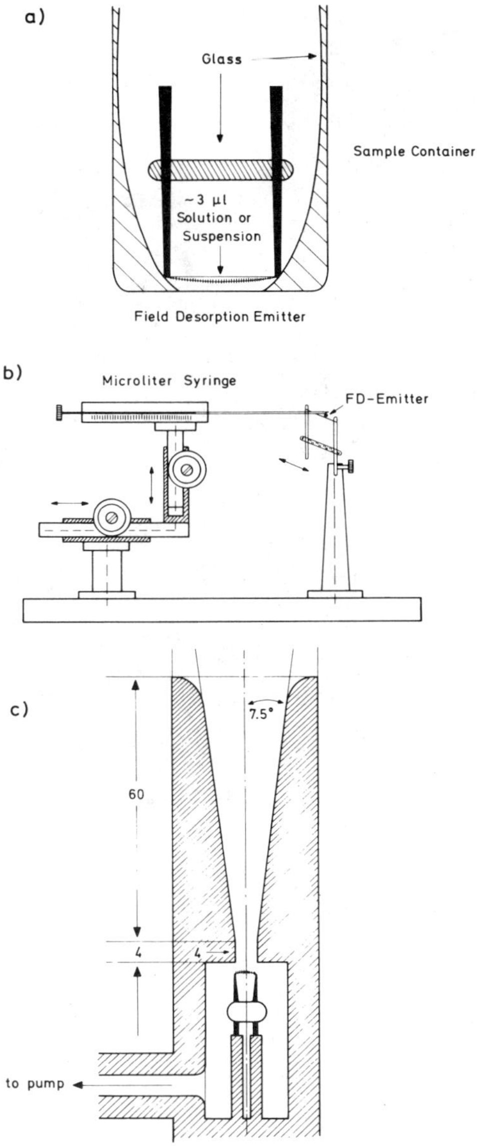

Figure 4. Sample supply techniques in FD-MS. (*a*) Emitter dipping technique. (*b*) Syringe technique. (*c*) Jet sampler (21,22); dimensions in millimeters.

Figure 5. (a) Scanning electron miscroscope photograph of an FD emitter after dipping into a 0.01*M* aqueous solution of D-glucose. Secton of tips of microneedles at a magnification of approximately 1:19,000. Clearly the adsorbed sample is deposited on and between the microneedles (slighter darkening). The ends of the fine needles are brought closer together by some of the glucose threads (formed after evaporation of the solvent water).

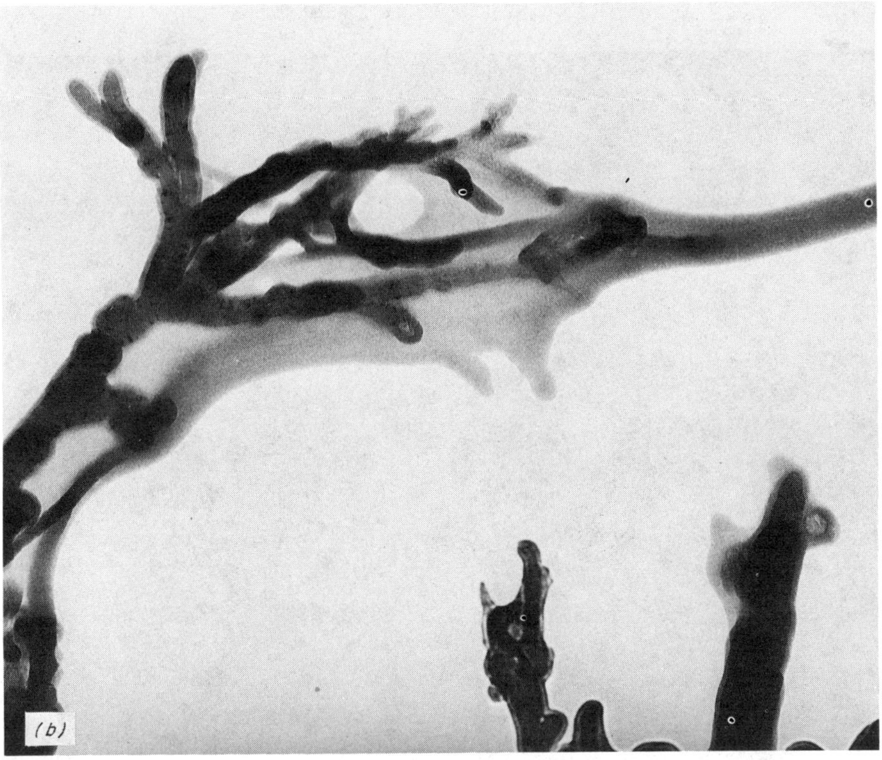

Figure 5. (b) Same object photographed at an approximate magnification of 1:92,000 in a transmission electron microscope. The thickness of the adsorbed layer of the sample in this part of the emitter microneedle is in the order of 0.01 μm. The total adsorbed sample amount on the whole FD emitter is about 100 ng.

Figure 5. (c) Scanning electron micrographs of an FD emitter coated by dipping into a concentrated solution of a nucleoside. Estimated sample amount 2 to 10 μg. (Photomicrograph courtesy of JEOL Ltd., Tokyo, Japan.)

have been developed to overcome the poor wettability of the emitter surface for some solvents (19,20).

Third, a procedure has been used to adsorb aerosols on the FD emitter (21). This jet technique (Figure 4*c*) employs a simple impactor, and the particulate matter in an air stream is (to some extent) adsorbed from the bushy microneedles of the FD emitter (22).

3. Registration of the Spectra

In principle, the FD spectra are recorded with the *electric* or *photographic* detection system of the mass spectrometer, as are the spectra obtained with conventional ion sources. Since the ion currents produced by the FD technique are low and often fluctuating, for electric detection it is essential to keep the recording time considerably shorter than the desorption time of the

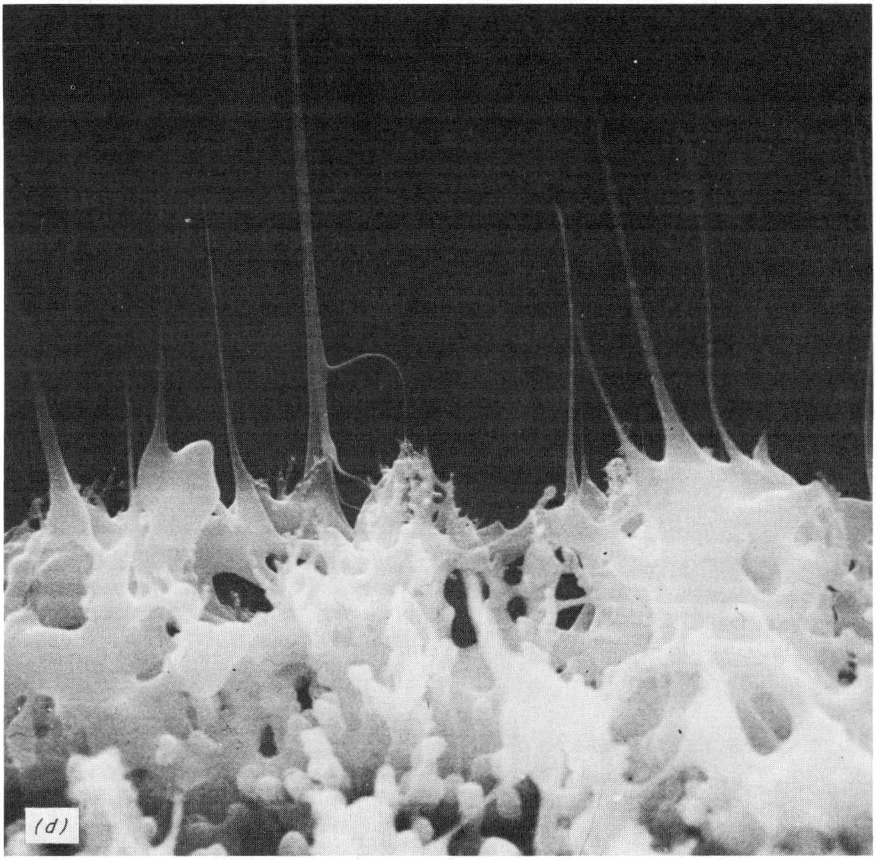

Figure 5. (d) Polymerization products formed during FD-MS of a nucleotide. The original microneedles produced by HT activation appear *below,* the spear; shaped, long protrusions (*above*) are generated at higher emitter temperatures from the sample molecules.

sample. In addition, it is advantageous to use a highly sensitive and integrating device for the registration of FD ions.

Most of the low-resolution work with magnetic and quadrupole instruments has been done with electric detection, using a computer or a multichannel analyzer for time averaging of the signal. The electric detection mode has the advantages of high sensitivity and flexible operation within a wide variety of analytical and fundamental applications; moreover, it gives results fast and relatively easily. However in FD-MS the photographic detection mode appears to have some decisive advantages. The photoplate as a "real" integrating device is capable of recording all ions that are produced during desorption of the sample simultaneously. Photographic detection allows the option of desorbing the entire sample either in one exposure or as desired. In the case of mixtures, fractionated desorption (23) can be performed. Furthermore, since the entire mass spectrum usually has been recorded at high resolution, it is convenient to do exact mass measurements, thus determine the elemental composition. On the other side, the photographic detection system suffers from two major drawbacks. The processing and data evaluation are rather tedious, particularly when a comparator and data system are not available. Also, in comparison with electric detection, the photoplate has a low dynamic range and lower precision in ion abundance measurements. Quantitative data are more difficult to obtain because of the need to determine emulsion response curves and because of the low dynamic range.

4. Special Problems of the FD Technique

A. EMITTER PREPARATION

For the period, 1969 to 1972, a widespread application of the FD method was hampered by the difficulty preparing chemically and mechanically resistant emitters. One essential requirement of FD emitters is that they yield ion currents intense enough for detection in conventional mass spectrometers. On the one hand the FD emitter should possess a surface large enough for the adsorption of the sample molecules, and on the other it should enable the production of the extremely high field strengths required for FD. The development of high-temperature activation (11) is one solution to the problem. Single and multiple activation devices are available, and the preparation of efficient FD emitters is now a routine procedure, although considerable technical proficiency is required for the routine production of acceptable emitters.

B. EMITTER ADJUSTMENT

The exact positioning of the emitter in the ion-optical axis of the mass spectrometer (formed by the slotted cathode plate, entrance slit, and energy resolving slit) appears to be an important prerequisite for optimal FI or FD ion-current intensities. The first high-resolution data in FD-MS were obtained using a mechanical device for three-dimensional adjusting of the emitter (24) (micromanipulator). The design of the micromanipulator is shown in Figure 6. The position of the emitter is first optimized in all three directions in the field ionization mode, using the m/e 58 peak of acetone (introduced by the gas inlet system), the fine axial screw, and four micrometers. After disconnecting the emitter high-voltage supply, the pushrod is withdrawn by the aid of the vacuum lock.

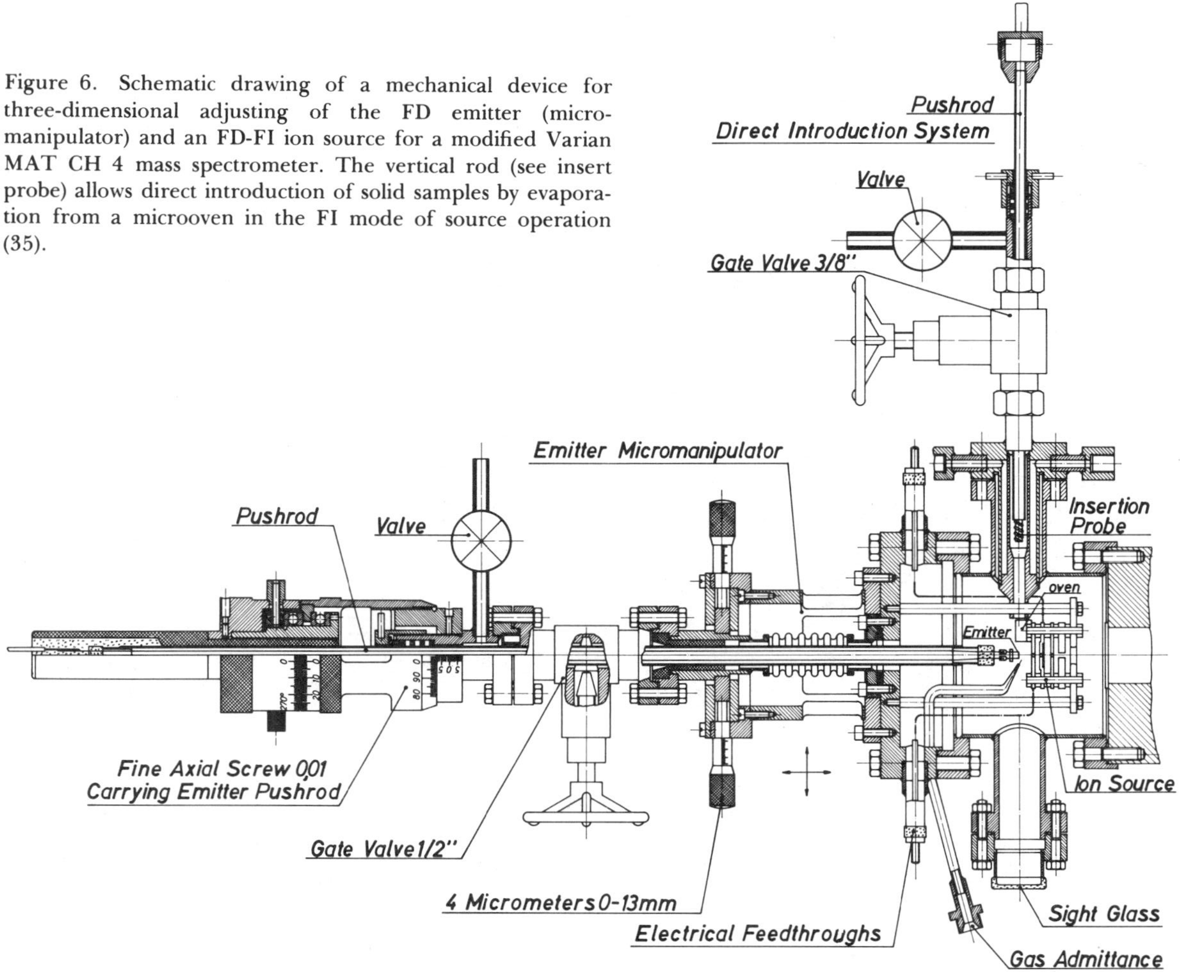

Figure 6. Schematic drawing of a mechanical device for three-dimensional adjusting of the FD emitter (micromanipulator) and an FD-FI ion source for a modified Varian MAT CH 4 mass spectrometer. The vertical rod (see insert probe) allows direct introduction of solid samples by evaporation from a microoven in the FI mode of source operation (35).

Next the emitter is coated with the sample and pushed back to the predetermined position. Further small corrections to the emitter position may now be made quickly (to avoid loss of the sample) in the FI mode before starting the desorption measurements. Slight deviations from the optimum position can have a dramatic effect on the signal intensity.

C. EMITTER TEMPERATURE

The FD spectra obtained are governed by the temperature of the sample and the emitter during desorption. In principle, five different temperature ranges for investigations by FD-MS are available.

1. It has been shown that by cooling of the emitter ($T < -50°C$) and pulsed FD-MS, ions with extremely low internal energy can be produced (25).
2. At ambient temperature (mostly 20–30 °C) a number of substances that have sufficiently high volatility desorb from the high-temperature emitter [e.g, phenanthrene, endrin (26), cyclophosphamide (27)].
3. With increasing polarity of the molecules, however, it is necessary to heat the emitter for desorption of the sample. A convenient way to transfer the thermal energy to the sample is by direct electric heating of the emitter (tungsten) wire (10,19). The FD mass spectra in this mode of operation are characterized by the emitter heating current (ehc). Indirect heating of the emitter (and the sample) by infrared radiation has been performed, and higher relative molecular ion intensities have been achieved (28). A best anode temperature (BAT in mA ehc or T^* in degree Celsius) has been defined as the temperature of the field anode that generates maximal molecular ion currents and minimal fragmentation (29). Thus for molecular weight determination and detection of a compound, the FD mass spectra obtained at BAT, expressed in milliamperes, are particularly important.
4. At temperatures above BAT (up to 30 to 40 mA ehc), thermally and field-induced fragmentation can be performed (30). Since this procedure allows the production of structurally significant fragments, FD-MS can be used for structure elucidation and identification of biochemicals. Pyrolytic products are easily generated, however, and *in situ* polymerization has been observed for hydroxycarboxylic acids (168) and oligonucleotides (49; Figure 5 *d*).
5. At a temperature range considerably above BAT (30–80 mA ehc), fast and controlled thermal degradation of materials, such as biopolymers, macromolecules, tissue, and microorganisms, can be achieved. This technique, pyrolysis FD-MS, (31) requires an integrating device for ion detection and a defined time-temperature profile for desorption.

Since the temperature of an FD emitter depends on the variation of the wire diameter and length, the length and distribution of the carbon microneedles, the solvent used to apply the sample, the field strength, and other parameters, it is difficult to establish an exact correlation between the ehc and the actual temperature. However two different methods have been reported to establish this correlation (32,33), and these can be used in a first approximation to compare the emitter heating currents with the temperatures (Figure 7).

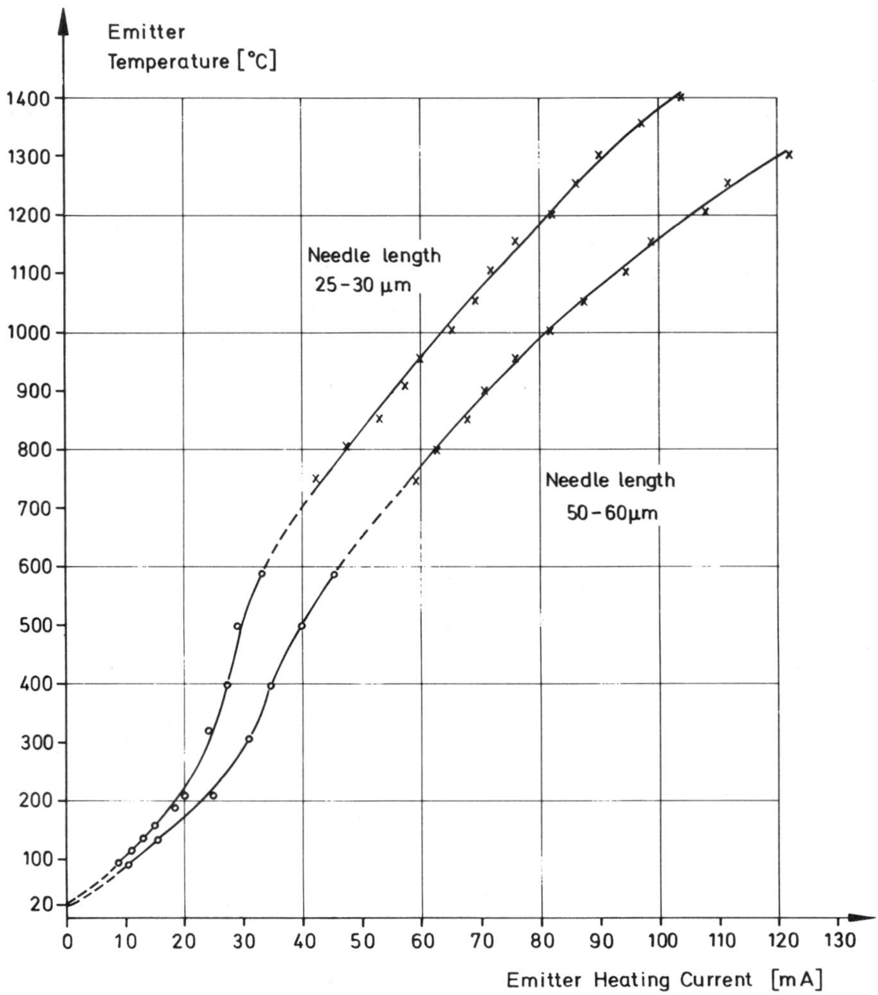

Figure 7. Graphic plot of ehc versus emitter temperature. The dashed portions of the curves were not accessible by the methods employed and have been interpolated (32).

D. MONO- AND COMBINED ION SOURCES

The ion source connected to a conventional mass spectrometer for the operation in the FI/FD mode is represented in Figure 6. Although EI and CI mass spectra cannot be obtained from this simple, "open" construction, an ion source for all four techniques described has been reported (34).

Commercially available instruments (e.g., from Varian MAT, Bremen, West Germany) have the following advantage: with their combined EI/FI/FD ion sources, the mass calibration can be obtained easily in the EI mode. However it is clear that the design of a combined ion source must be a compromise, and it cannot be expected that the optimal conditions can be met either in EI-MS or in FI/FD-MS.

E. EMISSION-CONTROLLED FD-MS

In addition to the absolute temperature of the sample, the rate of desorption from the emitter surface has a strong influence on the appearance of the FD spectrum. Therefore a device has been designed (Scheme 1) that allows an automatically controlled desorption. Reproducible time-temperature desorption profiles are obtained by setting a threshold measured at the cathode for the total emission of the FI/FD emitter. Emission-controlled FD-MS allows three modes of operation (36,37) for electric and photographic detection.

For minimal fragmentation, the threshold current is set at the lowest value ($\sim 1 \times 10^{-9}$ A for a CEC 21-110B instrument, 10 kV accelerating voltage, and 2 mm distance between the FD emitter and cathode) producing a

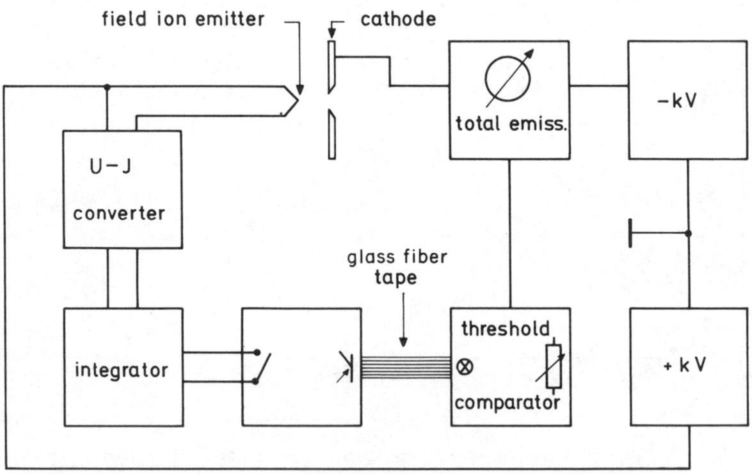

Scheme 1

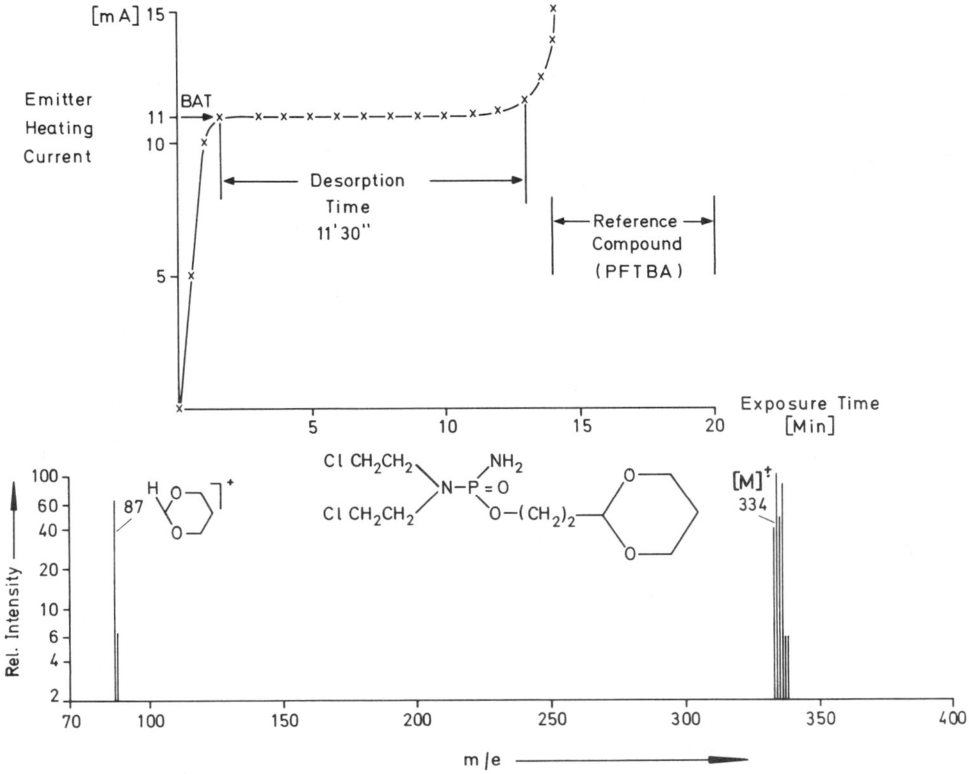

Figure 8. Thermogram of aldophosphamide propanediol-1,3- acetal and FD mass spectrum at a threshold emission of 1.2×10^{-9} A; BAT, 11 mA, total exposure time, 20 min. The actual desorption time of the sample was 11.5 min: 30 sec was consumed for the emitter current regulation to reach 11 mA ehc, and the rest of the total exposure time was used for the field ionization (supply from the gas phase) mass spectrum of the reference compound perfluorotributylamine (PFTBA), which is needed for mass calibration (36).

detectable ion current. The thermogram recorded for the desorption of a cancer drug conjugate appears in the upper part of Figure 8. The relationship between the ehc and the desorption time is clearly indicated. This thermogram also shows that the characteristic BAT for the compound is easily derived and that a rough estimate of the amount of sample used in the experiment can be made from the plateau in this curve. The FD mass spectrum (lower part of Figure 8) produced in this manner shows that the conditions for desorption were indeed close to BAT. Except for the molecular ions, only one fragment is found in the low-mass range. When the threshold current was set at values 1×10^{-9} A, more fragmentation occurred, but the spectra are still well reproducible for a given threshold current. The intensities of each line of a photographically recorded FD-MS for 10 determinations of the drug conjugate were reproducible within 10%. (36)

A third mode of operation, very suitable for quantitative work, is linear programming of the ehc with time. However the procedure of linear programming can be used only when sufficient sample material is available and the physicochemical properties of the compound under investigation are favorable for FD. For electric detection of the FD spectra, the use of an emitter current programmer (38) or a computer-controlled heating device has been reported (39).

F. HIGH-RESOLUTION AND ACCURATE MASS MEASUREMENTS

In general, high-resolution (HR) mass spectrometry is performed to establish the elemental composition of the ions produced. The HR data in FD-MS were almost exclusively reported for photographic detection (21,22,40–43). This is founded on the versatility and sensitivity of the photoplate as a detector for relatively weak or strongly fluctuating ion currents. However, peak matching of FD ions has also been described (44,176).

For accurate mass measurements in FD-MS, external as well as internal mass references can be used. The external mass calibration can be established using a different mode of ionization (e.g., EI or FI). Cluster FD ions can be used as internal standard, as has been shown for phosphate cluster ions (31). Two main problems are associated with HR FD-MS. It is difficult to produce suitable ions for mass reference in the mass range above m/e 1000. In addition, the accuracy of mass measurements in FD-MS is in the range of 10 ppm; thus a number of meaningful elemental compositions may be in agreement with the data obtained. Nevertheless, for pyrolysis FD experiments and for the analysis of complex mixtures and accurate molecular weight determination of natural products, HR FD-MS is a powerful tool.

G. REPRODUCIBILITY AND QUANTITATION

The characteristic features of the method for the *qualitative* analysis of a wide variety of model compounds is demonstrated in the following. However quantitative determinations by FD-MS are handicapped for the following reasons:

1. The quality of the field desorption emitter.
2. The adjustment of reproducible desorption conditions.

3. The transfer of the sample onto the FD emitter.
4. The influence of the chemical properties of the compound under investigation on the field desorption process.

The problems arising from the variable factors 1 to 4 can be solved in the following way.

1. The production of emitters of similar field desorption properties can be achieved by high-temperature activation in a multiple-activation apparatus, control of the shape, distribution, and length of the microneedles under a light microscope, and testing of the ionization effiency of the emitter in the FI mode.
2. Reproducible desorption conditions can be achieved when emission-controlled FD-MS is employed as described previously (37).
3. It is general experience that the syringe technique enables the reliable transfer of defined amounts of sample onto the emitter.
4. Quantitation in FD-MS is feasible using stable isotope dilution (36). Since sample molecules and their isotopically labeled analogs show virtually identical chemical properties but can be distinguished easily by mass spectrometric techniques, these compounds can be used successfully as internal standards for quantitation (177 and references cited).

IV. APPLICATION OF FIELD DESORPTION MASS SPECTROMETRY TO BIOCHEMICALS

At least four essential demands are made on the analytical methods for environmental, medical, and biochemical research.

1. The detection sensitivity of a technique should be within or below the nanogram range. This is of importance for the chemical-analytical problems of structure elucidation of natural products, for the trace analysis of biochemicals and, in particular, for the identification of products in biotransformation processes.
2. The selectivity of the method should be such that even substances in prepurified crude extracts of biological material can be detected unequivocally despite of the unavoidable admixtures.
3. With respect to the significance of the MS data, it is required that apart from characteristic fragment ions, the highest possible molecular ion intensities be produced, since frequently in the above-described areas the determination of the molecular weight of an unknown compound, or its elemental composition, is a main aim.
4. Reliability of the procedure in qualitative and quantitative analysis is

required. This is especially difficult because products from the biological environment tend to have little thermal stability (besides being sensitive to oxidizers, changes in pH, etc.). Moreover, a great number of free functional groups are usually present and coupled with strong intermolecular forces, which lead to low volatility. It is often very difficult, or even impossible, to convert the undecomposed intact molecule to the gas phase. This is, however, a basic requirement of other mass spectrometric methods such as EI, CI, and FI.

To demonstrate the contributions of the unique features of FD-MS and its analytical capabilities, we present four classes of compound—biochemicals, drugs and drug metabolites, biomacromolecules, and inorganic salts—for which the FD technique has been successfully applied.

1. Carbohydrates and Related Compounds

A. MONOSACCHARIDES

The first substance measured by FD-MS was D-glucose (Figure 9*c*). In 1969 Beckey introduced with the mass spectrum of this compound the FD technique "for the study of thermally unstable substances of low volatility" (10). In contrast to EI and FI mass spectra (Figures 9*a* and 9*b*), the fragmentation of D-glucose is strongly reduced with field desorption. Even in this early publication, four essential facts emerged.

First, the transferred excitation energies in the ionization processes are considerably smaller in the sequence EI > FI > FD.

Second, under the conditions selected, the EI mass spectrum of underivatized D-glucose yields no structural information and is useless for the determination of molecular weight. Although the FI mass spectrum shows intense fragments, due mainly to thermal processes such as water elimination, the method can be regarded as a "soft" ionization method because the $[M + 1]^+$ ion is the base peak of the spectrum. The portion of D-glucose molecules that survives the thermal stress of the evaporation and reaches the gas phase near the emitter is ionized so that most of these D-glucose molecules are detected as molecular ions or protonated molecules.

Third, the FD method is unique in that no evaporation is required and ionization takes place in the adsorbed layers. The FD spectrum of D-glucose (Figure 9*c*) is virtually a "one-peak spectrum"; only the molecular ion group is displayed. Thus the method appears to be ideally suited for molecular weight determination of thermally unstable substances. The fragmentation is drastically reduced, however, and this implies that the potential for structural investigations is limited.

Fourth, the first FD publication emphasizes that the type, preparation, and adjustment of the FD emitter is a decisive prerequisite of FD-MS, and this is completely valid to date.

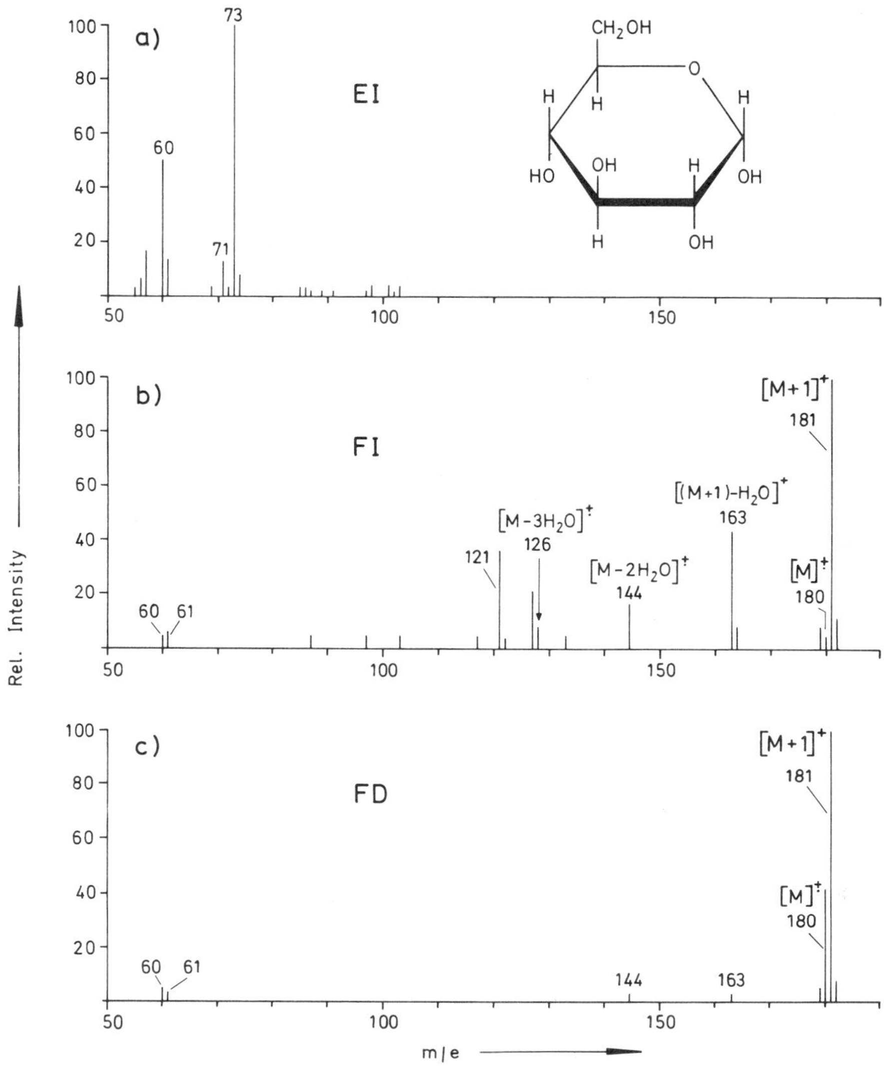

Figure 9. Mass spectra of D-glucose (10).

B. OLIGOSACCHARIDES (DIRECT BOND CLEAVAGES)

a. Disaccharides. Krone and Beckey (46) reported the FD spectrum of cellobiose. The $[M + H]^+$ ion at *m/e* 343 was displayed as the base peak. In general, this ion has been described as $[M + H]^+$, since no molecular ion $[M]^+$ is present and there is no isotopic contribution (^{13}C, D, etc.) from this ion. In cases where $[M]^+$ was found, the ion at a mass of one unit higher in *m/e* value was defined as $[M + 1]^+$. This also holds for $[M + 1]^+$ ions that

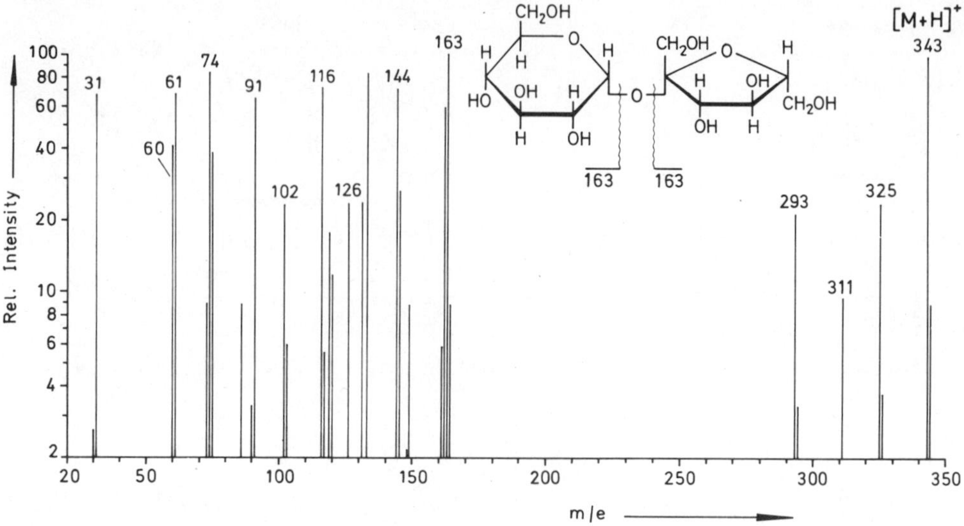

Figure 10. FD-MS of sucrose: solvent, water: BAT, 16 to 21 mA ehc; photographic detection, exposure time, 12 min (49).

apparently are generated in proton transfer reactions because of their high relative abundance (47,48).

A number of fragments are listed in the electrically recorded FD mass spectrum of cellobiose. Most prominent is the ion formed by direct bond cleavage at the glycosidic oxygen (*m/e* 163). The following sections demonstrate that the formation of this type of ion is a typical feature in FD-MS. The FD spectrum of sucrose, recorded photographically at high resolution (Figure 10) shows the same fragment at *m/e* 163.061 ($C_6H_{11}O_5$) and the $[M + H]^+$ ions as base peaks (49). For this disaccharide, however, two different pathways may lead to the production of the ion at *m/e* 163 (Figure 10). It is noteworthy that the FD spectrum is obtained in an ehc program between ambient temperature and approximately 300 °C (~ 30 mA ehc) within 12 min desorption time. The best anode temperature is reached at 16 mA ehc, and strong thermally and field-induced fragmentation is generated in the temperature range above T^*. Typical FD fragments of mono- and disaccharides are at *m/e* 30.016 (CH_2O), *m/e* 60.021 ($C_2H_4O_2$), *m/e* 73.029 ($C_3H_5O_2$), *m/e* 90.032 ($C_3H_6O_3$), and their protonated species at *m/e* 31, 61, 74, and 91. The upper mass range for these sugars is often characterized by loss of water from the $[M + H]^+$ ion (e.g., *m/e* 325), loss of CH_4O (e.g., *m/e* 311), and water elimination (e.g., *m/e* 293). Also the ions at *m/e* 102, 119, 126, and 144 are significant for oligosaccharides.

b. Trisaccharides. As an example of field desorption of this class of compounds, consider the mass spectrum of raffinose given in Figure 11. In general, the intensities of the fragments and the molecular ions for trisaccharides are considerably smaller than for disaccharides reflecting the higher polarity, because of a larger number of free hydroxyl groups and the possibility of the formation of strong hydrogen bonds. The $[M + H]^+$ ion at *m/e* 505 is still of 50% relative abundance and the determination of the molecular weight of trisaccharides by FD-MS should be no problem. In addition to the typical sugar fragments at *m/e* 30, 31; 60, 61; 91; 119; 126; and 144, and the monosaccharidyl cation at *m/e* 163, an intense, structurally significant ion is found at *m/e* 343. A direct bond fission accompanied by hydrogen and proton transfer reactions leads to the $[II + 2H]^+$ ion, which indicates the disaccharide unit of the molecule (Figure 11). Moreover, since

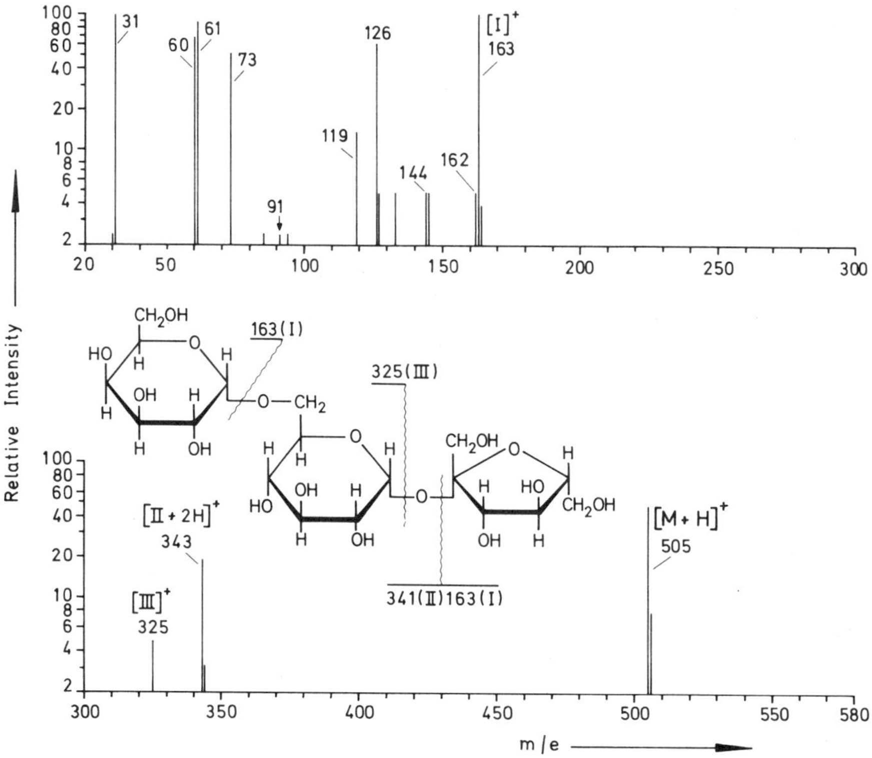

Figure 11. FD-MS of raffinose: solvent, dimethylsulfoxide; BAT 19 to 25 mA ehc; emission-controlled FD; total exposure time 28 min—15 min at a threshold of 3×10^{-9} A, 13 min at 2×10^{-8} A for the ion emission (49).

the typical cleavage at the glycosidic oxygen yields the ion at *m/e* 163 for the terminal sugar unit, useful structural data are obtained from the information of the molecular weight and these two complementary, charged parts of the molecule (40). (These complementary ions are very frequently observed with FD-MS for a wide variety of substances, such as pesticides (26,43), pesticide metabolites (50,51), drugs (52), and a large number of biochemicals, as described in this chapter.)

Moor and Waight studied a number of di-, tri-, and tetrasaccharides, and some of their permethyl and peracetyl derivatives, by field desorption (53). With the exception of stachyose, a tetrasaccharide, intense $[M + H]^+$ ions were reported for all sugars, thus enabling the determination of the molecular weights. Also the effect of variation of ehc on the relative intensities of fragment ions was investigated. Permethylated oligosaccharides showed intense molecular ions. In contrast, peracetylated substances of this class showed only weak molecular ions but strong $[M - 60]^+$ ions. Owing to the increased volatility, the derivatized sugars give better mass spectra with EI, FI, or CI (54). FD-MS is advantageous for the analysis of free sugars and it is not suited for substances of high volatility because the desorption rate cannot be controlled reproducibly.

However the interglycosidic cleavages already described for sucrose (*m/e* 163) and raffinose (*m/e* 163 and 325) were confirmed by these authors for all 14 sugars investigated. For trisaccharides, the fission leading to the disaccharidyloxy moiety of the molecule $[II + 2H]^+$ ion for raffinose (Figure 11) was also observed.

The conclusion follows that the few fragments of free oligosaccharides produced in FD-MS are highly informative as far as structure is concerned, but there is no doubt that the main advantage lies in the use of the method for molecular weight determination.

C. GLYCOSIDES (HIGH-RESOLUTION MS AND FD-MS)

For all spectroscopic methods there is an approximately reciprocal relationship between the signal intensity and the resolution. Because of the relatively small and fluctuating ion currents generated in field desorption, high-resolution measurements make severe demands on the transmission of the mass spectrometer and on the quality of the emitter (see Section III.1).

High resolution provides precise determination of the molecular weight, thus of the elemental composition, as well as important information for the interpretation of fragments. High-resolution data for some aryl glycosides and for glycosides of the coumarin and flavone series have been published (40). In the HR FD spectrum of rutin (Figure 12*b*), the protonated molecule $[M + H]^+$ is certainly recognizable at *m/e* 611.161 ($C_{27}H_{31}O_{16}$). Since HR FD-MS also provides the elemental composition of the fragments, *m/e*

302.043 ($C_{15}H_{10}O_7$) can be interpreted as $[A + H]^+$, and *m/e* 147.066 ($C_6H_{11}O_4$) can be attributed to direct bond fission at the glycosidic oxygen of the terminal sugar unit (*d*). It is of advantage that the structurally diagnostic ion (*e*) at *m/e* 463.088 ($C_{21}H_{19}O_{12}$) for the other part of the rutin molecule is also found. Thus the interglycosidic cleavages yield complementary charged parts of the molecule as described for oligosaccharides. In addition to the $[A]^+$ and $[A + H]^+$ ions, the aglycone appears as a doubly charged ion at *m/e* 151.021 (for $[A + H]^{2+}$) and *m/e* 151.525 (for $[A + 2H]^{2+}$). Protonation of the rutin molecule preferably occurs at the aglycone. This is indicated by the very intense ion at *m/e* 303, which has a relative intensity considerably exceeding the ^{13}C satellite of *m/e* 302 (A + H).

The EI mass spectrum of rutin (Figure 12*a*) does not give much structural information. Only the aglycone moiety of rutin is found, and signals in the mass range above *m/e* 304 are missing. Since EI-MS was performed at low resolution, it is not clear whether the weak signal at *m/e* 147 corresponds to the ion defined as (*d*) in the FD spectrum.

If enough sample is available for high-resolution study by MS and the special (and costly) instrumental requirements for the HR mass spectrometer

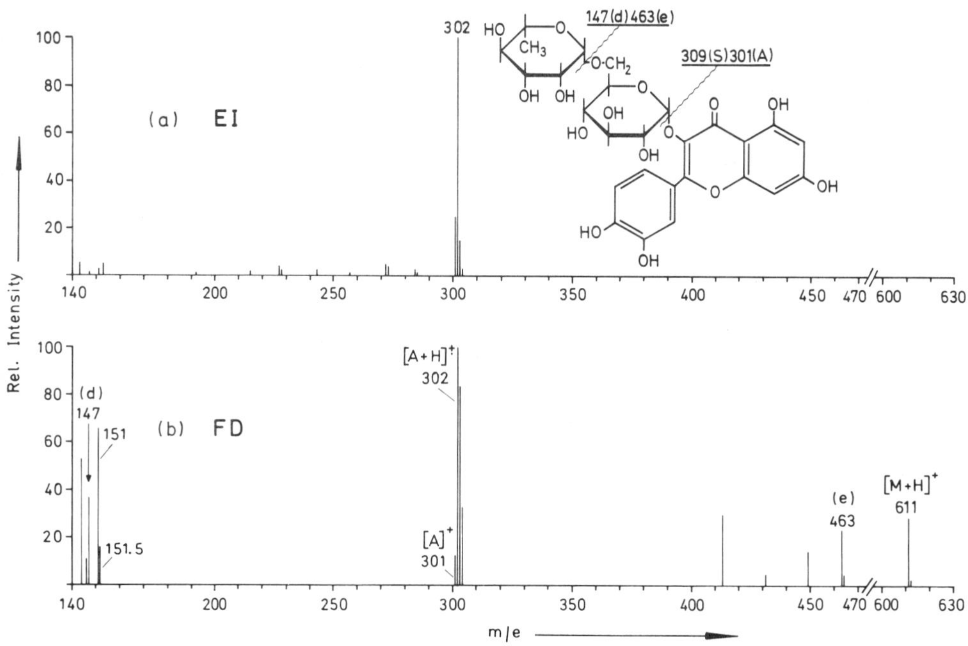

Figure 12. (*a*) Low-resolution (electric detection) EI mass spectrum of rutin. (*b*) High-resolution (photographic detection) FD mass spectrum of rutin (40).

can be met, HR FD-MS can be a powerful analytical tool. First, the specific features in the fragmentation pattern of FD can be evaluated using known model compounds. Second, this knowledge, together with information of the origin of the sample and data of other analytical methods, enables the elucidation of the elemental composition and (to some extent) of the structure of unknown substances.

As an example, the determination of the molecular weight and the confirmation of the elemental composition of cascaroside B (55), a naturally occurring glycoside, was performed by HR FD-MS (49). Although most FD work at high resolution has been recorded photographically, electrical detection and the peak matching technique have recently been used for an HR FD study (44). However accurate peak matching requires an ion current that lasts long enough to align the reference and the sample peaks and is intense enough to align the peak center. When the FD ions are produced only in intervals considerably shorter than 1 min, photographic recording is advantageous. On the other hand, photographic recording is limited to

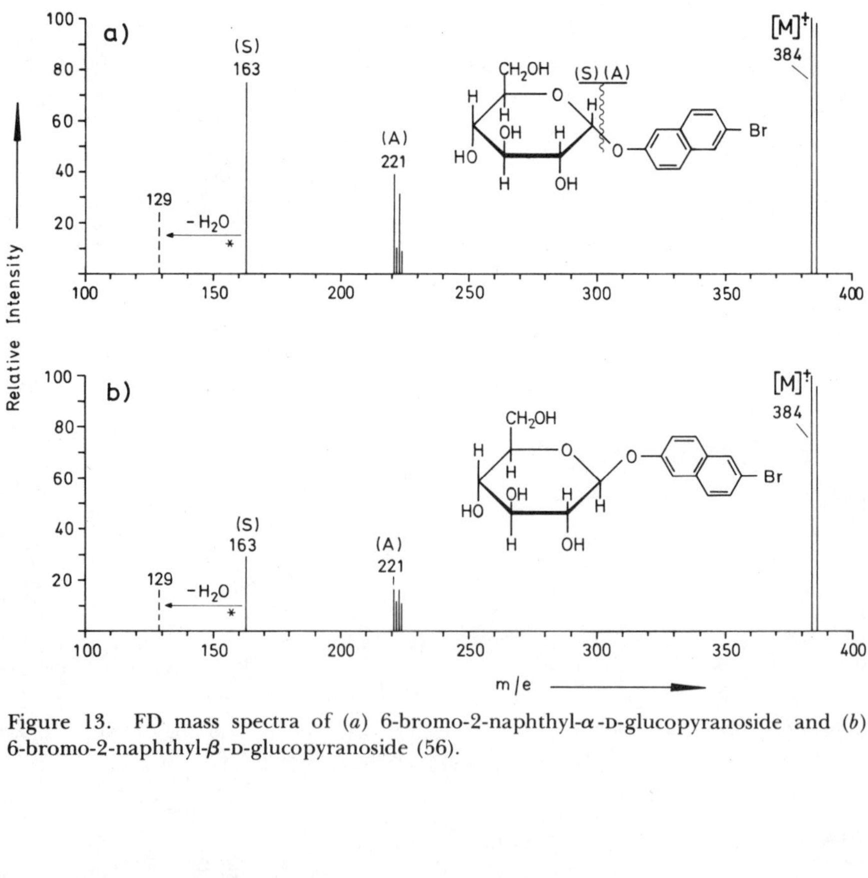

Figure 13. FD mass spectra of (*a*) 6-bromo-2-naphthyl-α-D-glucopyranoside and (*b*) 6-bromo-2-naphthyl-β-D-glucopyranoside (56).

instruments with Mattauch-Herzog geometry, whereas peak matching can be used with all high-resolution mass spectrometers.

Investigations on stereoisomeric glycosides with aromatic aglycon showed that the FD method gives distinct differences between α- and β-linked glycosides. The comparison of EI with FI spectra revealed that the simplest spectra containing the greatest differences for stereoisomers can be obtained with FD-MS (56).

Figure 13 compares the FD mass spectra of 6-bromo-2-naphthyl-α-D-glucopyranoside (a) and 6-bromo-2-naphthyl-β-D-glucopyranoside (b). The spectra exhibit few, but significant, signals for the sugar moiety (S), the aglycone (A), and the molecular ion as base peak. The α-isomer yields a more intense fragmentation consistent with the higher thermodynamic stability of the β-isomer. As can be derived from FI studies of stereoisomers (57), however, differences in the fragmentation pattern by direct bond cleavages are preponderantly directed by the orientation of the molecules in the high field. Therefore it is assumed that the observed differences between the α- and β-glycosides are due to an orientation effect rather than to differences in the thermodynamic stability.

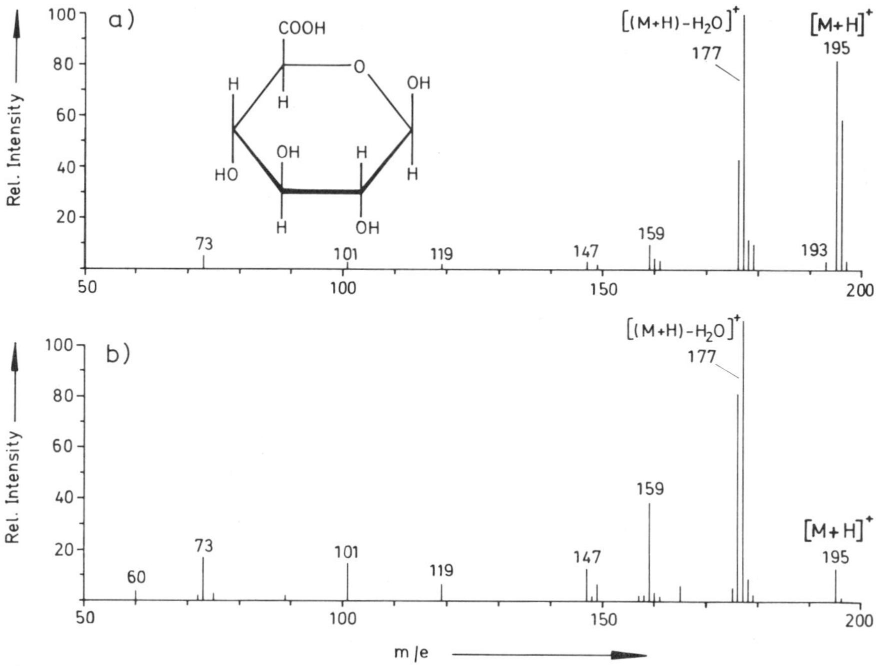

Figure 14. FD mass spectra of glucuronic acid: solvent, water; electric detection. (*a*) At 13 mA ehc. (*b*) At 15 mA ehc (59).

A stereochemical study by EI and FD mass spectrometry of the photodimers of cinnamic acid and related compounds indicated that FD-MS allows a clear differentiation between isomers linked head to head and head to tail (58). Thus the application of the FD technique to problems in stereochemistry appears to be promising.

D. GLUCURONIDES (TEMPERATURE INFLUENCE IN FD-MS)

One of the most intriguing effects in FD-MS is the strong influence of the emitter (and sample) temperature. When the FD mass spectrum of glucuronic acid is electrically recorded at 13 mA ehc, corresponding to an emitter temperature slightly above T^*, the $[M + H]^+$ ion at *m/e* 195 has a relative intensity of 82% (Figure 14*a*). The base peak is formed by the elimination of water from the protonated molecule. The prominence of this ion species clearly indicates that the FD spectrum was run at an emitter temperature exceeding T^* ($\sim$ 11 mA). Failure of the molecular ion to appear confirms the prominence of the energetically and kinetically favored even-electron species in FD-MS (59). When the FD spectrum is recorded at an emitter heating current that is 2 mA higher ($\sim$ 15 mA), there are dramatic changes in the relative intensities of the ions obtained (Figure 14*b*). The relative abundance of the $[M + H]^+$ ion falls from 82 to 13%, and the $[M - H]^+$ is missing; but the signal due to the loss of two water molecules from the protonated glucuronic acid at *m/e* 159 increases from 9 to 29%. Although the $[(M + H) - H_2O]^+$ ion is still the base peak of the spectrum, the generally enhanced abundance of the decompition products (e.g., *m/e* 60, *m/e* 73, as in FD MS of mono- and oligosacharides) indicates a thermally and field-induced fragmentation.

It is worthwhile to emphasize the decisive influence of the emitter temperature on the FD spectra obtained. Note the striking alteration of the spectra caused by small changes in the ehc, as also shown in Section V.6.B for the thermally labile metabolic oxydation products of the antineoplastic drug cyclophosphamide. To obtain reproducible spectra with the electric detection system at a constant field strength, with standard emitters (10-μ tungsten wires with 30-μ high-temperature-activated microneedles), using the same solvent and same sample size, the time-temperature program for desorption must be controlled very carefully. This can be achieved either directly by the computer of the data system, or by using the electronic device described in Section III.4.E that generates defined and reproducible time-temperature profiles for desorption (36).

Given the importance of glucuronic acid in the metabolic pathway of endogenous substances, and recognizing that the metabolites of drugs and pesticides are often excreted as conjugates, the behavior not only of the free

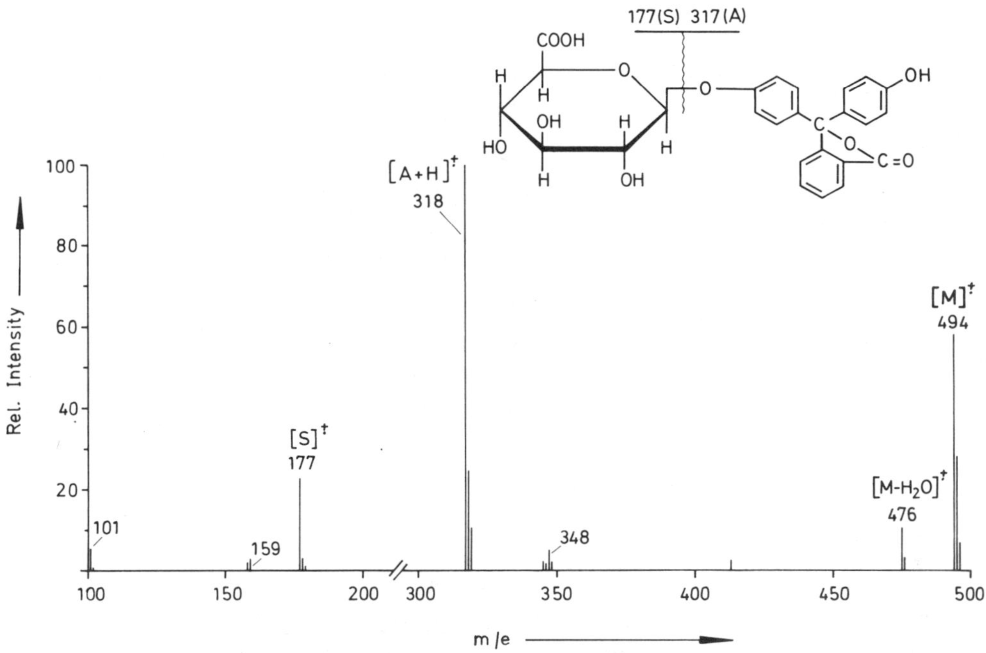

Figure 15. FD-MS of phenolphthalein-β-D-glucuronide; solvent, water; electric detection at 12 mA ehc (59).

acid but also of glucuronic acid bound to biochemically related molecules at the glycosidic oxygen in the C-1′ and C-4′ positions was examined under the conditions of field desorption.

The FD mass spectrum of phenolphthalein-β-D-glucuronide, a substrate for the assay of β-glucuronidase, was electrically recorded at a 12 mA ehc (Figure 15). As might be expected from the FD mass spectra of glycosides, the protonated aglycone at *m/e* 318 forms the base peak of the spectrum. Furthermore, the sugar moiety (S) at *m/e* 177 is clearly detected. Since the FD spectrum is obtained close to T^*, the loss of water from the parent molecule has only minor importance (16% relative intensity for *m/e* 476). These results support the assumption that the formation of ions that constitute complementary building blocks of the parent molecule is a more general feature of FD-MS and simplifies the interpretation of the mass spectra.

Steroid glucuronides are important in hormone excretion. Figure 16 presents a high-resolution FD spectrum of testosterone glucuronic acid sodium salt recorded on a vacuum-evaporated AgBr photoplate. As with

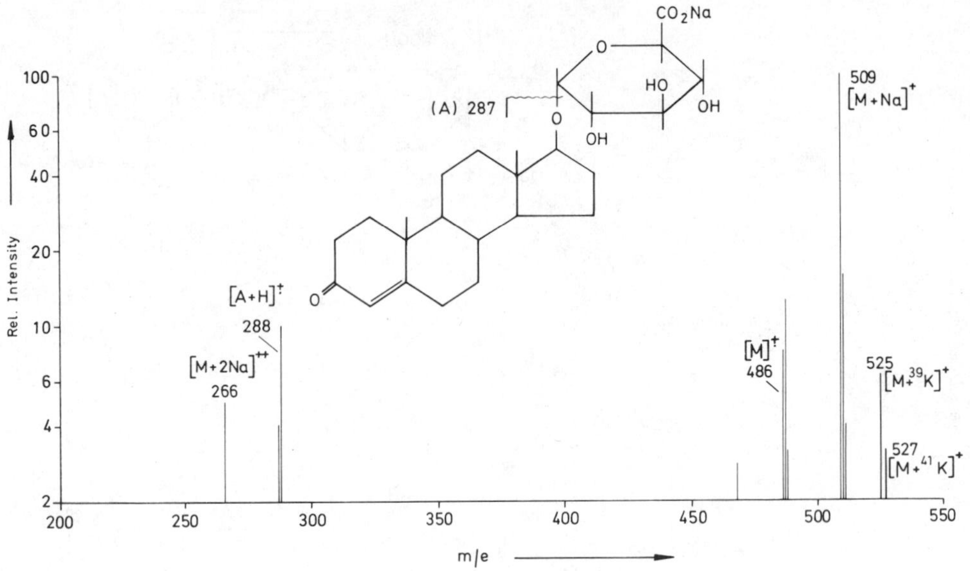

Figure 16. FD-MS of testosterone glucuronic acid sodium salt (40).

inorganic salts (60,61), the FD mass spectra of organic salts usually show molecular ions with no, or only minor intensity, but they also show the cation $[C]^+$ and the $[M + C]^+$ cluster ions with high relative abundances. Although the molecular ion at *m/e* 486.223 is found with about 8% relative intensity, the HR FD mass spectrum is clearly dominated by the $[M + Na]^+$ ion at *m/e* 509.213. In addition, *cluster formation* with accompanying potassium cations is recorded at *m/e* 525.187 for $[M + {}^{39}K]^+$ and *m/e* 527.185 for $[M + {}^{41}K]^+$. The $[Na]^+$ and $[K]^+$ ions are found with extremely high intensities (mostly with complete blackening of the photoplate). The direct bond cleavage between C-1 of the glucuronic acid and the glycosidic oxygen generates, concomitant with proton transfer, a $[A + H]^+$ ion at 288.209, which is ascribed to the testosterone part of the glucuronide. Of special interest is a doubly charged ion at *m/e* 266.101 formed by the clustering of two sodium cations with one molecule M of the testosterone glucuronic acid sodium salt: M = C + A (anion).

The work of Adlercreutz et al. (62) extends beyond investigations on model substances and brings field desorption into use for clinical chemistry and medicine. Thus oestriol-16-glucuronide, isolated from the urine of pregnant women, was successfully detected with FD-MS with high molecular ion intensity. This was achieved without derivatization at low resolution (*m/e* 500–600) and was supported by the FD measurement of standard synthetic compounds. For physiological studies of endogenous compounds and pharmacokinetic investigations, however, the quantitation of the FD results is basic. Preliminary investigations on the use of deuterated analogs of biogenic amines and a cancer drug for quantitative determinations by FD-MS are illustrated in Sections IV.9, and V.6.D.

To summarize the FD results obtained with oligosaccharides, glycosides, and glucuronides, we can say that FD-MS generally permits molecular weight determination based on high intensities of the molecular ions, the protonated molecules, or cluster ions. Similar conclusions were reported by Bursey and Sammons in a survey of applications of FD-MS in carbohydrate chemistry (63).

E. SUGAR PHOSPHATES

The alkali salts of sugar phosphates complete this list of sugar compounds of increasing polarity. As part of a program concerned with design of selective inhibitors of hexokinase isoenzymes (64), the substrate specificity of yeast hexokinase was established (65) using series of deoxyfluoro-D-glucoses in which the fluorine substituent was located at positions 2, 3, and 4. The high resolution FD-MS of these substances (42) which are the first large organic salts analysed with the technique give the $[M + H]^+$ ion as the base peak but also show products that result from exchange of $[Na]^+$ with $[H]^+$, elimination of water, and pyrolytic processes. These salts could only be recorded photographically, since they desorb within a very short time and heating current interval.

Tarry residues on the FD emitter at the end of the measurement (0 to 50 mA heating current, 10 min exposure time) greatly diminish the emission properties of the emitters. There are two reasons for this. First, the actual emission centers—the tips of the microneedles—are covered by sample residues are corroded; and second, the interstices between the needles are no longer completely available for the adsorption of the substance to be analyzed. The observed emission of these field anodes was always sufficient for satisfactory exposure of the reference compounds [perfluorokerosene, perfluorotributylamine, and tris-(perfluoroheptyl)-*S*-triazine] in the FI mode (i.e., sample in the gas phase), but results for repeated FD measurements of biological samples were disappointing.

3-Deoxyfluoroglucose 6-(disodium phosphate) gave the FD spectrum (Figure 17) in which the $[M + H]^+$ ion at *m/e* 307 is the base peak. Cation exchange in the high field leads to protonated phosphoric acid that is still linked to the sugar moiety. At higher emitter temperature, however, the breakdown of the organic part of the molecule proceeds rapidly, and FD spectra are characterized by inorganic salt clusters. Mixed clusters of phosphoric and metaphosphoric acids and alkaliphosphates similar to those

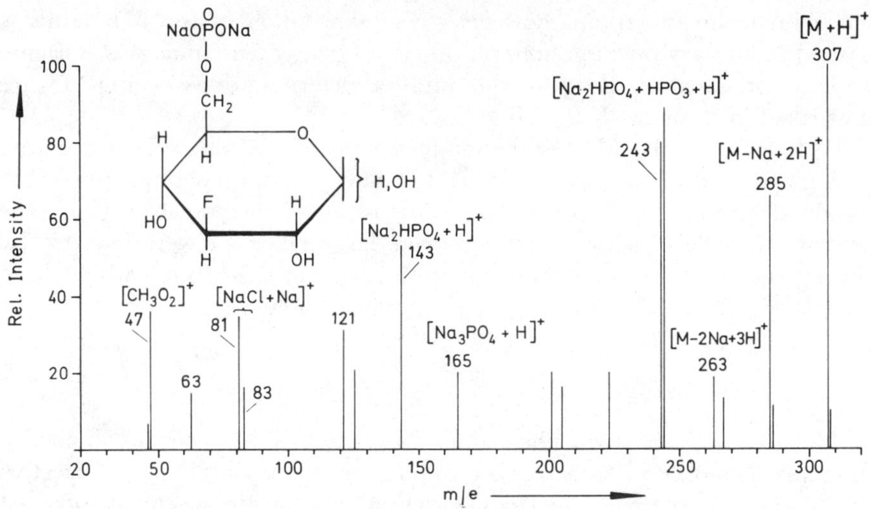

Figure 17. FD-MS of 3- deoxyfluoroglucose-6-(disodium phosphate) (III). For experimental data see Table I (42).

obtained by pyrolysis FD-MS of DNA (Section VI.1) were observed. Simultaneously, small impurities of salts such as sodium chloride are detected at *m/e* 81/83 for the cluster ion $[NaCl + Na]^+$. When the sugar phosphates were studied by HR FD-MS, the 2-, 3-, and 4-fluorine derivatives (II, III, IV) and D-glucose-6-(disodium phosphates) (I) gave accurate masses (Table I). Only peaks of a relative intensity of 5% of that of $[M + H]^+$ are included. As mentioned before, this ion is recorded with 100% relative intensity for all sugar phosphates. Thus the capability of FD-MS for determining the molecular formulas of sugar phosphates indicates that the technique may have considerable value for the identification of intermediates in carbohydrate metabolism.

It was of interest to explore the possibilities of the FD technique for defining the position of the fluorine substituent in the sugar moiety of substances II to IV. An EI-MS study of a series of deoxyfluorosugar acetates (66) and related hexitol derivatives (67) has shown that cleavage of the C-C bond in a C(F)-C grouping is relatively disfavored. On the basis of the available data, it was not possible to establish a similar effect in the fragmentation of compounds II to IV under FD conditions. This might be explained by two reasons. First, the short desorption time of the sample from the emitter surface at higher emitter temperatures hinders the detection, particularly of these smaller fragments from the organic moiety of the molecule. Second, the protonation of the molecule and the generation of ions

TABLE I

Field Desorption Mass Spectra[a] of D-Glucose 6-Phosphate (Disodium Salt, I) and Deoxyfluoroglucose 6-Phosphates (Disodium Salts, II–IV)[b] (42)

Mass		Relative intensity (%)						
Found[d,e]	Theory	I	II	III	IV	Composition	Origin	Ion type[c]
308.007	308.010	—	7	10	9	$^{13}CC_5H_{11}O_8FPNa_2$	(M + 2)	Isotope
307.004	307.006	—	$\underline{100}$	$\underline{100}$	$\underline{100}$	$C_6H_{11}O_8FPNa_2$	(M + 1)	Quasi-molecular
306.017	306.015	8	—	—	—	$^{13}CC_5H_{12}O_9PNa_2$	(M + 2)	Isotope
305.014	305.012	$\underline{100}$	—	—	—	$C_6H_{12}O_9PNa_2$	(M + 1)	Quasi-molecular
286.028	286.029	—	14	11	15	$^{13}CC_5H_{12}O_8FPNa$		Isotope
285.025	285.025	—	82	66	82	$C_6H_{12}0_8FPNa$	[(M − Na + H) + H]	
284.036	284.034	10	—	—	—	$^{13}CC_5H_{13}O_{10}PNa$		Isotope
283.031	283.030	78	—	—	—	$C_6H_{13}O_{10}PNa$	[(M − Na + H) + H]	
267.008	267.013	—	18	13	18	$C_6H_{10}O_7FPNa$	285 − H_2O	
263.042	263.044	—	20	19	20	$C_6H_{13}O_8FP$	[(M − 2Na + 2H) + H]	
261.045	261.049	10	—	—	—	$C_6H_{14}O_9P$	[(M − 2Na + 2H) + H]	
245.031	245.032	—	40	36	40	$C_6H_{11}O_7FP$	263 − H_2O	
244.898	244.899	15	52	48	52	$H_2O_7P_2Na_3$	$(Na_3PO_4)(HPO_3)H$	Inorganic cluster
243.035	243.037	40	85	80	85	$C_6H_{11}O_8P$	263 − HF (I,261 − H_2O)	
222.925	222.917	10	28	20	22	$H_3O_7P_2Na_2$	$(Na_2HPO_4)(HPO_3)H$	Inorganic cluster
204.906	204.905	30	50	16	43	$HO_6P_2Na_2$	$(NaPO_3)(NaPO_3)H$	Inorganic cluster
200.936	200.936	18	31	20	27	$H_4O_7P_2Na$	$(NaH_2PO_4)(HPO_3)H$	Inorganic cluster
164.933	164.931	30	26	20	35	HO_4PNa_3	$(Na_3PO_4)H$	Inorganic cluster
142.942	142.950	52	70	56	22	$H_2O_4PNa_2$	$(Na_2HPO_4)H$	Inorganic cluster

TABLE I (*Continued*)

Mass		Relative intensity (%)						
Found[d,e]	Theory	I	II	III	IV	Composition	Origin	Ion type[c]
124.939	124.938	16	60	10	7	O_3PNa_2	Na_2PO_3	Fragment
120.967	120.969	40	65	16	13	H_3O_4PNa	$(NaH_2PO_4)H$	
83.051	83.056	12	17	16	21	C_5H_7O	[Methylfuran + H]	Pyrolysis product
80.974	80.975	20	40	35	56	H_2O_3P	$(HPO_3)H$	
63.025	63.027	—	12	15	—	C_2H_3OF		
61.030	61.033	32	—	—	35	C_2H_5O		
60.025	60.024	31	—	—	8	C_2H_4O	Fragments	
47.016[f]	47.016	50	—	36	—	CH_3O_2		
46.006[f]	46.007	—	—	7	—	CH_2O_2		

[a] Measured at high resolution (15,000–20,000, 10% valley definition) on a modified CEC21-110B spectrometer (Mattach-Herzog geometry, photoplate detection) with a heating current of 0 to 50 mA. Other operating conditions were as given in ref. 42.
[b] Compounds adsorbed on the emitter from aqueous solution.
[c] Apart from m/e 125 (Na_2PO_3), all ions occured preponderantly as the protonated forms.
[d] From the spectrum of I and the 2-fluoro derivative II.
[e] High relative intensities are observed at *m/e* 22.990 for Na^+ and 38.964 and 40.962 for $^{39}K^+$ and $^{41}K^+$, the latter probably introduced when the sugar phosphates were eluted from an anion-exchange resin in the borate form which had been generated using potassium borate.
[f] From the spectrum of I and the 3-fluoro derivative III.

from the adsorbed surface layer in the extremely high field is expected to result in different fragmentation pathways.

It seems likely that C-2 and C-3 are present in the 2-C fragments in the FD spectra of the fluorosugar phosphates. These fragments have *m/e* values of 63 for the 2-(II) and the 3-fluoro derivative (III), and 61 for the 4-fluoro analog (IV) (Table I). This is supported by the occurrence of the fragments at *m/e* 46, 47 and *m/e* 60, 61 in the spectra (low and high resolution) of glucose-6-phosphate disodium salt (I). Moreover, the FD-MS of free nucleotides (24) display signals at *m/e* 60.021 $[(CHOH)_2]^+$ and 61.029 $[(CHOH)_2 + H]^+$, probably indicating a ring opening between C-1 and C-2 following the field-induced proton transfer to the ring oxygen.

Thus low- and high-resolution FD-MS is of particular value for the determination of molecular weights and molecular formula of sugar phosphates on small amounts ($\sim 5 \times 10^{-8}$g) of material without derivatization. Since data on fragmentation pathways of carbohydrates under FD conditions and on the relative contribution of pyrolytic processes (31) are limited, the scope of the method for defining the position of functional groups needs further investigation.

2. Sulfates and Related Compounds

A. SULFATE ESTERS AND SULFATE SALTS (MULTIPLY CHARGED IONS)

Macromolecular sulfate compounds, sulfate esters used as artificial substrates for sulfatase assay, sulfates of drugs, and biochemical reagents occupy an important place in the application to biochemical analysis. In general, the introduction of the–SO_4C grouping (where C may be a proton or a metal cation) increases the water solubility and the nonvolatility of a compound. Whereas other polar functional groups can be protected by derivatization, no successful procedure to make sulfates amenable to MS analysis has been reported.

The initial investigations of large organic sulfate esters by Games et al. (68) constitute a good demonstration of the utility of FD-MS. The potassium salts of *n*-hexyl, *n*-decyl, and *n*-undecyl sulfates showed $[M + K]^+$ cluster ions at *m/e* 259, 315, and 329, respectively, and no fragmentions. Similar behavior was observed for the potassium salt of cyclohexylphenyl-4-sulfate. When the sodium salt of cortisone sulfate was electrically recorded at different emitter heating currents, the compound showed unusual behavior at low ehc. The base peak in its spectrum was at *m/e* 254 and is ascribed to the doubly charged ion $[M + 2Na]^{2+}$; other ions were present at *m/e* 254.5 (13% rel. int.), 255 (21% rel. int.), and 485 (8% rel. int.) $[M + Na]^+$. Higher ehc increased the intensity of *m/e* 485 relative to 254. However at 20 mA, the highest ehc used, the only ions were at *m/e* 360, (33% rel. int.) and 342 (100%

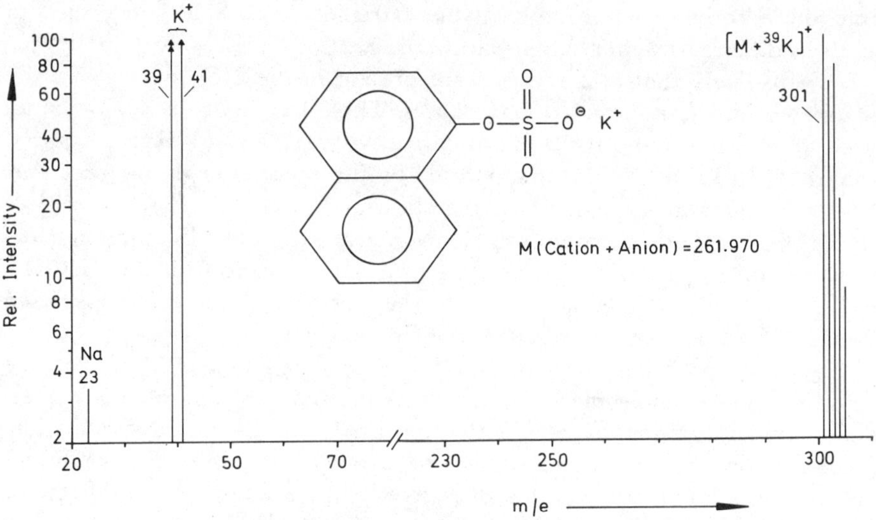

Figure 18. FD-MS of the potassium salt of α-naphthylsulfate ester (69).

rel. int.), and they are ascribed in the first case to a loss of the sulfate group with hydrogen transfer and to dehydration and hydrogen transfer in the latter case.

Recently, a number of organic sulfate esters and salts have been studied by FD-MS (69). As shown for inorganic substances such as alkali halides (Section VII), cluster formation (60) is the outstanding feature of small sulfate esters. The FD mass spectrum of α-naphthylsulfate potassium salt (Figure 18) illustrates clearly the high cation intensities obtained at *m/e* 39 and *m/e* 41 for the potassium isotopes. The FD method possesses extremely high sensitivity for traces of alkali cations. Although the α-naphthalenesulfate potassium salt was of analytical grade, an impurity due to sodium (*m/e* 23) was found. A comparison of the analytical data obtained with atomic absorption spectrometry revealed that FD-MS is more sensitive by orders of magnitude for $[Li]^+$, $[Na]^+$, and $[K]^+$ (49). Of equal importance, however, is the occurrence of the intense $[M + K]^+$ cluster ions from which the molecular weight is easily derived. No other signals were recorded between *m/e* 18 and 560 with the photographic detection system.

The FD spectra of seven organic sulfate salts with large organic cations have been reported (69). Sulfate salts, quinine-, strychnine-, atropine-, isoprenaline-, bamethane-, orciprenaline-, and amphetamine sulfates, all showed very intense cation $[C]^+$ and $[C - H]^+$ ion signals. In every case one of these ions is the base peak of the FD mass spectrum. Obviously the FD method is a suitable technique for the determination of cations of large organic amines that are present as sulfate salts. In all cases, elimination of water from the $[C]^+$ and/or $[C - H]^+$ is observed. At *m/e* values much above the nominal mass of the cation intense cluster ions are found, one fraction of them organic, the other purely inorganic, with sulfuric acid. The cluster ions serve in analyzing sulfate salts because the compound under investigation can be recognized as a sulfate salt, and some cluster ions (e.g., $[C + H_2SO_4]^+$) permit determination of the molecular weight (2C + anion).

The FD spectra of the sulfate salts in all cases revealed intense doubly charged ions. It appears that the formation of very intense doubly charged ions is a typical feature in FD-MS of large organic salts of such polyvalent acids, as sulfuric acid and phosphoric acid (31). Table II lists the prominent multiply charged ions in the FD mass spectrum of isoprenaline sulfate. Accurate mass measurements enabled the tentative assignment of elemental composition. On the basis of these data, an abbreviated description of the cluster ions, using the cation and fragments as building blocks, was used. It is noteworthy that in these investigations (69) intense triply charged ions were recorded, a novel type of ion in FD-MS not reported previously. The use of doubly charged ions to give additional analytic information for the interpretation of FD mass spectra was described earlier (30).

It is remarkable that all structural units of the multiply charged cluster ions, as formulated in Table II, are also observed as very intense singly

TABLE II
Doubly and Triply Charged Ions in the FD Spectrum of Isoprenaline Sulfate (69)

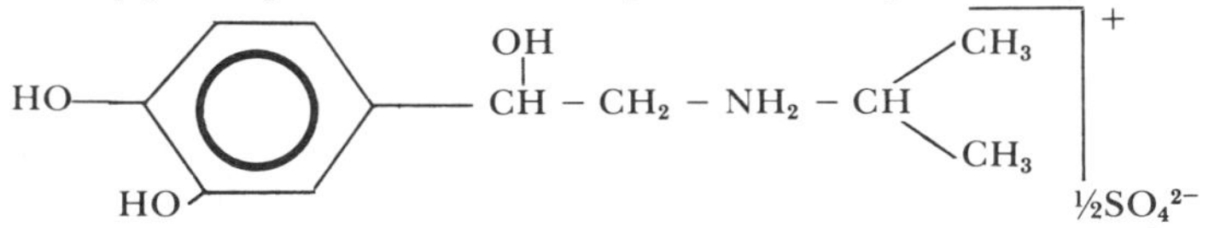

Doubly charged ions	Relative intensity[a]	Triply charged ions	Relative intensity[a]
$[2C - H_2O]^{2+}$	++	$[2(C - H_2O) + C]^{3+}$	+++
$[2C - H]^{2+}$	+++	$[3(C - H_2O) + (C - H) + H_2SO_4]^{3+}$	++
$[2(C - H_2O) + H_2SO_4]^{2+}$	+++		
$[(2C - H_2O) + H_2SO_4]^{2+}$	++		
$[2(C - H) + H_2SO_4]^{2+}$	++		
$[2C + H_2SO_4]^{2+}$	++		

[a] Relative intensities of the signals given in four degrees of blackening of the photoplate: high = ++++, medium = +++, low = ++, weak = +.

charged ions. In field ionization the formation of doubly charged ions is explained by ionization of singly charged species that are bound intermediately to the emitter surface (70). It is assumed that multiply charged cluster ions of large organic sulfate salts are generated under FD conditions, by way of aggregation of singly charged ions bound in the adsorbed layer of the emitter surface.

B. SULFONIC ACIDS (FD ANALYSIS OF COMPOUNDS OF TECHNICAL PURITY)

Another functional group that increases water solubility and decreases volatility of biochemicals is–SO_3 C (where C is a proton) in sulfonic acids and a metal cation in sulfonates. Apart from preliminary reports on the FD mass spectra of 7-iodo-8-hydroxyquinoline-sulfonic acid (50), the potassium salt of 1-hydroxy-2-methoxybenzene-sulfonic acid (71), and the sodium salt of benzenesulfonic acid (72), little is known about the mass spectrometric behavior of underivatized sulfonic acids and sulfonates. Because of the low volatility of these classes of compound, it is difficult or impossible to obtain useful and complete spectra with the conventional mass spectrometric methods. Since these substances play an important role as dyestuff intermediates and dyestuffs, detergents, pharmaceuticals, environmental chemicals, and biochemicals, a series of representative compounds was investigated by FD-MS (23,73).

In Table III the sulfonic acids examined and their major FD signals are listed. In general the FD spectra of free sulfonic acids have four characteristic features.

1. All compounds give high molecular ion intensities. In most cases complete photographic blackening was obtained for the $[M]^+$. With the time-temperature program employed for the desorption of the samples, $[M + 1]^+$ intensities were smaller than $[M]^+$ intensities in most cases.

2. Elimination of SO_3 from the parent molecule is often observed. Higher emitter heating currents (> BAT) used for desorption apparently favor this process. The resulting $[M - SO_3]^+$ ion is detected in some cases as a doubly charged ion $[M - SO_3]^{2+}$.

3. The FD spectra of the technical grade free sulfonic acids show contamination, due mainly to sulfonates and alkali salts. The intensities for the $[Na]^+$ and $[^{39}K]^+$ ions are given in Table III. In view of the extremely high sensitivity of FD-MS for alkali ions, contributions from solvent impurities and emitter material cannot be excluded.

Of particular interest are signals produced by *cationization*, a process that includes the attachment of a cation to a neutral particle (74; Section IV.3). In general the corresponding $[M + Na]^+$ ions are formed when the $[Na]^+$ ions are recorded with high relative abundance. However in a number of cases high alkali ion intensities do not lead to detectable signals for the

TABLE III
Major Types of Ion Observed in the FD Spectra of Sulfonic Acids (23)

Compound	Mol. Mol. wt.	Relative Intensities[a] [M]$^+$	[M + 1]$^+$	[M − SO_3]$^+$	[Na]$^+$	[K]$^+$	[M + Na]$^+$	Structurally significant fragments
Aminosulfonic acid	96,983	++++	+++	+++				
n–Ethylsulfonic acid	110,004		+++					
n–Butylsulfonic acid	138,035		+++					
Benzenesulfonic acid	158,004	++++	++		++	++		
2–Amino-benzene-sulfonic acid	173,015	++++	++++	++	++++	++++	++	[M + 1 − NH_3]$^{2+}$
3–Amino–benzene-sulfonic acid	173,015	++++	+++		+++			
4–Amino–benzene-sulfonic acid	173,015	++++	++++					
3–Camphorsulfonic acid	232,077	+++	++++	+				[M + 1 − H_2O]$^+$
6–Hydroxy–7–amino–3–naphthalene-sulfonic acid	239,025	++++	++++	+	++++	++++		[M + 1 − NH_3]$^+$
4–Hydroxy–6–amino–2–naphthalene-sulfonic acid	239,025	++++	++++	+++	++++	++++	+++	[M + 1 − NH_3]$^+$
4–Amino–5–hydroxy–1–naphthalenesulfonic acid	239,025	++++	+++	++++	++++	++++	+	[M + 1 − NH_3]$^+$
5–Nitro–2–naphthalenesulfonic acid	253,005	++++	++++		+++	+++		
4–Hydroxy–6–phenylamino–2–naphthalenesulfonic acid	315,057	++++	+++	+++				
5,6–Dihydroxy–1–anthraquinonesulfonic acid	319,999	++	+	+	++++			[M + 1 − H_2O]$^+$
1,4–Dihydroxy–2–anthraquinonesulfonic acid	319,999	+++	+++	++	++++	++++		
5–Nitro–3–anthraquinonesulfonic acid	332,994	+++	+		++++	++++	+	[R − SO_3Na + Na]$^+$, [R − SO_3Na + 2Na]$^+$ [R − SO_3Na + K]$^+$, [3R − SO_3Na + 2Na]$^{2+}$
5–Nitro–1–anthraquinonesulfonic acid	332,994	+++	+		++++	+++,+		
7–Iodo–8–hydroxy–5–quinoline-sulfonic acid	350,906	++++	+++	+	++++	+++	+	[I]$^+$, [R − SO_3Na + 2Na]$^{2+}$

[a] Relative intensities as in Table II.

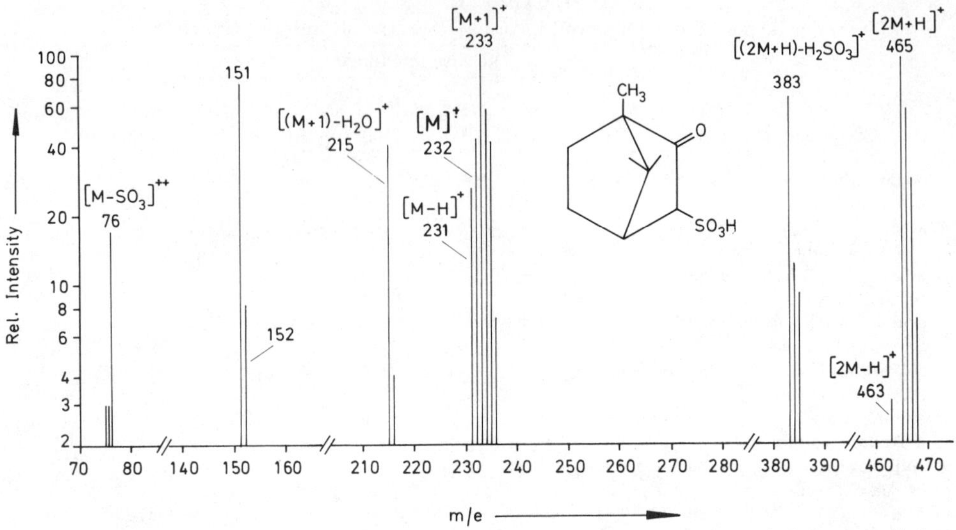

Figure 19. FD-MS of 3-camphorsulfonic acid: solvent, water; emission-controlled FD-MS; threshold, 5×10^{-8} A; BAT, 18 mA ehc.; exposure time, 4 min; temperature range 0–20 mA ehc (23).

cationized parent molecules. As the discussion of the parameters for cationization (74) indicates, the capacity for the formation of cationized molecules decreases in the first group of the periodic system from lithium to cesium. Furthermore, the presence of $[M + Na]^+$ ions allows the investigator to discriminate between $[M]^{+}$ or $[M + 1]^+$ ions if they are also recorded, thus providing helpful analytical information (75).

4. In addition to the reliable use of FD-MS for molecular weight determination of sulfonic acids, in some cases structurally significant fragmentation can be obtained. As may be inferred from Table III, the elimination of small neutral chemical species such as ammonia and water from the protonated molecule is observed preferentially. It has been shown that this phenomenon occurs with amino acids (29), sultams carrying carboxylic substituents (30), bile acids (491), and glucuronic acids (59), and apparently it is common in FD-MS of free acids.

The foregoing characteristic points are clearly illustrated in the FD mass spectrum of 3-camphorsulfonic acid (Figure 19).

1. The base peak of the spectrum is the $[M + 1]^+$ ion at *m/e* 233.077. The molecular ion is displayed with 42% relative intensity. The correct assignment of the molecular weight is aided by the occurrence of intense cluster ions $[2M + H]^+$ and their fragments in the upper mass range.

2. A relatively small signal found at m/e 152.120 can be interpreted as loss of SO_3 from the parent molecule. Further evidence for this process is given by the pronounced ion at m/e 76.060 for $[M - SO_3]^{2+}$. More details regarding the usefulness of the doubly charged ions in furnishing structural information and for the determination of the molecular weight of unknown compounds have become available recently (75,169).

C. SULFONATES (CLUSTER FORMATION IN FD-MS)

Of particular interest in the application of FD-MS is the analysis of organic and inorganic salts. The lack of volatility of the sulfonates carrying large organic substituents makes it almost impossible to convert the molecular unit M [= cation (C) + anion (A)] into the gas phase without thermal decomposition of the organic moiety. Therefore the mass spectrometric analysis by conventional methods, such as EI, CI, and FI, is hampered. Table IV gives the major signals obtained from basic sulfonates, dyestuff intermediates, and some widely used dyestuffs.

Five common features are observed:

1. The $[nM + C]^+$ cluster ion with $n = 1$ in most cases is the prominent ion. This process of cluster formation is energetically and kinetically favored and has been observed both for organic and inorganic salts.

2. As might be expected from the behavior of organic salts in FD-MS, the FD spectra of sulfonates often display smaller relative abundances of the molecular ion $[M]^+$ than for the first cluster $[M + C]^+$. However there is a tendency for the $[M]^+$ intensities to increase with increasing molecular size as compared with the intensities of the cluster ions. This is particularly pronounced for the two commercial dyestuffs listed at the end of Table IV. It is assumed that better charge stabilization due to mesomeric effects contributes to the production of molecular ions of large-molecule salts.

3. Structurally diagnostic ions are recorded in the FD spectra of sulfonates (e.g., doubly charged ions are more frequently demonstrated than with free sulfonic acids). Most information about the structure, however, is obtained by the outstanding and commonly observed field dissociation phenomenon of direct bond cleavages in the molecule. The resulting ions reflect straightforward building blocks in the molecule. Additionally, the frequent appearence of complementary charged particles simplefies the interpretation of the FD spectrum even for nonexperts in mass spectrometry.

These characteristic properties of the FD technique for the analysis of sulfonates are illustrated in Figure 20*a*, showing the FD spectrum of a common, commercial dyestuff. The strongest peak of the spectrum is produced by the $[M + C]^+$ cluster ion. It is noteworthy that when using HR signals due to a formation of $[2M + 2Na]^{2+}$, ions are observed in this mass region.

TABLE IV
Major Types of Ion Observed in the FD- Spectra of Sulfonates (23)

Compound	Mol. wt.	Relative intensities[a] $[M]^+$	$[M + C]^+$	$[2M + C]^+$	Significant ions
1-Chloroethyl-2-sulfonic acid, sodium salt	165,947		++++	+++	
Benzenesulfonic acid, sodium salt	179,986	+	++++	++++	
3-Nitrobenzene-1-sulfonic acid, sodium salt	224,971		++++	++	
Naphthalene-1-sulfonic acid, sodium salt	230,001	++	++++	++++	$[R - SO_3H]^+$
7-Hydroxynaphthalene-1-sulfonic acid, potassium salt	261,970		++++	+++	$[M + Na]^+$, $[2M + Na]^+$
Benzene-1,3-disulfonic acid, disodium salt	281,925		+++	+	$[M + 2Na]^{2+}$
Anthraquinone-2-sulfonic acid, sodium salt	309,991		++++	++	
5,6-Dihydroxy-1,3-benzenedisulfonic acid, disodium salt	313,914		++++	++	$[M + 2Na]^{2+}$ $[(M - H_2O) + Na]^+$
Dodecyl-benzenesulfonic acid, sodium salt	348,174	++	++++	+	$[M + 1]^+$, $[R - SO_3H + 1]^+$
6-*p*-Toluidino-3-methylanthrapyridone-2′-sulfonic acid, sodium salt	468,076	++++	++++	+	$[M + Na]^{2+}$ $[M + 2Na]^{2+}$ $[M + 1]^+$
1-Amino-4-(3′-4″-methoxybenzoylamino-anilino)-anthraquinone-2-sulfonic acid, sodium salt	565,092	++++	+++	+	$[M]^{2+}$, $[M + Na]^{2+}$ $[M + 1 + Na]^{2+}$ $[(R - SO_3H) - SO_3]^+$ $[M + 1]^+$

[a] Relative intensities as in Table II.

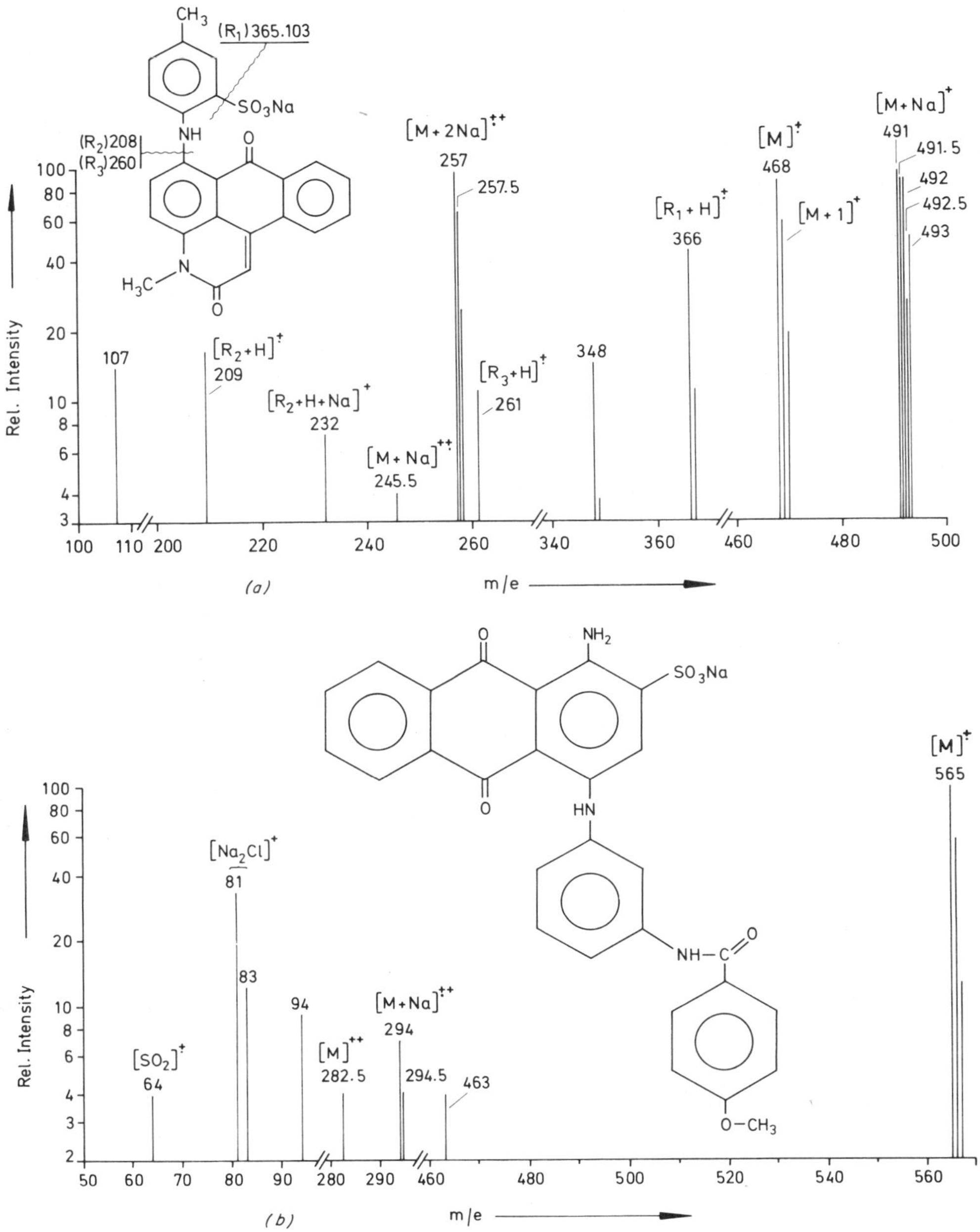

Figure 20. (*a*) FD-MS of 6-*p*-toluidino-3-methylanthrapyridone-2-sulfonic acid, sodium salt (23). (*b*) FD-MS of 1-amino-4-(3′-4″-methoxy-benzoylaminoanilino)-anthraquinone-2-sulfonic acid, sodium salt: solvent, water; emission-controlled FD-MS; threshold 5×10^{-8} A; BAT, 26 mA ehc, exposure time 15 min; temperature range 0–50 mA ehc (23).

4. Ions due to the corresponding acid of the sulfonate are of minor importance. For example, the FD spectrum of a blue anthraquinone-2-sulfonate dyestuff is exhibited (Figure 20*b*). To obtain the greatest sensitivity, the mass range was limited in this case from *m/e* 20 to 570. Therefore the first cluster ion ($n = 1$) was not recorded and the $[M]^+$ ion at *m/e* 565.092 is the base peak of the spectrum. As Figure 20*b* indicates, significant doubly charged ions are found for $[M + Na]^{2+}$ at *m/e* 294.041 and $[M]^{2+}$ at *m/e* 282.546. Although the free sulfonic acid is not detected at *m/e* 463.153, a relatively intense ion is found that possibly is generated in a solvolysis reaction. The $[SO_2]^+$ ion is often observed in the FD spectra of sulfonic acid and sulfonates and is apparently generated at high emitter heating currents at the end of the exposure time. The ion at *m/e* 94.042 (C_6H_6O) is also assumed to be such a decomposition product or an impurity.

It is remarkable that the cluster ion for sodium chloride stands out from the other organic ions, except the molecular ion. To the best of the author's knowledge, this is the first mass spectrum of a mixture of a large-molecule organic salt and an inorganic salt that reveals the cluster ions as the most intense ions of the spectrum. Obviously FD-MS is able to record large organic as well as inorganic salts in one sample if a *fractionated desorption* can be performed. This means that one compound after another desorbs from the FD emitter according to its best anode temperature in the mixture, and these compounds then can be recorded in "one run" when an integrating ion-recording system is employed. Thus all ions that were desorbed during the whole desorption time were recorded in one track of the photoplate or, alternatively, a number of different spectra from one sample can be recorded according to different heating current-time intervals. First examples of FD spectra of these preselected intervals of the whole desorption time on separate tracks of the photoplate are given in ref. 23. Clearly, partial fractionation of components of differing volatilities (e.g., NaCl and benzenesulfonate) occurs and gives useful analytical information. This mode of recording FD spectra photographically contributes to a better understanding of the basic processes in field desorption.

5. Typical for sulfonates is formation of large-cluster ions where $n \geqslant 1$. Cluster ions with $n = 1, 2, 3, 4, 5$, and 6 were described for sodium acetate (60). The relative abundance for these ions decreased gradually toward the higher mass range when an integrating recording device (photoplate) was employed. Similar spectra were obtained for sulfonates. When the mass range extended from *m/e* 38 to 1400 with photographic detection, the FD spectrum of the benzenesulfonic acid sodium salt (in short, benzenesulfonate) is characterized by the prevailing cluster ions. Moreover $[nM + Na]^+$ ions for $n = 1, 2, 3, 4, 5$ and $6[nM + {}^{39}K]^+$ ions for $n = 1, 2$, and 3 are found (23). In contrast to the results of Large and Knof obtained by electric recording (72), it was possible to record the $[6M + Na]^+$ cluster at *m/e* 1103 photo-

graphically (with 9% rel. int.). For benzenesulfonate (and the compounds listed in Table IV) a gradual and continuous decrease in the relative ion cluster intensities with increasing *m/e* values was observed. Integrating recording yields ion currents for the $[n\text{M} + \text{Na}]^+$ ions that drop for ions with $n = 1$ to $n = 6$ and for $[n\text{M} + {}^{39}\text{K}]$ for $n = 1$ to $n = 3$.

3. Nucleic Acid Derivatives

A. NUCLEOSIDES (SOLVENT AND SALT INFLUENCE IN FD-MS)

The application of mass spectrometry to nucleic acid derivatives has been severely limited by the very low volatility of these compounds (77). The first mass spectra of free bases and nucleosides were obtained by Biemann and McCloskey in 1962 (78), and later a number of new techniques for forming volatile derivatives were developed. In view of the biochemical importance of nucleosides and the large number of derivatives of biologically active nucleosides that have been synthesized (79), it was of interest to determine the feasibility of FD analysis of the native compounds. Therefore the next step after the FD-MS investigation of purine bases (80) was the study of unprotected nucleosides. Most members of this class of compound do not display larger intensities of the molecular ions under EI conditions (78), but do so with FI-MS (81). Guanosine constitutes a remarkable exception because it exhibits no signal for the molecular ion, nor in the EI or FI spectra (81). It is striking that all FI-MS are obtained at probe temperatures on the average 40 °C higher than for the EI spectra, which suggests an attempt to compensate for lower FI intensities by higher probe temperatures.

The FD-MS of guanosine displays the $[\text{M}]^+$ peak with 20% intensity of the base peak $[\text{B} + \text{H}]^+$ at *m/e* 151.049 (Figure 21*a*). The ehc of 18 to 20 mA for the 10-μm (diameter), high-temperature-activated tungsten wire corresponds to a desorption temperature between 180 and 200 °C. The FI-MS was obtained at a probe temperature of 250 °C (81). Thus we have confirmation of the view that in the comparison of FI and FD, the salient parameter is the heat of ionic evaporation that can be lowered by the high electric field in the case of FD. This is especially important for thermally unstable or insufficiently volatile compounds [guanosine melts at 240 °C, with decomposition (82)]. The FD-MS of adenosine (m.p. 220 °C) and thymidine (m.p. 186 °C) are similar to the FI spectra (81), displaying molecular peaks with high intensities and smaller, but important, fragmentation, which yields information about the sugar (S) and base (B) moieties of the molecules.

The FD-MS of guanosine (Figure 21*b*) was obtained from a mixture of acetone and 1*N* aqueous KOH, supporting previous findings (11) that the choice of a suitable solvent is decisive for the quality of the desorption spectrum. The intensities of the $[\text{M}]^+$ and $[\text{M} + 1]^+$ signals are remarkably enhanced, the latter being the base peak. Precise mass measurements give a

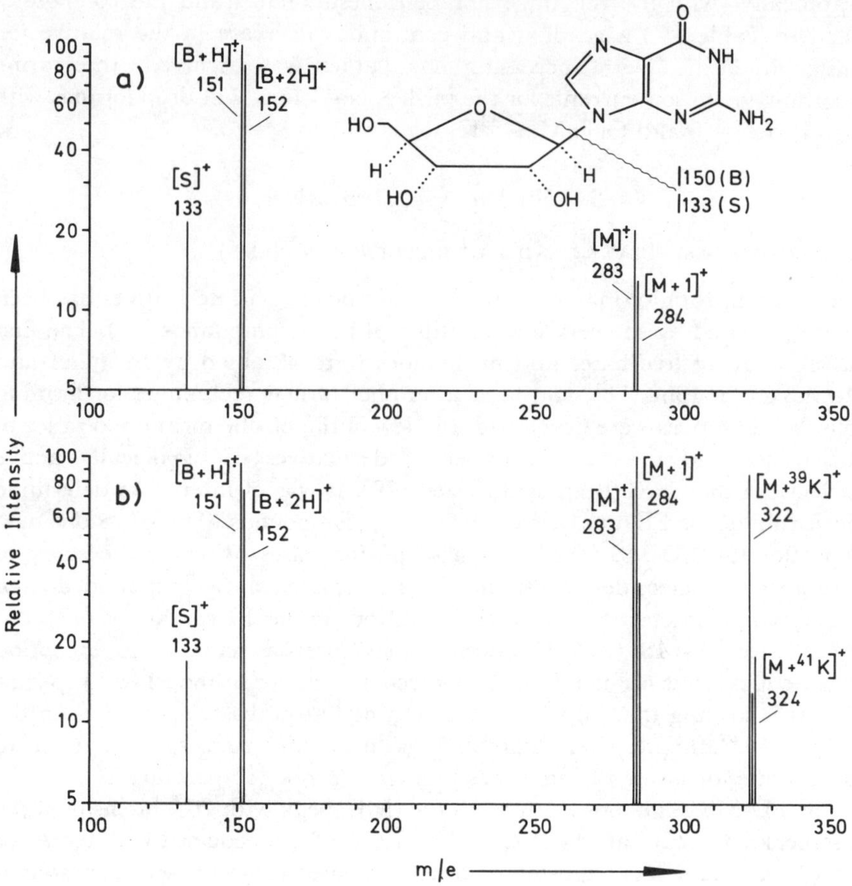

Figure 21. FD mass spectra of guanosine in solvents. (*a*) In acetone. (*b*) In a mixture of acetone and 1*N* aqueous KOH (11).

signal at *m/e* 322.055, making it certain that the complex ion consists of $[M + {}^{39}K]^+$ and is due to field-induced ion-molecule reactions in the surface layers. Like the high proportion of $[M + H]^+$ in the $[M + 1]^+$ signal, this intense $[M + {}^{39}K]^+$ signal may be ascribed to the energetically favorable even electron system. In this respect, the attachment of a positive ion to a neutral molecule and the detection of an intense signal from the complex provides an interesting parallel between FD and CI mass spectrometry.

General experience shows that solvents such as acetone and dimethylsulfoxide give good results when used in FD. Water and carbon tetrachloride are less suitable. It is assumed that chemically bound solvents, such as water, cause sputtering of the sample molecules during the desorption process. On the other hand, one drawback of aprotic solvents is that they cannot

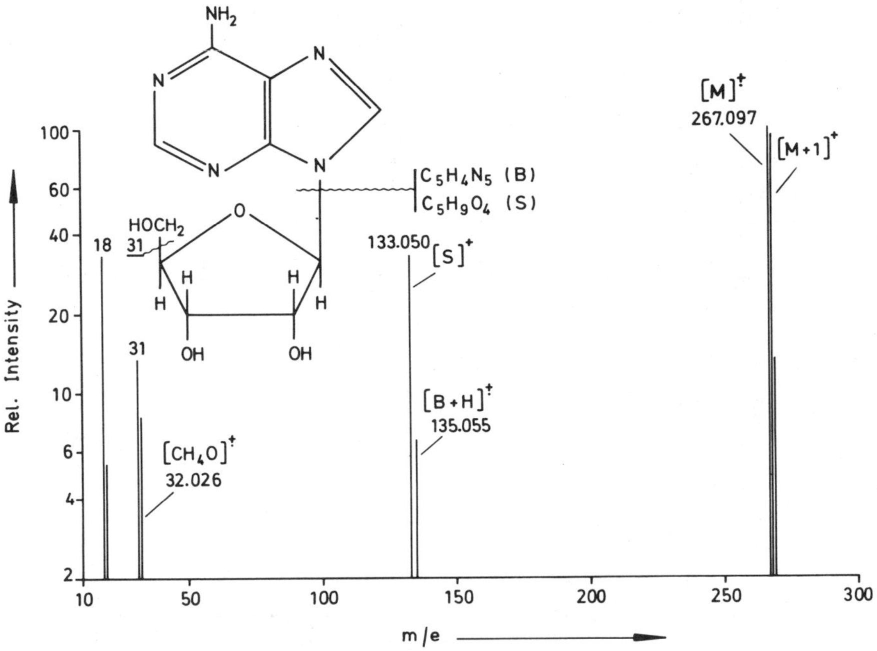

Figure 22. FD-MS of adenosine: solvent, water; sample consumption, 1 μg; photographic detection (24).

contribute to protonation reactions leading to $[M + H]^+$ ions. In general, solutions of the sample gave better results than suspensions. However it is possible to transfer a completely insoluble sample onto the emitter from a fine-grained suspension (e.g., by means of ultrasonic waves).

CATIONIZATION

The high-resolution FD mass spectrum of adenosine as recorded photographically appears in Figure 22. The signals for $[B + H]^+$ at *m/e* 135.054 and $[S]^+$ at *m/e* 133.050 are characteristic, as well as the base peaks for $[M]^+$ and $[M + H]^+$. The solvent used for this FD spectrum was twice-distilled water, and the adenosine was of analytical grade purity. The fragmentation was induced in an ehc program between 0 and 20 mA and at a relatively high emission threshold (5×10^{-8} A). Owing to the procedure employed, one obtains the typical FD mass spectrum of a pure model substance, with the characteristic fragmentation pattern.

Three FD spectra of adenosine (Figure 23) illustrate the dependence of FD-MS on the concentration of inorganic salts for the attachment of a cation

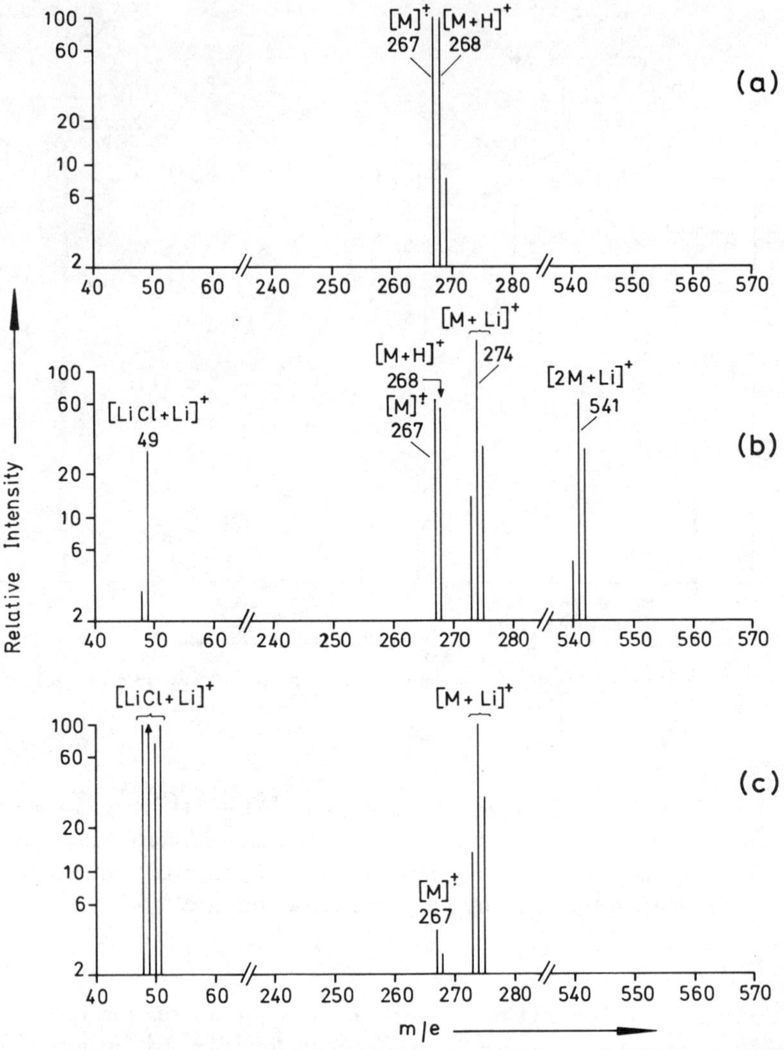

Figure 23. FD-MS of a mixture of adenosine (1*M*) and lithium chloride at different concentrations. Solvent, water; photographic detection (60). (*a*) $10^{-5}M$. (*b*) $5 \times 10^{-5}M$. (*c*) 1*M*.

(C) to a molecule (M) to form a $[M + C]^+$ quasimolecular ion. This process of cationization is of great importance for the molecular weight determination of polar organic substances (74,83,84,169).

To obtain the mass spectra of adenosine, a molar aqueous adenosine solution was used in which LiCl was present at different concentrations.

Under the experimental conditions used in Figure 23*a*, no cationized molecules could be detected. At higher salt concentrations (Figure 23*b*), $[M + Li]^+$ ions appeared. In addition to the $[M + Li]^+$ ions in this spectrum, cationized dimers are also present. In Figure 23*c* the intensities of the $[M]^+$ and $[M + H]^+$ ions are very small compared with those of the ions produced by cationization. Thus a comparison of the spectra in Figure 23 demonstrates the strong influence of alkali salts on the FD spectra of biochemicals. In general, the formation of quasi-molecular ions using other alkali ions than $[Li]^+$ showed a reduction of the quasi-molecular ion intensity from Li to Cs, as is to be expected from the affinity values (85).

Investigations revealed that the use of small-molecule inorganic salts such as alkali chlorides is not always advantageous for cationizaton. A relatively large thermal excitation is necessary to release the cations from such salts, and this demands high cationization temperatures at which the thermolabile, organic molecules can decompose before cationization is achieved. When using larger molecule salts with organic anions for cationization, work can proceed at lower emitter temperatures, since the cations become available at these temperatures by decomposition of the anion. When analyzing unstable organic molecules containing acidic groups, it seems advantageous to convert at least some of the molecules into the salt form by exchange of the acidic hydrogen with Li or Na, allowing these salt molecules to be used as cation donors during field desorption. The conditions for this are often automatically fulfilled, in that the samples from biological sources are applied to the emitter in salt-containing solutions, and consequently part of the substance is present in the form of a salt (169).

The main aspects of cationization (attachment of inorganic or organic cations to a neutral molecule) can be summarized as follows. (1) In general, cationized molecules are more stable than radical molecular ions or protonated molecules. (2) The fragmentation of the polar molecules resulting from a cationization (e.g., by alkali ions), has a higher activation energy than splitting off functional groups after a protonation. Cationization is therefore a useful tool for the determination of molecular weights. Three different methods for achieving cationization have been given (74): (*a*) field-induced cationization a low anode temperature using a heterogeneous reaction of the molecules in the gas phase with molecules of a salt in the adsorption layer on the field anode, (*b*) largely thermally induced cationization at high emitter temperatures and low electric fields, and (*c*) cationization, causing less thermal excitation to the molecules, using the field desorption technique.

B. NUCLEOTIDES (SEQUENCE INFORMATION)

Following the work of McCloskey et al. (86), the class of compound referred to as nucleotides has been subject to mass spectroscopic investigation.

To circumvent the extremely low volatility, the trimethylsilyl (TMS) derivatives (which are sufficiently volatile) have been studied, and EI-MS of mononucleotides have been obtained. This method yields valuable structural information but has some disadvantages (multiple derivatization, sensitivity of the products to traces of water, and increased molecular weight from additional protecting groups).

The first mass spectrum of a free nucleotide, adenosine-5′-monophosphate (Figure 24), permits the elucidation of the following basic structural information.

1. The molecular weight is indicated by the quasi-molecular ion $[M + H]^+$ at *m/e* 348.071 with 73% relative intensity of the base peak.

2. The $[N + H]^+$ ion for the corresponding nucleoside (adenosine), may be identified by precise mass measurements at *m/e* 267.097 (29% rel. int.).

3. The base signal of the FD-MS at *m/e* 136.062 is determined to be the $[B + H]^+$ peak for adenine.

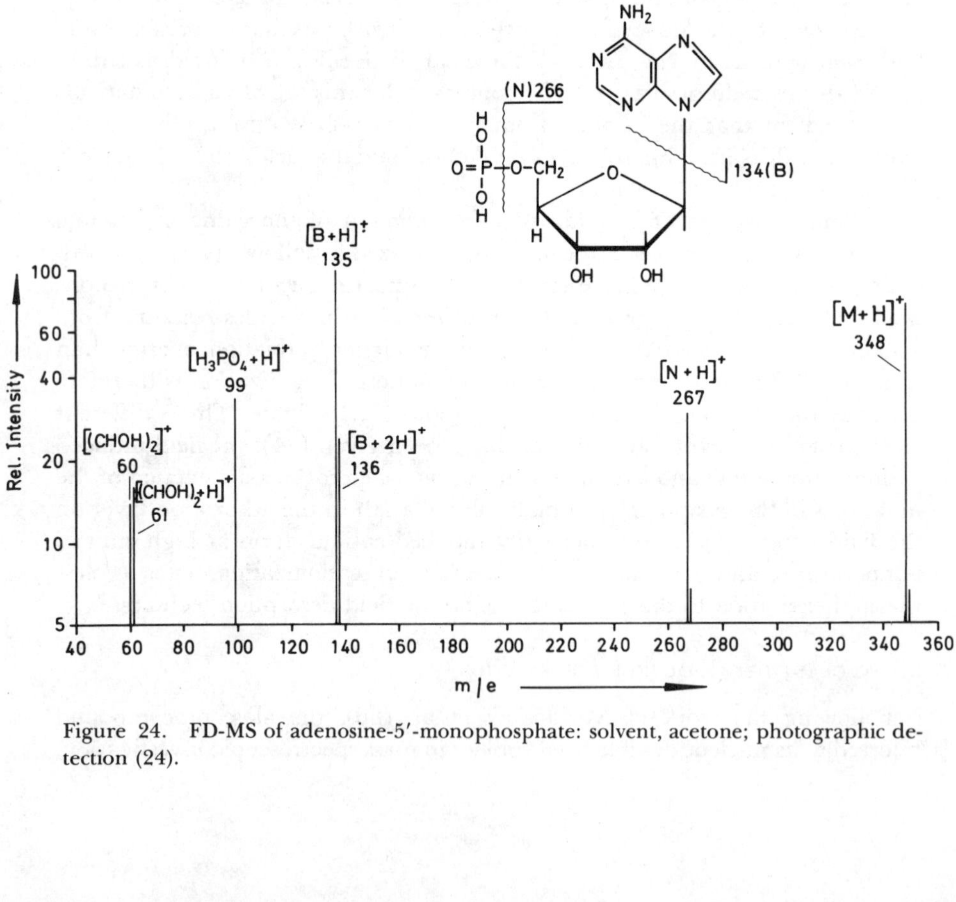

Figure 24. FD-MS of adenosine-5′-monophosphate: solvent, acetone; photographic detection (24).

4. Although there is no peak for the sugar (S) fragment, from the significant signals at *m/e* 60.021 $[(CHOH)_2]^+$ and 61.029 $[(CHOH)_2 + H]^+$, the decision between ribose and deoxyribose favors the ribose.

5. Though all these fragments appear at emitter heating currents up to 19 to 20mA, higher emitter temperatures (ehc up to 50 mA) display a distinct signal at 98.985 due to $[H_3PO_4 + H]^+$, protonated orthophosphoric acid.

Since other free nucleotides, 5′-thymidylic acid, 5′-deoxycytidylic acid, and 5′-deoxyguanylic acid, yield similar FD mass spectra, further investigations have been undertaken to apply this technique to oligonucleotides to obtain sequence information.

Recently the first results of the FD-MS of unprotected dinucleoside phosphates were published (41). For TpT, a deoxyribodinucleotide, the following significant FD signals are described. The most intense ion is generated by direct bond cleavage between the thymine portion (B) and the deoxyribose moiety of the free dinucleotide. The protonated thymine molecule $[B + 2H]^+$ at *m/e* 127.051 is the base peak of the spectrum. At *m/e* 323.064 the protonated thymidylic acid is detected. The most important feature, however, is the ion at *m/e* 547.144 for the protonated dinucleotide $[M + H]^+$, which is displayed with relatively high abundance (21%). Products of the thermal decomposition of the sugar part of the molecule are also detected. The ions at *m/e* 81.034 (C_5H_5O), at *m/e* 98.037 ($C_5H_6O_2$), and at *m/e* 117.055 ($C_5H_9O_3$) are interpreted as fragments of the deoxyribose.

The mechanism of the thermal formation of mass 98 ($C_5H_6O_2$) from DNA has been investigated by low-voltage ionization quadrupole MS and HR field ionization MS (87), and a structure (furfuryl alcohol) for this ion has been proposed. Further experiments by collision-activation MS (88) of the ion at *m/e* 98 in the pyrolysis products of DNA, confirmed the presence of furfuryl alcohol and α-angelicalactone. As has been reported for the FD-MS of adenosine monophosphate (24), protonated *o*-phosphoric acid (*m/e* 98.985) is also found in the FD-MS of TpT. In addition, a weak ion for protonated *m*-phosphoric acid is found at *m/e* 81.974 which also appeared in the HR FD-MS of D-glucose-6-disodium phosphates (42).

When the FD mass spectrum of GpU as a ribodinucleotide containing guanosine (i.e., the nucleoside with the lowest volatility was also taken in this case), the protonated molecular ion is clearly detected. In contrast to the FD-MS of the deoxyribodinucleotide TpT, the MS of GpU exhibits a reduced molecular ion intensity and considerably enhanced fragmentation. It is noteworthy that the base of lower volatility—namely, guanine—gives the base peak of the spectrum at *m/e* 152.057 $[B_1 + 2H]^+$, whereas uracil $[B_2 + H]^+$ at *m/e* 112.027 is displayed with 67% relative intensity. Ions with the composition of $[G + H]^+$ and $[GMP + H]^+$ are not found. However the products generated by thermally induced water elimination, such as the ions at *m/e* 266.089 for $[(G - H_2O) + H]^+$, at *m/e* 329.052 for $[(GMP - 2H_2O) + 2H]^+$, at *m/e* 346.055 for $[(GMP - H_2O) + H]^+$, and at *m/e* 461.095 for $[(M - uracil - H_2O) + H]^+$, are detected and reflect the low volatility of this nucleobase. Following the loss of guanine from the molecular ion, the species that contain uracil, such as U, UMP, and M-guanine, are displayed with high relative abundance. Also, in the case of uridylic acid (*m/e* 325.044) water elimination takes place, generating the ion at *m/e* 307.033. The ion at *m/e* 303.107 has been identified as the $[2M + H]^+$ ion of guanine. Neither thermal decomposition of the sugar moiety nor phosphate clusters have been observed.

The FD spectra of a pair of inverse dinucleotides CpA and ApC were investigated (89). It was of special interest to evaluate the differences in the FD spectra and possibly to find key fragments that give a clue to distinguishing the position of the bases B_1 and B_2. The results of HR FD-MS and accurate mass measurements of CpA appear in Table V. A resolution of 10.000, 10% valley definition, was achieved, the error in the accurate mass determination being in the 10-ppm range. The samples were introduced by way of the modified syringe technique, using water as the solvent. The sample amount consumed in one FD measurement was 1 μg. In the FD-MS of CpA (Figure 25) the peaks for the protonated molecule at *m/e* 573.146, and for the

TABLE V

Proposed Chemical Structure of Observed Ions in the FD-MS of CpA (41)

Compounds	Elemental composition	Calculated mass[a]
Cytosine + H	$C_4H_6N_3O$	112,051
Adenine + H	$C_5H_6N_5$	136,062
Cytidine − H_2O + H	$C_9H_{12}N_3O_4$	226,083
Cytidine + H	$C_9H_{14}N_3O_5$	244,093
Adenosine − H_2O + H	$C_{10}H_{12}N_5O_3$	250,094
Adenosine + H	$C_{10}H_{14}N_5O_4$	268,105
M + 2H[b]	$C_{19}H_{27}N_8O_{11}P$	287,077
Cyclo-CMP + H	$C_9H_{13}N_3O_7P$	306,049
CMP + H	$C_9H_{15}N_3O_8P$	324,060
		328,024[c]
Cyclo-AMP + H	$C_{10}H_{13}N_5O_6P$	330,060
AMP + H	$C_{10}H_{15}N_5O_7P$	348,071
		400,107[c]
M − adenine + H	$C_{14}H_{21}N_3O_{11}P$	438,091
M + H	$C_{19}H_{26}N_8O_{11}P$	573,146

[a] Protonated ions.
[b] Doubly charged.
[c] Not identified.

doubly protonated molecule at *m/e* 287.077, are found with 7 and 37% relative abundance, respectively. In addition, the FD-MS of this dinucleotide

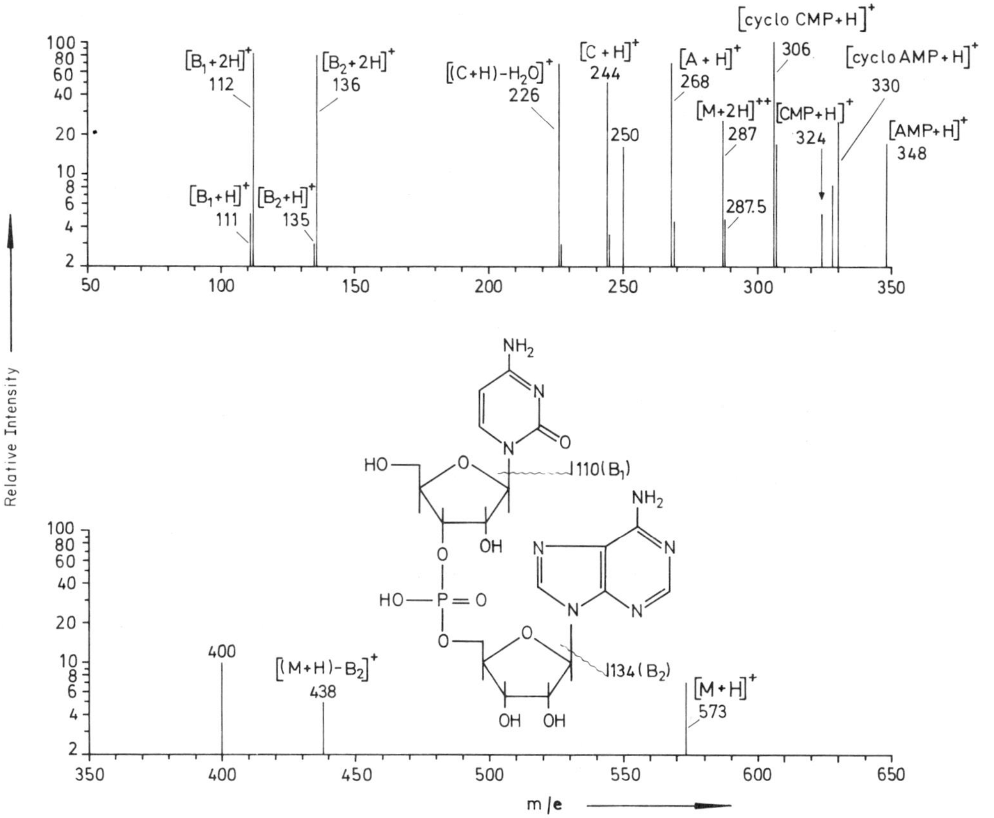

Figure 25. FD-MS of CpA: emission-controlled field desorption; exposure time of the photoplate, 8 min; threshold 1×10^{-8} A; solvent, water (41).

displays all essential building blocks of the CpA molecule. The protonated bases not only occur with high relative abundance, but also the corresponding nucleosides, and, with lower intensity, both mononucleotides, AMP and CMP. The base peak of the spectrum is produced by an ion at mass 306.049. In accordance with elemental composition and chemical criteria, the structure of a cyclo-CMP is assigned to this ion. An ion of structure analogous to that of cyclo-AMP is recorded at *m/e* 330.060. The signal at *m/e* 438.091 (5% rel. int.) is due to the loss of adenine from the protonated molecule. The origin of the ion at *m/e* 400.107 has not yet been explained.

The inverse dinucleotide ApC yields a significantly different mass spectrum. A number of ions are missing that are characteristic in the FD spectrum of CpA, and fragments and doubly charged ions are found with high intensity that do not appear in the spectrum (Figure 25). Most remarkable is the high intensity of the singly and doubly cationized molecules—for example, at *m/e* 573.146 $[M + H]^+$, at *m/e* 595.127 $[M + Na]^+$, at *m/e* 287.077 $[M + 2H]^{2+}$, at *m/e* 298.067 $[M + H + Na]^{2+}$, and at *m/e* 309.058 $[M + 2Na]^{2+}$. The attachment of sodium cations to the parent molecule indicates the presence of alkali salt impurity. The strong influence of accompanying inorganic salts on the FD-MS of organic molecules has been discussed in the previous section, and it was clear that the assignment of significant ions to distinguish between the two inverse isomers was not feasible.

Since the relative intensity of *m/e* 306 for $[\text{cyclo - CMP} + H]^+$ is greater than *m/e* 330 for $[\text{cyclo - AMP} + H]^+$ (Figure 25), the formation of the cyclophosphate-containing base B_1 appears to be favored rather than the cyclophosphate of B_2. Later experiments with other inverse dinucleotides confirmed these findings (89). Thus the first step in obtaining sequence information on unprotected oligonucleotides by mass spectrometry has been made, and future application of FD-MS to larger subunits of nucleic acids is one of the most promising areas of the technique.

4. Amino Acids and Peptides (Sequence and Mixture Analyses)

A. SINGLE COMPOUNDS

The amino acids in ester form were another of the earliest classes of organic compound to be successfully investigated by mass spectrometry (90). The moderate molecular weight and the relatively low polarity of the esters of most common amino acids allowed their vaporization for use in the direct introduction system. Intense efforts have been made to find appropriate reagents and procedures to produce derivatives of amino acids that make the compounds amenable to analysis by GC and MS. As a rule, any compound that passes through a GC column without suffering decomposition can also be handled by EI, FI, and CI mass spectrometry.

Several of the biologically important unprotected α-amino acids were investigated with FD-MS by Winkler and Beckey (29). The FD spectra obtained generally give the molecular ion or quasi-molecular ion with high relative intensity. With arginine and cystine these ions are also clearly shown with FD, although they are not detectable with EI, CI, and FI. In the FD mass spectrum of free arginine the $[M + H]^+$ ion appears with 30% relative intensity; the base peak of the spectrum at *m/e* 158 is marked by loss of ammonia from the protonated molecule $[(M + H) - NH_3]^+$. In addition, a direct bond rupture occurs between the guanidyl moiety and the C-5 rest of the molecule and gives *m/e* 116 or *m/e* 117, respectively. With the α-amino acids (e.g., Glu, Val, Pro, Ser, and Cys), one frequently gets a signal at $[M - 45]^+$ that can be interpreted as $[(M + H) - COOH_2]^+$. If the FD-MS are recorded a little above T^*, one observes water elimination (e.g., with Glu, Asp, and Cys-*S*-*S*-Cys).

Comparing the EI, CI, FI, and FD spectra of leucine, isoleucine, and norleucine (91), it appears that the highest molecular ion intensity and the smallest amount of fragmentation is obtained with CI. Compared to FD, however, the spectra reflect the different solubilities of the amino acids during preparation for the FD analysis more than the different transferred thermal energies.

The FD spectrum of glutamic acid (Figure 26*c*) is contrasted with the EI (Figure 26*a*) and FI (Figure 26*b*) mass spectra. Similar to the spectra of D-glucose (Section IV.1.A), these spectra are more useful in showing characteristic features of each method than in enabling a thorough comparison. However the unique value of FD for molecular weight determination is again illustrated. The $[M + H]^+$ ion of glutamic acid is virtually the only signal detected with FD-MS. Since a small proportion of the glutamic acid molecules can be evaporated without decomposition, the "soft ionization" mode FI, consistent with the small energy transferred in the field ionization process, yields a clearly detectable $[M + H]^+$ ion at *m/e* 148. This ion is missing with EI-MS, and thermal water elimination produces the $[M - H_2O]^+$ ion at *m/e* 129 as the highest signal in the upper mass range.

One particular difficulty that occurs in FD-MS when working with unknowns arises because either $[M]^+$ or $[M + H]^+$ (as shown for glutamic acid in Figure 26*c*), or both, may occur in a given FD spectrum and cannot be predicted. Since the ability to identify these ions correctly is critical for interpretation, the following general observations can be of use (75).

1. If the number of nitrogen atoms in the molecule is known, the molecular weight can, as usual, be predicted to be odd or even according to the nitrogen rule.
2. With few exceptions, the intensity of the $[M + H]^+$ ion decreases relative to that of $[M]^+$ with increasing emitter temperature.
3. Although the relationship between the real field strength at the site of ionization and the production of the $[M + H]^+$ ions is complicated, the intensities of $[M + H]^+$ ions decrease with increasing field strength.
4. If an even electron fragment ion corresponding to loss of a small neutral molecule such as H_2O, NH_3, or CH_2O_2, is seen in the FD mass spectrum, the precursor ion in this reaction is more likely to be $[M + H]^+$ than $[M]^+$. The presence of the metastable ion of the transition in question is very useful as a means of proving that the fragment ion does not result from some impurity. Furthermore, processes starting from the $[M + H]^+$ ion—for

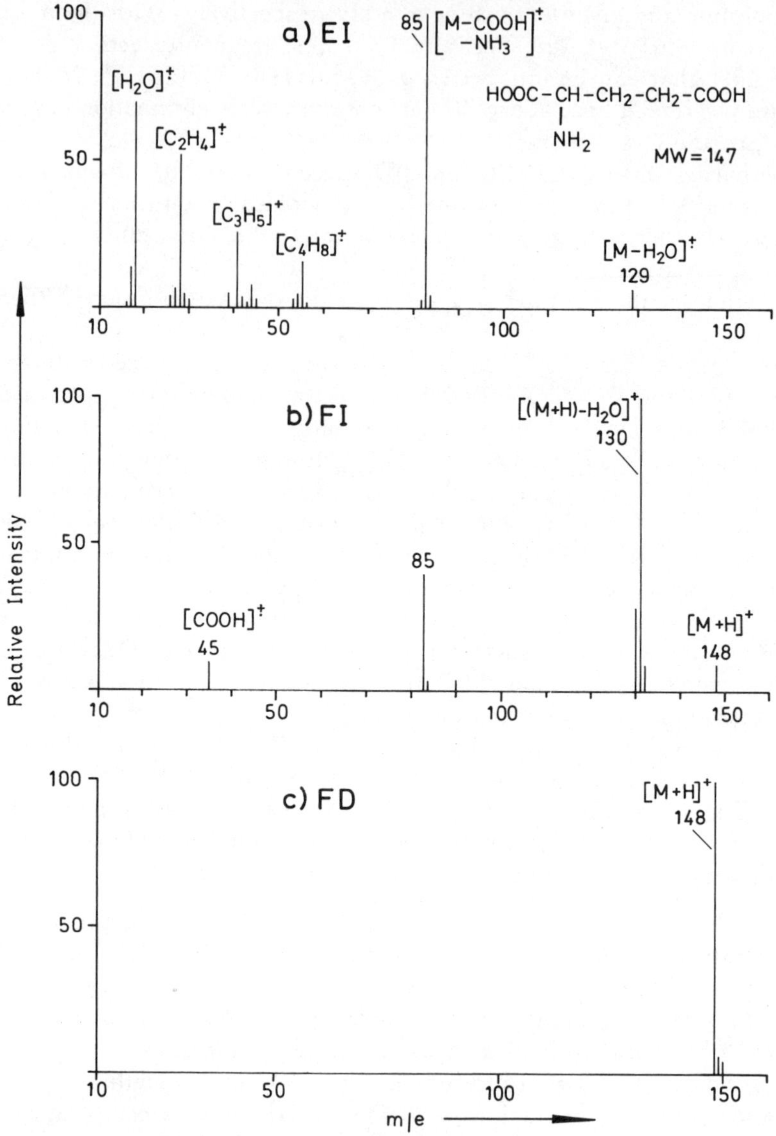

Figure 26. Electrically recorded mass spectra of glutamic acid (18). (*a*) EI-MS (70 eV). (*b*) FI-MS (lr). (*c*) FD-MS (lr).

example, $[(M + H) - H_2O]^+$—result in sharp peaks at the appropriate *m/e* value for the fragment ion. However elimination from the parent molecule—for example, $[M - H_2O]^+$—always yields broadened asymmetrical peaks in which the peak centroid is shifted toward the lower end of the mass scale.

5. The absolute intensity of the $[M + H]^+$ ion increases more rapidly than that of $[M]^+$ from increasing acidity of the solvent used to dissolve or suspend the sample for coating the emitter.

6. In addition to reactions involving the solvent, the $[M + H]^+$ ion is also generated in a disproportional reaction [1] in the high electric field, that is,

$$2M \xrightarrow[-e]{\text{field}} \longrightarrow [M + H]^+ + [M - H] \quad [1]$$

Compounds containing nucleophilic functions such as carbonyl-, amino-, polyhydroxy-, or acid groups are particularly susceptible to such field-induced proton transfer reactions, therefore exhibit intense $[M + H]^+$ ions. For a detailed discussion of protonation and field reactions see refs. 92 to 94.

7. For polar compounds, an intense ion at an m/e value corresponding to $[M - H]^+$ is generally diagnostic of intermolecular proton transfer reactions in the high field that also lead to intense $[M + H]^+$ ions, as indicated by [1] and [2].

$$[M - H] \xrightarrow[-e]{\text{field}} \longrightarrow [M - H]^+ \quad [2]$$

8. When multiple molecule complexes are formed, these will be protonated clusters $[nM + H]^+$ as opposed to dimers, trimers, and so on (27).

9. The FD mass spectra of nitrogen-containing substances such as amines and nitriles often show higher $[M + H]^{2+}$ intensities than $[M]^{2+}$. In general, the intensities of ions corresponding to $[M + H]^{2+}$ are invariably lower than those of $[M]^{2+}$.

10. Traces of inorganic salts present as contaminants often produce, in addition to the $[M]^+$ and/or $[M + H]^+$ ions, peaks resulting from the combination of the cation of the salt and a neutral molecule (cationization). This unequivocally defines the molecular weight of the compound (40, 74, 69).

It may not be difficult to find exceptions to any one of these rules alone. However careful application of as many as apply to any given case will greatly facilitate the assignment of the correct molecular weight to a compound from its FD mass spectrum. The serious problem of ambiguity in molecular weight may be alleviated in this way.

Peptides. Glycylycine, a dipeptide, was the first example indicating that underivatized peptides give complete mass spectra with FD-MS in which the molecular ions form the base signal using a 10 to 24 mA ehc (95). In a subsequent FD investigation, Winkler and Beckey (96) successfully recorded the spectra of partially protected and free oligopeptides with 3 to 9 units of amino acid (e.g., arginine and histidine), not only with high molecular ion intensities but also with partial information about the sequence.

In the FD mass spectrum of Ac-Gly-Arg-Arg-Gly-OMe, the following results were reported. Although the N-terminal end of the tetrapeptide is

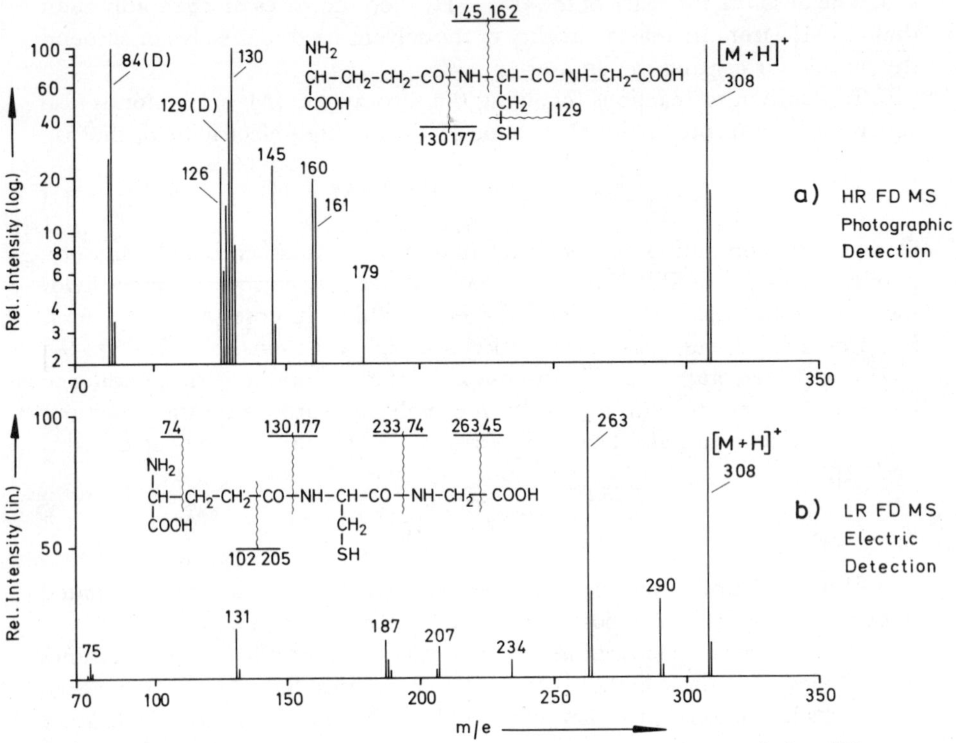

Figure 27. FD-M of glutathione. (*a*) High resolution (49). (*b*) Low resolution (97).

acetylated and the C-terminal carboxyl group methylated, the base intensity of the $[M + H]^+$ ion at *m/e* 501 is the salient feature of the spectrum, because among other groups the peptide contains two unprotected arginines that greatly reduce the volatility. The direct bond rupture (with hydrogen transfer) in the middle of the molecule leads to two sequence-specific fragments. Structure information is obtained from the molecular ion and the complementary signals at *m/e* 256 and *m/e* 244, which are relatively easy to interpret. However the qualification applies that in this case the identity of the individual amino acids and their arrangement require additional data of EI-MS. Hence if the complete sequence is to be determined, a combination of EI and (CI) FD-MS would probably be most successful.

From the HR FD-MS of glutathione (Figure 27*a*) and the fragmentation depicted in this spectrum, it becomes clear that photographic detection gives only the overall (or rapid motion) picture for the decomposition of the unprotected tripeptide. Although sequence-specific cleavages are found [e.g., at *m/e* 130.050 for $(C_5H_8O_3N)$ and at *m/e* 179.049 for $(C_5H_{11}O_3N_2S)$], the complete structure cannot be elucidated on the basis of the HR data. In electric detection the ehc can be modified specifically, and on line with the

LR FD spectra to produce the desired thermally and field-induced fragmentation containing most of the sequence-specific signals. This time-resolved (or slow-motion) picture of the desorption of glutathione at 25 mA is shown in Figure 27*b* (97). Sufficient fragmentation of the peptide backbone occurred to allow sequence determination of the tripeptide. Typically, most of the sequence peaks arise from ions formed by cleavages on either side of the carbonyl group of the free peptide.

Asante-Poku et al. (98) reported the FD spectra of a tripeptide Pro-Leu-Gly-NH_2, a pentapeptide Cbz-Gly-Pro-Leu-Gly-Pro, and a nonapeptide Arg-Pro-Pro-Gly-Phe-Ser-Pro-Phe-Arg. The FD mass spectrum of the free nonapeptide, bradykinin, appears in Figure 28. Again the molecular ion and sequence-specific cleavages on both sides of the carbonyl group were observed. Although a series of overlapping peptide fragments is produced, the determination of the complete sequence from these data would be difficult. A promising approach for sequence information of small peptides has been reported by Frick et al. (178) using methylation-methanolysis product mixtures of oligopeptides and FD-MS. However, an FD study of underivatized glucagon, a peptide with 29 amino acid residues, by Winkler et al. (179) showed the upper limit for the detection of large underivatized peptides probably to be about 15 to 20 amino acid residues.

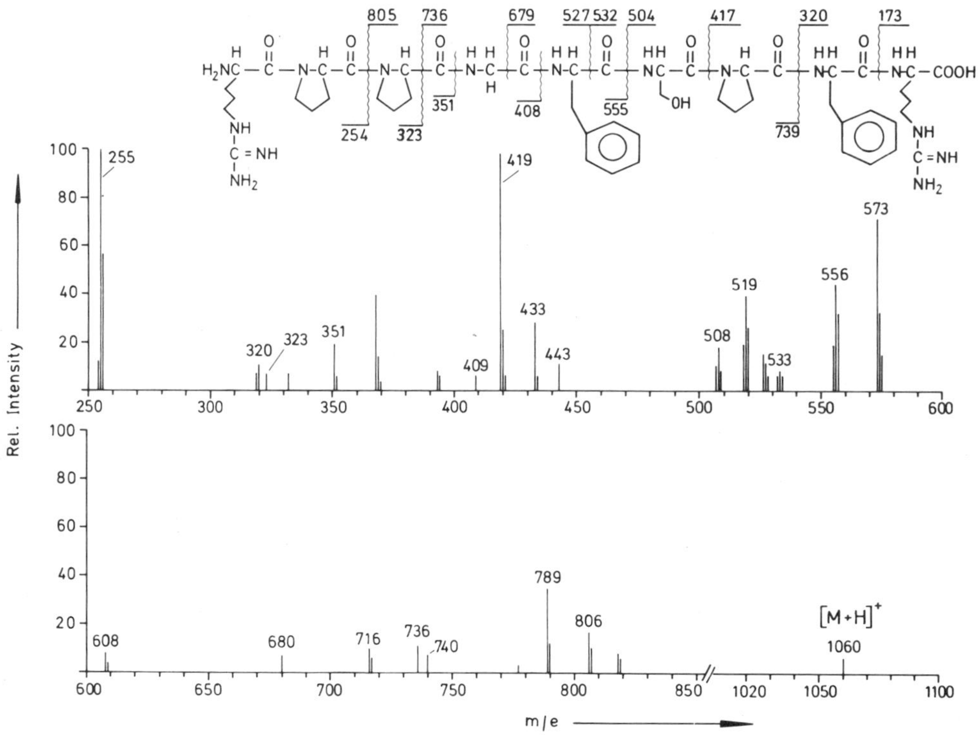

Figure 28. FD-MS of Arg-Pro-Pro-Gly-Phe-Ser-Pro-Phe-Arg (bradykinin) (98).

B. MIXTURES OF OLIGOPEPTIDES

An important analytical problem in biochemistry and medicine is the analysis of mixtures of peptides. Nearly all living tissues appear to contain appreciable quantities of low molecular weight peptides, but the most interesting group of biologically active large peptides and proteins is out of the range of direct mass spectrometric investigation. Therefore the members of this group are split chemically or enzymatically into subunits such as amino acids or small peptides, which are amenable to mass spectrometry. One essential problem connected with MS is that the free peptides cannot be introduced into the mass spectrometer as such but must be derivatized to gain sufficient volatility for vaporization. A number of protecting groups and special techniques have been introduced for this purpose (99), but there is considerable difficulty when the mixtures obtained have a very complex composition and the material of interest is available only in minute amounts.

To explore the possibilities of FD-MS for this analytical problem, model mixtures of underivatized oligopeptides have been studied (59). Figure 29 presents the FD spectrum of the mixture of Gly-Ala, Gly-Val, and Gly-Leu. Except for Gly-Leu, which gives only the $[M + H]^+$ ion and no $[M]^+$, the molecular ions and protonated molecules of all three peptides give prominent

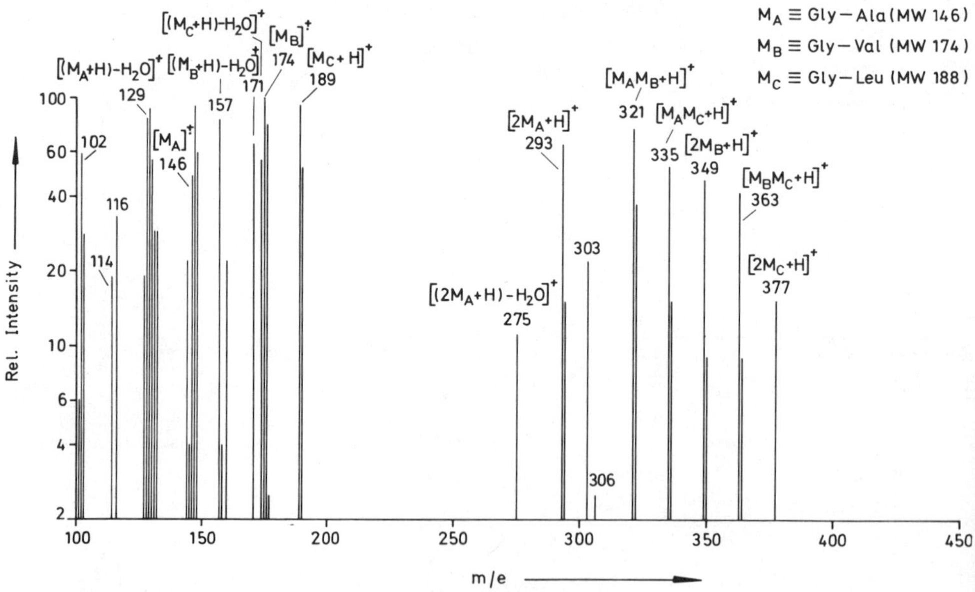

Figure 29. FD-MS of a mixture of three dipeptides: approximately equimolar concentration in aqueous solution; photographic detection; exposure time, 11 min; BAT 22 to 30 mA ehc (59).

signals in the spectrum. The process of water elimination from the protonated molecules is usually very pronounced, indicating that the time-temperature program employed was chosen for desorption above T^* to induce thermal fragmentation. The ehc was raised at a constant rate from 0 to 50 mA during the 11-min exposure time.

The photoplate as an integrating recording device collected all the ions generated during this desorption procedure. This also holds true for the peaks in the upper mass range (*m/e* 275–377). At first glance this part of the spectrum may be puzzling, but HR FD-MS reveals that these signals contain very useful information for the identification of the different chemical species in the mixture. For all three free peptides, the $[2M + H]^+$ ions are easily found, and these cluster ions can be used to confirm the molecular weight of the individual peptides. In addition, all mixed clusters of Gly-Ala, Gly-Val, and Gly-Leu are detected; thus in this first example of a peptide mixture, the three dipeptides can be assigned unambiguously.

When the synthetic, equimolar mixture of a tripeptide (Leu-Gly-Leu), a tetrapeptide (Try-Met-Asp-PheNH_2), and a pentapeptide (Met-Asp-Val-Gly-Leu) was desorbed under extremely soft experimental conditions, the FD spectrum presented in Figure 30 was obtained. In this case the time-temperature program for FD-MS was automatically controlled. The total emission was kept at the lowest measurable emission threshold ($\sim$ 0.5

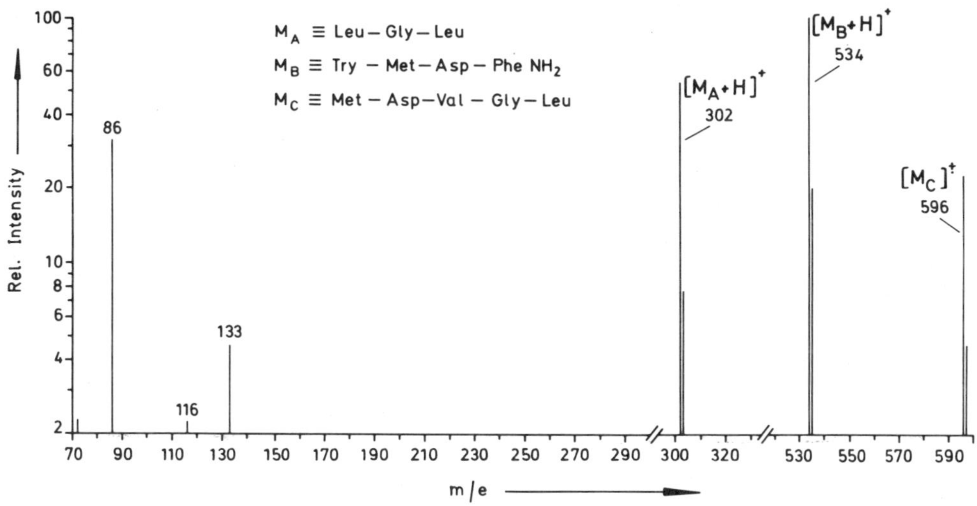

Figure 30. FD-MS of a mixture of a tripeptide, a tetrapeptide, and a pentapeptide: approximately equimolar concentration in aqueous solution; emission-controlled field desorption; threshold, 4×10^{-9} A (11 min) and 1×10^{-8} A (7 min); total exposure time of the photoplate, 24 min (165).

$\times 10^{-9}$ A) for more than 10 min desorption time, strongly reducing the thermally and field-induced fragmentation. Only four fragment ions of minor relative abundance were found. The most intense signal at *m/e* 86 was determined with accurate mass measurement of its elemental composition (Figure 30). It is assumed to originate from the leucine part of the peptides. With the exception of these fragment peaks, the resulting mass spectrum contains virtually only the molecular ions or protonated molecules of the oligopeptides, and it gives a clear picture of the components of the mixture.

Since the synthetic mixtures of underivatized peptides gave such meaningful and easily understandable mass spectra, FD-MS appears to be a promising technique for analyzing peptide mixtures and thus might apply to the problem previously mentioned.

C. MIXTURES OF PTH AMINO ACIDS AND EDMAN DEGRADATION PRODUCTS

As part of a continuing effort to explore systematically the analytical potential of FD-MS for mixture analysis, we investigated (100) the FD spectra of single PTH amino acids, synthetic mixtures of these compounds, and samples from automated Edman degradation steps of a ribosomal protein (100).

The behavior of the PTH-derivatives of the amino acids glycine, alanine, serine, proline, valine, threonine, leucine, isoleucine, asparagine, aspartic acid, glutamine, glutamic acid, methionine, histidine, phenylalanine, arginine, carboxylmethyl cysteine, tyrosine, tryptophan, and lysine was studied under FD conditions. In general, these substances showed the molecular ion $[M]^+$ or the protonated molecule as the base peak of the spectrum. In many cases no fragment ions were detected above *m/e* 50 as for example in Figure 31*b* for PTH-aspartic acid. Only the peaks for the solvent acetone at *m/e* 58.042 (C_3H_6O) appeared (Figures 31*a*, 31*b* and 31*e*). However some thermally and field-induced fragmentation was found and was attributed to the standardized ehc program employed and the integrating function of the photographic recording system.

The PTH amino acids of lower volatility showed this particular type of fragmentation. In the FD spectrum of PTH-arginine (Figure 31*d*) the elimination of ammonia from the parent molecule led to an ion at *m/e* 274.089 ($C_{13}H_{14}N_4OS$). Additional intense signals recorded at *m/e* 119.037 (C_7H_5NO) and *m/e* 155.062 ($C_6H_{11}N_4O$) could be explained by bond cleavages and proton transfer (Figure 31*d*). The FD spectra were characterized by pronounced doubly charged ions. The $[M]^{2+}$ ions for PTH-histidine (Figure 31*e*) at *m/e* 136.037) and PTH-tryptophan (Figure 31*e*, at *m/e* 160.547) and the $[M - CO]^{2+}$ ions for PTH-proline (Figure 31*a*, at *m/e* 102.036) and PTH-His (at *m/e* 122.039) were typical in this respect.

A sample was taken from an equimolar solution of PTH-proline, PTH-aspartic acid, PTH-histidine, PTH-arginine, and PTH-tryptophan. The FD spectrum obtained for this mixture (Figure 32) had three significant features.

1. With the exception of PTH-arginine, which underwent the strongest fragmentation (*m/e* 119, *m/e* 155/156, Figure 31*d*), the molecular ions of the PTH derivatives gave the base peaks of the spectrum.

2. Although ionization in FD-MS takes place in adsorbed layers, which means that a plurality of intermolecular reactions is possible, the FD

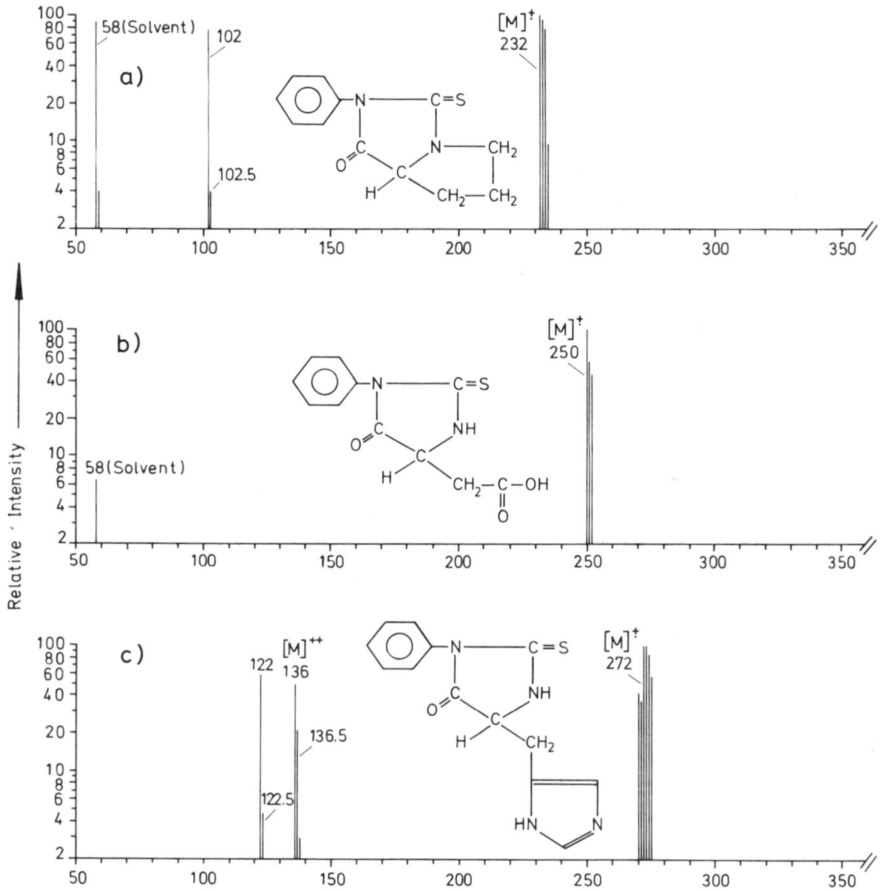

Figure 31. FD-MS of five PTH amino acids: Emission-controlled desorption; ehc 0 to 30 mA within about 12 min exposure time (100). (*a*) PTH-Proline. (*b*) PTH-Aspartic acid. (*c*) PTH-Histidine. (*d*) PTH-Arginine. (*e*) PTH-Tryptophan.

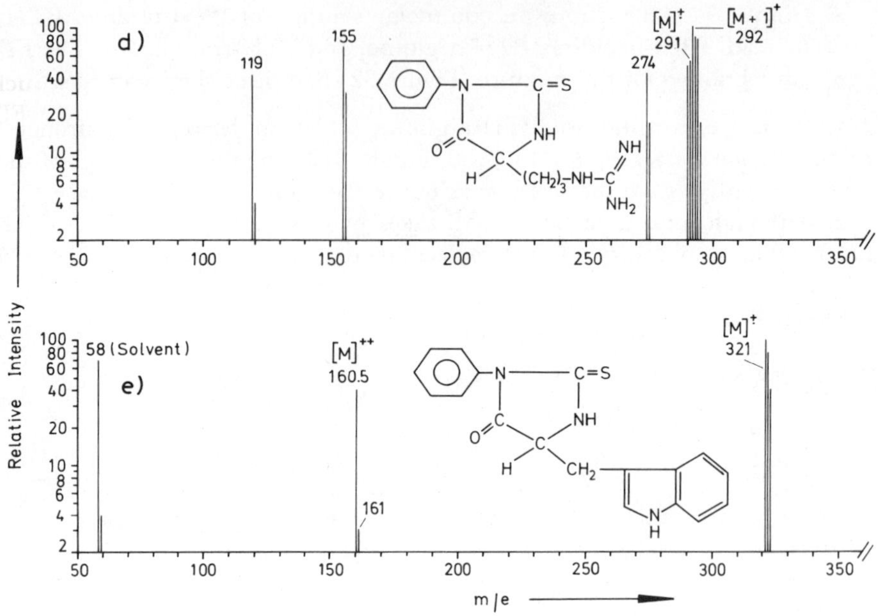

Figure 31. Continued.

spectrum of the mixture showed virtually the same fragments as the sum of the spectra of the individual components. It is noteworthy that with only two exceptions (at *m/e* 135 and *m/e* 206), under the experimental conditions described, no product of a chemical interaction between components of the mixture was detected.

3. Within the polarity range of the thiohydantoin derivatives studied, FD-MS appeared to have an approximately equal response and enabled the molecular weight determination of all components of the mixture.

To evaluate this capacity of the method for a more complex mixture, the FD spectrum of a synthetic mixture of 15 PTH amino acids was taken. An equimolar solution of PTH-glycine, PTH-serine, PTH-proline, PTH-valine, PTH-threonine, PTH-leucine, PTH-aspartic acid, PTH-glutamic acid, PTH-methionine, PTH-histidine, PTH-phenylalanine, PTH-arginine, PTH-cysteic acid, PTH-tyrosine, and PTH-tryptophan was prepared and a total of 1 μg of solid substance was transferred to the FD emitter.

As shown in the FD mass spectrum (Figure 33), the molecular ions of these phenylthiohydantoin derivatives clearly dominated the upper mass range (*m/e* 190). In fact the molecular ion intensities for 13 PTH amino acids were between 85 and 100% relative intensity. The two exceptions were PTH-

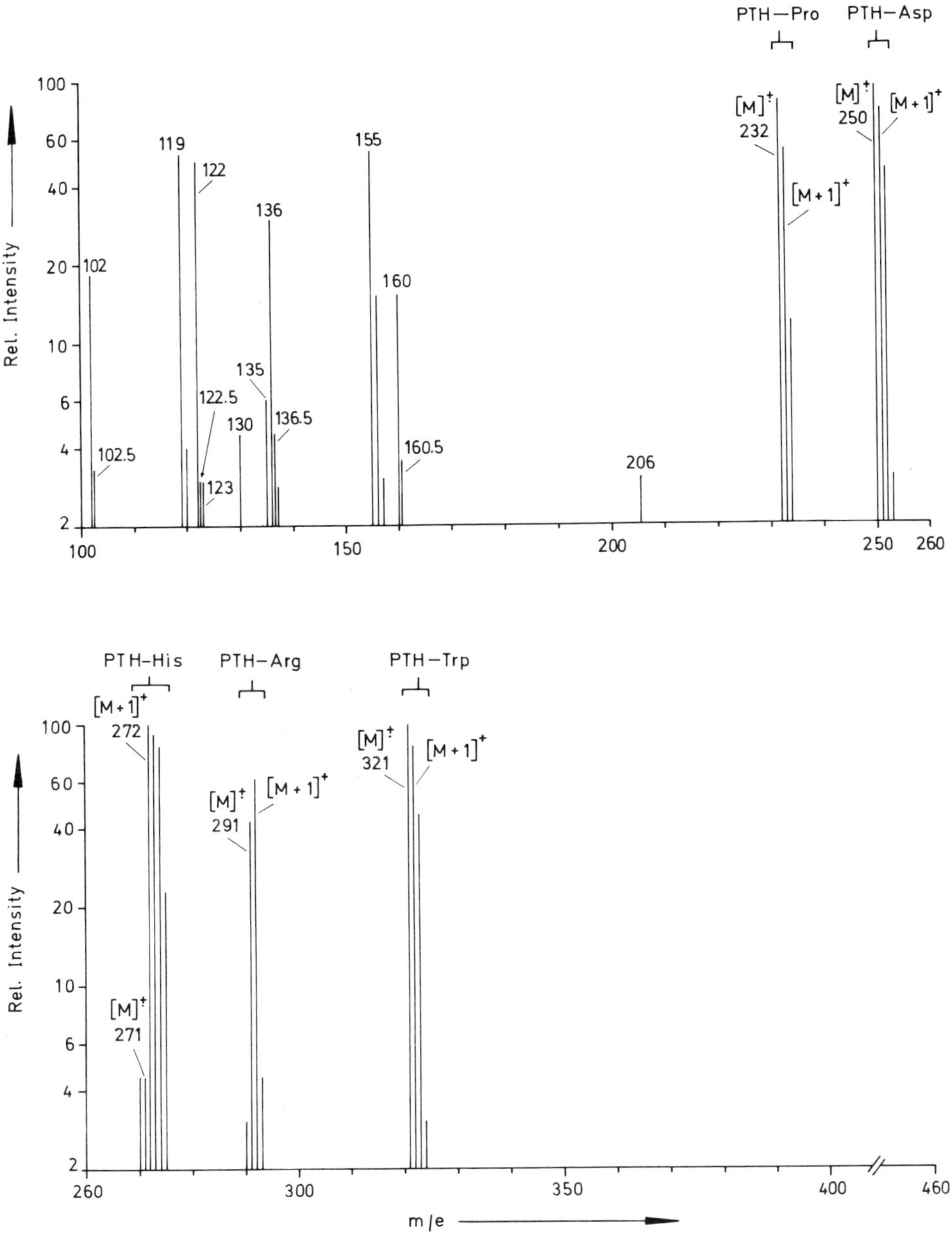

Figure 32. FD-MS of a synthetic, equimolar mixture of PTH-proline, PTH-aspartic acid, PTH-histidine, and PTH-tryptophan; desorption program as described for Figure 31 (100).

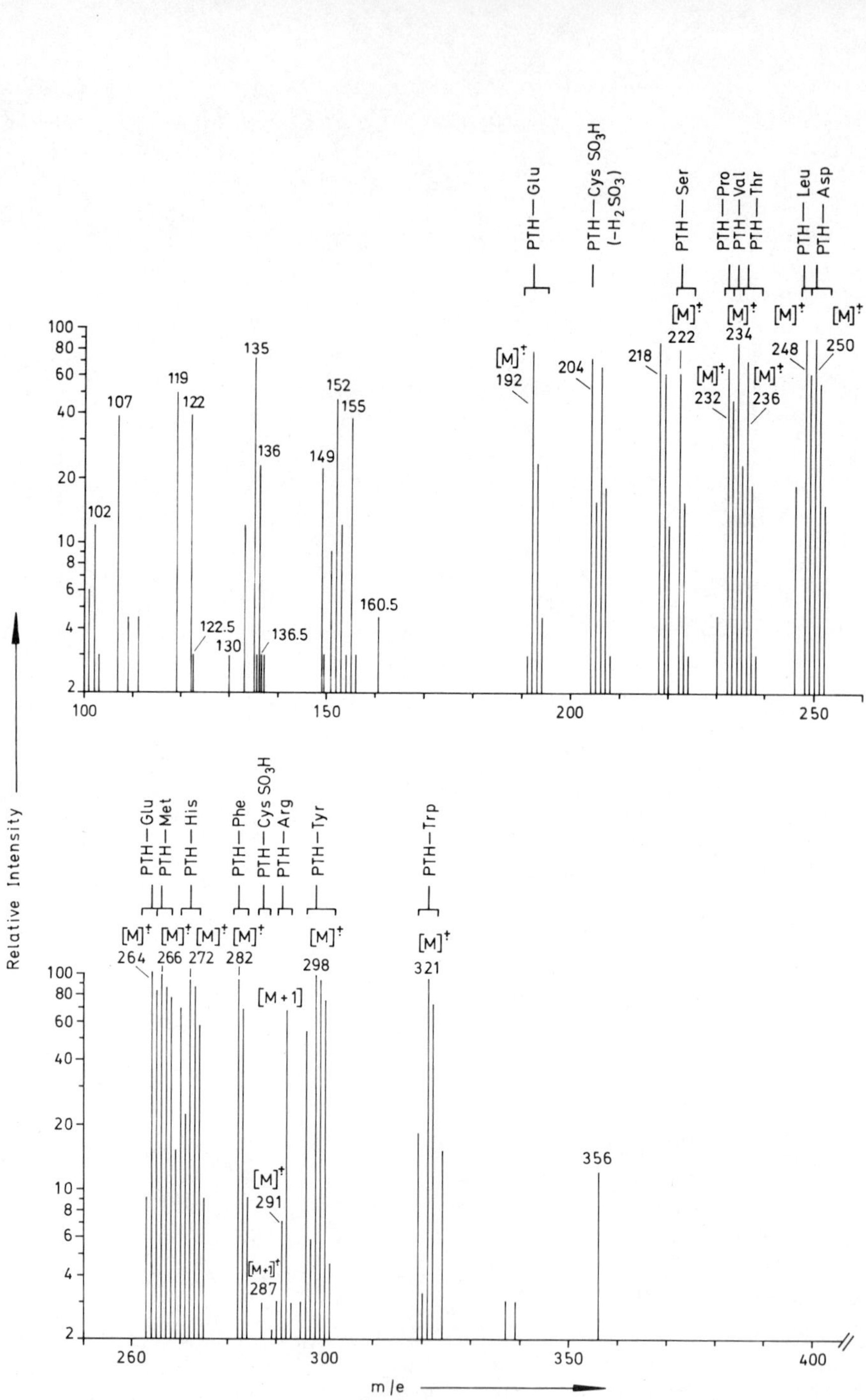

Figure 33. FD-MS of a synthetic mixture of 15 PTH amino acids: emission-controlled desorption; ehc 0 to 30 mA within approximately 20 min exposure time; emission starts at 7 mA ehc, best emission at 22 mA ehc (threshold of 1×10^{-7} A) (100).

arginine and PTH-cysteic acid. According to its high proton affinity, PTH-arginine gave a more intense $[M + H]^+$ ion that $[M]^+$, and because of its low volatility, the compound underwent strong fragmentation as indicated in Figure 31*d* (*m/e* 119,155,156). The free sulfonic acid PTH-cysteic acid, however, gave only a weak $[M + H]^+$ ion at *m/e* 287.016 ($C_{10}H_{11}O_4N_2S_2$) and no molecular ion. A very intense fragment at *m/e* 204.036 ($C_{10}H_8ON_2S$) for the loss of sulfonic acid from the protonated molecule was confirmed by high resolution and accurate mass measurements.

Thus the identification of PTH-cysteic acid in the mixture was possible when emission-controlled desorption of all phenylthiohydantoin derivatives was performed. However the experimental conditions selected do not favor the protonation and/or cluster formation of the extremely polar free sulfonic acid. These results with synthetic mixtures of PTH amino acids suggested that FD-MS could be applied to the analysis of products of the Edman degradation.

In principle, when mass spectrometry is applied to the automatic sequential degradation of proteins or peptide chains, the most significant information is the knowledge of the molecular weight of the N-terminal amino acid that has been reacted with phenyl isothiocyanate in each degradation step. In this case it is the relative abundance of the molecular ion that is particularly important, since the structures of the commonly encountered PTH-derivatives are known. The fragments are used only as additional evidence for the correct assignment of the molecular ion.

In the first qualitative approach, the FD mass spectra were taken of samples from the first 10 degradation steps of the ribosomal protein L17 (methionine-arginine-histidine-arginine-lysine-serine-glycine-arginine-glutamine-leucine-), whose N-terminal sequence had been established (101) up to position 35 with a modified Beckman sequenator (102). With EI mass spectrometry, four of the first 10 degradation steps (i.e., the lysine and arginine steps) gave no molecular ions and only characteristic fragment ions were found.

Figure 34*b* gives the FD mass spectrum of step 5, PTH-(ϵ-PTC)-lysine. The amount of sample necessary (at a resolution of 15,000, 50% peak valley) was approximately 1% of the total output of the automatic sequenator or about 200 ng. The FD spectrum of 500 ng of authentic PTH-(ϵ-PTC)-lysine obtained under the same experimental condition is shown in Figure 34*c*. Comparison of both spectra enabled an unambiguous identification of PTH-(ϵ-PTC)-lysine in the actual sample from step 5. Accurate mass measurements at *m/e* 398.124 ($C_{20}H_{22}N_4OS_2$), for $[M]^+$, and the three most intense fragments at *m/e* 305.066 ($C_{14}H_{15}N_3OS_2$), *m/e* 263.109, and *m/e* 135.014, were in good agreement within the standard error of accurate mass measurements (about 10 ppm). The ion at *m/e* 199.062 for $[M]^{2+}$ gave a good demonstration of the use of doubly charged ions for molecular weight determination. However a number of peaks appeared in the FD spectrum in

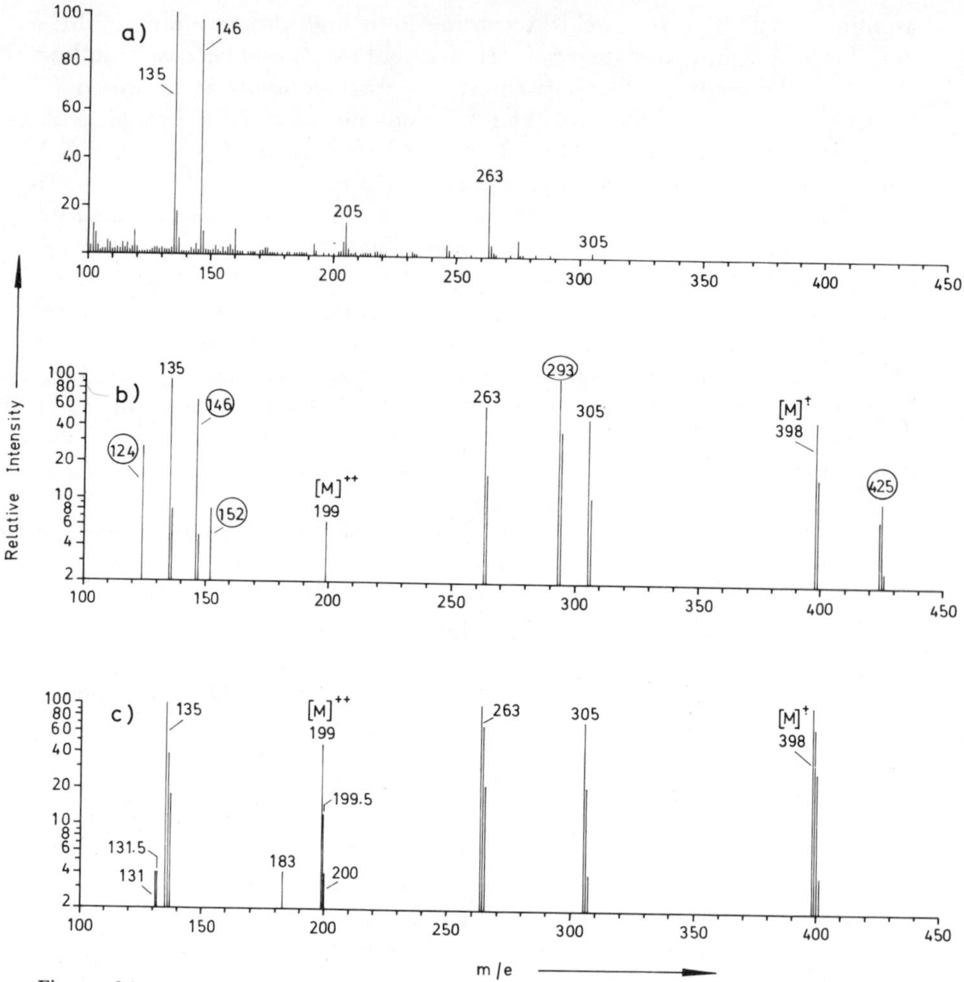

Figure 34. (*a*) EI-MS of step 5 (PTH-(ε-PTC)-lysine of the sequenator: sample amount between 400 and 600 ng; electric detection on a Varian CH-7 instrument (linear intensity scale); resolution (10% valley definition) about 500. (*b*) FD-MS of step 5 (PTH-(ε-PTC)-lysine) of the sequenator: sample amount approximately 200 ng; photographic detection (logarithmic intensity scale); resolution (50% valley definition), approximately 15,000, emission-controlled desorption; exposure time 20 min. (*c*) FD-MS of an authentic sample of PTH-(ε-PTC)-lysine: sample amount, approximately 500 ng; photographic detection; same conditions as described for (*b*) (100).

Figure 34*b* (circled mass numbers) that were not found in the spectrum of the authentic sample of PTH-(ε-PTC)-lysine.

A partial explanation is possible when the FD mass spectrum is contrasted with the EI spectrum of the same sample of the sequenator. It was found that most of these signals are due to sequenator chemicals. For example, *m/e* 293 in the FD spectrum in Figure 34*b* is the $[M + H]^+$ ion of *N,N,N′,N′*-tetra-(2-hydroxypropyl)ethylenediamine, a reagent used in the sequenator. However when the FD mass spectrum of the wash cycle is subtracted from the FD spectrum (Figure 34*b*), all signals with circled mass numbers are eliminated.

In summary, when the results with FD-MS are compared with those of EI-MS, two facts are outstanding. First, since the resolution of the FD spectra is about 30 times greater, and the sample amount needed is approximately 2 times less than for EI-MS, the field desorption method has excellent sensitivity for PTH derivatives. Second, regarding the high molecular ion intensities, and the uncomplicated fragmentation and reliability of the FD data, FD-MS seems to be ideally suited for PTH amino acids.

The use of FD mass spectrometry as a complementary technique for the confirmation of PTH amino acids released in Edman degradation steps appears to be advantageous. This is especially true for PTH amino acids (and amides) of low volatility. In comparison with EI-MS, the analysis of samples containing mixtures of these substances is facilitated. Also in CI-MS the FD data might be helpful as an additional proof for the correct assignment of PTH amino acids in peptide sequencing.

5. Steroids

A. HORMONES (HIGH-PRESSURE LIQUID CHROMATOGRAPHY AND FD-MS)

Although the EI spectra of steroids have been extensively documented (103), to date few reports have been published of their CI (104,105) and FD spectra (62,106). A comparison of the FD, CI, and EI mass spectra of seven steroids displaying varying stability to electron impact has been performed (107), with the following results.

In the EI spectra only estrone showed the molecular ion as its base peak (Table VI). Cortisone, cortisol, tetrahydrocortisone, aldadiene, and predni-

TABLE VI
Molecular Ion Abundances by EI, FD, and CI-Mass Spectrometry of Seven Steroid Hormones (107)

Compound	Mol. wt.	EI $[M]^{+\cdot}$	FD $[M]^{+\cdot}$	CI $[M + H]^+$
Estrone	270	100	100	100
Cortisone	360	13	100	100
Cortisol	262	9	100	100
Tetrahydrocortisone	364	9	100	20
Aldadiene	340	22	100	100
Prednisone	358	4	100	100
Pregnanetriol	336	0	100	0

sone all had weak molecular ions, and pregnanetriol showed no molecular ion. The presence of weak molecular ions or their total absence can make identification of steroids difficult, particularly if the sample under examination has been obtained from natural sources and contain trace contaminants. With the exception of tetrahydrocortisone and pregnanetriol, all the steroids had $[M + H]^+$ ions as the base peaks in their CI spectra; $[M]^+$ ions were base peaks in the FD spectra of all the steroids at optimal emitter temperature.

For the detection and molecular weight determination of steroids, FD appears to be the method of choice, particularly for steroids carrying three or more free hydroxyl functions. However when a mixture of thermally labile compounds cannot be analyzed directly by FD-MS (because of low sample concentration or disturbing influences from contaminants, e.g., inorganic salts) and preseparation is necessary, there is one ideal combination available: liquid chromatography and FD-MS.

Many substances of biological interest are stable only in their solid phase or in solution. This is one reason for the extensive application of liquid chromatography in biomedical work, where mass spectrometry is much less frequently employed. Also soft ionization methods, such as FI, CI and low-voltage electron impact, require volatilization of the sample before ionization takes place. Therefore the molecular ion can be obtained only if the molecule in question is stable in the vapor phase. This precondition also applies if, for a compound mixture, preseparation is performed by gas chromatography.

Although the coupling of a gas chromatograph to the mass spectrometer has been an extremely useful combination, compounds are amenable to this system only if they themselves are sufficiently volatile, or if this volatility can be produced by chemical pretreatment (i.e., by the preparation of derivatives). In view of the rapid progress in high-pressure liquid chromatography and in field desorption, both of which avoid the necessity of vaporizing the sample, the two methods have now been combined.

The first experiments in this direction were made by a two-step technique (106). The eluate from the liquid chromatograph was collected, concentrated on the emitter, and analyzed by FD spectrometry. In this sequence the FD mass spectrometer serves as a (second) molecule-specific detector. The use of column liquid chromatography in combination with field desorption mass spectrometry was demonstrated by the separation and identification of the components of a steroid mixture extracted from rat serum (106). The liquid chromatogram of this mixture appears in Figure 35*a*. The effluent of the high-pressure chromatograph was collected and transferred to the surface of the field desorption emitter. The FD spectrum of prednisolone (Figure 35*b*) shows the molecular ions as the base peak of the spectrum. Only one fragment is observed because of a common cleavage in the side chain between C-20 and C-21.

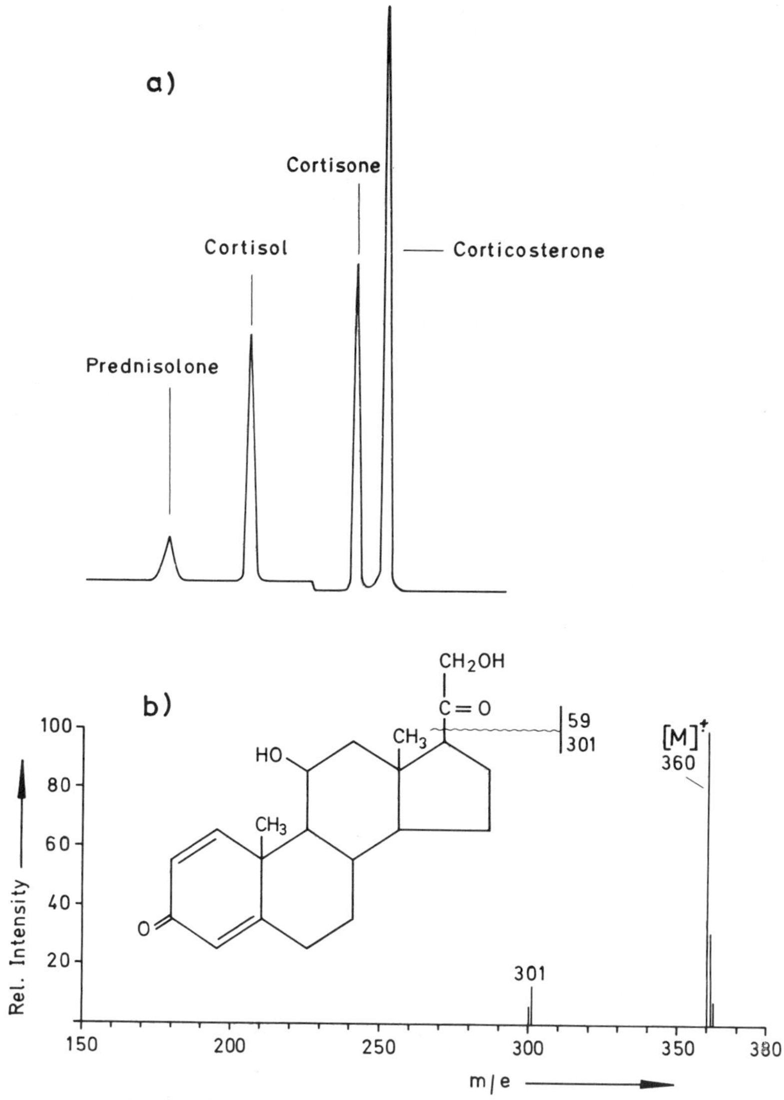

Figure 35. (*a*) Column liquid chromatogram of a steroid mixture from rat serum. (*b*) High-resolution FD-MS of prednisolone (106).

Progesterone and corticosterone yield similar FD spectra consistent with the results given in Table VI. Steroids displayed relatively intense doubly charged ions, mostly of the $[M]^{2+}$ and $[M + H]^{2+}$ type. The occurrence of doubly charged ions is a characteristic phenomenon of FD-MS and is closely related to the formation of ions in the adsorbed layers (30). In view of the size of the molecule and the functional groups located far apart in the molecule, stabilization of a second charge is apparent for steroids.

B. BILE ACIDS

Bile acids have been the subject of extensive mass spectral study, and for optimal results they are normally examined as their methyl esters (108). Dean and Aplin (109) have shown that good EI spectra are obtainable from underivatized bile acids; however the molecular ions produced were usually of low intensity. In contrast, Games et al. (68) obtained the complete FD spectra of four bile acids, two bile salts, and five bile acid conjugate salts. From these spectra the information of the molecular weight was readily found.

The high-resolution FD mass spectrum of deoxycholic acid (Figure 36) gives a good demonstration of the utility of the technique for analysis of single bile acids. The $[M + H]^+$ ion at *m/e* 393 represents the base peak of the spectrum. As might be expected, the corresponding proton transfer reactions lead to intense $[M - H]^+$ ions at *m/e* 391; the molecular ion, typical in the FD spectra of free acids, is only about 30% relative intensity. The desorption of the sample was performed at T^* and above. Therefore water elimination yields prominent singly (*m/e* 375, 357) and doubly charged (*m/e* 188, 179) ions.

The FD spectra of cholic, chenodeoxycholic, and lithocholic acids gave similar results (49). In addition, the complete FD spectra of the sodium salts

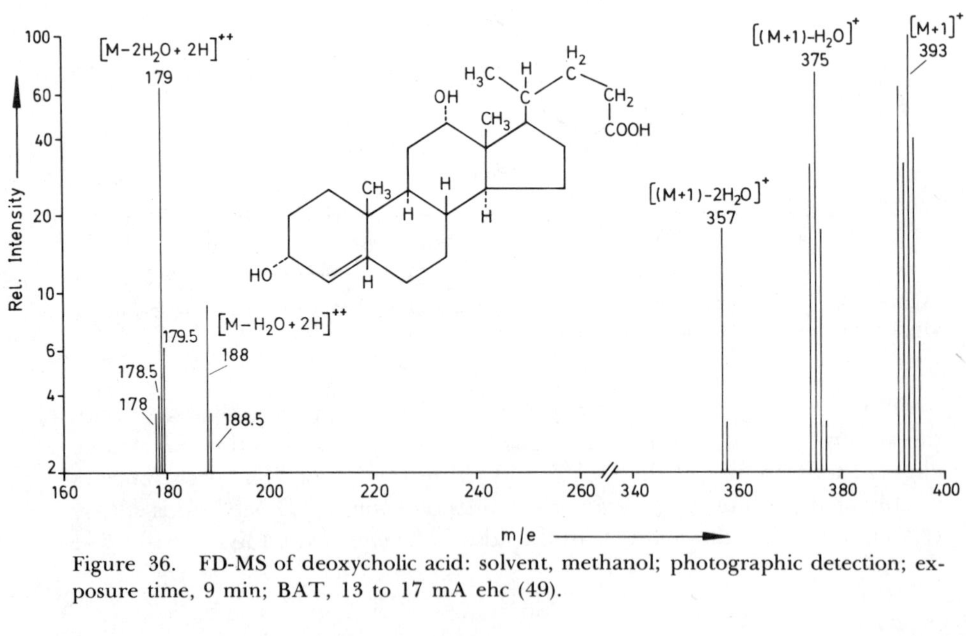

Figure 36. FD-MS of deoxycholic acid: solvent, methanol; photographic detection; exposure time, 9 min; BAT, 13 to 17 mA ehc (49).

of deoxycholic, taurocholic, taurodeoxycholic, taurochenodeoxycholic, dehydrocholic, and glycocholic acids were obtained.

Thus FD mass spectrometry should provide a convenient and useful method for the identification of bile acids, their alkali salts, and their conjugates, in crude biological samples. As a first approach to the examination of natural mixtures of bile acids, the FD spectra of the supernatant of methanol-treated bile were reported (68). Figure 37 gives two of the electrically recorded FD spectra obtained from one of the bile samples. At lower emitter heating currents (Figure 37*a*) the $[M + 2Na]^{2+}$ and $[M + Na]^+$ ions, corresponding to some of the expected conjugates, are present. At higher emitter heating currents only the $[M + Na]^+$ ions were observed. Other samples of bile gave similar results.

6. Porphyrins

The utility of the combination of HPLC and FD-MS has already been shown for studies of steroids (106), vitamins (110), alkaloids (111), dyestuffs (180) and PTH-amino acids (177). Recently natural porphyrins and chlorophyll derivatives were investigated by this combination technique (112). Mixtures of porphyrins from various biological sources, especially urine and feces of porphyric subjects, were separated by HPLC, and the FD emitter wire was dipped directly into the eluates.

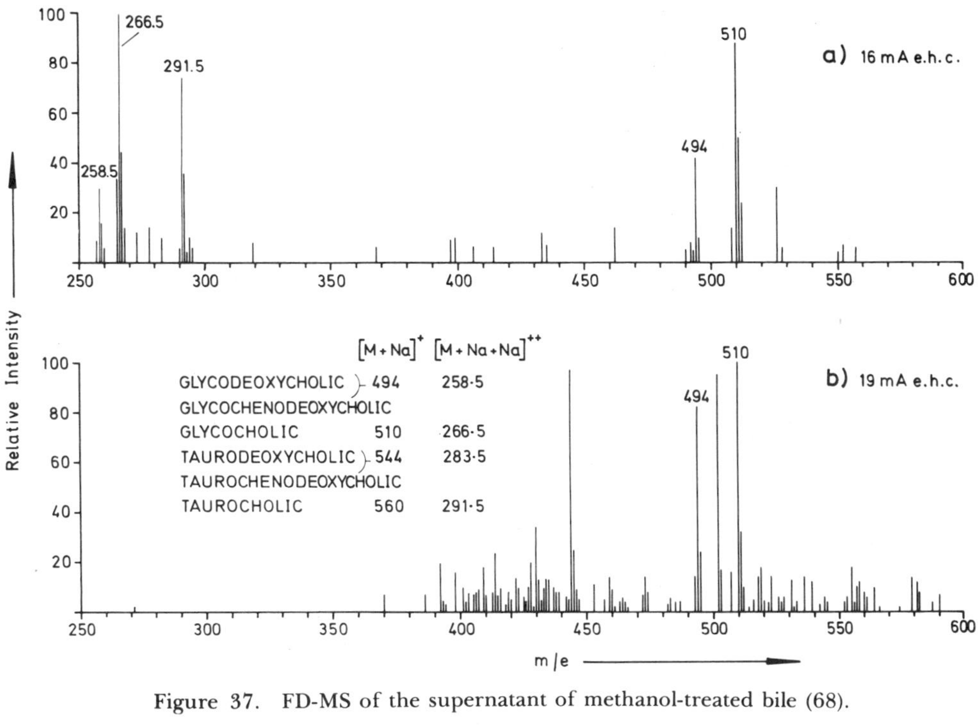

Figure 37. FD-MS of the supernatant of methanol-treated bile (68).

For the complete identification of a complex mixture of porphyrins, a combined HPLC-MS technique was employed. A field desorption spectrum of the crude mixture was taken first, to provide information of the number and type of components. These were separated by HPLC, and further mass spectra of the individual isolated fractions, together with retention data from the chromatograms, allowed complete characterization of the constituents to be made by direct comparisons with authentic samples wherever possible.

The field desorption spectra of the various tetrapyrrolic compounds are given in Table VII. With the exception of a hematoporphyrin and hemin, $[M]^+$ or $[M + 1]^+$ ions are the base peaks in the spectra and there is little or no fragmentation.

Hematoporphyrin has as its base peak an ion at *m/e* 562 attributed to the loss of two molecules of water from the parent compound, and in the case of hemin the base peak in its spectrum is at *m/e* 616, consistent with the chlorine-free ligand.

7. Carotenoids

The mass spectra of a variety of pure carotenoids covering a range of structural types have been studied with the field desorption technique (113). All the compounds examined exhibited the molecular ion as the base peak, even when this ion is of very low intensity in the EI spectra. Little fragmentation was observed, but thermal decomposition could be induced in certain cases by using higher emitter currents. Application to a simple mixture of known compounds, and a more complex mixture of sedimentary origin, revealed the potential of the method in the rapid qualitative screening of carotenoids in crude extracts. Similar results using FD-MS as a screening technique have been obtained by Sphon et al. (181) in the FD spectra of mycotoxins, mycotoxin mixtures, and extracts from foodstuffs.

Table VIII lists the characteristic carotenoid ions in the FD spectra of a series of pure compounds, with the corresponding EI data for comparison. Each spectrum shows a molecular ion as the base peak with the expected reduced fragmentation compared with the corresponding EI spectrum.

8. Lipids

EI mass spectrometry has proved particularly useful in the determination of structure of new lipids and in the identification and estimation of known lipids from natural sources (114). However some of these compounds are not amenable to conventional mass spectrometric methods unless in the form of a derivative. Problems arise for the determination of the molecular weight, particularly if only small amounts of material are available. Evans et al. (115) showed that the FD mass spectra of triglycerides and phosphoglycerides enabled ready identification of the molecular species. The relative intensities of the $[M]^+$ and $[M + 1]^+$ ions obtained were reported for glycerol trimyris-

TABLE VII

Field Desorption Mass Spectra of Porphyrins (112)

Compound	Mol. wt.	*m/e* (relative intensity, %, in parentheses)
Mesoporphyrin IX	566	568 (45), 567 (100), 566 (76)
Hematoporphyrin	598	564 (39), 563 (73), 562 (100)
Coproporphyrin III	654	(*a*) 658 (2), 657 (9), 656 (32), 655 (100), 653 (1), 609 (2)
		654 (2), 653 (3), 611 (2), 609 (3), 541 (3)
Hemin	652	618(17), 617 (50), 616 (100)
Aetioporphyrin	478	480 (11), 479 (35), 478 (100)
Octaethylporphyrin	534	536 (13), 535 (40), 534 (100)
Protoporphyrin IX dimethyl ester	590	591 (41), 590 (100)
Mesoporphyrin IX dimethyl ester	594	597 (1), 596 (10), 595 (45), 594 (100) 297.5 (8), 297 (10)
Coproporphyrin III tetramethyl ester	710	713 (5), 712 (19), 711 (52), 710 (100), 651 (3)
Pentacarboxylic porphyrin pentamethyl ester	768	770 (14), 769 (53), 768 (100), 709 (8)
Hexacarboxylic porphyrin hexamethyl ester	826	830 (3), 829 (5), 828 (18), 827 (36), 826 (100), 767 (1), 725(1), 413 (0.5)
Heptacarcoxylic porphyrin heptamethyl ester	884	885 (41), 884 (100), 882 (8)
Uroporphyrin III octamethyl ester	942	945 (2), 944 (6), 943 (12), 942 (100), 941 (8), 940 (8), 913 (10), 912 (7), 884 (6)

[a] Higher emitter current.

TABLE VIII
Comparison of Relative Intensities of Ions in the FD and EI Mass Spectra of Carotenoids (113)

Carotenoids (semisystematic names in parentheses)	Mol. wt.	Relative intensity (% FD/EI)	Wire current (m/A)
β-Carotene (II) (*β*,*β*-carotene)	536	$[M]^+$ (100/100), $[M-2]^+$ (5/4.2)	15
Lycopene (III) (*ψ*,*ψ*-carotene)	536	$[M]^+$ (100/14.7), $[M-2]^+$ (5/1.4), $[M]^{2+}$ (3.5/1.8)	14
Bixin (IV) (methyl hydrogen 9′-*cis*-6,6′-diapocarotene-6,6′-dioate)	394	$[M]^+$ (100/9.6), $[M-2]^+$ (2.6/0.6)	12
β-Apo-8′-carotenal (V) (8′-Apo-*β*-Caroten-8′-al)	416	$[M]^+$ (100/100), $[M]^{2+}$ (7/7.4)	10
Echinenone (VI) (*β*,*β*-caroten-4-one)	550	$[M]^+$ (100/100), $[M-2]^+$ (3/2.7)	10
Canthaxanthin (VII) (*β*,*β*-caroten-4,4′-dione)	564	$[M]^+$ (100/100), $[M]^{2+}$ (25.8/8.5)	17
Lutein (VIII) (*β*,*ε*-carotene-3,3′-diol)	568	$[M]^+$ (100/74.9), $[M-2]^+$ (4/6), $[M-18]^+$(40/86.5), $[M-36]^+$ (3.6/4), $[M]^{2+}$ (3.6/4.1)	16
Zeaxanthin (IX) (*β*,*β*-carotene-3,3′-diol)	568	$[M]^+$ (100/100), $[M-2]^+$ (1.3/7), $[M]^{2+}$ (0.7/7)	15
Anhydrorhodovibrin (X) (1-methoxy-3,4-didehydro-1,2-dihydro-*ψ*,*ψ*-carotene)	566	$[M]^+$ (100/4.7)	13
Spirilloxanthin (1) (1,1′-dimethoxy, 3,4,3′,4′-tetradehydro-1,2,1′,2′-tetrahydro-*ψ*,*ψ*-carotene)	596	$[M]^+$ (100/1.3)	16
Spheroidenone (XI) (1-methoxy-3,4-didehydro-1,2,7′,8′-tetrahydro- *ψ*,*ψ*-caroten-2-one)	582	$[M]^+$ (100/10), $[M-2]^+$ (0.2/0.1), $[M]^{2+}$ (0.1/0)	13
2,2′-Diketo-spirilloxanthin (XII) (1,1′-dimethoxy-3,4,3′,4′-tetradehydro-1,2,1′,2′-tetrahydro-*ψ*,*ψ*-carotene-2,2′-dione)	624	$[M]^+$ (100/1.6), $[M-2]^+$ (0.4/0.1), $[M]^{2+}$ (3/0.4)	15
Peridinin (XIII) (5′,6′-epoxy-3,5,3′-trihydroxy-6,7-didehydro-5,6,5′,6′-tetrahydro-10,11,20-trinor-*β*,*β*-caroten-19′,11′-olide 3-acetate)	630	$[M]^+$ (100/7.4), $[M-18]^+$ (5/20.6), $[M-78]^+$ (10/23.6), $[M-92]^+$ (7/6.5)	12
Fucoxanthin (XIV) (5,6-epoxy-3,3′,5′-trihdroxy-6′,7′-didehydro-5,6,7,8,5′,6′-hexahydro-*β*,*β*-caroten-8-one 3′acetate)	658	$[M]^+$ (100/0), $[M-18]^+$ (3.6/7)	19
Myxoxanthophyll tetraacetate (XV) (2′-(*β*-*L*-rhamnopyranosyloxy)- 3′,4′-didehydro-1′,2′-dihydro-*β*,*ψ*-carotene-3,1′-diol tetraacetate)	898	$[M]^+$ (100/5)[a]	19
4-Ketomyxol 2′-(methylpentoside) tetraacetate (XVI) (3,1′-dihydroxy-2′-(5-*C*-methylpentosyloxy)-3′,4′-didehydro-1′,2′-dihydro-*β*,*ψ*-caroten-4-one tetraacetate)	912	$[M]^+$ (100/–)[b]	14

[a] Literature value for EI data.

[b] No available EI intensity data in literature.

tate (mol. wt. 722) with 26 and 55%; glycerol tripalmitate (mol. wt. 806) with 13 and 19%; glycerol trioleate (mol. wt. 884) with 100 and 56%; 1,2-dipalmityl-Sn-glycero-3-phosphoryl-ethanolamine (mol. wt. 691) with 0 and 100%; 1,2-dipalmityl-Sn-glycero-3-phosphoryl-*N,N*-dimethylethanolamine (mol. wt. 719) with 10 and 100%; and 1,2-dipalmityl-Sn-glycero-3-phosphorylcholine (mol. wt. 733) with 87 and 100%. Though the FD spectra contained fewer fragment ions than the EI spectra, ions were present for the key structural fragmentations, particularly at higher emitter currents.

Natural mixtures of olive and safflower oil were studied, producing molecular ions of the different components in relatively high abundance. Qualitative analysis of a mixture of phosphatidylcholine from egg yolk by FD-MS reflected the composition of the different long-chain fatty acids. Wood and Lau (116) reported the analysis of intact phospholipids by FD-MS. Diacyl glycerophosphorylcholines (lecithins) gave prominent molecular ions. For 1,2-dipalmitoyl-glycero-3-phosphorylcholine, a $[M]^{+}$ ion was obtained with 90% relative abundance.

Figure 38 gives the FD spectrum of cholesterylcaprylate. Here the molecular ion has only 12% relative intensity. The direct bond cleavage between the fatty acid and cholesteryl moieties gives a base peak of the spectrum at *m/e* 368. Apparently this cleavage is accompanied by hydrogen transfer to caprylic acid. A minor process is loss of C_3H_7 from the side chain that gives a peak at *m/e* 326. Since the spectrum is recorded electrically at low resolution, and proton transfer reactions produce five signals in the molecular ion region, an unambiguous interpretation is possible only for an authentic

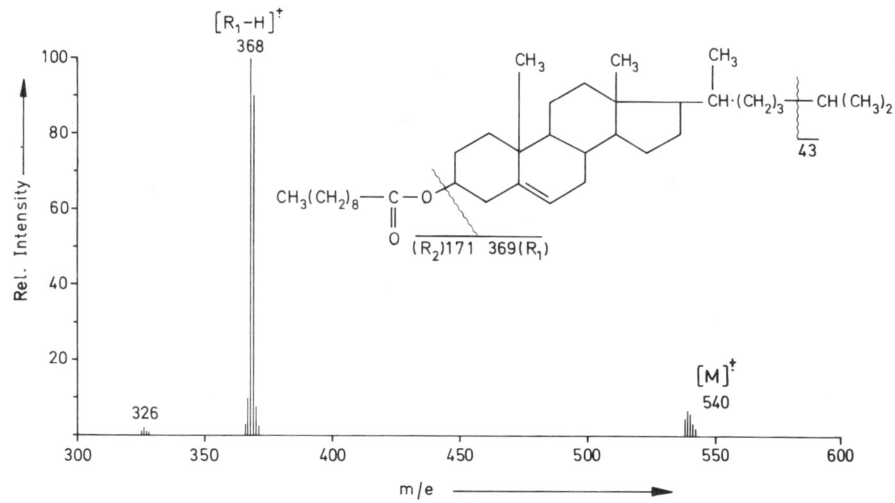

Figure 38. FD-MS of cholesteryl caprylate: solvent, acetic acid ethylester; electric detection at 9 mA ehc; ion source, Varian MAT 731 mass spectrometer; temperature, 80°C (49).

model substance. For the identification of an unknown compound, additional analytical information from other techniques is essential.

9. Biogenic Amines (Thin-Layer Chromatography and FD-MS)

Aliphatic amines (i.e., quaternary ammonium bases) such as acetylcholine and phenol or catechol amines such as adrenaline and dopamine, are of particular interest, and sensitive methods for their qualitative and quantitative investigation have been developed (117,118). In general two procedures are used for the identification and quantitation of aromatic amines in body fluids or tissue: (1) Separation of the amines by column- and thin-layer chromatography (TLC) and formation of highly fluorescent derivatives that are quantitated by fluorometry (117,119). (2) Relatively volatile derivatives of the amines are produced which are estimated by a coupled GC-MS unit (120,121). The sensitivity of both methods is sufficient to determine biogenic amines at concentrations in natural products (1 to 100 ng/ml).

Very recently, a new method for qualitative and quantitative analysis of dopamine (DA), and its conjugates 3-*O*-methyl-dopamine (3-MDA), and 4-*O*-methyldopamine (4-MDA) was introduced. Defined, fluorescent derivatives of high molecular weight were generated by reaction with 5-di-*n*-butylaminonaphthalene sulfonyl chloride (BANS-Cl) (122,123). The BANS-derivatives were isolated by TLC followed by identification and quantitation by two independent methods: EI-MS and FD-MS.

A. DOPAMINE

The reaction of DA with BANS-Cl produces a mixture of four derivatives: a monobansyl-DA, two constitutionally isomeric bisbansyl-DAs, and a trisbansyl-DS. Their differences in polarity allow the reaction products to be easily separated by TLC and clearly localized by their strong ultraviolet fluorescence (Figure 39). Under the conditions selected (reaction with a hundredfold molar excess of BANS-Cl at room temperature for about 3 hr) the tris-BANS represents the main product. As indicated on the left in Figure 39, the EI-spectra of all four derivatives yield the $[M - 43]^+$ ion as the base peak. The molecular ion is recorded with approximately 40% relative intensity. In contrast, the FD spectra exhibit only the molecular ion group. In this preliminary study into the use of combined TLC and EI/FD-MS in physiological and pharmacokinetic studies, the bisbansyl- and trisbansyl-DS derivatives have been isolated from human urine samples that were spiked with DA at a level of 10 ng/ml. This value is of the order of magnitude of the concentration of unconjugated DA excreted in the urine of Parkinson's disease patients undergoing L-dopa therapy (124).

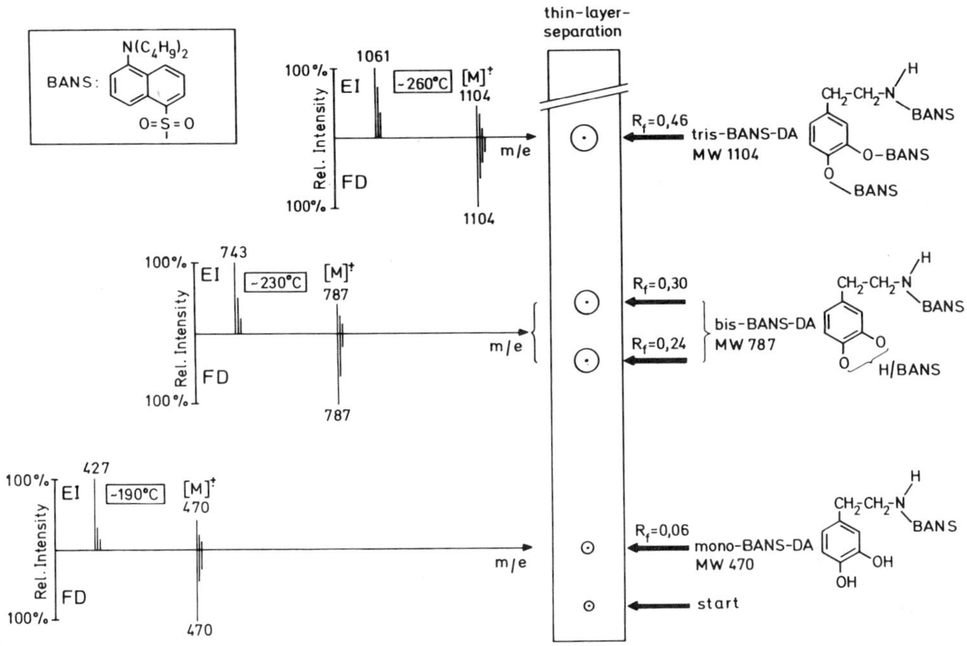

Figure 39. Thin-layer separation of the four derivatives that are formed by reaction of BANS-Cl with dopamine and partial EI and FD mass spectra of the separated reaction products. The mass range from *m/e* 300 to 1200 was recorded electrically with EI-MS. For photographic detection of the FD spectra the mass range from *m/e* 40 to 1400 was registered (125).

B. QUANTITATION OF BANS DERIVATIVES

For quantitative determination it appears to be advantageous to use the trisbansyl derivative of DA and bisbansyl derivatives of 3-MDA or 4-MDA. Two novel mass spectrometric methods were used for their measurement (125): (1) a twin direct introduction system for calibration in the electron impact mode (external standard), and (2) stable isotope dilution in connection with EI and FD mass spectrometry (internal standard). The method using an external standard yields quantitative results for concentrations between 0.1 and 10 μg/ml of urine, with average error of ±8%. To obtain reliable quantitative results using an internal standard, the molecular ion group of the bansyl derivatives was scanned with the electric detection system in the EI mode, or high-resolution FD-MS with photographic detection was employed. With EI-MS at low resolution, the estimation of the trisbansyl derivative in the range just described could be performed with an error of about ±5%. Photographic detection with FD-MS of the molecular ion group of a mixture of trisbansyl-DA-d_0/d_4 in the ratio 1:0.95 gave good results at a resolution of 10.000. The sample consumption was about 200 ng (180 pmole).

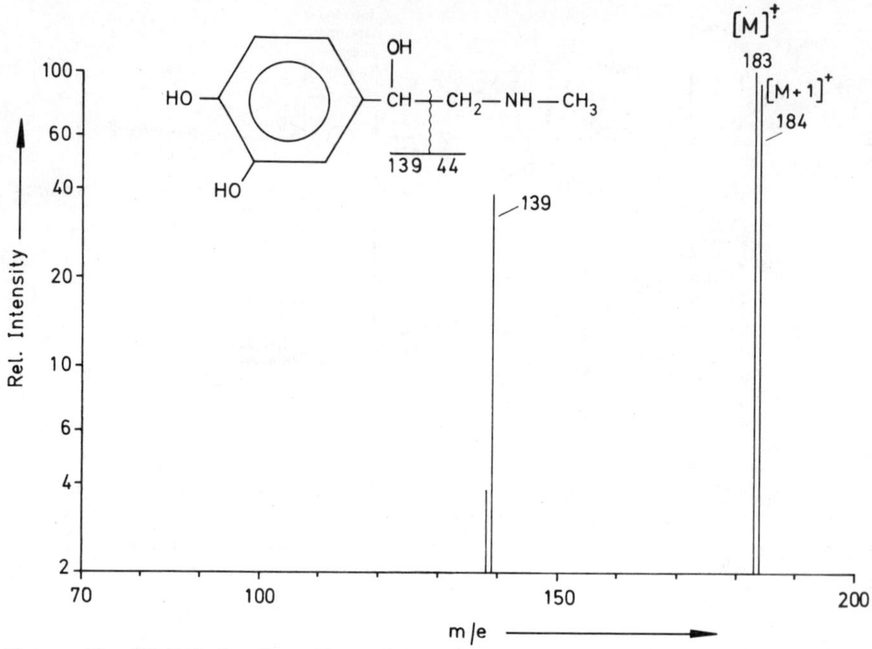

Figure 40. FD-MS of L-adrenaline: solvent, water; BAT, 22 to 24 mA ehc; exposure time of the photoplate, 4 min (49).

The theoretical molar ratio was reproducible within an error of 5%, which is consistent with a determination of cyclophosphamide-d_0/d_6 with FD-MS and photographic detection, reported in Section V.6.B.

In conclusion, we can note three main points.

1. The reaction of catechol amines with BANS-Cl leads to derivatives of high molecular weight, good chemical stability, strong fluorescence, and low polarity.
2. Extraction with a nonpolar solvent and TLC separation yields relatively pure derivatives in quantities that are sufficient for mass spectrometric investigation.
3. To produce quantitative data that are highly reliable, cross-checks with independent methods such as EI-MS *and* FD-MS are very useful.

The potential of FD-MS for underivatized biogenic amines has been demonstrated (111, 126). The FD mass spectrum of adrenaline is given in Figure 40. With gentle desorption the compound gives the molecular ion as base peak and only two complementary signals at m/e 139 and m/e 44, which are due to a direct bond cleavage in the side chain. Accurate mass measurements established the elemental composition and confirmed the

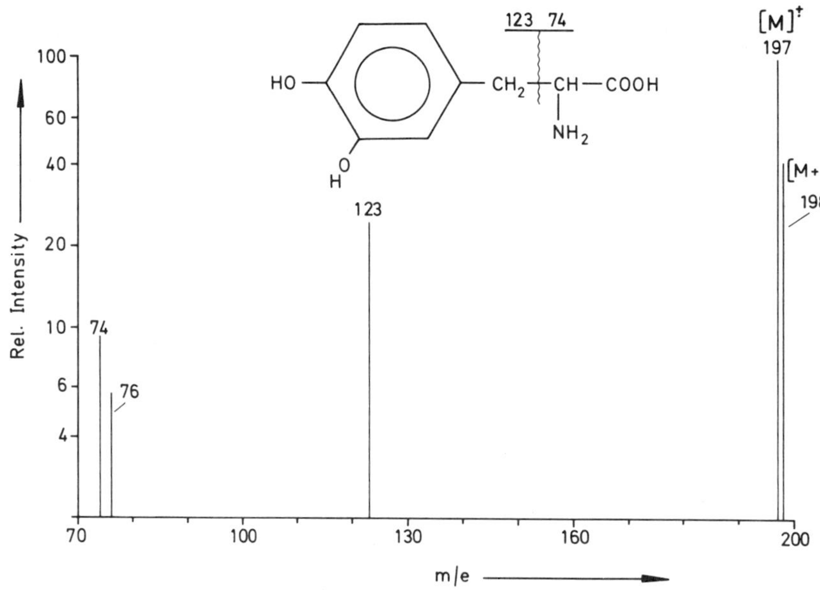

Figure 41. FD-MS of L-β-3,4-(dihydroxy)-phenylalanine (dopa): solvent, water; BAT, 25 to 28 mA ehc; emission-controlled field desorption threshold, 1×10^{-8} A; exposure time, 5 min. For the photoplate detection of the FD spectra, the mass range from *m/e* 40 to 1400 was registered (49).

correct assignment of these fragments (49). When free L-β-(3,4-dihydroxyphenyl)-alanine (dopa) was desorbed under the same experimental conditions, the FD spectrum in Figure 41 was obtained. As shown previously, the characteristic feature is the high intensity of the molecular ion with the straightforward and easily interpreted fission in the side chain.

For any mass spectrometric technique to assume practical use in the analysis of body fluids or biological extracts, several difficulties must be overcome. FD-MS has a major advantage in its ability to give useful spectra of nonvolatile and underivatized compounds. The results indicate that the method has promise for qualitative and quantitative analysis of biogenic free amines when the problems of sample concentration and contamination with inorganic salts are solved.

10. Vitamins and Coenzymes

With conventional EI-MS (70-eV electrons, 250 °C source temperature) the tetrakis-trimethylsilyl (TMS) derivative of vitamin C gave only a very minor molecular ion and intense fragmentation (127). When the ion source temperature was lowered to 160 °C, an EI spectrum (Figure 42*a*) was

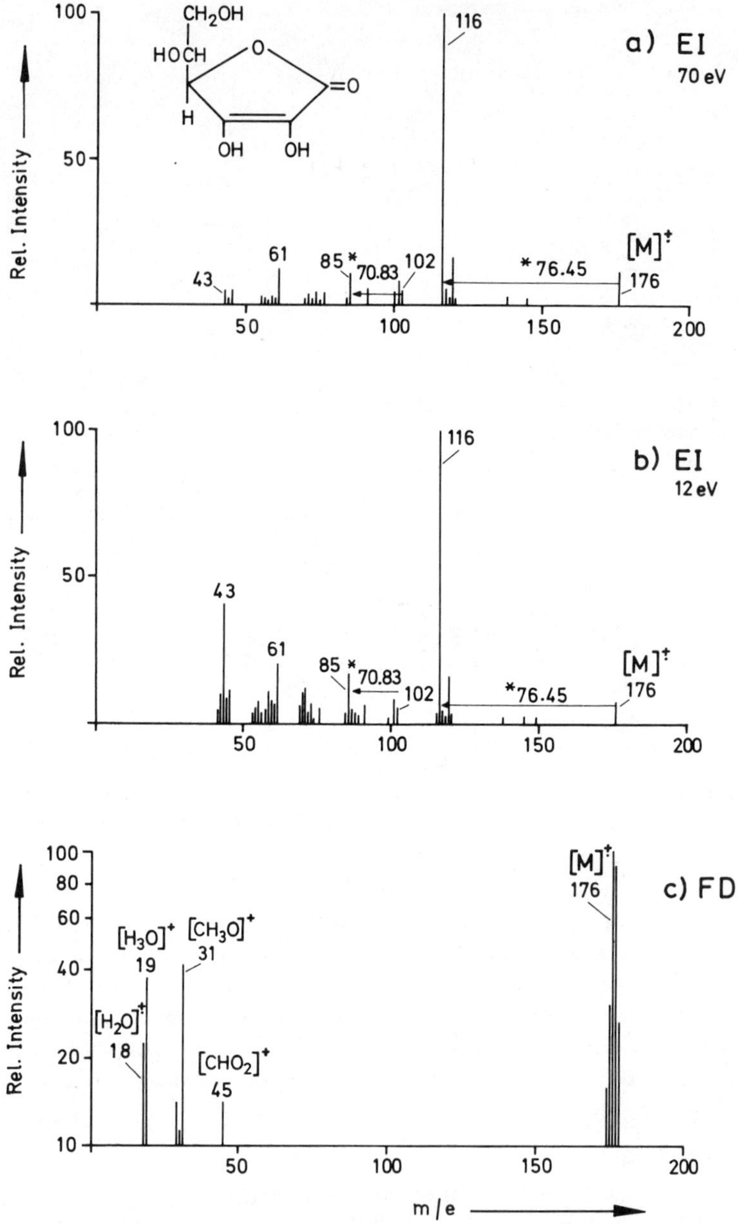

Figure 42. Mass spectra of ascorbic acid (vitamin C). (*a*) EI-MS: electron energy, 70 eV; spectrometer, AEI MS 12; probe temperature, 160°C (electric recording). (*b*) EI-MS: electron energy, 12 eV; spectrometer, AEI MS 12; probe temperature, 160°C (electric recording). (*c*) FD-MS: solvent, water; BAT, 20 mA ehc, photographic recording, CEC 21-110B spectrometer (50).

obtained from underivatized ascorbic acid (50). The molecular ion at *m/e* 176 is displayed with 6% relative intensity. The base peak at *m/e* 116 is due to loss of side chain and scrambling of one hydrogen ($M - C_2H_4O_2$), which is confirmed by a metastable ion at *m/e* 76.45. Reducing the electron energy to 12 eV (and the ionization efficiency of the electrons), the molecular ion intensity is increased to 11% (Figure 42*b*). This reflects the phenomena of enhanced relative abundances of the molecular ions at low electron energy, despite reduced absolute intensity. In sharp contrast, the FD-MS of ascorbic acid (Figure 42*c*) recorded photographically and at high resolution (resolution > 15,000, 10% valley definition), exhibits the molecular ion as base peak of the spectrum. Because of field-induced proton transfer reactions (48,92), either from the acidic moiety of the molecule or from the solvent as proton donor, the $[M + 1]^+$ signal has about the same intensity.

This high protonation is generally observed with partially nucleophilic compounds carrying (1) keto functions in the molecule (e.g., in free aldehydes or free carboxylic acids), (2) nitrogen functions of high basicity (e.g., in amines and amino acids), (3) polyhydroxy functions (e.g., in sugars and polyalcohols), and (4) inorganic acids (e.g., phosphoric and sulfonic acids).

Fragments with minor intensities are observed in the FD-MS of ascorbic acid only at *m/e* 18 $[H_2O]^+$, *m/e* 19 $[H_3O]^+$, *m/e* 30 $[CH_2O]^+$, *m/e* 31 $[CH_3O]^+$, and *m/e* 45 $[CHO_2]^+$.

As demonstrated for other classes of polar compounds, a distinct advantage of FD-MS is the possibility of fast and direct detection of unprotected polar vitamins. Recently Sammons et al. reported on the characterization of the B vitamins pyridoxal, pyridoxamine, their phosphates, and thiamine (128). In connection with FD spectra of the free phosphoric acid ester, pyridoxamine-5′-phosphate and pyridoxal-5′-phosphate, as hydrochlorides, give the $[M + H]^+$ ion as the base peak at T^*. Furthermore, the bond rupture between the ester oxygen and the phosphorus atoms characterizes the fragmentation as with FD-MS of the nucleosides, nucleotides, and glycosides. Riboflavin has not been found to be amenable to direct mass spectrometric analysis. The high-resolution FD spectrum (Figure 43) recorded with gentle desorption program (at T^*) is characterized by an intense molecular ion. When stronger thermal stress is applied, a pronounced water elimination—for $[M - H_2O]^+$ at *m/e* 358.128 and for $[M - 2H_2O]^+$ at *m/e* 340.117—is found. An additional signal at *m/e* 348.143 is explained by loss of carbon monoxide from the parent molecule. The structurally most significant bond cleavage between the isoalloxazine ring and the ribityl moiety accompanied by proton transfer is clearly indicated at *m/e* 242.080.

In summary, the FD method seems to be feasible for the detection and identification of very polar vitamins and should be a helpful tool in the structural elucidation of metabolic products and synthetic intermediates of vitamins.

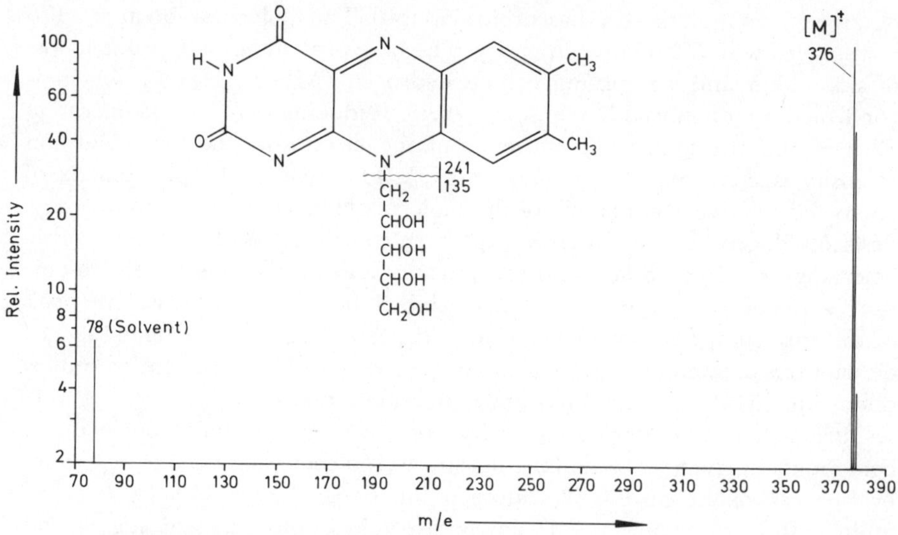

Figure 43. FD-MS of riboflavin (vitamin B_2): solvent, dimethylsulfoxide, BAT, 19 to 22 mA ehc; emission-controlled field desorption; threshold, 4×10^{-9}A; exposure time, 10 min (49).

V. APPLICATION OF FD-MS TO THE ANALYSIS OF DRUGS AND DRUG METABOLITES

To explore the potential of the field desorption method for synthetic drugs and drug metabolites, characteristic features of FD-MS were obtained and some rules for the interpretation of the spectra were established. These studies were applied to antineoplastic, immunosuppressive, and bacteriostatic agents, tranquilizers, hypnotics, and antibiotics were also investigated.

1. Antibiotics

Rinehart et al. (129) showed that FD-MS is the method of choice for determining molecular weights of a number of nonvolatile or thermally unstable antibiotics. Examples were neomycin, streptolydigin, and novobiocin, which do not give molecular ions by electron impact or chemical ionization mass spectrometry. In addition, field desorption mass spectrometry provided data of the composition of complexes of antibiotics, as illustrated by streptovaricin, filipin, and dermostatin. Since many antibiotics are found as complexes, the use of FD mass spectrometry in determining the composition of mixtures of related antibiotics appears to be very promising.

A broad spectrum antibiotic that affects the cell membrane and possibly inhibits protein synthesis is kanamycin. This compound carries seven free hydroxyl groups and four free amino functions and is therefore a good example of an extremely nonvolatile drug. The high-resolution FD spectrum of kanamycin (Figure 44) shows that the molecular ion is the base peak. As for glycosides and oligosaccharides, the direct bond cleavage at the glycosidic oxygen and proton transfer produce the most intense fragment at *m/e* 323 for $[II + H]^+$. Another structurally- interesting finding is the ions at *m/e* 163 $[I + H]^+$ for one amino sugar unit. The signals at *m/e* 142, 125, and 109 are probably due to the loss of water and ammonia from part III of the molecule. It should be emphasized that the doubly charged $[M + H]^{2+}$ ion is very prominent (40% rel. int.) and can be useful for the confirmation of the molecular ion.

A commercial penicillin available as a potassium salt is phenethicillin. As for glucuronides and sugar phosphates, large-molecule organic salts are of particular interest for FD investigations because this class of compound is not amenable to other mass spectrometric techniques. As might be expected from previous studies (60), the molecular ion of phenethicillin at *m/e* 402 does not appear, and cluster formation leads to the $[M + {}^{39}K]^+$ ion at *m/e* 441, from which the molecular weight is derived (Figure 45). The isotopic pattern of potassium is also clearly discerned for the cluster ions at *m/e* 236/238 and it

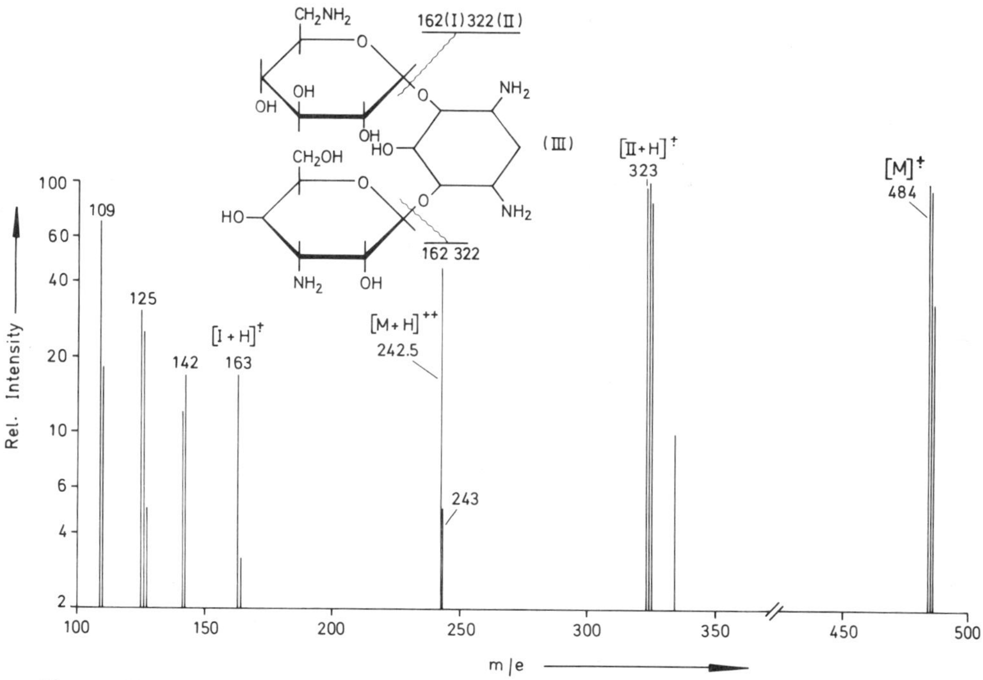

Figure 44. FD-MS of kanamycin: solvent, methanol; BAT, 18 to 22 mA ehc; exposure time, 12 min (49).

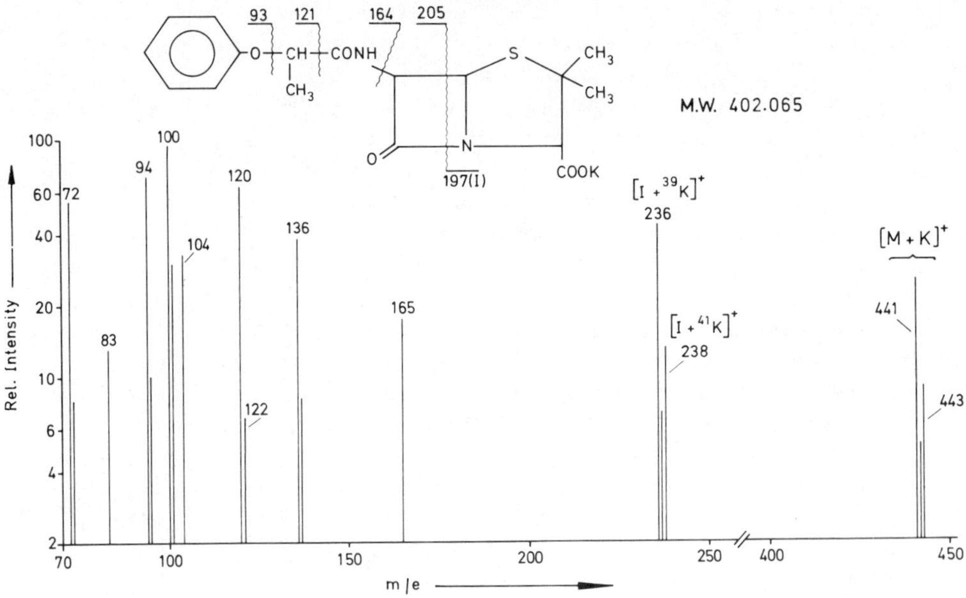

Figure 45. FD-MS of phenethicillin potassium salt: solvent, methanol; BAT, 25 to 29 mA ehc; exposure time, 10 min (164).

indicates the β-lactam moiety of phenethicillin. Three bond cleavages in the side chain give relevant structural information. Signals at *m/e* 94 and 122 give accurate masses, hence elemental composition. This enables a clear distinction to be made from other penicillins (e.g., benzylpenicillin, phenoxymethylpenicillin, propicillin, and phenbenicillin).

Recently Zerilli et al. (130) have shown the FD spectra of a number of antibiotics of the rifamycin type, obtained by fermentation and chemical modification, which are active against Gram-positive and Gram-negative bacteria and mycobacteria. The basic structure of rifamycins consists of a naphthoquinone moiety, usually referred to as the chromophore, which is spanned by an aliphatic bridge, called the ansa. When compared with the EI spectra, the FD spectra revealed that with FD-MS all the compounds gave abundant molecular or quasi-molecular ions, whereas no significant fragmentation was observed in the low-mass region of the spectra. It appears that this technique is superior for determining molecular weights of rifamycins, and it is recommended for new derivatives.

2. Bacteriostatic Agents

The EI mass spectrum of the sulfonamide aristamid (Figure 46*a*) is contrasted with the FD spectrum (Figure 46*b*) (49). The relative intensity of

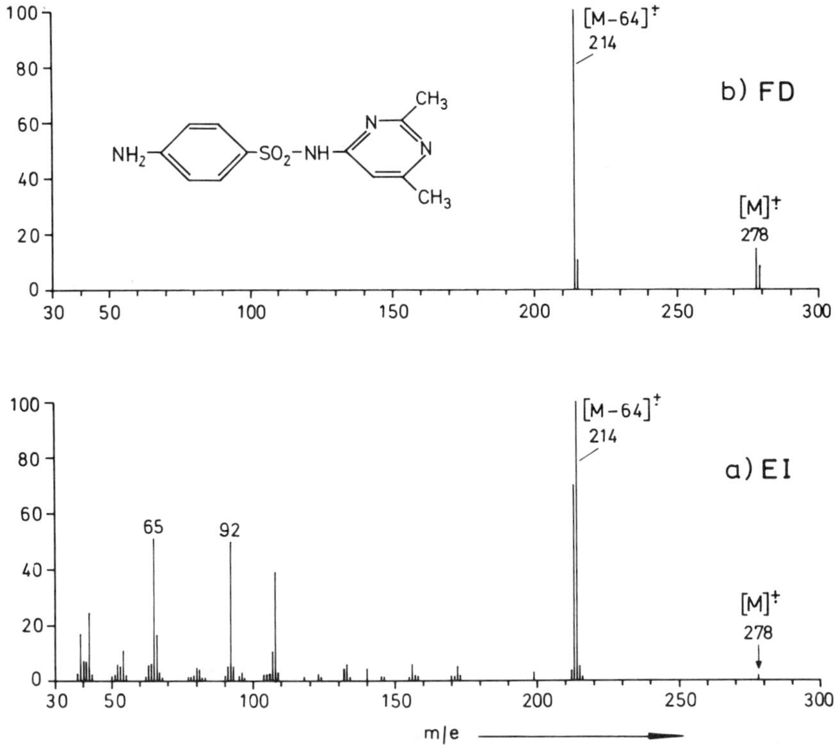

Figure 46. Mass spectra of aristamid. (*a*) LR EI-MS: 70 eV. (*b*) FD-MS: solvent, 0.1*N* HCl; BAT, 18 mA ehc; exposure time, 11 min (49).

the molecular ion is about 1% in EI-MS and 14% in FD-MS when the spectra are recorded electrically at low resolution. It is unique in that the base peak with both methods is obtained for an ion at *m/e* 214 because of a rearrangement and elimination of SO_2—a process that has been observed in the EI mass spectra of diaryl sulfones. Although with FD no more fragmentation is observed, there are a number of structurally significant ions in the EI spectrum: for example, *m/e* 65 for $[SO_2H]^+$, *m/e* 92 for the phenylamine ion, and *m/e* 108 for protonated 2,4-dimethyl-pyrimidine. As long as the molecular ion of this sulfonamide is clearly detected, Figure 46 shows that EI-MS yields more information and for this compound is obviously the better method.

However this advantage does not apply when the EI and ED mass spectra of badional are compared. The $[M]^+$ was very weak in EI-MS, but this ion at *m/e* 231 is of about 75% relative intensity in the FD mass spectrum (Figure 47). The common direct bond cleavage between the *p*-amino-benzenesulfone and the thiocarbamate moieties of the molecule shows the two essential parts

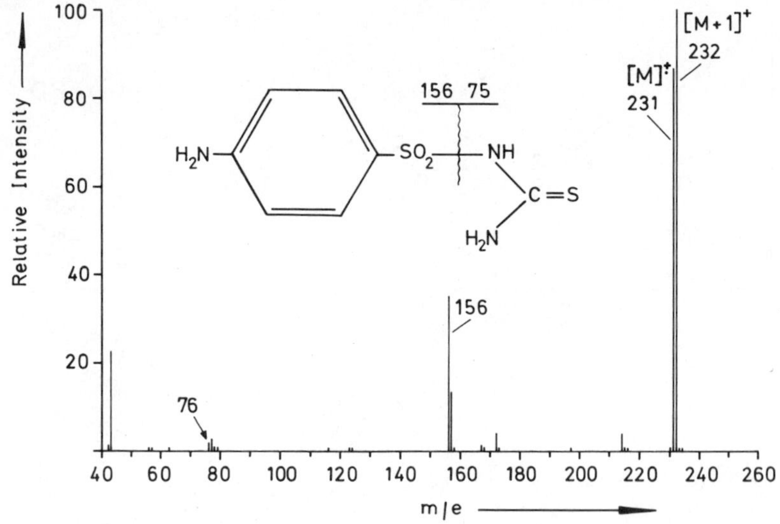

Figure 47. FD-MS of badional: solvent, acetone; BAT, 20 mA ehc; exposure time, 13 min (49).

of the drug. Hence for highly polar substances, FD-MS yields not only information of molecular weight but also structurally important details by way of significant fragmentation. As for glycosides (40) and sultams (37), in some cases these straightforward and easily interpreted fragments are produced by FD only, not EI.

3. Immunosuppressive Agents (Drug Conjugates)

Azathioprine (I), [(1-methyl-4-nitroimidazol-5-yl) thio] purine, is the most widely used immunosuppressive agent in clinical organ transplantation. The metabolism of I has been the subject of extensive studies (131). When the metabolic fate of the methylnitroimidazole moiety of ^{14}C-azathioprine, labeled in the imidazole ring, was investigated (132), evidence was found that nucleophilic attack occurred on the nitromethylimidazole ring, *in vivo*, leading to the formation of 5-substituted derivatives. In recent studies on the rat (132,133), the following metabolites of azathioprine were characterized: 1-methyl-4-nitro-5 (*S*-glutathionyl)imidazole (II), 1-methyl-4-nitro-5(*N*-acetyl-*S*-cysteinyl)imidazole (III), and 1-methyl-4-nirto-5-carboxymethylaminoimidazole (IV).

The predominant analytical methods used to identify II to IV were liquid chromatography and ultraviolet spectroscopy. Conventional EI mass spectrometry was of limited value because the resulting spectra had small to nonexistent molecular ions and weak high-mass fragments. Therefore molecular weights and/or elemental compositions of these metabolites were difficult or impossible to obtain by EI-MS.

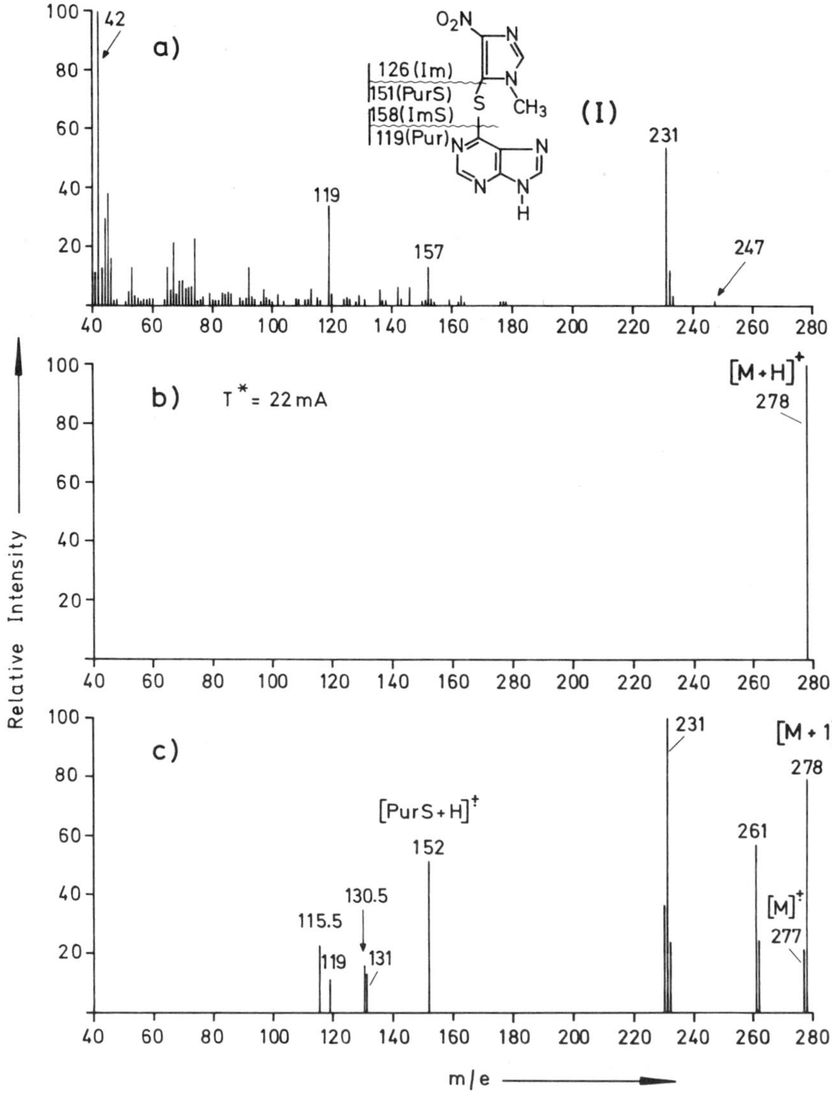

Figure 48. Mass spectra of azathioprine (1). (*a*) EI-MS, using a direct probe at 210°C. (*b*) FD-MS, using 22 mA ehc. (*c*) FD-MS, using 25 mA ehc (52).

The EI spectrum of azathioprine (Figure 48*a*) (52) shows that the relative intensity of the $[M]^+$ ion is below 1%. A small peak at *m/e* 247 ($C_9H_7N_6OS$) represents the loss of NO from the molecular ion. The most intense high-mass ion, *m/e* 231 ($C_9H_7N_6S$), is formed by the loss of NO_2. Fission of the sulfurimidazole bond with rearrangement of a hydrogen to the purine moiety, $[PurS + H]^+$, results in an ion isomeric with 6-mercaptopurine at *m/e* 152 ($C_5H_4N_4S$). Cleavage between sulfur and the purine ring with retention of charge on the purine ring $[Pur]^+$ accounts for the formation of the ion at *m/e* 119 ($C_5H_3N_4$). The base peak of the spectrum is *m/e* 42 (C_2H_4N).

In contrast, the FD mass spectrum gives an intense $[M + H]^+$ ion, *m/e* 278 ($C_9H_8N_7O_2S$) at *T** (Figure 48*b*). When the emitter current is raised (Figure 48*c*), fragmentation is induced. The FD mass spectrum of azathioprine at 25 mA emitter current is: *m/e* 278, 79%, $[M + 1]^+$; *m/e* 277, 22% $[M]^+$; *m/e* 262, 24% $[M - CH_3]^+$; *m/e* 152, 51%, $[PurS + H]^+$; *m/e* 130.5, 16%, doubly charged mass 261; *m/e* 131, 13%, doubly charged mass 262; *m/e* 119, 11% $[Pur]^+$; and *m/e* 115.5, 22%, doubly charged mass 231. The fragments at *m/e* 231, 152, and 119 are common to both the EI and FD spectra of azathioprine. They have the same elemental composition and probably the same structure. Although both techniques offer fragments that can aid in characterizing (I), the quasimolecular ion is available in the field desorption spectra only.

The EI mass spectrum of a metabolite of azathioprine, a gluthathionylconjugate (II), results primarily from pyrolytic decomposition (Figure 49*a*). This spectrum is remarkably similar to the low-resolution field desorption spectrum taken at 27 mA emitter heating current (Figure 49*c*), with the exception of the molecular ion. The *T** for II is 24 mA (Figure 49*b*). Under this mild thermal condition, only the $[M + H]^+$ signal is seen. As the emitter temperature is increased, fragments are generated, yielding valuable structural information (Figure 49*c*). The $[M + H]^+$ peak is still large (26% rel. int.), and the base peak is *m/e* 159, $C_4H_5N_3O_2S$, $[ImS + H]^+$, *m/e* 158, and $[ImS + 2H]^+$, *m/e* 160, are present. The high-resolution FD mass spectrum has an ion at *m/e* 274 ($C_{10}H_{16}N_3O_6$) because of the loss of (ImS) from the molecule. Thus the $[M + 1]^+$ and *m/e* 159 and 274 ions characterize this molecule.

The observation that unprotected amino acids and oligopeptides exhibit high molecular intensities (Section IV.4) prompted this exploration of the FD method for the study of metabolites that are conjugates of drugs with amino acids and oligopeptides.

The first attempt to explore the potential of FD-MS for the detection of a drug metabolite in an actual sample was performed with the *N*-acetyl-*S*-cysteinyl conjugate (III) of azathioprine (52). This sample was isolated from the urine of rats given an oral dose of azathioprine. Compound III was isolated by column chromatography and purified by paper chromatography. Low-resolution FD-MS revealed the molecular weight of this metabolite. In an interval of the ehc between 0 and 24 mA, only intense signals were obtained at *m/e* 288 (100% rel. int.) and 289 (40% rel. int.).

After this field desorption measurement was completed, observation of the emitter in a light microscope, 100 × magnification, showed that the microneedles were still covered with a considerable amount of the sample.

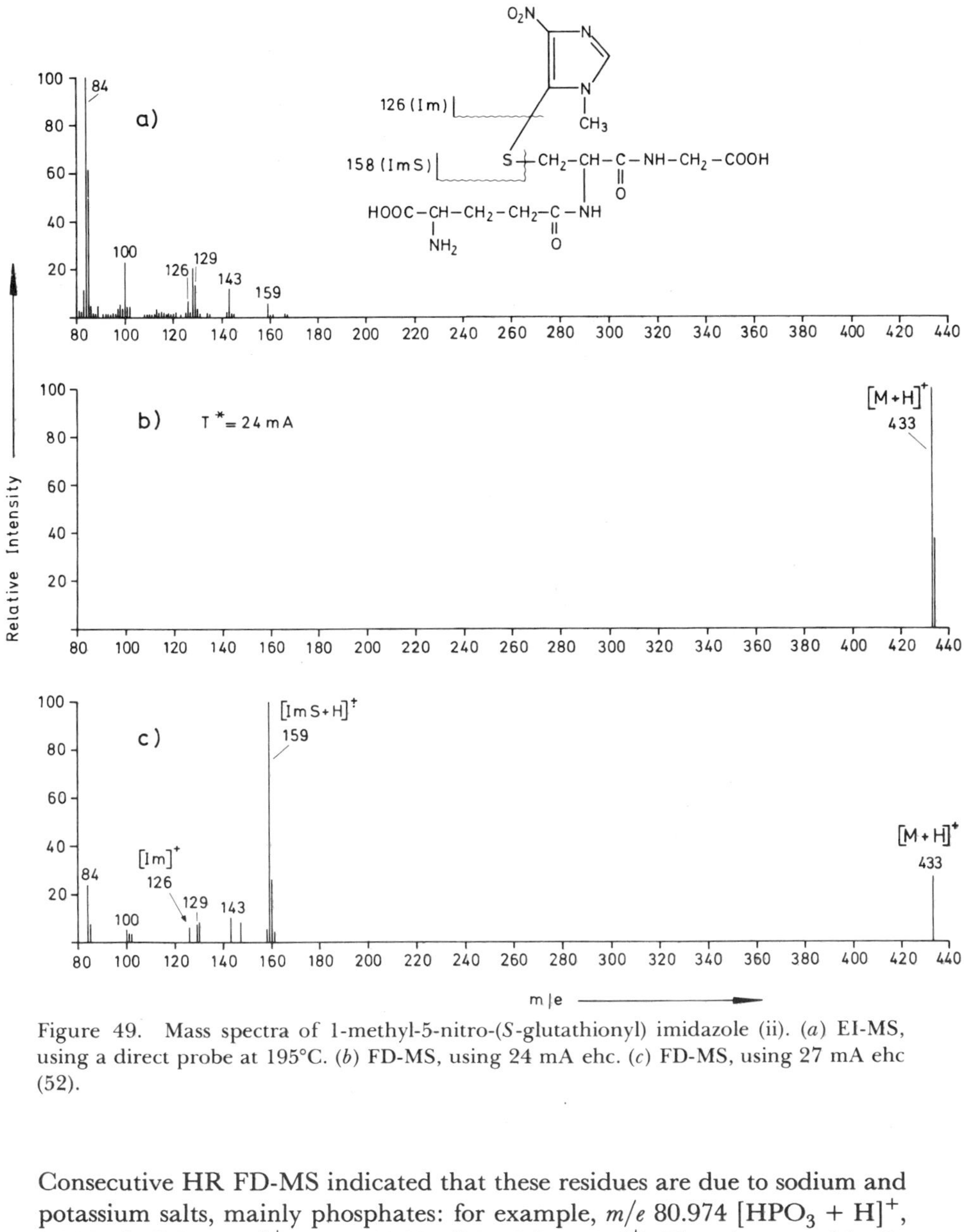

Figure 49. Mass spectra of 1-methyl-5-nitro-(*S*-glutathionyl) imidazole (ii). (*a*) EI-MS, using a direct probe at 195°C. (*b*) FD-MS, using 24 mA ehc. (*c*) FD-MS, using 27 mA ehc (52).

Consecutive HR FD-MS indicated that these residues are due to sodium and potassium salts, mainly phosphates: for example, *m/e* 80.974 $[HPO_3 + H]^+$, 96.969 $[HPO_4 + H]^+$, and 164.931 $[Na_3PO_4 + H]^+$. The photographically recorded spectrum showed $[Na]^+$ and $[^{39,41}K]^+$ peaks with saturated blackening.

Similar to results from FD-MS of biochemicals and drugs that are conjugated as glucuronides (Section IV.1.D), phosphates (Section IV.1.E) and sulfates (Section IV.2.A), it was shown that drug conjugates of amino acids and peptides give useful spectra with field desorption.

In summary, the FD method proved especially useful for the determination of molecular weights in drug metabolism studies of the immunosuppressive agents. This was confirmed by FD investigations of biotransformation products of pesticides (26,51) and other environmental chemicals (50,59).

4. Tranquilizers

The field desorption mass spectra of 10 medicinal carbamates (meprobamate, mebutamate, carisoprodol, emylcamate, bethanechol chloride, styramate, hydroxyphenamate, mephenesin carbamate, methocarbamol, and chlorphenesin carbamate) were run and compared with their published EI spectra (135).

To evaluate the potential of EI-MS in characterizing these compounds, Coutts studied their EI mass spectra and found that the molecular ion was either absent or of low intensity (134). In contrast, of the 10 compounds just listed , emylcamate did not give a field desorption mass spectrum. All others gave abundant molecular ions, with the exception of bethanechol chloride, which has a quaternary ammonium group. No general characteristic fragmentation was attributed to the carbamate function in the field desorption mode.

5. Hypnotics (Quadrupole Mass Filters and FD-MS)

Barbiturates, their metabolites, and derivatives have been investigated by EI-MS (139), FI-MS (140), CI-MS (141), and low-resolution FD-MS (142). To give an estimate of the merits of a new analytical method such as field desorption mass spectrometry, not only the advantages of the available information but also the costs of an FD analysis should be mentioned. The photographically recorded FD spectra obtained with double-focusing mass spectrometers are expensive, and the evalution of the spectra is tedious and time-consuming. An alternative is the combination of a quadrupole mass filter and the field desorption ion source. The physical and technical problems involved in this combination have been solved (136). Recent investigations revealed the versatility of the FD-quadrupole mass spectrometer (137,138). Wide ranges of the mass spectrum can be displayed quasi-simultaneously on the oscilloscope screen because of the facility for rapid mass scan. Hence changes in the operating mode for resolution, mass range, emitter temperature, and field strength can be visualized directly during the FD process. If low-resolution mass analysis proves not to be sufficient and a high-resolution FD mass spectrum with a double-focusing instrument becomes necessary, a preinvestigation of the substance with the quadrupole mass spectrometer is of advantage to exploit the desorption behavior of one or several components.

The utility of FD-MS for mixture analysis has been demonstrated (Sections

IV.4. and IV.5). If the determination of the molecular weight of known substances is sufficient for routine work, the quadrupole-FD combination not only offers a highly time-resolved observation of the FD spectrum but also a distinction between components with different desorption behaviors. This is demonstrated by an example from forensic medicine where the chloroform urine extract from a case of overdose of hypnotics was examined by FD (137).

It was suspected that the patient had taken an overdose of the hypnotic "trisomnin," a combination of the barbiturates propanal (5,5-dipropylbarbituric acid), luminal (5-ethyl- 5-phenyl-barbituric acid), and eldoral [5-ethyl, 5(1-piperidyl)-barbituric acid], and attention was paid only to the molecular ion regions of these barbiturates. Figures 50*a* to 50*c* show the FD mass spectra of the extract at different emitter temperatures. Clearly desorption patterns of the barbiturates present in the extract differed: eldoral desorbed completely at low emitter temperatures with low intensities, whereas luminal desorbed over a wide temperature range with high intensities. The desorption behavior of propanal resembled that of eldoral, with higher ion intensities at a higher emitter temperature. Even though the concentration of eldoral was very small, the consumption of "trisomnin" was confirmed.

Since the whole recording procedure for the mass spectrum takes only a few minutes, the application of a quadrupole-FD mass spectrometer for fast identification of drugs in extracts of blood, gastric contents, and urine appears very promising. Because a rough estimate of the costs for a completely

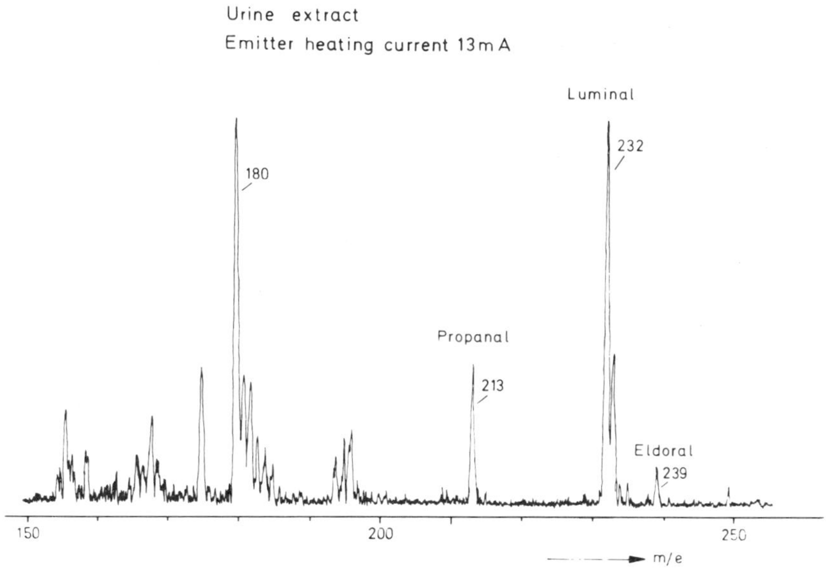

a

Figure 50. FD-MS of the chloroformic extract of a case of overdose of hypnotics obtained with electric detection at different emitter heating currents (137).

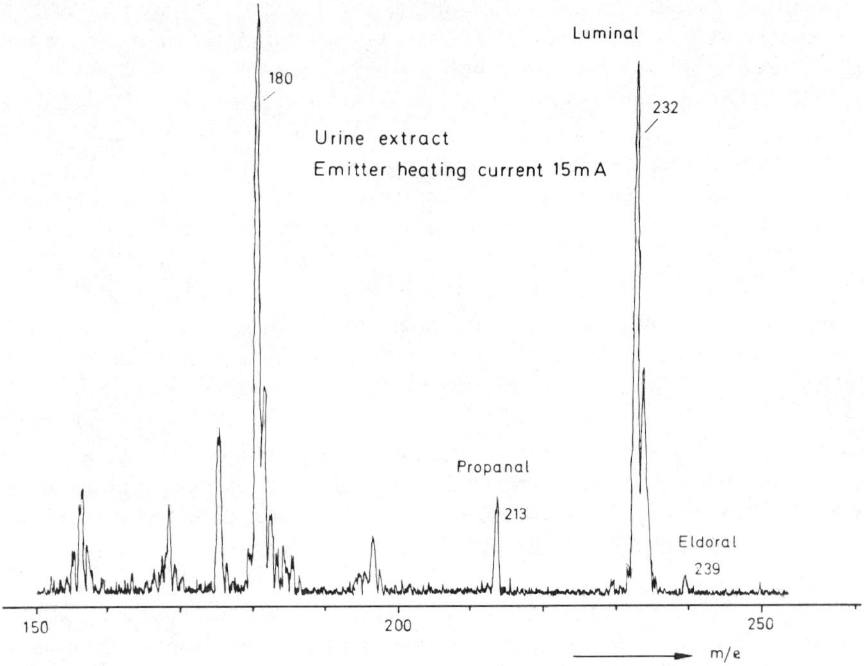

b

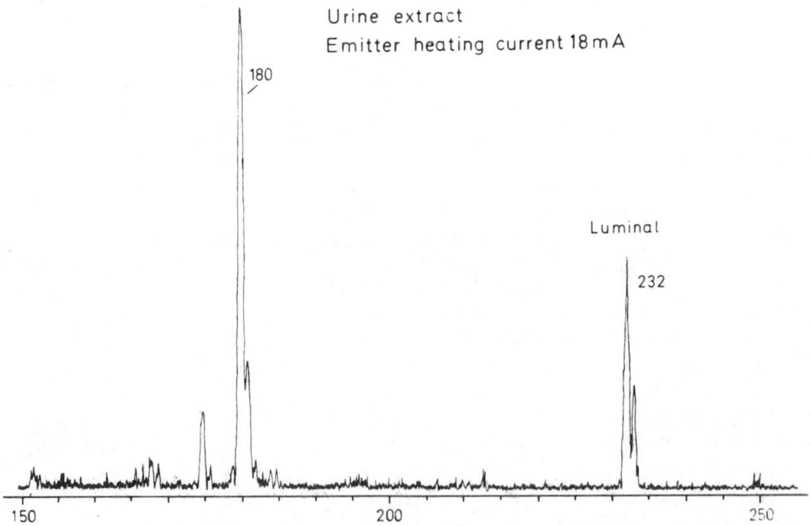

c

Figure 50. Continued.

equipped quadrupole instrument (with data system) is only about a third of the price of a double-focusing mass spectrometer for highly resolved, electrically and photographically recorded spectra, the combination FD and quadrupole mass spectrometer is a new tool for clinical and forensic analysis.

6. Antineoplastic Agents

Most of the earlier work with FD-MS of drugs was carried out with anticancer drugs and their metabolites (11,27,36,80). Two outstanding features of field desorption were considered. First, a comparison of the EI and FD spectra of a large number of substituted sultams showed the significant fragmentation pattern for each method and the critical differences for the formation of ion in both techniques (30). Second, the potential of FD for studies of biotransformation was shown for a common anticancer drug and most of its known metabolites.

A. SUBSTITUTED SULTAMS

It was reported that some sultam derivatives have antitumor effects on Ehrlich ascites tumors (143). The LR and HR EI and FD mass spectra of similarly substituted 5- or 6-membered ring sultams were reported, and the main fragmentation pathways were interpreted (30). Sultams carrying polar substituents, such as dicarboxylic acids and their mono- and diamides, showed either small or no molecular ions with EI-MS. In contrast, FD-MS showed in all cases the quasi-molecular ions $[M + 1]^+$ as the base peaks of the spectra.

a. Electron Impact Mass Spectra. Of the compounds **1** to **5** (structures in Scheme 3) only **4** showed a peak corresponding to the molecular ion in its EI mass spectrum. The position of the molecular ions in the remaining compounds could be found, however, by metastable defocusing. The EI-MS fragmentation pathways for substituted butanesultams are given in Scheme 2, showing quite similar spectra for the glutaric acid derivatives 1-3. The fragmentation of **1** (Figure 51*a*) is initiated by α-cleavage, leading to ions **a**, which then fragment further by elimination of stable molecules. α -Fission between the α-C atom and the adjacent $(CH_2)_nCOR'$-group, which would afford unfavorable primary radicals, does not occur in the EI spectra of **2** to **5** and only to a minor extent for **1** (*m/e* 192, 6% rel. int.). As was to be expected, bond rupture between the ring nitrogen and the substituent, which is important in FD-MS, was not observed with the EI technique.

The most characteristic feature of the succinic acid derivative **4** on electron impact is loss of SO_3 or SO_3H from the molecular ion, leading to ions $[C_8H_{13}NO_3]^+$ (*m/e* 171: 7% rel. int.) and $[C_8H_{12}NO_3]^+$ (*m/e* 170; 8% rel. int.), and from the product of α-cleavage $[C_7H_{12}NO_4S]^+$ (*m/e* 206 base peak) leading to the ions $[C_7H_{12}NO]^+$, (*m/e* 126; 47% rel. int.) and $[C_7H_{11}NO]^+$ (*m/e* 125, 20% rel. int.). This can be explained by a transfer of

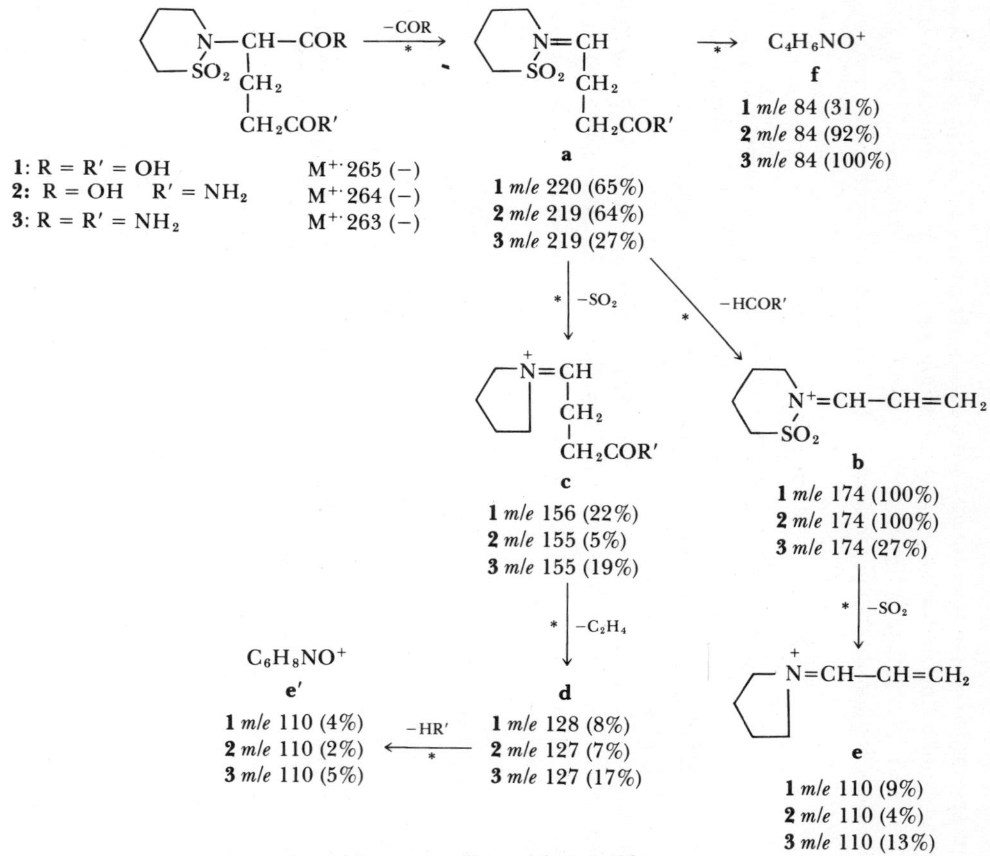

Scheme 2. Main EI fragmentation pathways (30).

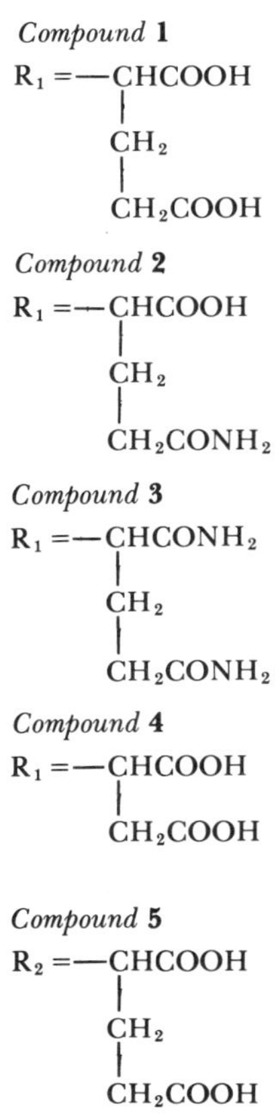

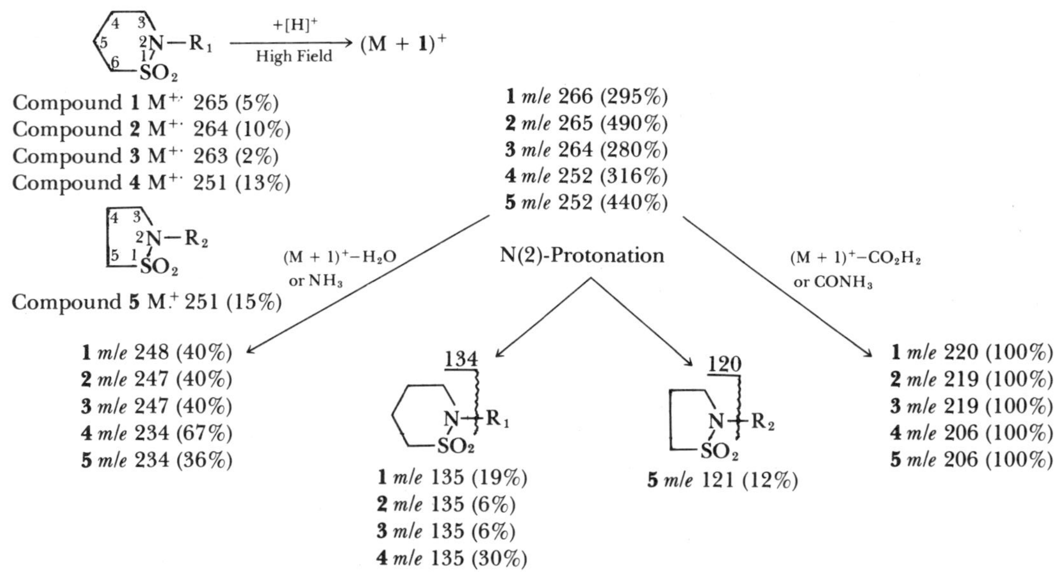

Scheme 3. Main FD fragmentation pathways (30).

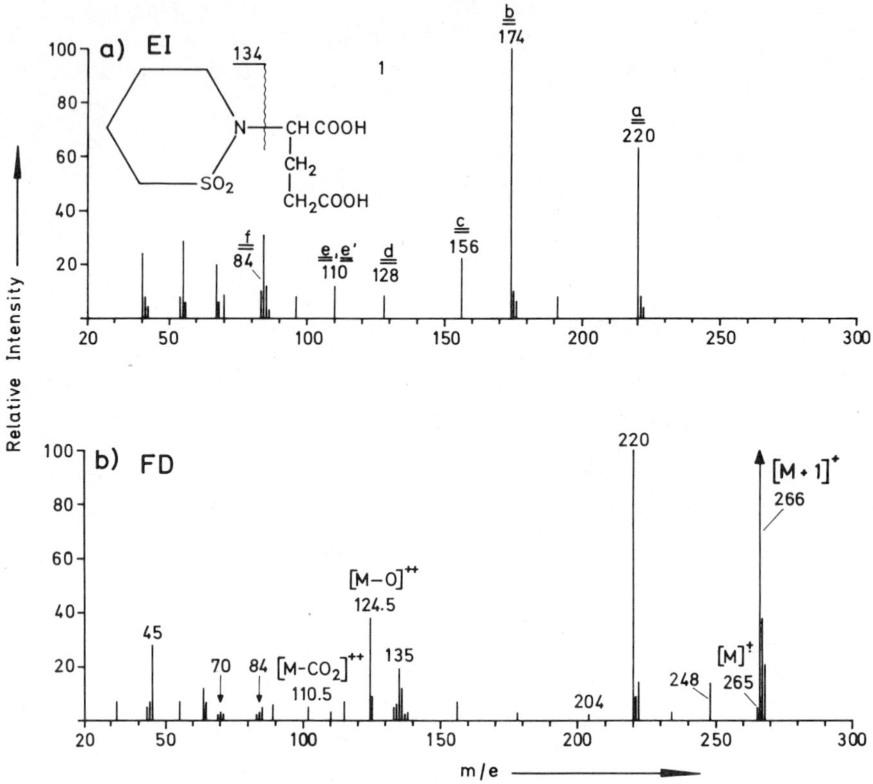

Figure 51. (*a*) EI-Ms of the substituted sultam **1**. Lowercase letters indicate ions interpreted in Scheme 2. (*b*) FD-MS of **1**. Main fragments are explained in Scheme 3 (30).

the OH function from the β-carboxyl group to the S atom, by way of a 6-membered transition state. The transfer of the OH group in **1** to **3** requires a 7-membered transition state. Although possible in principle, there is no positive evidence for this occurrence, since at no fragmentation step was a loss of SO_3 observed. The fragmentation behavior of **5** is closely related to that of **1** to **3**, the main difference being that after α-cleavage leading to the base peak of the spectrum (*m/e* 206), COOH instead of HCOOH is eliminated, yielding $[C_6H_{11}NO_2S]^+$ (*m/e* 161; 78% rel. int.). The other main fragments are analogous to those described in Scheme 2.

b. Field Desorption Mass Spectra. Utilizing the high ionization efficiency and extreme thermal stability of high-temperature-activated emitters (11), FD spectra were obtained by inducing a controlled thermal fragmentation of the compounds examined. The basic conditions for this procedure are as follows.

First, a thick (5 to 15 μ) layer of the adsorbed solid substance is applied by the emitter dipping technique to the surface and into the gaps between the microneedles of the field ion emitter. From electron scanning micrographs the maximal amount of sample deposited on a 10-μ high-temperature-activated tungsten wire (needle length 15 μ average) is estimated to be approximately 1×10^{-6} g. For adjusting the FD emitter before the desorption measurements in the FI mode, it is important that a number of needle tops protrude from the adsorbed compound layer (see Fig. 5*c*).

Second, highly polar samples desorb rapidly and, because their FD mass spectra are strongly dependent on the ehc, it is necessary to have an integrating detection system. One solution to the problem is the time-averaging technique, in cases where on-line computer facilities and a fast scan of the magnetic field are available. The other solution, photographic recording, has been preferred for this work. The use of photoplates (especially vacuum-evaporated AgBr plates) seems to offer crucial advantages, particularly when, despite extremely short desorption times and small ion current intensities, high-resolution data are required.

Third, since molecules of volatile substances escape from the emitter surface without accepting much thermal energy, the FD-MS show only molecular ion peaks and fail to give fragment ions. This is no disadvantage, since most of these compounds yield good FI, CI, or EI spectra.

Five general features of the production and interpretation of FD-MS were described:

1. Sultams carrying free amino acids or their mono- or diamides as functional groups exhibit the quasi-molecular ions as the most intense peaks in all observed spectra. To visualize more clearly the relative intensities of the fragments, the actual intensity of the $[M + 1]^+$ ion is given in Scheme 3. In the FD-MS of **1** and **4**, the peak with the second highest intensity was chosen as the 100% relative abundance (base) peak and the spectrum normalized to its intensity. This was necessary because at low emitter temperatures (about 5 to 10 mA ehc) the fragments are scarcely detectable and only the quasi-molecular ion can be recorded unambiguously.

2. It is clear from Scheme 3 that another prevailing feature of FD is the single-step formation of fragment ions of higher intensity in the upper mass range. Field-induced protonation of the heteroatom followed by the elimination of small, stable molecules from the quasi-molecular ion was observed for a wide variety of different compounds (59). The loss of CO_2H_2 from carboxylic acids has already been described (Section IV.4.A), and it produces the most intense signal in the FD-MS of **1**, **2**, **4**, and **5** apart from the $[M + 1]^+$ ion. The analogous expulsion of $CONH_3$ from the protonate diamide **3** is observed with comparable intensity.

3. The compounds investigated show elimination of water (Scheme 3) from the quasi-molecular ions of **1**, **2**, **4**, and **5** or NH_3 from 3. Decarboxylation is

observed with lower intensity (*m/e* 220, **1**, Figure 51*b*; *m/e* 206, **4**). In addition, the overall loss of H_2O and CO_2 from $[M + 1]^+$ is a minor process only: *m/e* 204 for **1**; Figure 51*b*; *m/e* 190 for **4**. These reactions are critically dependent on the emitter heating current; that is, the decay of the $[M + 1]^+$ ion with rising emitter temperature is accompanied by an increase in the relative abundance of signals due to these eliminations.

4. As a result of field-induced proton transfer reactions on the emitter surface, either from the acidic moiety of the substance or from the solvent molecules, a simple cleavage between N (**2**) and the substituent is observed in all cases (Scheme 3). The signals obtained at *m/e* 135 for **1** to **4** and 121 for **5** came from the 5- or 6-membered sultam ring, consistent with this being a major structural unit. The decay of the compounds into stabilized, even-electron building blocks has also been observed for nucleosides, nucleotides (24,41), and the pyrolysis FD products of DNA (31). The resulting FD mass spectra yield easily identifiable and relevant structure information.

5. Doubly charged ions were first reported by FD-MS of sultams. With high resolution and photographic detection, two species of these ions are recognized by their mass position at approximately half nominal masses. Precise mass determinations revealed that the more prominent ones may be described formally by $[M - 0]^{2+}$ at *m/e* 124.5 (Figure 51*b*) and *m/e* 117.5 (**4**), the minor one, by $[M - CO_2]^{2+}$ at *m/e* 110.5 (Figure 51*b*) and *m/e* 103.5 (**4**).

c. Comparison of EI and FD Mass Spectra. It is a general observation that FD mass spectra display greater intensities of molecular or quasi-molecular ions compared to EI mass spectra. This is the most significant difference between the two. Only minimal energy is transferred in the ion-generating process, since cationization (Section IV.3) in the adsorbed surface layer results in $[M + H]^+$ even-electron species. Regarding the high intensities of quasi-molecular ions obtained, FD shares the advantage of chemical ionization (CI) in contrast to EI-MS, where primary ionization of molecules in the gas phase results in odd-electron molecular ions $[M]^{+}$. Whereas in EI-MS α-cleavage is the prevailing feature for large fragments, with FD-MS the ions of the upper-mass range are formed by field-induced proton transfer followed by the elimination of small, neutral particles. Furthermore, protonation of heteroatoms (according to their proton affinity) leads to direct bond rupture in the sample molecules, yielding easily interpreted and valuable structure information. Although these single-step processes are dominant in FD-MS, signals of smaller intensity indicate multistep processes such as association and rearrangement reactions similar to EI-MS.

B. CYCLOPHOSPHAMIDE (FD-MS AND METABOLISM STUDIES)

Mass spectrometry has played a major role in recent investigations into the mechanism of action of antitumor agents (144–146). Many of these substances, particularly their pharmacologically active forms, are either nonvolatile and/or thermally unstable and must be converted into derivatives before examination by EI-MS. Also many of these agents possess reactive groupings, such as the bis-(2-chloroethyl) amino function of the nitrogen mustards, which can be affected by chemicals used to derivatize polar functional groups present elsewhere in the molecule. The formation of polar metabolites of drugs is also a common detoxification mechanism. A knowledge of the structure of these metabolic compounds can illuminate the mechanism of action of the drug. These points were illustrated in an FD study of the antitumor drug cyclophosphamide (**1**), which is part of a wider investigation into the mechanism of action of this agent.

Cyclophosphamide is a widely used antitumor agent that requires metabolic activation. The mechanism of activation and the detoxification processes have been extensively investigated (183 and references cited). The principal urinary metabolites of cyclophosphamide are 4-ketocyclophosphamide [2-(bis-(2-chloroethyl)amino) tetrahydro-2H-1,3,2-oxazaphosphorine-4-one-2-oxide (**2**)], and carboxyphosphamide [2-carboxyethyl-*N,N*-bis-(2-chlorethyl)phosphorodiamate (**3**)]. In contrast to carboxyphosphamide, which is a detoxification product of **1**, an additional polar metabolite *N,N*-bis-(2-chloroethyl) phosphorodiamidic acid (**4**), which has been characterized as its methyl ester by EI-MS, may be one of the important intermediate metabolites responsible for the cytotoxic effects of the drug (144,147).

Parent Drugs. The FD mass spectra of **1** and its metabolites' **2, 3,** and **4a** structures (Figure 54) and the isomeric cancer drugs isophosphamide (**5**) and trophosphamide (**6**) were examined by FD-MS.

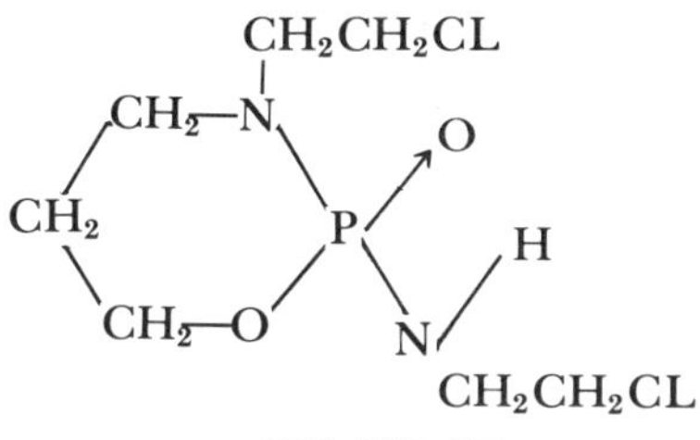

5

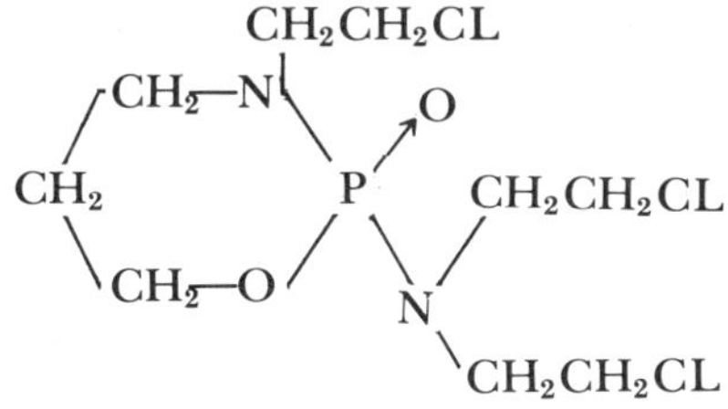

6

It was of interest to determine whether this technique is better suited than conventional EI-MS to the characterization of these compounds. The FD

TABLE IX
High-Resolution[a] Field Desorption Mass Spectra: Elemental Composition of Structurally Diagnostic Ions[b] (27)

m/e	Elemental composition	Tentative interpretation
	a. Cyclophosphamide (**1**)	
521.057	$2(C_7H_{15}O_2N_2P^{35}Cl_2) + H$	(2M + 1)
260.025	$C_7H_{15}O_2N_2P^{35}Cl_2$	M
224.048	$C_7H_{14}O_2N_2P^{35}Cl$	M − ($H^{35}Cl$)
211.040	$C_6H_{13}O_2N_2P^{35}Cl$	M − ($CH_2{}^{35}Cl$)
198.032	$C_5H_{12}O_2N_2P^{35}Cl$	(M + 1) − ($CH_2CH_2{}^{35}Cl$)
141.011	$C_4H_9N^{35}Cl_2$	$HN(CH_2-CH_2Cl)_2$
63.000	$C_2H_4{}^{35}Cl$	$-CH_2CH_2Cl$ direct bond cleavage
48.985	$CH_2{}^{35}Cl$	$-CH_2Cl$
35.977	$H^{35}Cl$	Hydrochloric acid
	b. 4-Ketocyclophosphamide (**2**)	
549.016	$2(C_7H_{13}O_3N_2P^{35}Cl_2) + H$	(2M + 1)
274.004	$C_7H_{13}O_3N_2P^{35}Cl_2$	M
238.027	$C_7H_{12}O_3N_2P^{35}Cl$	M − ($H^{35}Cl$)
225.020	$C_6H_{11}O_3N_2P^{35}Cl$	M − ($CH_2{}^{35}Cl$)
	c. Carboxyphosphamide (**3**)	
585.037	$2(C_7H_{15}O_4N_2P^{35}Cl_2) + H$	(2M + 1)
293.022	$C_7H_{16}O_4N_2P^{35}Cl_2$	(M + 1)
275.012	$C_7H_{14}O_3N_2P^{35}Cl_2$	(M + 1) − H_2O
248.025	$C_6H_{15}O_2N_{2P}{}^{35}Cl_2$	(M + 1) − COOH (or M − CO_2)
152.011	$C_3H_7O_4NP$	Direct bond cleavage N − P
142.019	$C_4H_{10}N^{35}Cl_2$	Bis-(2-chloroethyl)amine (protonated)
98.985	H_4PO_4	Phosphoric acid (protonated)
72.081	$C_4H_{10}N$	Pyrrolidine (protonated)
71.073	C_4H_9N	Pyrrolidine
46.005	CH_2O_2	Formic acid (protonated)
35.977	$H^{35}Cl$	Hydrochloric acid
	d. Cyclohexylammonium salt of *N,N*-bis-(2-chloroethyl)phosphorodiamidic acid (**4a**)	
419.211	$C_{10}H_{24}O_2N_3P^{35}Cl_2) + (C_6H_{14}N)$	(M + cation)
320.106	$C_{10}H_{25}O_2N_3P^{35}Cl_2$	(M + 1)
284.129	$C_{10}H_{24}O_2N_3P^{35}Cl$	(M + 1) − $H^{35}Cl$
235.144	$C_9H_{22}O_2N_3P$	{(M − CH_2Cl)} − Cl
185.024	$C_4H_{12}O_2N_2P^{35}Cl$	(Anion + H) − ^{35}Cl
184.016	$C_4H_{11}O_2N_2P^{35}Cl$	(Anion + H) − $H^{35}Cl$ or anion − ^{35}Cl
141.011	$C_4H_9N^{35}Cl_2$	$HN(CH_2-CH_2Cl)_2$
99.105	$C_6H_{13}N$	Cyclohexylamine
92.026	$C_3H_7N^{35}Cl$	$CH_2=NH-CH_2CH_2{}^{35}Cl$
83.086	C_6H_{11}	Cyclohexyl-radical

m/e	Elemental composition	Tentative interpretation
	e. Isophosphamide **(5)**	
521.057	$2(C_7H_{15}O_2N_2P^{35}Cl_2) + H$	(2M + 1)
520.049	$2(C_7H_{15}O_2N_2P^{35}Cl_2)$	(2M)
484.072	$C_{14}H_{29}O_4N_4P_2{}^{35}Cl_3$	$\{(M - H^{35}Cl) + M\}$
260.025	$C_7H_{15}O_2N_2P^{35}Cl_2$	M
224.048	$C_7H_{14}O_2N_2P^{35}Cl$	$M - (H^{35}Cl)$
211.040	$C_6H_{13}O_2N_2P^{35}Cl$	$M - (CH_3{}^{35}Cl)$
198.032	$C_5H_{12}O_2N_2P^{35}Cl$	$(M + 1) - (CH_2CH_2{}^{35}Cl)$
	f. Trophosphamide **(6)**	
645.042	$2(C_9H_{18}O_2N_2P^{35}Cl_3) + H$	(2M + 1)
322.017	$C_9H_{18}O_2N_2P^{35}Cl_3$	M
286.040	$C_9H_{17}O_2N_2P^{35}Cl_2$	$M - (H^{35}Cl)$
273.032	$C_8H_{16}O_2N_2P^{35}Cl_2$	$M - (CH_2{}^{35}Cl)$
260.025	$C_7H_{15}O_2N_2P^{35}Cl_2$	$(M + 1) - (CH_2CH_2{}^{35}Cl)$
252.079	$C_9H_{18}O_2N_2P^{35}Cl$	$M - 2^{35}Cl$

[a] The obtained resolution was > 15,000 (10% valley definition), the error in the determination of the precise masses was in the 10-ppm range.

[b] In general isotope peaks containing heavier nuclides were not listed here.

mass spectra of cyclophosphamide obtained at low and high resolution showed the molecular ions only when emitter heating currents between 0 and 5 mA were applied. This indication of a relatively high volatility for FD measurements is supported by the observation of a $[M]^+$ ion in EI-MS at a relative intensity of approximately 1% (27,50). In comparing FD and EI spectra, it was found that if molecular ions occur with small abundances in EI-MS, then FD-MS normally yields these ions as very intense signals, usually the base peaks. At higher emitter heating currents, 10 to 15 mA, a small but structurally diagnostic fragmentation was induced.

These spectra were all recorded with an integrating recording system to collect all the ions, especially those generated in fast decomposition and desorption processes. In the FD mass spectrum of **1** (HR data, Table IXa) four typical fragmentation pathways appeared: (**a**) loss of hydrochloric acid fr m the parent molecule (*m/e* 224); (**b**) α-cleavage in the chloroethyl side chain (*m/e* 211); (**c**) elimination of a chlorethyl radical from the protonated molecule (*m/e* 198); (**d**) direct bond rupture between P-2 and the *N*-mustard group. This led to the most significant signal of *N*,*N*-bis-(2-chloroethyl) amine at *m/e* 141 (in addition to the protonated species).

The isomeric compound isophosphamide [3-(2-chloroethyl)-2-(2-chloroethylamino)tetrahydro-1,3,2,-oxazaphosphorine-2-oxide, (**5**)] (HR data in Table IXe) showed only the molecular ions between 0 and 8 mA emitter heating currents, and similar fragmentation between 10 and 15 mA: (**a**) loss of HCl (*m/e* 224; 11% rel. int.); (**b**) α-cleavage (*m/e* 211, 18% rel. int.); (**c**) elimination of CH_2CH_2Cl (*m/e* 198, 10% rel. int.); (**d**) direct cleavage between P-2 and the (2-chloroethyl)amino group was not observed in this case, helping to distinguish between **1** and **5**. The FD spectrum of **5** in comparison with **1** is unique in that the spectrum contains a $[2M]^+$ ion at *m/e* 520 (9% rel. int.) and a $[2M - HCl]^+$ ion at *m/e* 484 (5% rel. int.). The rarely observed $[2M]^+$ may arise from favorable hydrogen bonding available only to **5**, that is, an accessible secondary nitrogen as opposed to a tertiary nitrogen. As expected, a pronounced $[2M + 1]^+$ ion (16% rel. int.) at *m/e* 521 is also observed.

It is sometimes difficult to determine the molecular weight of unknowns because $[M]^+$ or $[M + H]^+$ ions, or both, may occur in given FD spectra. A survey of methods to distinguish between these signals in FD-MS has therefore been given in section IV.4.A.1. However the $[2M + H]^+$ ion displayed in the FD spectra of polar compounds can aid in resolving this, particularly when an integrating recording system is available and high-resolution mass measurements can be performed. When multiple molecule complexes are formed, they are usually protonated clusters as opposed to dimers, trimers, and so on (75).

The third parent drug, trophosphamide (**6**), exhibits fragmentation consistent with the pathways mentioned (HR data, Table IXf); (**a**), (**b**), and (**c**) are intense fragments, whereas (**d**) is weak (1% rel. int.) at 10 to 15 mA ehc. The $[2M]^+$ and $[2M - HCl]^+$ ions are not present, which is consonant with the secondary amine postulate. A key peak for **6** at *m/e* 252 (5% rel. int.) involves the net loss of two chlorines from the molecule. This fragment is not observed in the FD spectra of **1** to **5**.

Drug Metabolites. In the FD mass spectra of 4-ketocyclophosphamide (**2**) (HR data, Table IXb), the molecular ion was the base peak. This is in marked contrast to the EI mass spectra of this compound, in which the relative intensity of $[M]^+$ is less than 1%. Fragments from paths (**a**) and (**b**) are present in the FD mode. In addition, the diagnostic $[2M + 1]^+$ ion is seen. The EI spectrum of carboxyphosphamide (**3**) contains no molecular ion. A very weak $[M - H_2O]^+$ is present at *m/e* 274/276 (2Cl), and a weak $[M - CH_2Cl]$ peak at *m/e* 243/245 (Cl).

In the FD mass spectrum obtained at room temperature with a single-focusing instrument and electric detection (0 mA ehc), the protonated intact molecule (*m/e* 293/297) is the only significant feature (Figure 52, top). The thermal instability of carboxyphosphamide, implicit in the low intensity of the higher mass fragments in the EI mass spectrum, was confirmed for the FD determination by the observation that the total ion current fell when the emitter temperature was raised from 40 °C to about 70% of the value at 20 °C (Figure 52, bottom). This illustrates the capability of FD-MS, especially for analysis of thermally unstable compounds.

When the photoplate exposure in a double-focusing instrument took place during a gradual (6-min exposure) rise of the ehc from 0 to 30 mA, a

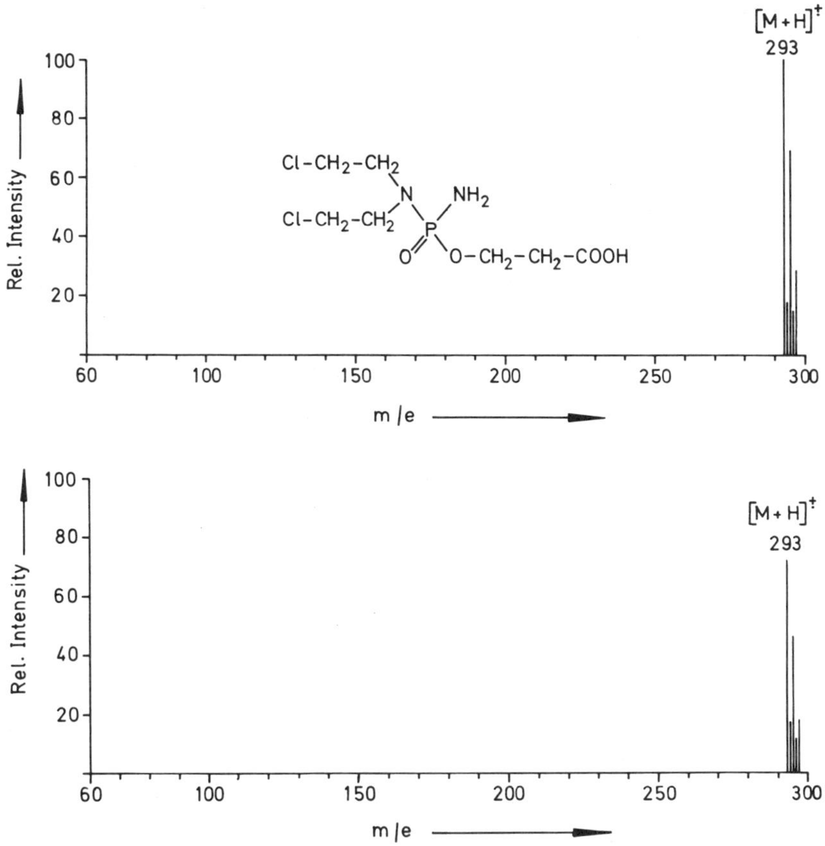

Figure 52. FD-MS of carboxycyclophosphamide, recorded with electric detection. Upper spectrum obtained at 20°C, lower spectrum at approximately 40°C emitter (and sample) temperature (49).

thermally and field-induced fragmentation, yielding some structurally diagnostic fragments (**c**), was observed. An intense $[M + 1]^+$ ion is found. Direct bond rupture (**d**) leads to the formation of the protonated *N*-mustard, and *m/e* 152 is the conjugate fragment. A pronounced ion at *m/e* 275 represents the thermal elimination of water. At the reasonably higher emitter temperatures involved in this procedure of thermally and field-induced fragmentation, pyrolytic products such as $[H_3PO_4 + H]^+$ and $[CHO_2 + H]^+$ are generated.

Another metabolite **4** is stabilized in the form of a cyclohexylammonium salt **4a**. The FD mass spectrum of this compound gave the HR data shown in Table IXd. This salt is easily analyzed with FD-MS, giving a relatively intense protonated molecular ion even at higher emitter heating currents used to generate fragments. Fragments from paths **a** and **d** are observed, and other ions are present that can be used to characterize this compound.

It appears that one important intermediate in the *in vivo* activation of cyclophosphamide is 4-hydroxycyclophosphamide. Therefore the high-resolution FD spectra of the synthetic compound were studied (Figure 53). The molecular ion at *m/e* 276.020 ($C_7H_{15}O_3PCl_2$) is the base peak of the FD spectrum. Since an aprotic solvent was used (CCl_4), protonation from the solvent can be excluded. However the relatively high abundance of the $[M + H]^+$ ion (*m/e* 277), approximately 21% relative intensity, indicates that intermolecular protonation takes place. This is confirmed by the occurrence of an $[M - H]^+$ ion (*m/e* 275) and a strongly pronounced $[M - 2H]^+$ ion at *m/e* 274, which are probably produced in a high field process. From the FD spectrum alone, it is impossible to decide whether small impurities of 4-ketocyclophosphamide (mol. wt. 274) were present in the synthetic samples or whether oxidation occurred during sample handling and transfer to the emitter. Other ions in the FD spectrum of the 4-hydroxy derivative (Figure

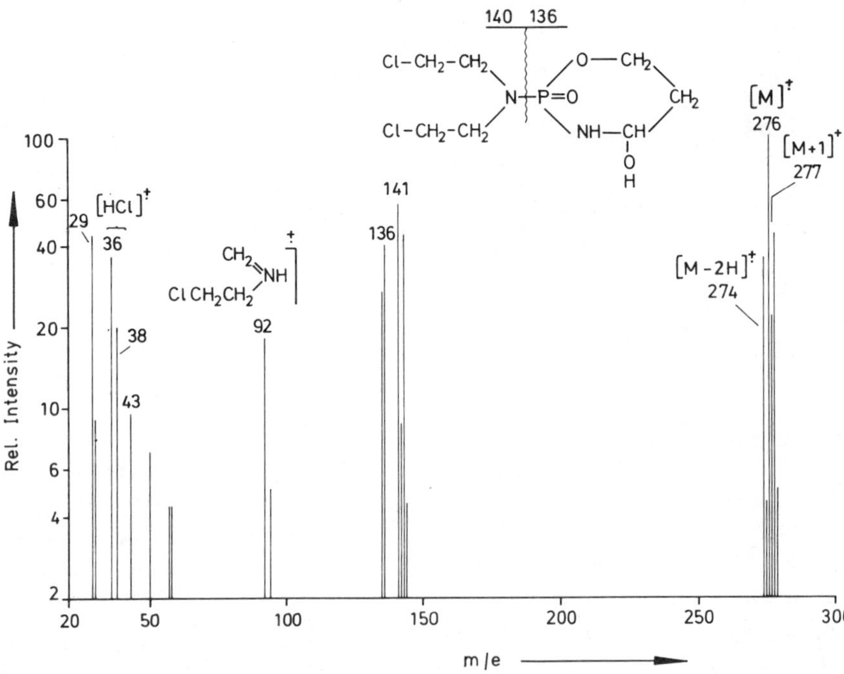

Figure 53. FD-MS of 4-hydroxycyclophosphamide: solvent, carbon tetrachloride; emission-controlled FD; threshold, 5×10^{-8} A; BAT, 11 mA ehc; exposure time, 10 min; temperature range 0–30 mA ehc (36).

53) are readily assigned. Thus *m/e* 136 and 141 are generated by direct bond cleavages and protonation. This fission of a molecular ion into two complementary, charged particles is indeed the most frequently observed phenomenon in FD-MS. A feature of the FD spectra of the other cyclophosphamide metabolites and analogs that has greatly simplified the interpretation of these spectra is P-N cleavage to give the *N*-mustard and oxazaphosphorine moieties. Some ions of minor significance, such as $[HCl]^+$ and $[C_3H_7NCl]^+$, are generally recorded in the lower mass range (see Table IX).

Drug Conjugates. In addition to oxidation products of drugs, conjugates of cancer drugs are of particular interest. One of the usual reactions of alkylating agents in the body is conjugation with cysteine or *N*-acetylcysteine. An example of analysis of these conjugates *without* derivation is the FD spectrum of *S*-[*N*-(2,4-dinitrophenyl)] mercapturic acid (Figure 54). This compound occurs as a urinary metabolite of the antitumor agent 1-(1-arizidinyl)-2,4-dinitrobenzene in the rat (148). The molecular ion (*m/e* 372.074) of the free acid is displayed with 70% relative intensity in the FD

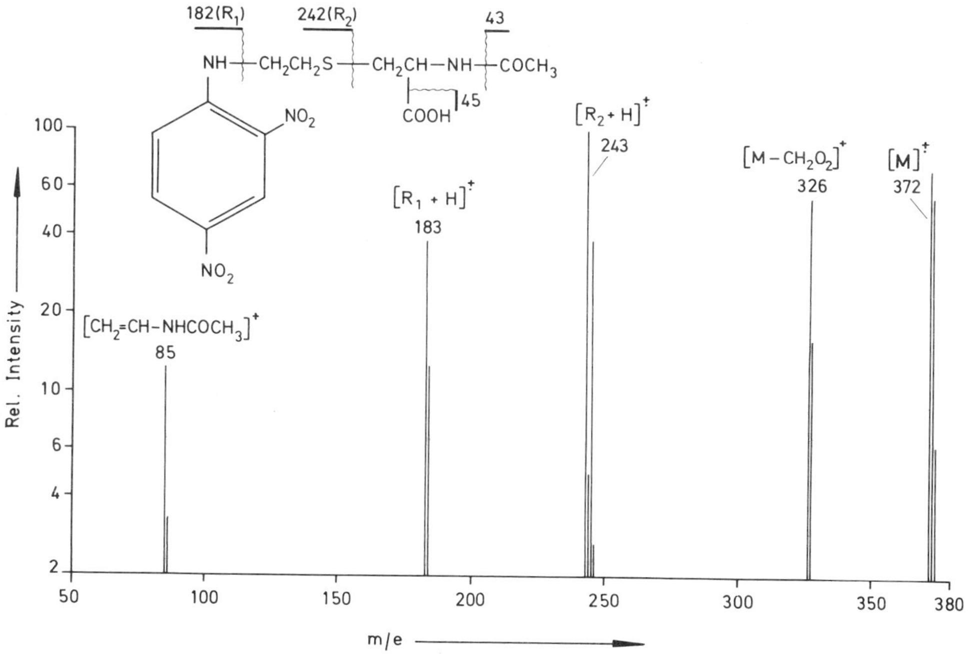

Figure 54. HR FD-MS of *S*-[*N*-(2,4-dinitrophenyl)] mercapturic acid. The average error in the accurate mass determination was ±3 mmu (49).

spectrum. Direct bond cleavages and protonation yield two structurally diagnostic ions at *m/e* 183.028 $[R_1 + H]^+$ and 243.031 $[R_2 + H]^+$. In addition, the generally observed loss of formic acid in the FD spectra of free carboxylic acids generates a prominent ion at *m/e* 326.068. Fragments of smaller relative intensity are recorded at *m/e* 85.053 $[C_4H_7NO]^+$ and 43.018 $[C_2H_3O]^+$. The cysteine conjugate gave similar results with FD-MS. The molecular ion intensities obtained, however, are smaller (about 30% rel. int.) when the spectrum is recorded on a Varian CH4 instrument with electric detection.

C. FD-MS OF DRUG MIXTURES

a. Cyclophosphamide and Some of Its Metabolites. It is important to determine whether FD-MS can be applied practically to mixtures of drugs and their metabolites, especially when the components are of widely differing polarities. For this an equimolar mixture of **1, 2, 3,** and **4a** was tested. One exposure of the photoplate was made for emitter heating currents of 0 to 30 mA. The spectrum (Figure 55) was almost the sum of each FD spectrum of these compounds. Although desorption occurs from the solid, adsorbed state, interaction between the different compounds of the mixture is amazingly small. The molecular ions $[M_1]^+$ or protonated molecules $[M_{2,3,4a} + H]^+$ are the most prominent ions in the spectrum. With accurate mass measurement, the spectrum can be easily interpreted. Again this interpretation is aided by the presence of $[2M + H]^+$ ions, which can be used to distinguish between $[M]^+$ and $[M + H]^+$ under low-resolution conditions.

Although the FD spectrum (Figure 55) reflects qualitatively the composition of the synthetic mixture, there are clearly marked differences in the ionization efficiencies in FD for the components of this equimolar mixture. FD-MS for quantitative biomedical and pharmacokinetic studies depends on the development of a reliable calibration method (e.g., stable isotope dilution). This involves also careful study of the parameters that influence the reproducibility of FD spectra, as well as improvements of the recording instrumentation (70) (Section III). Recently a comprehensive review on the results and possibilities of quantitative MS in biochemistry and medicine has been given (175) and reports the present state of art in quantitative FD-MS.

b. Methotrexate and Sodium Chloride. Methotrexate (*N*-[*p*-[[2,4-diamino-6-pteridinyl)methyl] methylamino] benzyl]-glutamic acid disodium salt, or MTX) is one of the most useful anticancer agents so far developed. Recently an FD study of methotrexate and some folic acid analogs gave high molecular ion intensities and, at higher emitter currents, structur-

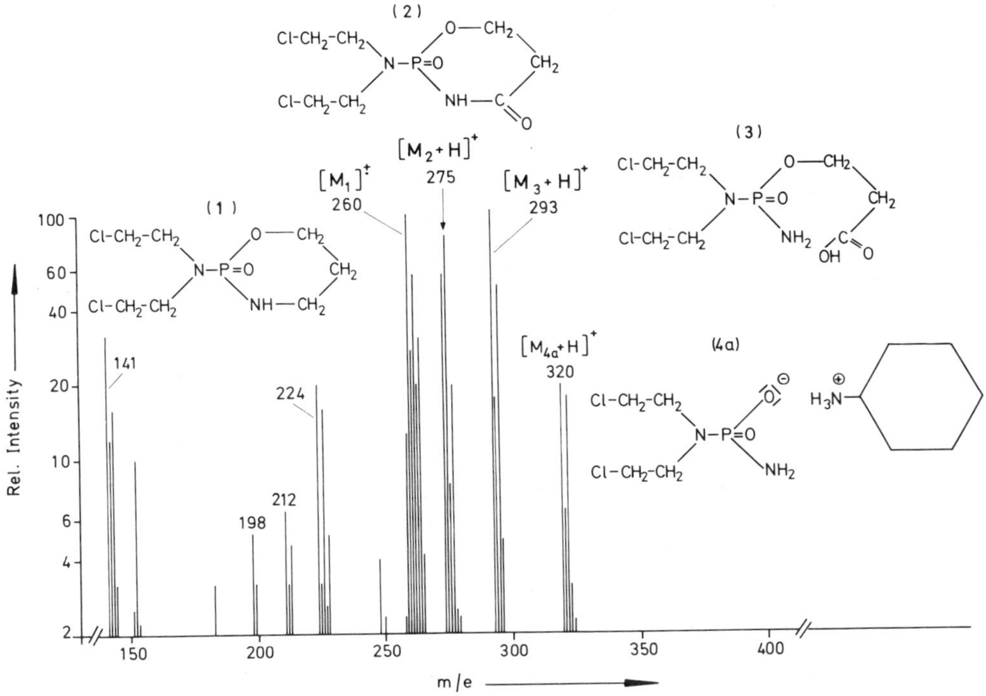

Figure 55. FD-MS of a mixture of **1, 2, 3,** and **4a** A solution/suspension of 50 μg of each compound in 10 μl of acetone was used in this experiment. The amount of sample adsorbed on the emitter surface and consumed in one high-resolution (>15,000) FD mass spectrum was estimated and calculated to be less than 10^{-7} g (27).

ally significant fragmentation (149). In view of the strong influence of inorganic salts on the FD spectra of nonvolatile organic substances (Section IV.3), MTX was investigated in its commercially available form. The injection vial contains 5.48 mg of MTX and 17.2 mg of sodium chloride, which are dissolved in 5 ml of aqueous solution for medical application. About 1 μl of this solution was transferred to the FD emitter by syringe (18). The FD spectrum recorded with high resolution on vacuum-evaporated AgBr photoplates (49) appears in Figure 56. Although the molar ratio or MTX disodium salt to NaCl is about 1:30, the spectrum can be interpreted when the processes of cluster formation (60) are taken into consideration. At *m/e* 521 the $[M + Na]^+$ ion is shown with 30% relative intensity; thus the molecular weight of the MTX disodium salt is derived easily. The $[M + 2Na]^{2+}$ ion at *m/e* 272 gives a helpful indication of the correct assignment of the molecular cluster ion. The components of the MTX molecule R_1 (as $[R_1 + H + Na]^+$ at *m/e* 185), R_2 (as $[R_2 + H]^+$ with 8% rel. int. at *m/e* 176) were found. When accurate mass measurements are

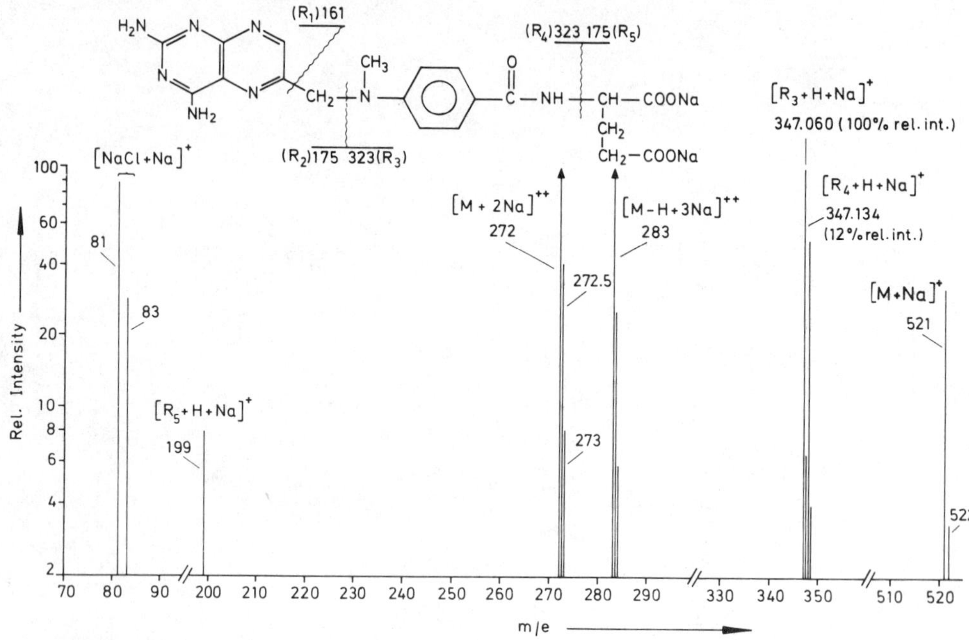

Figure 56. FD-MS of methotrexate: solvent, water; 1 μg of sample material was applied via the syringe technique; emission-controlled FD; threshold, 4×10^{-9} A; BAT, 29 to 31 mA ehc; exposure time, 26 min (167).

performed, the R_3, R_4, and R_5 moieties of the MTX salt are identified as $[R_3 + H + Na]^+$ at *m/e* 347.060, $[R_4 + H + Na]^+$ at *m/e* 347.134, and $[R_5 + H + Na]^+$ at *m/e* 198.996. Thus structurally significant information is obtained from a "practical" drug sample when the appropriate desorption program is chosen. However in FD-MS few experimental data have been reported for the limits of inorganic salt concentrations in mixtures with large organic molecules except a very recent study on steroid- and triterpene saponins (169).

D. CALIBRATION STUDIES BY STABLE ISOTOPE DILUTION AND QUANTITATION BY FIELD DESORPTION

The choice of an internal standard for quantitation of cyclophosphamide by FD-MS stable isotope dilution (36) was governed by the existence of $[M + H]^+$ ions containing two ^{37}Cl atoms. The molecular ion cluster for cyclophosphamide contained ions of *m/e* 259 to 265, but there was no detectable ion at *m/e* 266 (detection limit 2% of base peak at 15.000 resolution).

Therefore a suitable internal standard is the d_6-analog, cyclophospham-

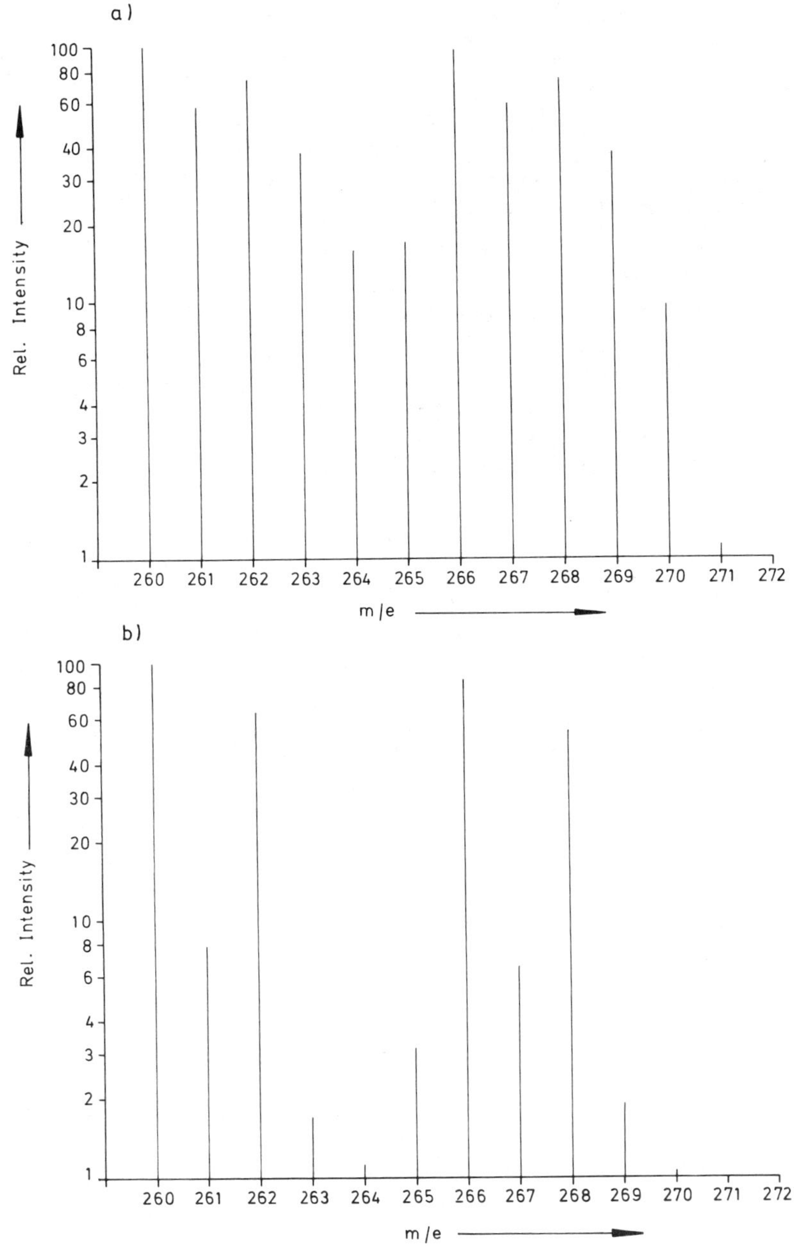

Figure 57. FD-MS of a mixture of cyclophosphamide and cyclophosphamide-d_6 (1.00 : 0.85) solvent, acetone; threshold, 1×10^{-8} A; exposure time, 30 min (36). (*a*) 1 μg Cyclophosphamide. (*b*) 0.1 μg Cyclophosphamide.

ide-4,5,6-d_6 (150). Figure 57 presents the molecular ion groups for mixtures of the cyclophosphamide and the d_6-analog containing the d_0-and d_6-forms in the molar ratio 1:0.85 (1:1 with respect to protium and deuterium forms, but the latter contains d_5- and d_6-analogs in the ratio 15:85). One microgram of each component is present on the emitter and is recorded by FD-MS with photographic detection (Figure 57*a*). The resulting densitogram was overexposed in relation to the more abundant isotopic peaks. The results with 100 ng of each component are given in Figure 57*b*. The ratio *m/e* 260:266 of two ^{35}Cl-containing $[M]^+$ for d_0- and d_6-analog was 100:86 (theory, 100:85), and the ^{35}Cl ^{37}CL-containing ions (*m/e* 262, 268) are approximately in the same ratio.

Greater departures from theoretical values are observed for the less intense ions, particularly *m/e* 264 and 265. At the present resolution (15.000) 100 ng represents a near-optimal value for quantitative determination of cyclophosphamide, but the sensitivity would be enhanced by the use of lower resolution or by a different detection system. In the latter connection, the use of an FD source in combination with a quadrupole mass spectrometer and a time-averaging computer appears promising. Recent investigations using stable isotope dilution confirmed the utility of quantitative FD-MS in a comparison of EI and FD data of dopamine derivatives from human urine samples (125,151). First results on the instrumental precision and the accuracy of the technique have been described (45,152).

A survey of the advantages and problems of FD-MS in metabolism studies has been given (153,154) and clearly illustrates one focal point in the application of the method for physiological and pharmacokinetic studies.

VI. BIOMACROMOLECULES

1. Pyrolysis Field Desorption Mass Spectrometry of DNA

A new technique, pyrolysis field desorption mass spectrometry, Py-FD-MS (31), consists of the direct submicropyrolysis on the surface of a high-temperature-activated tungsten wire emitter. This method is necessary for obtaining structural information of compounds with a molecular weight considerably greater than *m/e* 1500. Some of the most important biological compounds, such as polysaccharides, polynuleotides, and polypeptides, have very high molecular weights ($>$ 20,000). Currently it is impossible to detect these macromolecules in a routine mass spectrometer because they cannot be ionized without decomposition. Controlled thermal degradation and identification of the structural units by mass spectrometric methods has promise in two directions.

1. Rapid, accurate, and reproducible analysis of the pyrolysis products (e.g., with a combination of Curie point pyrolyzer, quadrupole MS, and data

system) leading to fingerprint spectra (155). Use of this system, with comparison of spectra, can provide unambiguous identification of known polymers.

2. Identification of ionized primary fragments in the mass spectrum provides indications of the composition and structure of known and unknown polymers. Py-MS of such material naturally yields extremely complex mixtures of pyrolysis products. However, the qualitative determination of the components is greatly facilitated by the use of ionization methods displaying enhanced molecular ion intensities and minimal or no fragmentation. In line with these considerations, Meuzelaar et al. (156) examined the Curie point pyrolysis products of biological samples with low-voltage EI ionization. Schulten et al. (157) reported the use of high-resolution FI mass spectrometry for the analysis of mixtures of bacterial pyrolysis products, identifying more than 140 chemical species.

To establish a correlation between the pyrolysis products and the structure of a (bio- or technical) polymer it is advantageous to produce large ionized fractions in the process of thermal degradation. As a rule, the larger the weight of the detected ions, the more valuable the information obtained.

In Py-FD-MS, the observation of these large primary products is favored by four important conditions: minimal difference in time and place between pyrolysis and ionization, minimal sample qualities, minimal residence time of the free ions in the pyrolysis zone, and minimal excess energy transferred by ionization.

Py-FD-MS fulfills all these requirements: pyrolysis and ionization coincide in place and time, the sample quantity is 10^{-8} to 10^{-9} g, the residence time of the free ion in the pyrolysis zone is of the order of 10^{-11} sec, and the average excess energy does not exceed 0.2 eV.

Since investigators were familiar with the behavior of underivatized, highly polar components of DNA, such as purine bases (11), nucleosides, and nucleotides (24,41), under conditions of FD-MS, DNA was chosen for studying the feasibility of Py-FD-MS.

The only mass spectral data thus far available on thermal degradation products of DNA were obtained by Charnock and Loo (158), using conventional EI ionization at 70 eV and sample quantities of 25 to 50 μg. They reported high-resolution mass measurements on 26 peaks up to m/e 391 and suggested elemental compositions for 17 of these peaks. No nucleotides or other phosphate-containing units of DNA were found.

For the Py-FD mass spectrum (Figure 58) herring DNA was suspended in acetone and transferred to the 10-μ high-temperature-activated tungsten wire emitter by the dipping technique. The sample consumption was estimated from electron scanning micrographs before and after Py-FD-MS to be about 5×10^{-8} g. With a magnetic field strength of 12 kG and an accelerating voltage of + 8 kV on the field anode, the photographically recorded mass

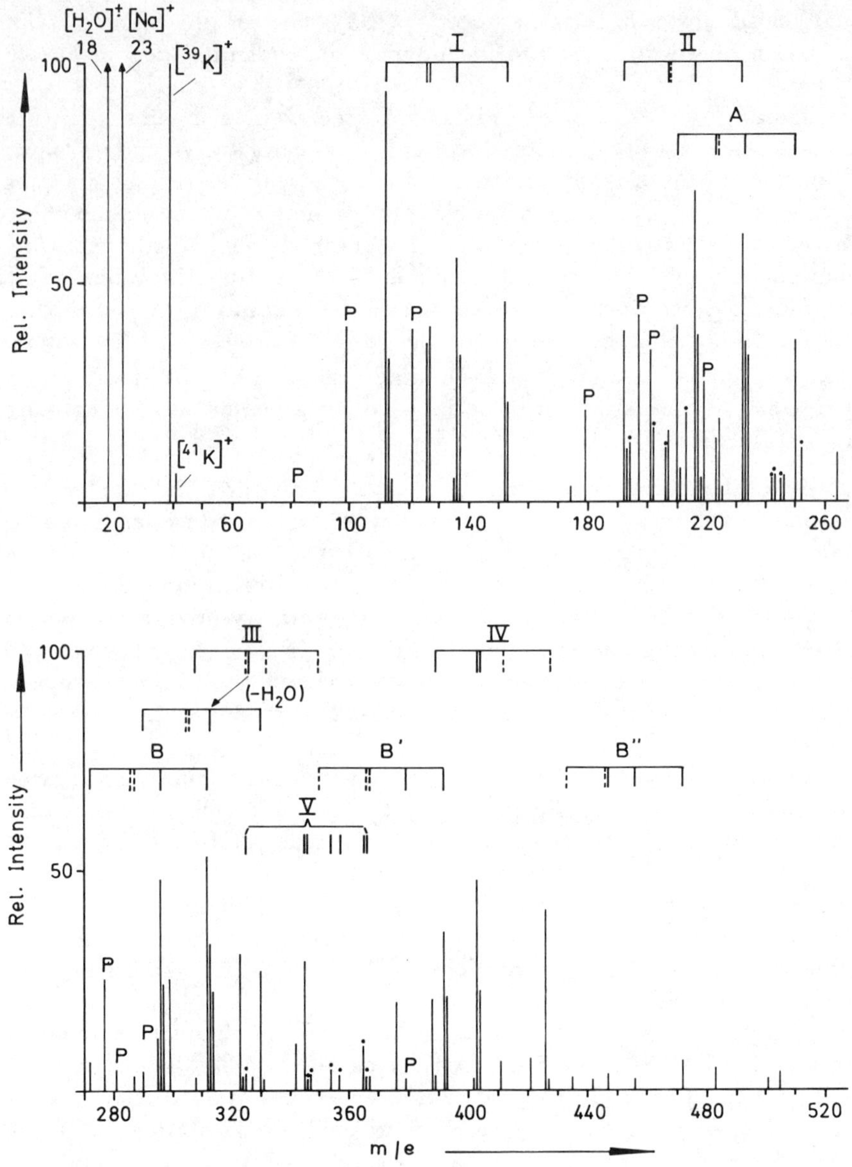

Figure 58. Py-FD-MS of herring DNA: solid dots, doubly charged ions; P, phosphate or phosphoric acid (cluster) ions used as internal mass reference, I, bases + H^+; II, nucleosides − $2H_2O$ + H^+; III, nucleosides + H^+ (monophosphates); IV, nucleotides + H^+ (diphosphates); V, doubly charged dinucleotides; A, (I + H_3PO_4) clusters; B, II + 1 · (C_5H_4O); B′, II + 2 · (C_5H_4O); B″, II + 3 · (C_5H_4O); relative ion intensity 160 corresponds to the saturation blackening of the vacuum evaporated AgBr photoplate (31).

range extended from *m/e* 18 to 665. For pyrolysis, the tungsten wire was heated with a stabilized dc current of 0 to 80 mA for an exposure of 6 min. The total emission of the field anode measured on the slotted cathode plate rose from 5×10^{-10} to 1×10^{-6} A while the ehc was increased to 28 mA (about 1 min). After 3 min the total emission fell slowly to 1×10^{-7} A (50 mA heating current) and, at the end of the exposure time, 1×10^{-9} A emission and 80 mA heating current were reached.

Two prominent features were directly observed in the Py-FD mass spectrum (Figure 58). First, there are no organic ions detected below *m/e* 112. This is in sharp contrast to all previously mentioned pyrolysis studies, indicating a different pathway of thermal degradation in the extremely high field and virtual absence of ion fragmentation. Second, the high relative abundance of the ions in the upper mass range (*m/e* 200 to 400) is remarkable, especially when compared to the high-resolution EI-MS spectrum of DNA pyrolysis products (158).

A closer examination of the mass spectrum revealed the presence of intense $[M + 1]^+$ signals for the five essential bases, adenine, guanine, cytosine, thymine, and methylcytosine of herring DNA. A typical base pattern (Figure 58) is also observed among the heavier ions and strongly suggests the presence of larger structural units. To establish their chemical identity, the high-resolution data obtained were subsequently used for accurate mass determination. This was greatly facilitated by the appearance of inorganic phosphate cluster ions throughout the Py-FD-MS, which were used as internal mass references. Because these ions are easily discerned by their characteristic mass defects, it became unnecessary to introduce the usual reference substances, perfluorokerosene or perfluorotributylamine. The addition of such substances during the pyrolysis field desorption process might influence the reactions unpredictably. Introducing these substances after termination of the pyrolysis would create an additional source of error, caused, for example, by drifting of the magnetic field or the focusing potentials.

The phosphate cluster ions identified in the mass spectrum (Figure 58) are listed in Table Xa. Precise mass determination of organic ions directly adjacent to the phosphate peaks confirmed the existence of several "base patterns" throughout the mass spectrum, as is apparent from the nominal mass values. Iterative interpolation between the newly established mass values then enabled precision mass determination of the remaining organic ions. The average error in the accurate mass measurement in this range is less than ± 2 millimass units, equivalent to 10 ppm.

The combined presence of different classes of compound arranged in typical ion patterns appears to provide a basis for the chemical identification of these compounds, minimizing the chance of accidental misinterpretation of individual ions. The chemical identities listed (Tables Xb, Xc) show groups of compounds containing bases associated with phosphoric acid or chemically bound to dehydrated deoxyribose units. The compounds in Table Xc may be regarded as building blocks of DNA. Whether these components originate from direct fragmentation or arise by accidental combinations of smaller pyrolysis-products is difficult to know. Data in Table Xd reveal an interesting group of doubly charged dinucleotide ions. The presence of such large ions could be of great importance because it indicates the feasibility of eventually obtaining base sequence information by Py-FD-MS. No phosphate-containing ions were found by Charnock and Loo (158), a result that minimizes chances of obtaining such information by their technique. Moreover, the small sample consumption in Py-FD-MS would be especially useful in analytical biochemistry.

As evident from Table Xd, all doubly charged dinucleotide ions are quasi-molecular $[M + 2]^+$ species, and several of them contain sodium. Similar ions were observed for the cytidine and thymidine diphosphates, but no doubly charged ions were observed for any of the atypical DNA products listed in Tables Xa and Xb. The results described in Section IV.3 for FD-MS of oligonucleotides are consistent with these findings; for example, ApC and CpA give intense $[M + 2H]^{2+}$ ions (41). Some of the reaction mechanisms and the potential of the Py-FD-MS method for analysis of biopolymers require further study.

TABLE Xa
Singly and Doubly Charged Ions in the Pyrolysis Field Desorption Mass Spectrum of Herring DNA (31)

Inorganic Cluster Ions Identified in the Py-FD Mass Spectrum of DNA: All Species Appeared as Protonated $[M + H]^+$ Ions

HPO_3	$HPO_3 \cdot H_3PO_4$	$HPO_3 \cdot (H_3PO_4)_2$
H_3PO_4	$(H_3PO_4)_2$	$(H_3PO_4)_3$
NaH_2PO_4	$NaH_2PO_4 \cdot H_3PO_4$	
$NaH_2PO_4 \cdot HPO_3$	$NaH_2PO_4 \cdot HPO_3 \cdot H_3PO_4$	
$NaH_2PO_4 \cdot (HPO_3)_2$	$NaH_2PO_4 \cdot (HPO_3)_2 \cdot H_3PO_4$	

Group	Compounds	Elemental composition	Calculated mass (protonated ions)
	b. Proposed Chemical Structure of Observed Ions, Not Representing Typical Building Blocks of DNA		
A	Cytosine + phosphoric acid	$C_4H_8O_5N_3P$	210,028
	Methylcytosine + phosphoric acid	$C_5H_{10}O_5N_3P$	224,044
	Adenine + phosphoric acid	$C_5H_8O_4N_5P$	234,039

TABLE X (*Continued*)
Singly and Doubly Charged Ions in the Pyrolysis Field Desorption Mass Spectrum of Herring DNA (31)

	Guanine + phosphoric acid	$C_5H_8O_5N_5P$	250,034
B	Cytosine + 2 methylfuran − 4H	$C_{14}H_{13}O_3N_3$	272,104
	Adenine + 2 methylfuran − 4H	$C_{15}H_{13}O_2N_5$	296,115
	Guanine + 2 methylfuran − 4H	$C_{15}H_{13}O_3N_5$	312,110
B′	Adenine + 3 methylfuran − 4H	$C_{20}H_{17}O_3N_5$	376,141
	Guanine + 3 methylfuran − 6H	$C_{20}H_{17}O_4N_5$	392,136
B″	Thymine + 4 methylfuran − 8H	$C_{25}H_{22}O_6N_2$	447,156
	Adenine + 4 methylfuran − 8H	$C_{25}H_{21}O_4N_5$	456,167
	Guanine + 4 methylfuran − 8H	$C_{25}H_{21}O_5N_5$	472,162

c. Proposed Chemical Structure of Observed Ions, Representing Typical Building Blocks of DNA

I	Cytosine	$C_4H_5ON_3$	112,051
	5-Methylcytosine	$C_5H_7ON_3$	126,067
	Thymine	$C_5H_6O_2N_2$	127,051
	Adenine	$C_5H_5N_5$	136,062
	Guanine	$C_5H_5ON_5$	152,057
II	(2′-Deoxycytidine)-$2H_2O$	$C_9H_9O_2N_3$	192,077
	(5-Methyl-2′-deoxycytidine)-$2H_2O$	$C_{10}H_{11}O_2N_3$	206,093
	(2′-Deoxyadenosine)-$2H_2O$	$C_{10}H_9ON_5$	216,089
	(2′-Deoxyguanosine)-$2H_2O$	$C_{10}H_9O_2N_5$	232,083
III	(2′-Deoxycytidylic acid)-H_2O	$C_9H_{12}O_6N_3P$	290,054
	(2′-Deoxyguanylic acid)-H_2O	$C_{10}H_{12}O_6N_5P$	314,065
	2′-Deoxycytidylic acid	$C_9H_{14}O_7N_3P$	308,065
	2′-Deoxythymidylic acid	$C_{10}H_{15}O_8N_2P$	323,064

d. Proposed Chemical Structure of Observed Doubly Charged Ions Representing Structural Units of DNA. All Ions Appear as Quasi-molecular $[M + 2]^{2+}$ Ions

Group	Ion structure	Elemental composition	Calculated mass
IV	(deoxycytidine diphosphate + $2H)^{2+}$	$C_9H_{17}O_{10}N_3P_2$	390,038
	(deoxythymidine diphosphate + $2H)^{2+}$	$C_{10}H_{18}O_{11}N_2P_2$	404,038
	(Na-deoxycytidine diphosphate + $2H)^{2+}$	$C_9H_{16}O_{10}N_3P_2Na$	411,020
	(Na-deoxythymidine diphosphate + $2H)^{2+}$	$C_{10}H_{17}O_{11}N_2P_2Na$	426,020
V	(Na-dithymidine diphosphate + $2H)^{2+}$	$C_{20}H_{29}O_{15}N_4P_2Na$	650,100
	(cytidine-thymidine triphosphate + $2H)^{2+}$	$C_{19}H_{30}O_{17}N_5P_3$	693,085
	(dithymidine triphosphate + $2H)^{2+}$	$C_{20}H_{31}O_{18}N_4P_3$	708,085
	(Na-cytidine-thymidine triphosphate + $2H)^{2+}$	$C_{19}H_{29}O_{17}N_5P_3Na$	715,067
	(Na-dithymidine triphosphate + $2H)^{2+}$	$C_{20}H_{30}O_{18}N_4P_3Na$	730,067
	(Na-^{13}C-dithymidine triphosphate + $2H)^{2+}$	$^{13}CC_{19}H_{30}O_{18}N_4P_3Na$	731,070

2. Curie Point Pyrolysis of Nucleic Acids

Recently two mass spectrometric investigations have shed some light on pyrolysis of biomacromolecules. First, the products resulting from Curie point pyrolysis of nucleic acids were studied by low-voltage EI ionization quadrupole and HR FI-MS (87). The spectra from both techniques are in good agreement. As already indicated for low resolution, the high-resolution spectra proved that most peaks originate from the carbohydrate moiety of the nucleic acid. Micrograms of DNA/RNA can be distinguished well by use of these key fragments, although the bases themselves are not found. Significant ions in the Py mass spectra of DNA are observed at m/e 68, 70, 82, 98, and 116. For RNA, typical pyrolysis products give ions at m/e 84, 86, 98, and 114. It is suggested that most of these peaks with even mass numbers are due almost exclusively to pyrolysis products of deoxyribose and ribose. Some of the peaks in the pyrolysis mass spectra of DNA and RNA can be interpreted on the assumption that pyrolysis tends to produce rather stable substances, such as compounds with conjugated double bonds and aromatic compounds. Thus the relatively intense peaks at m/e 68, 82, 98, and 116 in the DNA spectrum, and at 84, 96, 98, and 114 in the RNA spectrum, might originate from initial thermal degradation.

A. MASS 96 (RNA)

Irrespective of the reaction sequence, removal of the 5′-phosphate ester and the base by simple 1,2-elimination and of the 3′-phosphate ester through an epoxide intermediate, may finally generate a 4-pyrone (mol. wt. 96, b.p. 215 °C) from RNA. For DNA, the corresponding mass 80 is not found and only two 1,2-eliminations are possible, thus obviating analogous thermal degradation for DNA.

B. MASS 82 (DNA) AND 98 (RNA)

For both DNA and RNA, interaction between the base and the 5′-phosphate ester may result in a thermolytic elimination of the base as a salt, although removal of the 3′-phosphate ester through 1,2-elimination, or through an epoxide intermediate (for RNA only), also seems possible (Scheme 4). These thermal degradations may finally generate 2-methylfuran (mol. wt. 82, b.p. 63 °C) from DNA, and its analog from RNA, containing the additional oxygen and partly contributing to mass 98 (see below). Regarding elimination of the base as a salt, it is noteworthy that in the pyrolysis experiments with DNA described (Table Xb), peaks with masses corresponding to the association of various bases with phosphoric acid were observed.

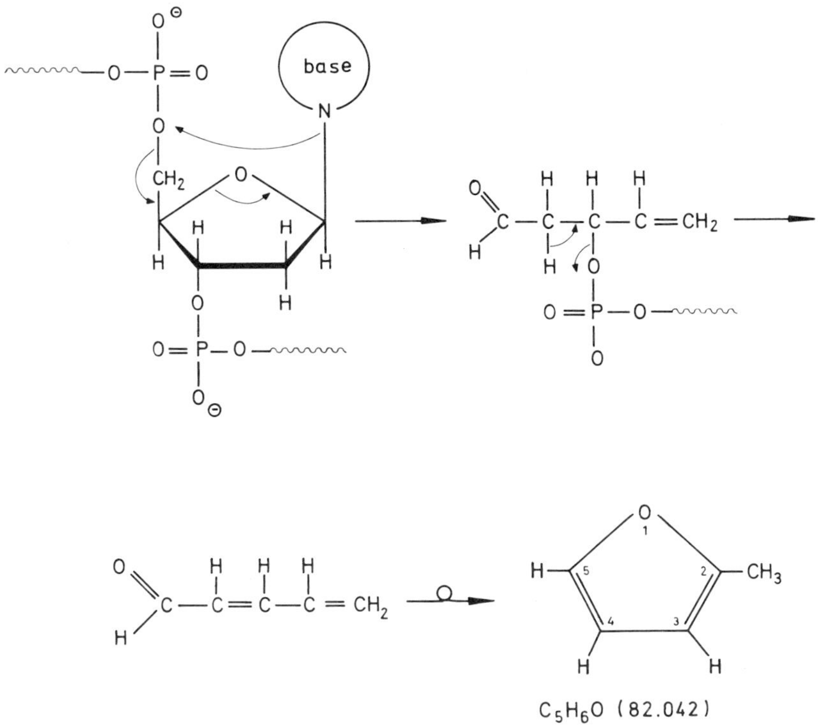

Scheme 4. Proposed mechanism for the formation of 2-methylfuran in the pyrolysis of DNA (87).

c. MASSES 68, 98, AND 116 (DNA) AND 84 AND 114 (RNA)

Free phosphoric acid is undoubtedly formed during pyrolysis (31). This may protonate and induce the cleavage of the ribo-ester linkages, producing metaphosphoric acid, which recently was identified as such and also occurred in cluster ions formed during pyrolysis field desorption of DNA.

Successive elimination of the 3′-phosphate ester and the base by 1,2-elimination may then lead to furfuryl alcohol (mol. wt., 98; b.p., 172 °C). Analogous reactions may account for the formation of mass 116 from DNA and mass 114 from RNA. Removal of CH_2O from the furan ring may result in mass 68 from DNA and mass 84 from RNA. The presence of ribose fragments sharply contrasts with the total absence of intact bases. The latter phenomenon is not entirely due to adsorption of the bases on the walls of the inlet system, as indicated by the pyrolysis of the various bases themselves. The pyrolysis mass spectra then exhibit, almost exclusively, intense molecular ion peaks originating from evaporated, unpyrolyzed molecules of the bases.

The presence of nitrogen-containing compounds of lower molecular weight, as found in the high resolution FI spectra (e.g., HCN, HCNO, NH_3, nitriles) suggests a degradation of the base after elimination from the nucleic acid skeleton. The first step in this degradation could be a reaction of the base with phosphoric acid (Scheme 4) or trapping of the free base by pyrolytically generated free phosphoric acid to form a nonvolatile salt. These complexes have been reported (Table Xb). Further arguments for this follow from the large peak at *m/e* 17 (NH_3), which is observed in the pyrolysis mass spectra of nucleic acids but not in those of deoxyribo- and ribonucleosides. The ammonia molecules are possibly generated from the reaction

$$[RNH_3]^+[H_2PO_4]^- \rightarrow [R]^+[H_2PO_4]^- + NH_3\uparrow$$

The resulting nonvolatile salt may then further decompose thermally to generate other gaseous molecules.

In conclusion, Curie point pyrolysis, in conjunction with low-resolution quadrupole and/or high-resolution FI mass spectrometry, appears to be a suitable tool for analyzing biopolymers at the microgram level, and it contributes to understanding the fundamental processes in pyrolysis.

3. Collisional Activation Mass Spectrometry of Pyrolysis Products

The second approach to confirm the chemical identity of the pyrolysis products of RNA and DNA involves collisional activation (CA) mass spectrometry. If ions of high kinetic energies (> 1 keV) collide with neutral target atoms, part of the translational energy is converted to internal energy, leading to subsequent collision-induced fragmentation. If the collision occurs in the second field-free region of a double-focusing instrument of reversed geometry, the collision-induced fragments may be energy-analyzed, thus mass-analyzed conveniently by scanning the electric sector. The fragmentation pattern of such a CA spectrum (88 and refs. cited) or a given molecular ion qualitatively resembles that obtained by electron impact.

It is significant that the relative abundances of such collision-induced fragments reflect the ion structure predominantly (not the internal energy distribution). This makes CA spectra useful for structure elucidation, not only of molecular ions but especially of fragment ions. Another promising application of CA spectra is the direct analysis of mixtures. The identification of the components of a mixture can be achieved by matching the CA spectrum of the unknown compound with that of a known reference, or by interpretation of the fragmentation pattern. This interpretation is based on the close similarity between the fragmentation patterns of an EI and a CA spectrum.

By successively recording the CA spectra of the peaks in the normal mass spectrum that characterize the various components of the mixture, it should be possible to identify these components without prior separation (e.g., by gas chromatography). Since fragments do not reflect a given precursor unambiguously, and various components may contribute to a given fragment peak, it is desirable to produce the primary spectrum of the mixture under conditions that favor the occurrence of molecular ions. FI and FD-MS are especially suited to this. The FI spectra of complex multicomponent mixtures of bacterial pyrolysis products (157) and the pyrolysis FD spectra of products such as polyethylene (59), glycogen (159), and DNA (Section VI.1), were recorded at high resolution and enabled determination of the elemental composition of the ions produced. The practical usefulness of high-resolution MS analysis is limited by the inability to separate and identify isomers, unless incidental additional information is obtained from metastable processes or fragment ions, or doubly charged ions.

It appears that CA-MS is a useful method to establish chemical identity of isomers in mixtures. In a first CA study of the rather complex mixtures produced by pyrolysis of biological material (88), six major components of a mixture of pyrolysis products of deoxyribonucleic acid were analyzed by combined low-energy EI and CA spectra. The method allowed the assignment of structures of different components without prior separation, and it verified the pyrolysis mechanisms deduced from Py-FD-MS (Section VI.1) and Curie point pyrolysis in combination with EI and FI mass spectrometry (Section VI.2).

Assuming that the pyrolysis technique employed produces reasonably stable compounds, the 14-eV EI mass spectrum of the mixture of pyrolysis products should be dominated by molecular ions. CA spectra were taken of all the major peaks in the spectrum (m/e = 32, 41, 55, 68, 82, and 98). Table XI summarizes the elemental compositions and proposed structures of these peaks. When the CA spectra of these peaks were compared with the reference spectra of the proposed structures, as well as spectra of other possible structures, the results leave no doubt that the structures of m/e = 32 (methanol), 41 (acetonitrile), 68 (furan), and 82 (methylfuran) have been correctly assigned. As Table XII indicates, the CA spectra of the peak at m/e 82 generated in the Py-MS of DNA and the reference compound 2-(methylfuran) are in good agreement. Indeed the methylfuran isomer, with the methyl group the in 2-position, is formed as was postulated previously (Scheme 4, ref. 87).

The foregoing interpretation is based on the assumption that all peaks discussed represent molecular ions, and this is supported by the following: (*a*) a rather similar spectrum is observed under FI conditions when ionization is even softer than with low-energy electrons; (*b*) in the pyrolysis spectrum produced under the conditions described, there are no significant peaks above m/e 120 from which the assumed fragments may originate; (*c*) the compounds that have been finally identified show little fragmentation at 14 eV.; and (*d*) all the assumed molecular ions are odd-electron species.

TABLE XI
Important Peaks in the Pyrolysis Mass Spectrum of DNA (87)

m/e	Elemental composition	Proposed structure
32	CH_4O	Methanol
41	C_2H_3N	Acetonitrile
55	—[a]	—[a]
68	C_4H_4O	Furan
82	C_5H_6O	Methylfuran
98	$C_5H_6O_2$	Furfuryl alcohol

[a] Not previously published.

An exception exists at *m/e* 55 (Table XI). No elemental structure had been proposed for this peak, which is abundant in the EI spectrum but absent in the FI spectrum published (87). The CA spectra of the only conceivable compounds with a molecular ion at *m/e* 55 (ethylisocyanide and propiontrile) differ considerably with respect to type and relative abundance of the fragments, suggesting that this is not a molecular ion but a fragment ion, possibly formed by hydrogen elimination from *m/e* 56. Comparison with reference spectra suggests that propargyl alcohol may be a candidate. Even at low electron energy, propargyl alcohol readily loses hydrogen to give an abundant *m/e* 55. A comparison of the CA spectra of *m/e* 55 affirms this postulated structure.

The results demonstrate that CA spectra in conjunction with low-energy EI spectra allow direct analysis of fairly complex mixtures, making gas chromatographic separation unnecessary. The analysis is facilitated when the low-energy EI spectrum of the mixture is dominated by the molecular ions of the components, as expected for rather stable compounds. This technique seems to be especially suited to analysis of pyrolysis products, since the

TABLE XII
CA Mass Spectra of Mass 82 in the Pyrolysis Mass Spectrum of DNA and the Reference Spectrum[a] (88)

m/e	DNA	2-Methylfuran
15	0.2	0.2
27	3.1	2.9
28	1.3	1.1
39	4.3	4.3
43	2.0	2.1
51	4.3	3.8
53	9.8	9.5
67	0.1	0.1
81	75	76

[a] Abundance relative to the sum of all fragments.

thermal degradation of organic molecules usually leads to stable compounds. The identification of propargyl alcohol as a component of the DNA pyrolysate demonstrates that fragment ions are also useful for identification of components.

CA-MS analysis of mixtures might not be applicable for three main reasons. (1) The mixture analysis may be complicated by the occurrence of isomers, leading to a superposition of several CA spectra. (2) When many plausible structures exist, reliable, pure standard compounds for the reference CA spectra might be hard to obtain. (3) For instrumental reasons it is difficult to obtain highly resolved and significant CA spectra of heavier ions (e.g., $> m/e$ 200).

4. Pyrolysis FD-MS of Polysaccharides

Further applications of Py-FD-MS to the analysis of polysaccharides and polypeptides have been reported (159,160). Studies have been undertaken to clarify some of the reaction mechanisms in Py-FD-MS as well as to explore the potential of the method for the structural analysis of biopolymers. The first attempt was the Py-FD mass spectrum of glycogen (159), a branched polymer of amylose chains joined through $1 \rightarrow 6$ linkages (Figure 59). Although this is a complex molecule, the Py-FD mass spectrum shows significant molecular and quasi-molecular ions derived from mono- and disaccharide units. High-resolution data of the peaks at m/e 162, 324, and 340 directly indicate structural units of the biopolymer (Figure 59). The ions at m/e 117 and m/e 126/127 are typical of FD spectra of oligosaccharides (Section IV.1.B) containing glucose units. Similarly, other structural homopolysaccharides such as cellulose and starches (amyloses and amylopectins) gave intense, high molecular weight fragments with Py-FD-MS (49).

Other signals are due to compounds generated by pyrolytic bond rupture, water elimination, and subsequent protonation of the dehydrated molecule. The peak at m/e 99.042 ($C_5H_7O_2$) is apparent as the $[M + H]^+$ peak of furfuryl alcohol or its isomers. This identification was proved by accurate mass measurements, by reference to our library of pyrolysis products of sugar-containing materials, and by the identification of the compound in bacterial pyrolysis (157). It is still uncertain whether branching and other structural characteristics of the polymer chain can be directly correlated with the ion species obtained. Investigations of such technical polymer products as polyethylene (59), polyacrylonitrile, and polyacrylamide showed that with photographic detection and a controlled time-temperature program for the ehc, significant FD spectra can be obtained. The polyethylene was ground to a fine powder at liquid N_2 temperature before it was applied to the emitter from a suspension in acetone. Thus not only solutions but also suspensions can be analyzed with FD-MS.

A crucial problem in Py-FD-MS is reliable and reproducible control of the

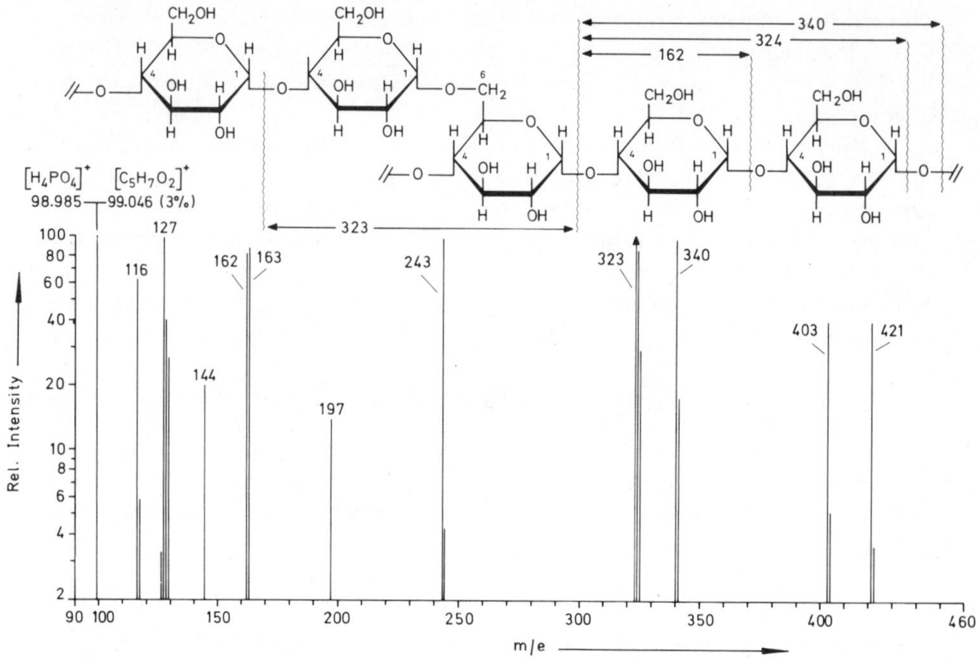

Figure 59. Py-FD-MS of glycogen: photographic detection with vacuum-evaporated AgBr plate; approximate resolution, 20,000 (10% valley definition). A part of the D(+)-glycogen molecule $(C_6H_{10}O_5)_n$, molecular weight 27,000 to 3,500,000, appears in the upper section of the figure, and characteristic bond cleavages are tentatively assigned (159).

time-temperature profile of the pyrolysis. Since the pyrolysis-products depend strongly on the time-heating current program, an optimal and reproducible procedure controlled by an automatic heating device is necessary (Section III.4.E). Repeated Py-FD mass spectra of polymers have indicated that both shorter and longer pyrolysis times drastically reduce the appearance of larger building units. It must be emphasized that Py-FD-MS can be applied only when two fundamental prerequisites are met. First, the mass spectrometer must have sufficient resolving power. Without this, assignment of elemental composition of peaks becomes increasingly difficult as mass number increases. Second, the extremely complicated spectra of multicomponent mixtures can be evaluated only if electronic data -processing is available.

5. Pyrolysis FD-MS of Microorganisms, Cells, and Tissues

At present, very few results have been reported with FD, although direct fingerprinting by pyrolysis of microorganisms such as bacteria (155) has received considerable attention. Figure 60 illustrates the behavior of a small droplet (1 μl) of human blood on a field desorption emitter wire. The ehc (and sample temperature) rise gradually in the first 20 min (0 to 20 mA ehc,

1 to 5 × 10^{-9} A, total emission). At this stage volatile components are desorbed and the FD spectrum mainly contains ions at *m/e* 18 $[H_2O]^+$ and *m/e* 19 $[H_3O]^+$. The next 10 min of desorption is characterized by strong, fluctuating outbursts of ions from the FD emitter. Within a very narrow ehc interval (19.5 to 20.5 mA), the sample "explodes."

Under the conditions used, Py-FD-MS generates large particle pyrolysis products. The difficulty in this part of the desorption is recording the relatively weak ion currents (for an individual type of ion). Integrating photographic recording offers a solution, at least when a minimum of about 2000 ions of each type reach the photoplate. When ehc was kept constant at about 21 mA, and simultaneously the total emission was fairly stable between 9 × 10^{-9} and 1 × 10^{-8} A, the high-resolution Py-FD-MS data in Table XII were obtained. Since only small ions with organic elemental composition (⩽ *m/e*84) were found, most of the organic material was decomposed at this

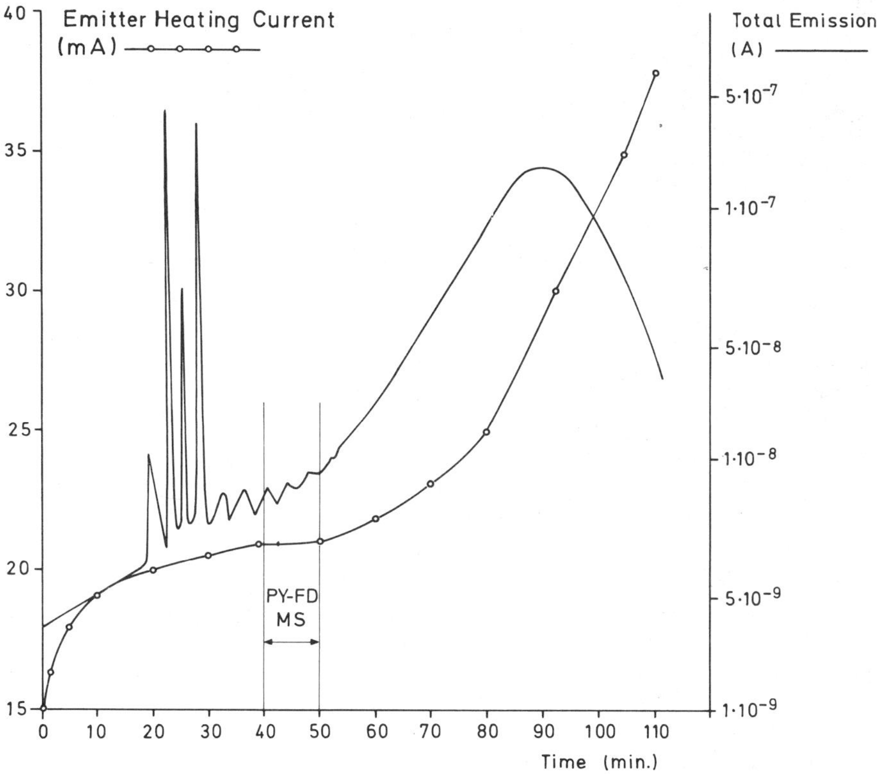

Figure 60. Schematic diagram correlating the emitter heating currents (thus the temperature) and the total emission (thus the number of produced ions) of a human blood sample as a function of time. The ions generated between the 40th and 50th minutes of the pyrolysis program are listed in Table XI (49).

TABLE XIII
Pyrolysis Field Desorption Mass Spectrum of Human Blood (49); Accurate Mass Determination (Resolution = 10,300 at *m/e* 86.909)

Mass		Error			Densitogram
Calculated	Measured	(mmu)	Formula	Interpretation	(log intensities)
17.027	17.027	0	NH_3	Ammonia	14
18.011	18.012	+1	H_2O	Water	36
18.035	18.035	0	NH_4	Ammonium	34
19.018	19.018	0	H_3O	Hydronium	20
22.989	22.990	+1	Na	Sodium	60
27.024	27.023	−1	C_2H_3		20
27.995	27.992	−3	CO	Carbon monoxide	2
28.031	28.033	+2	C_2H_4	Ethylene	12
29.039	29.041	+2	C_2H_5		10
30.010	30.007	−3	CH_2O	Formaldehyde	2
30.047	30.051	+4	C_2H_6	Ethane	4
38.964	38.964	0	K	^{39}Potassium	60
39.963	39.964	+1	Ca	^{40}Calcium	18
40.962	40.961	−1	K	^{41}Potassium	60
43.006	43.007	+1	CHNO		18
43.042	43.038	−4	C_2H_5N		20
43.055	43.057	+2	C_3H_7		24
44.026	44.024	−2	C_2H_4O	Ethylene oxide	20
44.063	44.063	0	C_3H_8	Propane	15
84.044	84.041	−3	C_4H_6NO		15
84.911	84.911	0	Rb	^{85}Rubidium	26
86.909	86.909	0	Rb	^{87}Rubidium	18
132.905	132.907	+2	Cs	Cesium	10

stage. However alkali cations were detected with high relative intensity. The high total emission values above 25 mA ehc (Table XIII) indicate these inorganic components of human blood almost exclusively. Very recently a determination of cesium in 0.2 μl of untreated human blood was performed by FD-MS using the single ion monitoring mode. The concentration of $[Cs]^+$ ions in 1μl blood was 18 pg. The precision of the method was found to be ± 10% and the accuracy of the determination was ± 20% (182). Whether Py-FD-MS is a useful method for the determination of salts in organic material, or whether a preseparation and fingerprinting of the organic components is the more promising approach, remains to be seen.

VII. APPLICATION OF FD-MS TO INORGANIC SUBSTANCES (INORGANIC SALTS AND METAL CHELATES)

The foregoing results indicate that an analytically important feature of field desorption mass spectrometry is its applicability in investigations of

TABLE XIV

Prominent Ions in the FD-MS of Inorganic Compounds (61)

Salt	Three most intense ions in FD spectra (except cations)
LiF	$(LiF)Li^+$, $(LiF)_2Li^+$, $(LiF)_3Li^+$
LiCl	$(LiCl)Li^+$, $(LiCl)_2Li^+$, $LiCl^{+\cdot}$
LiJ	$(LiJ)Li^+$, J^+, $(LiJ)_2Li^+$
Li_2CO_3	$(Li_2CO_3)Li^+$, $(Li_2CO_3)_2Li^+$
KF	$(KF)K^+$, $(KF)_2K^+$
KCl	$(KCl)K^+$, $(KCl)_2K^+$, $KCl^{+\cdot}$
KBr	$(KBr)K^+$, Br^+
KJ	$(KJ)K^+$, $(KJ)_2K^+$, J^+
NaCl	$(NaCl)Na^+$, $(NaCl)_2Na^+$, $NaCl^{+\cdot}$
$NaNO_3$	$(NaNO_3)Na^+$, $(NaNO_3)_2Na^+$, $(NaNO_3)_3Na^+$
Na_2SO_4	$(Na_2SO_4)Na^+$
$NaClO_3$	$(NaClO_3)Na^+$, $(NaClO_3)_2Na^+$, $(NaCl)Na^+$
$Na(CH_3CO_2)$	$[Na(CH_3CO_2)]\ Na^+$, $[Na(CH_3CO_2)]_2Na^+$
NaN_3	$(NaN_3)Na^+$
AgCl	$AgCl^{+\cdot}$, $(AgCl)Ag^+$
$AgNO_3$	$(AgNO_3)Ag^+$, $(AgNO_3)_2Ag^+$
$CaCl_2$	$CaCl^+$, $Ca^{+\cdot}$, $CaCl_2{}^{+\cdot}$
$(NH_4)_2SO_4$	$(NH_4)_2SO_4 + NH_4{}^+$, $[(NH_4)_2SO]_2 + NH_4{}^+$

inorganic salts. The mass spectrometric analysis of this class of compounds by other methods is more difficult because volatilization must occur before ionization. The first complete and consistent FD spectra of inorganic salts were obtained from alkali halides (161). In general, extraordinarily high relative intensities for the cations (C) were recorded. $[Li]^+$, $[Na]^+$, and $[K]^+$ ions gave ion current intensities that were orders of magnitude greater than those of all other types of ion observed in FD-MS. In addition, cluster formation generates ions from which the molecular structure can be easily derived (60).

Table XIV lists the three most abundant ions (except the cations) in the FD spectra of some representative inorganic salts with simple and complex anions (A). Typical ionic products in the FD of inorganic salts are $[C]^+$, $[M + C]^+$, or ions of the composition $[nM + C]^+$ (salt cluster), where C signifies the cation and M the salt molecule; $[M]^+$ ions of these salts are seldom recorded. The detection of inorganic salt clusters has been successful in the pyrolysis of biopolymers (Section VI.1), and in the chemical analysis of inorganic components in biochemicals (Section IV.2.C), pesticides (43), drugs (Section V.6.C), and natural aerosols (22). Metal chelates such as $Fe(acac)_3$ (Figure 61) give simple FD spectra that consist only of the base peak for the molecular ion and an intense signal for acetylacetone at *m/e* 100. This FD spectrum is a particularly clear example of the potential of the method for qualitative and quantitative analysis of metal ions and chelates.

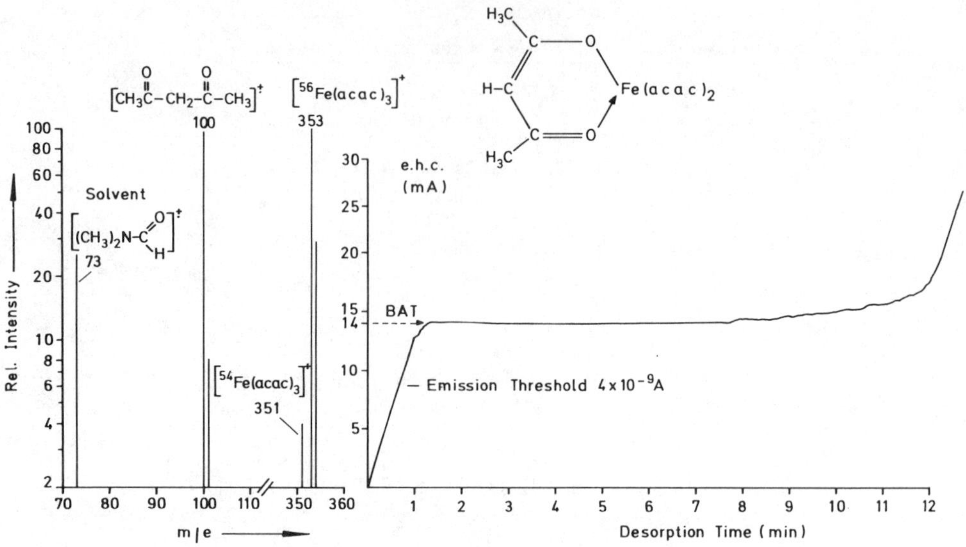

Figure 61. Emission-controlled FD-MS of a chelate: solvent, dimethylformamide; BAT, exactly 14 mA; threshold, 4×10^{-9} A for the total emission. *Right:* thermogram (time versus ehc); *left:* corresponding FD-MS (49).

The FD spectra of five tricarbonyl (π-hydrocarbon) metal salts and cyclopentadienyl (arene) metal salts have recently been reported by Games et al. (162). They also published the FD spectra of a number of phthalocyanines that normally gave the molecular ions as the base peak of the spectrum (163).

A survey of the application of FD-MS to the investigation of organic and inorganic salts has recently been given (69,164 and refs. cited therein).

As a method in biochemical analysis, FD-MS is promising for trace analysis of metal ions, isotope determination, detection and identification of inorganic salts, and for molecular weight determination plus structural information about complex, nonvolatile inorganic substances (184).

VIII. SUMMARY AND OUTLOOK

Summarizing the results that have been obtained by FD-MS (until July 1977), one can say that the field desorption technique is well on its way to being the method of choice for the mass spectrometric analysis of nonvolatile, thermally unstable compounds. This is not to say, however, that FD is yet an established tool of the analytical chemist (4).

As described in this chapter, the studies of model compounds have established a basic understanding of the utility and the specific features of FD in the field of biochemical, medical, and environmental analyses. Developments to improve sensitivity, to simplify the FD operation, and to make less costly commercial instruments available can be expected. In concert with other analytical methods, the unique information produced by FD-MS will thus contribute to the solution of problems in a wide area of analytical applications.

Acknowledgments

This work was supported financially by the Deutsche Forschungsgemeinschaft, the Fonds der Deutschen Chemischen Industrie, and the Ministerium für Wissenschaft und Forschung des Landes Nordrhein-Westfalen.

The constructive criticism and stimulating support throughout all phases of this research by Prof. H. D. Beckey is gratefully acknowledged.

References

1. H. Budzikiewicz, C. Djerassi, and D. H. Williams, *Structure and Elucidation of Natural Products by Mass Spectrometry*, Vol. I, *Alkaloids*, Vol. II, *Steroids, Terpenoids, Sugars, and Miscellaneous Classes*, Holden-Day, San Francisco, 1964; *Interpretation of Mass Spectra of Organic Compounds*, Holden-Day, San Francisco, 1964.
2. F. W. McLafferty, *Interpretation of Mass Spectra*, Benjamin, New York, 1966.
3. G. R. Waller, ed., *Biochemical Applications of Mass Spectrometry*, Wiley-Interscience, New York, 1972.
4. A. L. Burlingame, B. J. Kimble, and P. J. Derrick, *Anal. Chem. Ann. Rev.*, *48*: 5, 368R (1976).
5. E. Stenhagen, S. Abrahamsson, and F. W. McLafferty, eds., *Atlas of Mass Spectral Data*, Vols. 1 to 3, Wiley-Interscience, New York, 1969; *Eight-Peak Index of Mass Spectra*, Mass Spectrometry Data Centre, AWRE, Aldermaston, Reading, U.K., 1974.
6. R. G. Ridley, in *Biochemical Applications of Mass Spectrometry*, G. R. Waller, Ed., Wiley-Interscience, New York, 1972, p. 177.
7. G. W. A. Milne and M. J. Lacey, *Modern Ionization Techniques in Mass Spectrometry*, CRC. Critical Reviews in Analytical Chemistry, Cleveland, Ohio, 1974, p. 45.
8. F. H. Field, in *MPT International Review of Science*, Vol. 5, A. Maccoll, Ed., Butterworthes, London, 1972, p. 133.
9. H. D. Beckey, *Field Ionization Mass Spectrometry*, Pergamon Press, Oxford, 1971; H. D. Beckey, *Principles of Field Ionization and Field Desorption Mass Spectrometry*, Pergamon Press, Oxford, in press.
10. H. D. Beckey, *Int. J. Mass Spectrom. Ion Phys.*, *2*, 500 (1969).
11. H.-R. Schulten and H. D. Beckey, *Org. Mass Spectrom.*, *6*, 885 (1972).
12. H.-R. Schulten and H. D. Beckey, *Messtechnik*, *81*, 121 (1973).
13. H. D. Beckey, E. Hilt, and H.-R. Schulten, *J. Phys. E.: Sci. Instrum.*, *6*, 1043 (1973).
14. R. Ajeian, H. D. Beckey, A. Maas, and U. Nitschke, *Appl. Phys.*, *6*, 111 (1975).
15. M. Anbar and W. H. Aberth, *Anal. Chem.*, *46*, 59A (1974); M. E. Scolnick, W. H. Aberth, and M. Anbar, *Int. J. Mass Spectrom. Ion Phys.*, *17*, 139 (1975); H. D. Beckey, H. H. Gierlich, H. J. Heinen, and E. Hilt, *J. Phys. E.: Sci. Instrum.*, *8*, 355 (1975); F. W. Röllgen, U. Giessmann, H. J. Heinen, and S. J. Reddy, *Int. J. Mass Spectrom. Ion Phys.*, *24*, 235 (1977).
16. I. V. Goldenfeld, R. N. Bondarenko, and V. G. Golovatyi, *Prib. Tekh Eksp.*, *16*, 166 (1973);

V. G. Golovatyi, R. N. Bondarenko, and E. N. Korol, *Teor. Eksp. Khim.*, *11* 349 (1975); Yu. N. Artyukh, R. N. Bondarenko, V. G. Golovatyi, and G. M. Kozub, *Dopov. Akad. Nauk Ukr. RSR, Ser. B*, 121 (1975).

17. R. M. Wightman, D. M. Hinton, M. C. Sammons, and M. M. Bursey, *Int. J. Mass Spectrom. Ion Phys.*, *17*, 208 (1975); M. M. Bursey, C. E. Rechsteiner, M. C. Sammons, D. M. Hinton, T. S. Colpitts, and K. M. Tvaronas, *J. Phys. E.: Sci. Instrum.*, *9*, 145 (1976); C. E. Rechsteiner, D. E. Mathis, M. M. Bursey, and R. P. Buck, *Biomed. Mass Spectrom.*, *4*, 52 (1977).
18. H. D. Beckey, A. Heindrichs, and H. U. Winkler, *Int. J. Mass Spectrom. Ion Phys.*, *3*, App. 9–11 (1970).
19. H. D. Beckey and H.-R. Schulten, *Angew. Chem.*, *87* 425 (1975); *Angew. Chem. Int. Ed.*, *14*, 403 (1975).
20. K. L. Olson, J. C. Cook, and K. L. Rinehart, *Biomed. Mass Spectrom.*, *1*, 358 (1974).
21. H.-R. Schulten and U. Schurath, *J. Phys. Chem.*, *79*, 51 (1975).
22. H.-R. Schulten and U. Schurath, *Atmos. Environ.*, *9*, 1107 (1975).
23. H.-R. Schulten and D. Kümmler, *Z. Anal. Chem.*, *278*, 13 (1976).
24. H.-R. Schulten and H. D. Beckey, *Org. Mass Spectrom.*, *7*, 861 (1973).
25. F. W. Röllgen and H. D. Beckey, *Int. J. Mass Spectrom. Ion Phys.*, *12*, 465 (1973).
26. H.-R. Schulten and H. D. Beckey, *J. Agr. Food Chem.*, *21*, 372 (1973).
27. H.-R. Schulten, *Biomed. Mass Spectrom.*, *1*, 223 (1974).
28. H. U. Winkler and H. D. Beckey, *Org. Mass Spectrom.*, *7*, 1007 (1973).
29. H. U. Winkler and H. D. Beckey, *Org. Mass Spectrom.*, *6*, 655 (1972).
30. H.-R. Schulten, H. D. Beckey G. Eckhardt, and S. H. Doss, *Tetrahedron*, *29*, 3861 (1973).
31. H.-R. Schulten, H. D. Beckey, A. J. H. Boerboom, and H. L. C. Meuzelaar, *Anal. Chem.* *45*, 2358 (1973).
32. D. Kümmler and H.-R. Schulten, *Org. Mass Spectrom.*, *10*, 813 (1975).
33. H. U. Winkler and B. Linden, *Org. Mass Spectrom.* *11*, 327 (1976).
34. H. H. Gierlich, A. Heindrichs, and H. D. Beckey, *Rev. Sci. Instrum.*, *45*, 1208 (1974).
35. H. D. Beckey and H.-R. Schulten, *Z. Anal. Chem.*, *273*, 345 (1975).
36. H.-R. Schulten, *Cancer Treat. Rep.*, *60*, 501 (1976).
37. H.-R. Schulten and N. M. M. Nibbering, *Biomed. Mass Spectrom.*, *4*, 55 (1977).
38. J. W. Maine, B. Soltmann, J. F. Holland, N. D. Young, J. N. Gerber, and C. C. Sweeley, *Anal. Chem.*, *48*, 427 (1976).
39. H. U. Winkler, W. Neumann, and H. D. Beckey, *Int. J. Mass Spectrom. Ion Phys.*, *21*, 57 (1976).
40. H.-R. Schulten and D. E. Games, *Biomed. Mass Spectrom.*, *1*, 120 (1974).
41. H.-R. Schulten and H. M. Schiebel, *Z. Anal. Chem.*, *280*, 139 (1976).
42. H.-R. Schulten, H. D. Beckey, E. M. Bessel, A. B. Foster, M. Jarman, and J. H. Westwood, *J. Chem. Soc. Chem. Commun.*, *13*, 416 (1973).
43. H.-R. Schulten, *J. Agr. Food Chem.*, *24*, 743 (1976).
44. C. N. McEwen and A. G. Bolinski, *Biomed. Mass Spectrom.*, *2*, 112 (1975).
45. H.-R. Schulten, W. D. Lehmann, and M. Jarman, *Quantitative Mass Spectrometry in Life Sciences*, A. DeLeenheer and R. R. Roncucci, Eds., Elsevier P., Amsterdam, 77, (1977).
46. H. Krone and H. D. Beckey, *Org. Mass Spectrom.*, *5*, 983 (1971).
47. F. W. Röllgen and H. D. Beckey, *Surface Sci.*, *27*, 321 (1971).
48. F. W. Röllgen, in *Fast Reactions on Surfaces*, D. O. Hayward, Ed., Plenum Press, London, in press.
49. H.-R. Schulten, unpublished results.
50. H.-R. Schulten, in *New Methods in Environmental Chemistry and Toxicology*, F. Coulston, F. Korte, and M. Goto, Eds., International Academic Printing Co., Tokyo, 1973, pp. 31–42.
51. H.-R. Schulten, H. Prinz, H. D. Beckey, W. Tomberg, W. Klein, and F. Korte, *Chemosphere*, *2*, 23 (1973).

52. D. A. Brent, P. de Miranda, and H.-R. Schulten, *J. Pharm. Sci.*, *63*, 1370 (1974).
53. J. Moor and E. S. Waight, *Org. Mass Spectrom.*, *9*, 903 (1974).
54. T. Radford and D. C. DeJongh, in *Biochemical Applications of Mass Spectrometry*, G. R. Waller, Ed., Wiley-Interscience, New York, 1972, p. 313.
55. H. Wagner and G. Demuth, *Z. Naturforsch.*, *29c*, 444 (1974); *Z. Naturforsch*, *31b*, 267 (1975).
56. W. D. Lehmann, H.-R. Schulten, and H. D. Beckey, *Org. Mass Spectrom.*, *7*, 1103 (1973).
57. W. D. Lehmann and H. D. Beckey, *Org. Mass Spectrom.*, *9*, 1086 (1974).
58. S. Caccamese, G. Montaudo, and M. Przybylski, *Org. Mass Spectrom.*, *9*, 1114 (1974).
59. H. D. Beckey and H.-R. Schulten, in *Practical Spectroscopy* Series, C. N. McEwen and C. Merritt, Eds., Dekker., New York, in press.
60. H.-R. Schulten and F. W. Röllgen, *Org. Mass Spectrom.*, *10*, 649 (1975).
61. H.-R. Schulten and F. W. Röllgen, *Angew. Chem.* *87*, 544 (1975); *Int. Ed.*, *14*, 561 (1975).
62. H. Adlercreutz, B. Soltmann, and M. J. Tikkanen, *J. Steroid Biochem.*, *5*, 163 (1974).
63. M. M. Bursey and M. C. Sammons *Carbohydr. Res.*, *37*, 355 (1974).
64. J. Adamson and A. B. Foster, *Carbohydr. Res.*, *10*, 517 (1969).
65. E. M. Bessel, A. B. Foster, and J. H. Westwood, *Biochem. J.*, *128*, 199 (1972).
66. O. S. Chizhov, V. I. Kadentsev, V. M. Zolotarev, A. B. Foster, M. Jarman, and J. W. Westwood, *Org. Mass Spectrom.*, *5*, 437 (1971).
67. J. Adamson, A. D. Barford, E. M. Bessel, A. B. Foster, M. Jarman, and J. H. Westwood, *Org. Mass Spectrom.*, *5*, 865 (1971).
68. D. E. Games, M. P. Games, A. H. Jackson, A. H. Olavesen, M. Rossiter, and P. J. Winterburn, *Tetrahedron Lett.*, 2377 (1974).
69. H.-R. Schulten and W. D. Lehmann, *Anal. Chim. Acta*, *87*, 103 (1976).
70. F. W. Röllgen and H. D. Beckey, *Z. Phys. Chem.*, NF, *82*, 161 (1972).
71. H.-R. Schulten and H. D. Beckey, in *Advances in Mass Spectrometry*, Vol. 6, A. R. West, Ed., Applied Science, London, 1974, pp. 499–507.
72. R. Large and H. Knof, *J. Chem. Soc. Chem. Commun.*, 935 (1974).
73. A. Mathias, A. E. Williams, D. E. Games, and A. H. Jackson, *Org. Mass Spectrom.*, *11*, 266 (1976).
74. F. W. Röllgen and H.-R. Schulten, *Org. Mass Spectrom.*, *10*, 660 (1975).
75. H.-R. Schulten and H. D. Beckey, *Org. Mass Spectrom.*, *9*, 1154 (1974).
76. H. D. Beckey, F. W. Röllgen, K. Levsen, and H.-R. Schulten, *Surface Sci.*, in press.
77. C. Hignite, in *Biochemical Applications of Mass Spectrometry*, G. R. Waller, Ed., Wiley-Interscience, New York, 1972, p. 429.
78. K. Biemann, *Mass Spectrometry: Organic Chemical Applications*, McGraw-Hill, New York, 1962, p. 351; K. Biemann and J. A. McCloskey, *J. A. Chem. Soc.*, *84*, 2005 (1962).
79. A. Bloch, in *Medicinal Chemistry*, Vol. 4, E. J. Ariens, Ed., Academic Press, New York, 1973, p. 286.
80. H. D. Beckey, H.-R. Schulten, and H. U. Winkler, 18th Field Emission Symposium, August 23–27, 1971, Eindhoven, the Netherlands, *Conference Proceedings*, p. 81.
81. P. Brown, G. R. Pettit, and R. K. Robins, *Org. Mass Spectrom.*, *2*, 521 (1969).
82. R. M. Dawson, D. C. Elliott, W. H. Elliott, and K. M. Jones, in *Data for Biochemical Research*, Oxford University Press, New York, 1969.
83. F. W. Röllgen and H.-R. Schulten, *Z. Naturforsch.*, *30a*, 1685 (1975).
84. F. W. Röllgen, U. Giessmann, and H.-R. Schulten, in *Advances in Mass Spectrometry*, Vol. 7, N. Daly, Ed., Heyden o Son, London, in press.
85. U. Giessmann and F. W. Röllgen, *Org. Mass Spectrom.*, *11*, 1094 (1976).
86. J. A. McCloskey, A. M. Lawson, K. Tsuboyama, P. M. Krueger, and R. N. Stillwell, *J. Am. Chem. Soc.*, *90*, 4182 (1968).
87. M. A. Posthumus, N. M. M. Nibbering, A. J. H. Boerboom, and H.-R. Schulten, *Biomed. Mass Spectrom.*, *1*, 352 (1974).
88. K. Levsen and H.-R. Schulten, *Biomed. Mass Spectrom.*, *3*, 137 (1976).

89. H.-R. Schulten and H. M. Schiebel, *Nucl. Acids Res.*, *3*, 2027 (1976).
90. W. Vetter, in *Biochemical Applications of Mass Spectrometry*, G. R. Waller, Ed., Wiley-Interscience, New York, 1972, p. 387.
91. H. M. Fales, G. W. A. Milne, H. U. Winkler, H. D. Beckey, and J. N. Damico, *Anal. Chem.*, *47*, 207 (1975).
92. F. W. Röllgen and H. D. Beckey, *Surface Sci.*, *23*, 69 (1970).
93. F. W. Röllgen and H. D. Beckey, *Ber. Bunsenges. Phys. Chem.*, *75*, 988 (1971).
94. F. W. Röllgen and H. D. Beckey, *Z. Naturforsch.*, *29a*, 230 (1974).
95. H. D. Beckey, G. Hoffmann, K. H. Maurer, and H. U. Winkler, *Advances in Mass Spectrometry*, Vol. 5, N. Daly, Ed., Heyden o Son, London, 1971, p. 626.
96. H. U. Winkler and H. D. Beckey, *Biochem. Biophys. Res. Commun.*, *46*, 391 (1972).
97. H. U. Winkler and H. D. Beckey, 20th Annual Conference on Mass Spectrometry and Allied Topics, June 4–9, 1972, Dallas, Texas, Paper J 8.
98. S. Asante-Poku, G. W. Wood, and D. E. Schmidt, Jr., *Biomed. Mass Spectrom.*, *2*, 121 (1975).
99. K. Biemann, in *Biochemical Applications of Mass Spectrometry*, G. R. Waller, Ed., Wiley-Interscience, New York, 1972, p. 405.
100. H.-R. Schulten and B. Wittmann-Liebold, *Anal. Biochem.*, *76*, 300 (1976).
101. B. Wittmann-Liebold, A. W. Geissler, and E. Marzinzig, *J. Supramol. Struct.*, *3*, 426 (1975).
102. B. Wittmann-Liebold, *Hoppe-Seyler's Z. Physiol. Chem.*, *354*, 1415 (1973).
103. H. Budzikiewicz, in *Biochemical Applications of Mass Spectrometry*, G. R. Waller, Ed., Wiley-Interscience, New York, 1972, p. 251.
104. J. Michnowicz and B. Munson, *Org. Mass Spectrom.*, *6*, 765 (1972).
105. J. Michnowicz and B. Munson, *Org. Mass Spectrom.*, *8*, 49 (1974).
106. H.-R. Schulten and H. D. Beckey, *J. Chromatogr.*, *83*, 315 (1973).
107. N. J. Haskins, D. E. Games, and K. T. Taylor, *Biomed. Mass Spectrom.*, *1*, 423 (1974).
108. W. H. Elliott, in *Biochemical Applications of Mass Spectrometry*, G. R. Waller, Ed., Wiley-Interscience, New York, 1972, p. 291.
109. P. D. G. Dean and R. T. Aplin, *Steroids*, *8*, 565 (1966).
110. K. Habfast and H.-R. Schulten, 21st Annual Conference on Mass Spectrometry, San Francisco, May 20–25, 1973, Paper H 7.
111. J. F. J. Hughes, N. Evans, D. E. Games, M. J. E. Hewlins, A. H. Jackson, J. R. Jackson, N. A. Khan, S. A. Matlin, M. Rossiter, R. G. Saxton, H. A. Swaine, and K. T. Taylor, in *Advances in Mass Spectrometry in Biochemistry and Medicine*, Vol. I, A. Frigerio and N. Castagnoli, Eds., Spectrum Publications, New York, 1976, p. 357.
112. N. Evans, D. E. Games, A. H. Jackson, and S. A. Matlin, *J. Chromatogr.*, *115*, 325 (1975).
113. C. D. Watts, J. R. Maxwell, D. E. Games, and M. Rossiter, *Org. Mass Spectrom.*, *10*, 1102 (1975).
114. G. Odham and E. Stenhagen, in *Biochemical Applications of Mass Spectrometry*, G. R. Waller, Ed., Wiley-Interscience, New York, 1972, p. 229.
115. N. Evans, D. E. Games, J. L. Harwood, and A. H. Jackson, *Trans. Biochem. Soc.*, *2*, 1091 (1974).
116. G. W. Wood and P.-Y. Lau, *Biomed. Mass Spectrom.*, *1*, 154 (1974).
117. S. H. Snyder and K. M. Taylor, in *Research Methods in Neurochemistry*, N. Marks and R. Rodnight, Eds., Plenum Press, New York, 1972, p. 287.
118. N. Seiler, in *Methods of Biochemical Analysis*, Vol. 18, D. Glick, Ed., Wiley-Interscience, New York, 1970, p. 259.
119. A. H. Anton and D. F. Sayre, *J. Pharmacol. Exp. Ther.*, *145*, 326 (1964).
120. S. H. Koslow, F. Cattabeni, and E. Costa, *Science*, *176*, 177 (1972).
121. E. Änggard and G. Sedvall, *Anal. Chem.*, *41*, 1250 (1969).
122. N. Seiler, T. Schmidt-Glenewinkel, and H. H. Schneider, *J. Chromatogr.*, *84*, 95 (1973).
123. N. Seiler and H. H. Schneider, *Biomed. Mass Spectrom.*, *1*, 381 (1974).

124. N. Seiler, L. Demisch, and H. H. Schneider, *Angew. Chem., Int. Ed., 10*, 51 (1971).
125. W. D. Lehmann, H. D. Beckey, and H.-R. Schulten, *Anal. Chem., 48*, 1572 (1976).
126. G. W. Wood, N. Mak, and A. M. Hogg, *Anal. Chem., 48*, 981 (1976).
127. W. H. Elliott and G. R. Waller, in *Biochemical Applications of Mass Spectrometry*, G. W. Waller, Ed., Wiley-Interscience, New York, 1972, p. 532.
128. M. C. Sammons, M. M. Bursey, and D. A. Brent, *Biomed. Mass Spectrom., 1* 169 (1974).
129. K. L. Rinehart, Jr. C. Cook, Jr., K. H. Maurer, and U. Rapp, *J. Antibiot., 27*, 1 (1974).
130. L. F. Zerilli, M. Landi, G. G. Gallo, K. H. Maurer, and U. Rapp, *Biomed. Mass Spectrom., 2*, 307 (1975).
131. G. B. Elion, S. Callahan, S. Bieher, G. H. Hitchings, and R. W. Rundles, *Cancer Chemother. Rep., 14*, 93 (1961).
132. P. de Miranda, L. M. Beacham, III, T. H. Creagh, and G. B. Elion, *J. Pharmacol. Exp. Ther., 187*, 588 (1973).
133. P. de Miranda and L. C. Chu, *Fed. Proc., 29*, 608 (1970).
134. R. T. Coutts, *J. Pharm. Sci., 62*, 769 (1973).
135. D. J. Rouse and D. A. Brent, *Biomed. Mass Spectrom., 1*, 256 (1974).
136. H. J. Heinen, C. Hötzel, and H. D. Beckey, *Int. J. Mass Spectrom. Ion Phys., 13*, 55 (1974).
137. H. H. Gierlich, H. J. Heinen, and H. D. Beckey, *Biomed. Mass Spectrom., 2*, 31 (1975).
138. H. J. Heinen, H. H. Gierlich, and H. D. Beckey, *J. Phys. E: Sci. Instrum., 8*, 877 (1975).
139. W. Arnold and H. F. Grützmacher, *Arch. Toxikol., 25*, 200 (1969).
140. J. B. Callaghan, W. B. Askew, and W. E. Mochel, 21st Annual Conference on Mass Spectrometry and Allied Topics, San Francisco, May 20–25, 1973, Paper H 5.
141. H. M. Fales, G. W. A. Milne, and T. Axenrod, *Anal. Chem., 42*, 1432 (1970).
142. D. E. Games, A. H. Jackson, K. T. Taylor, and N. J. Haskins, in *Advances in Mass Spectrometry in Biochemistry and Medicine*, Vol. I, A. Frigerio and N. Castagnoli, Eds., Spectrum Publications, New York, 1976, p. 383.
143. S. Hayshi, H. Ueki, Y. Sako, K. Hayashi, and K. Takase, *Kumamoto Pharm. Bull., 5*, 327 (1962).
144. T. A. Connors, P. B. Farmer, A. B. Foster, A. M. Gilsenan, M. Jarman, and M. Tisdale, *Biochem. Pharmacol., 22*, 1971 (1973); T. A. Connors, P. J. Cox, P. B. Farmer, A. B. Foster, and M. Jarman, *Biochem. Pharmacol., 23*, 115 (1974). C. Fenselau, M. N. Kan, S. Billets, and M. Colvin, *Cancer Res., 35*, 1453 (1975). R. F. Struck, M. C. Kirk, M. H. Witt, and W. R. Laster, *Biomed. Mass Spectrom., 2*, 46 (1975).
145. N. Brock, in *Chemotherapy*, K. Hellmann and T. A. Connors, Eds., Plenum Publ. Corp., New York, Vol. 7, 19 (1976).
146. N. Brock and H.-J. Hohorst, *Z. Krebsforsch., 88*, 185 (1977).
147. M. Colvin, C. A. Padgett, and C. Fenselau, *Cancer Res., 33*, 915 (1973).
148. T. A. Connors, J. A. Hickman, M. Jarman, D. H. Melzack, and W. C. J. Ross, *Biochem. Pharmacol, 24*, 1665 (1965).
149. M. C. Kirk, W. C. Coburn, Jr., and J. R. Piper, *Biomed. Mass Spectrom., 3*, 245 (1976).
150. P. J. Cox, P. B. Farmer, A. B. Foster, E. D. Gilby, and M. Jarman, *Cancer Treat. Rep., 60*, 483 (1976).
151. W. D. Lehmann, H. D. Beckey, and H.-R. Schulten, *Proceedings of the First International Symposium on Quantitation in Life Sciences*, A. De Leenheer, Ed., Elsevier, Amsterdam, in press.
152. H.-R. Schulten, *Advances in Mass Spectrometry*, Vol. 7, N. Daly, Ed., Heyden o Son, London, in press.
153. H.-R. Schulten, in *Advances in Mass Spectrometry in Biochemistry and Medicine*, Vol. I, A. Frigerio and N. Castagnoli, Eds., Spectrum Publications, New York, 1976, p. 289.
154. H.-R. Schulten, in *Mass Spectrometry in Drug Metabolism*, A. Frigerio and E. L. Ghisalberti, Eds., Spectrum Publications, New York, in press.
155. H. L. C. Meuzelaar, P. Kistemaker, and M. A. Posthumus, *Biomed. Mass Spectrom., 1*, 312 (1974).

156. H. L. C. Meuzelaar, M. A. Posthumus, P. G. Kistemaker, and J. Kistemaker, *Anal. Chem.*, *45*, 1546 (1973).
157. H.-R. Schulten, H. D. Beckey, H. L. C. Meuzelaar, and A. J. H. Boerboom, *Anal. Chem.*, *45*, 191 (1973).
158. G. A. Charnock and J. L. Loo, *Anal. Biochem.*, *27*, 81 (1970).
159. H.-R. Schulten, in *New Approaches to the Identification of Microorganisms*, Part A: *New Technologies in the Automation of Microbiological Identification Routines*, Wiley, New York, 1975, p. 155.
160. H.-R. Schulten, in *Analytical Pyrolysis*, C. E. Roland Jones and C. A. Cramers, Eds. Elsevier Scientific Publ., Amsterdam, 1977, p. 17.
161. H.-R. Schulten and H. D. Beckey, 22nd Annual Conference on Mass Spectrometry and Allied Topics, May 19–24, 1974, Philadelphia, Paper O 7.
162. D. E. Games, A. H. Jackson, L. A. P. Kane-Maguire, and K. Taylor, *J. Org. Metals. Chem.*, *88*, 345 (1975).
163. D. E. Games, A. H. Jackson, and K. T. Taylor, *Org. Mass Spectrom.*, *9*, 1245 (1974).
164. W. D. Lehmann and H.-R. Schulten, in *Mass Spectrometry in Drug Metabolism*, A. Frigerio and E. L. Ghisalberti, Eds., Plenum Press, New York, 1977, p. 325.
165. H.-R. Schulten and H. D. Beckey, 24th Annual Conference on Mass Spectrometry and Allied Topics, May 9–14, 1976, San Diego, Paper K 11.
166. F. W. Röllgen and H.-R. Schulten, *Org. Mass Spectrom.*, *11*, 330 (1976).
167. H.-R. Schulten and H. D. Beckey, 24th Annual Conference on Mass Spectrometry and Allied Topics, May 9–14, 1976, San Diego, Paper K 12.
168. G. W. Wood and D. L. Wade, *Can. J. Chem.*, *53*, 3500 (1975).
169. H.-R. Schulten, T. Komori, and T. Kawasaki, *Tetrahedron*, in press.
170. S. Pfeifer, H. D. Beckey, and H.-R. Schulten, *Z. Anal. Chem.*, *284*, 193 (1977).
171. H. J. Veith, *Angew. Chem.*, *88*, 762 (1976); *Int. Ed.*, *15*, 695 (1976).
172. H.-R. Schulten and W. D. Lehmann, *Anal. Chim. Acta*, in press.
173. J.-C. Prome and G. Puzo, *Org. Mass Spectrom. 12*, 28 (1977).
174. W. D. Lehmann and H.-R. Schulten, *Angew. Chem.*, *89*, 180 (1977); *Int. Ed.*, *16*, 184 (1977).
175. W. D. Lehmann and H.-R. Schulten, *Angew. Chem.*, in press.
176. G. L. Peele and D. A. Brent, *Biomed. Mass Spectrom.*, in press.
177. W. D. Lehmann and H.-R. Schulten, *Biomed. Mass Spectrom.*, *4*, 152 (1977).
178. W. Frick, G. Doyle Daves Jr., D. F. Barofsky, E. Barofsky, G. H. Fisher, D. Chang, and K. Folkers, *Biomed. Mass Spectrom.*, *4*, 152 (1977).
179. H. U. Winkler, R. J. Beuhler, and L. Friedman, *Biomed. Mass Spectrom.*, *3*, 201 (1976).
180. C. N. McEwen, S. F. Layton, and S. K. Taylor, *Anal. Chem. 49*, 922 (1977).
181. J. A. Sphon, P. A. Dreifuss, and H.-R. Schulten, *J. Assoc. Off. Anal. Chem.*, *60*, 73 (1977).
182. W. D. Lehmann and H.-R. Schulten, *Anal. Chem.* in press.
183. M. Przybylski, H. Ringsdorf, U. Lenssen, G. Peter, G. Voelker, T. Wagner, and H. J. Hohorst, *Biomed. Mass Spectrom.* in press.
184. R. D. Smith, J. J. Decorpo, J. R. Wyatt, and F. E. Saalfeld, *Int. J. Mass Spectrom. Ion Phys.*, *21* 411 (1976).

AUTHOR INDEX

Numbers in parentheses are reference numbers and show that an Author's work is referred to although his name is not mentioned in the text. Numbers in *italics* indicate the page on which the full references appear.

SUBJECT INDEX

Methods of Biochemical Analysis

CUMULATIVE INDEX, VOLUMES 1–24 AND SUPPLEMENT

Author Index

CUMULATIVE INDEX, VOLUMES 1–24 AND SUPPLEMENT

Subject Index